Dead Sea Transform Fault System: Reviews

Modern Approaches in Solid Earth Sciences

VOLUME 6

For further volumes:
http://www.springer.com/series/7377

Zvi Garfunkel • Zvi Ben-Avraham
Elisa Kagan
Editors

Dead Sea Transform Fault System: Reviews

Springer

Editors
Zvi Garfunkel
The Fredy and Nadine Herrmann
Institute of Earth Sciences
The Hebrew University of Jerusalem
Jerusalem, Israel

Elisa Kagan
Department of Geophysical, Atmospheric
and Planetary Sciences
Tel Aviv University
Tel Aviv, Israel

Zvi Ben-Avraham
Department of Geophysical, Atmospheric
and Planetary Sciences
Tel Aviv University
Tel Aviv, Israel

Charney School of Marine Sciences
University of Haifa
Mount Carmel, Haifa, Israel

Responsible Series Editor: Y. Dilek

ISSN 1876-1682 ISSN 1876-1690 (electronic)
ISBN 978-94-017-8871-7 ISBN 978-94-017-8872-4 (eBook)
DOI 10.1007/978-94-017-8872-4
Springer Dordrecht Heidelberg New York London

Library of Congress Control Number: 2014943516

Credit Cover Figure: buabone/stockphoto.com

Printed on acid-free paper

Springer is part of Springer Science+Business Media (www.springer.com)

Preface

The ca. 1,000 km long Dead Sea Transform is a prominent geologic-physiographic feature, including the topographically lowest land area on Earth. It has attracted geologists' attention for more than 150 years, as it offers an excellent opportunity to study a variety of important topics. It is one of the largest and most accessible zones of strike slip faulting, associated with spectacular pull-apart basins which are some of the largest on earth. As well, the Dead Sea Transform has transpressional structures, which allow the studying of how such fault lines and associated secondary structures developed. The Dead Sea Transform was formed by the mid-Cenozoic breakup of a region with a long stable history; therefore, the breakup processes are uncomplicated by previous modifications of the lithosphere. It also offers an opportunity to study how seismic shaking affects sediments and contains a high-resolution paleoseismic archive. The sediments that accumulated continuously in the land-locked pull-apart basins also provide a detailed archive of paleoclimate variations and how they influenced the water chemistry of the lakes. The Dead Sea Transform also offered a major pathway for human migration out of Africa, evidenced by many prehistoric remains, thus providing an understanding of the ancient geographic and environmental settings as recorded by sediments and ancient structures. These are important for understanding the history of mankind. Because of these and other research opportunities, the Dead Sea Transform was, and still is, the object of extensive research in many directions. As geologic studies advanced, many papers and several collections of review articles on various geological aspects of the Dead Sea Transform were published.

The present collection of papers includes updated summaries of recent advances in the study of major aspects of the Dead Sea Transform. The need for this summarized review arose in view of the significant expansion and diversification of research in the last 10–20 years. There have been major advances in the understanding of previously studied topics, and also new vistas have opened on topics that were only modestly studied before. This book is intended for the use of those interested specifically in the Dead Sea Transform, particularly in geologic-geodynamic aspects, as well as for those who view the Dead Sea Transform in the broader framework of major strike-slip fault zones and continental breakup processes dominated by lateral motion, including other aspects such as environmental and seismic hazards.

The aim is to review and synthesize the present state of knowledge of some major topics in which significant advances were achieved. Obviously, a single volume cannot do justice to all the relevant directions of research; therefore, due to the limitations of space, this book includes only studies that are based on observations along the Dead Sea Transform. Within this framework, it is not possible to include studies regarding its relations with the bordering areas, the role of large-scale regional processes, and achievements in theoretical interpretations.

The papers represent the recent major increase in the application of geophysical and related methods, such as seismic, paleomagnetic, and GPS studies, as well as absolute dating and geochemical studies to the research of the Dead Sea Transform and the surrounding regions. The results greatly expanded the database regarding the deep structure along the Dead Sea Transform, its history, and the ongoing dynamic processes along it that supplement past and ongoing geological studies. Also represented are recent advances in the analysis of the historic and the paleoseismic record, one of the longest available on Earth, and on the development of the present-day Dead Sea Lake and of precursor lakes that existed along the transform. The findings of these studies provide important and unique insights into these issues which are not only of great scientific interest but are also important for the assessment of environmental and geological hazards which are of great significance to the population living near the transform. The papers collected in this book summarize and synthesize the major achievements of these studies.

We are very grateful to Y. Dilek who realized the significance of recent research and prompted us to undertake a compilation of papers reviewing some of the major advances. Special thanks go to Petra van Steenbergen, her staff, and Springer for the encouragement, advice, and guidance during the preparation of this book.

Givat Ram, Jerusalem, Israel — Zvi Garfunkel
Ramat Aviv, Tel Aviv, Israel — Zvi Ben-Avraham
Tel Aviv, Israel — Elisa Kagan

Contents

Contributors

Khalil Abu-Ayyash Natural Resources Authority, Amman, Jordan

Amotz Agnon The Fredy and Nadine Herrmann Institute of Earth Sciences, The Hebrew University of Jerusalem, Givat Ram, Jerusalem, Israel

Freddy Aldersons Seismological Consultant, Bellevue, Switzerland

Zvi Ben-Avraham Department of Geophysical, Atmospheric and Planetary Sciences, Tel Aviv University, Ramat Aviv, Tel Aviv, Israel

Charney School of Marine Sciences, University of Haifa, Mt. Carmel, Haifa, Israel

Sungchan Choi Christian-Albrechts-Universitaet zu Kiel, Kiel, Germany

Jaser Darwish Natural Resources Authority, Amman, Jordan

Catherine Dorbath Ecole et Observatoire des Sciences de la Terre, University of Strasbourg, Esplanade, Strasbourg, France

IRD, UMR 154, Toulouse, France

Louis Dorbath Ecole et Observatoire des Sciences de la Terre, University of Strasbourg, Strasbourg, France

Radwan El-Kelani An-Najah National University, Nablus, Palestine

Zvi Garfunkel The Fredy and Nadine Herrmann Institute of Earth Sciences, The Hebrew University of Jerusalem, Givat Ram, Jerusalem, Israel

Hans-Jürgen Götze Christian-Albrechts-Universitaet zu Kiel, Kiel, Germany

Abraham Hofstetter Geophysical Institute of Israel, Lod, Israel

Ecole et Observatoire des Sciences de la Terre, University of Strasbourg, Strasbourg, France

Amitai Katz The Fredy and Nadine Herrmann Institute of Earth Sciences, The Hebrew University of Jerusalem, Jerusalem, Israel

Yann Klinger Institut de Physique du Globe de Paris, Sorbonne Paris Cité, Univ. Paris Diderot, Paris, France

Ivan Koulakov Institute of Petroleum Geology and Geophysics, Novosibirsk, Russia

Gabi Laske University of California, San Diego, La Jolla, CA, USA

Shmuel Marco Department of Geophysical, Atmospheric, and Planetary Sciences, Tel Aviv University, Ramat Aviv, Tel Aviv, Israel

James Mechie Deutsches GeoForschungsZentrum, GFZ, Telegrafenberg, Potsdam, Germany

Uwe Meyer Bundesanstalt fuer Geowissenschaften und Rohstoffe, Hannover, Germany

Ayman Mohsen Deutsches GeoForschungsZentrum, GFZ, Telegrafenberg, Potsdam, Germany

Alexei Petrunin Deutsches GeoForschungsZentrum, GFZ, Telegrafenberg, Potsdam, Germany

Ernesto Meneses Rioseco Leibniz-Institut fuer Angewandte Geophysik, Hannover, Germany

Georg Rümpker Goethe-Universitaet, Frankfurt am Main, Germany

Trond Ryberg Deutsches GeoForschungsZentrum, GFZ, Telegrafenberg, Potsdam, Germany

Stephan V. Sobolev Deutsches GeoForschungsZentrum, GFZ, Telegrafenberg, Potsdam, Germany

Abraham Starinsky The Fredy and Nadine Herrmann Institute of Earth Sciences, The Hebrew University of Jerusalem, Jerusalem, Israel

Mordechai Stein Geological Survey of Israel, Jerusalem, Israel

Michael Weber Deutsches GeoForschungsZentrum, GFZ, Telegrafenberg, Potsdam, Germany

Ram Weinberger Division of Geological Mapping and Tectonics, Geological Survey of Israel, Jerusalem, Israel

Department of Geological and Environmental Sciences, Ben-Gurion University of the Negev, Beer-Sheva, Israel

Yishai Weinstein Department of Geography and Environment, Bar-Ilan University, Ramat-Gan, Israel

Chapter 1
Geophysical Studies of the Crustal Structure Along the Southern Dead Sea Fault

Zvi Ben-Avraham

Abstract The Dead Sea Fault (DSF) is an active transform fault linking opening in the Red Sea with collision in the Taurus/Zagros Mountains. Motion is left-lateral and estimated at approximately 5–7 mm/year. The fault is seismically active. The extensional regime combined with the dominant lateral motion along the DSF resulted in the formation of a series of deep pull-apart basins. These basins are among the largest and deepest in the world. The crustal structure of the DSF and its surroundings played a significant role in the development of the physiography and the architecture of the basins. The subsurface structure of the southern DSF has been studied quite intensively by various geophysical methods over the past few decades. This review summarizes the main geophysical findings of the crustal structure under the deep basins along the southern DSF, comparing the three deep basins against each other in contexts of various geophysical properties, and discussing their tectonic implications. A simulation of faulting processes along the northern DSF and the Levant margin suggested that the formation of the DSF could be explained as a result of simultaneous propagation from the north and south. Normally most basins are bordered only on one side by a strand of the DSF, leading to their asymmetry. Where asymmetry occurs basins do not extend to great depth. In areas where the deep basins occur, two strands of the DSF overlap in an en-echelon pattern. This situation is quite rare along the DSF. It is suggested that in these places an isolated block of lithosphere has dropped into the mantle. Simulations of this mechanism indicate that the resulting basin is rhomb-shaped and that with time it grows by the addition of distinct segments to its edges.

Z. Ben-Avraham (✉)
Department of Geophysical, Atmospheric and Planetary Sciences,
Tel Aviv University, Ramat Aviv, Tel Aviv 69978, Israel

Charney School of Marine Sciences, University of Haifa,
Mt. Carmel, Haifa 3190501, Israel
e-mail: zviba@tau.ac.il

Z. Garfunkel et al. (eds.), *Dead Sea Transform Fault System: Reviews*,
Modern Approaches in Solid Earth Sciences 6, DOI 10.1007/978-94-017-8872-4_1,

Keywords Dead Sea Fault • Crustal structure • Geophysics • Basin structure • Sedimentary basins

1.1 Introduction

The Dead Sea Fault (DSF), a plate boundary of the transform type, is characterized by a distinct topographic expression. It forms a narrow and elongated valley with uplifted shoulders showing asymmetry which changes along the length of the fault (Picard 1987; Wdowinski and Zilberman 1997).

The structure along the DSF varies greatly, as expressed in its physiography. Structurally, it can be divided into two segments: south and north of lat. 33°10′N. The southern segment, which is the topic of this paper, is marked by an almost continuous valley, most of which is underlain by a series of deep basins that are separated by less pronounced saddles (Garfunkel 1981; Garfunkel and Ben-Avraham 2001). These structures are controlled by longitudinal en-echelon faults on which lateral motion takes place. Normal faults produce the morphologic boundaries of that valley. Transverse faults are also present, but less conspicuous. In a map view, the segment has an arcuate shape, and the strike of the major longitudinal faults varies gradually from approximately 25°NE in the south to nearly N-S in the north, which is in accordance with the kinematics of Arabia-Africa plate separation (Ben-Avraham et al. 2008).

The main basinal areas along the DSF are the Gulf of Elat (Aqaba), the Dead Sea and the Sea of Galilee–Bet Shean (Fig. 1.1). In these areas the basins are quite deep and narrow and quite often the width/depth ratio is less than one. The central part of the southern DSF is associated with lower crustal seismicity and low heat flow.

Crustal structure of the DSF and its surroundings played a significant role in the development of the physiography and the architecture of the basins. The subsurface structure of the southern DSF has been studied quite intensively by various geophysical methods over the past few decades. Ben-Avraham et al. (2008) reviewed the subsurface structures of each of the deep basins along the DSF focused on specific properties for each basin. This review, however, is a summary of the main geophysical findings of the crustal structure under the deep basins along the southern DSF, presented in a comparative mode: a comprehensive review of the three deep basins against each other in contexts of several geophysical properties and discussing their tectonic implications.

1.2 Physiography

The DSF forms along most of its length a rather narrow valley with steep slopes on its margins. The development of the DSF was accompanied by uplifting and igneous activity on a regional scale. Uplifting was greatest away from the plate boundary itself – maximal at distances of several tens of kilometers on both sides of the transform. Elevations are quite variable, but often reach 0.7–1.5 km, while in places topography rises to 2–3 km above sea level. Igneous activity, mainly basaltic volcanism, occurred on a regional scale, mostly east of the transform

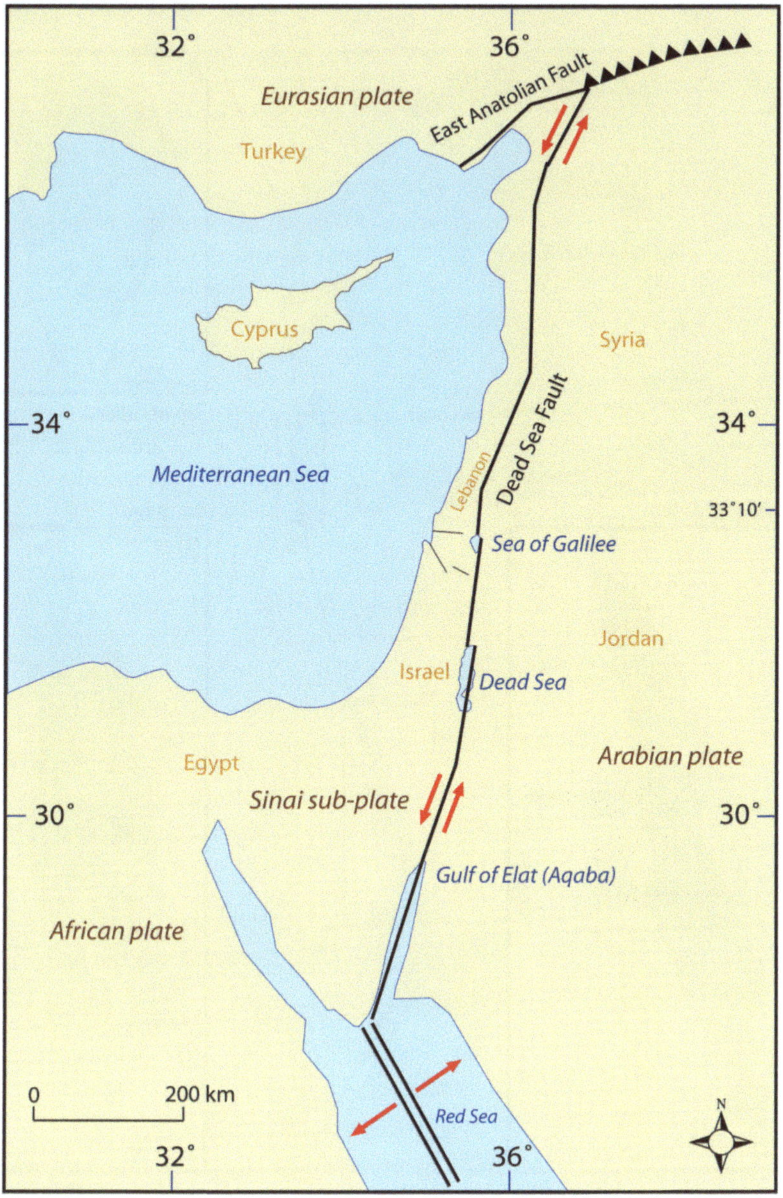

Fig. 1.1 Location and tectonic setting of the Dead Sea Fault (DSF) (Note the basinal areas of the Gulf of Elat (Aqaba), the Dead Sea and the Sea of Galilee). *Red arrows* mark regional tectonic plate movement

(Ben-Avraham et al. 2008; Garfunkel 1989). Relatively large basinal areas exist along the DSF. These are, from south to north, the Gulf of Elat (Aqaba), the Dead Sea basin and the Kinneret (Sea of Galilee)–Bet Shean basin (Fig. 1.2). To the north lays another smaller basin, the Hula Basin.

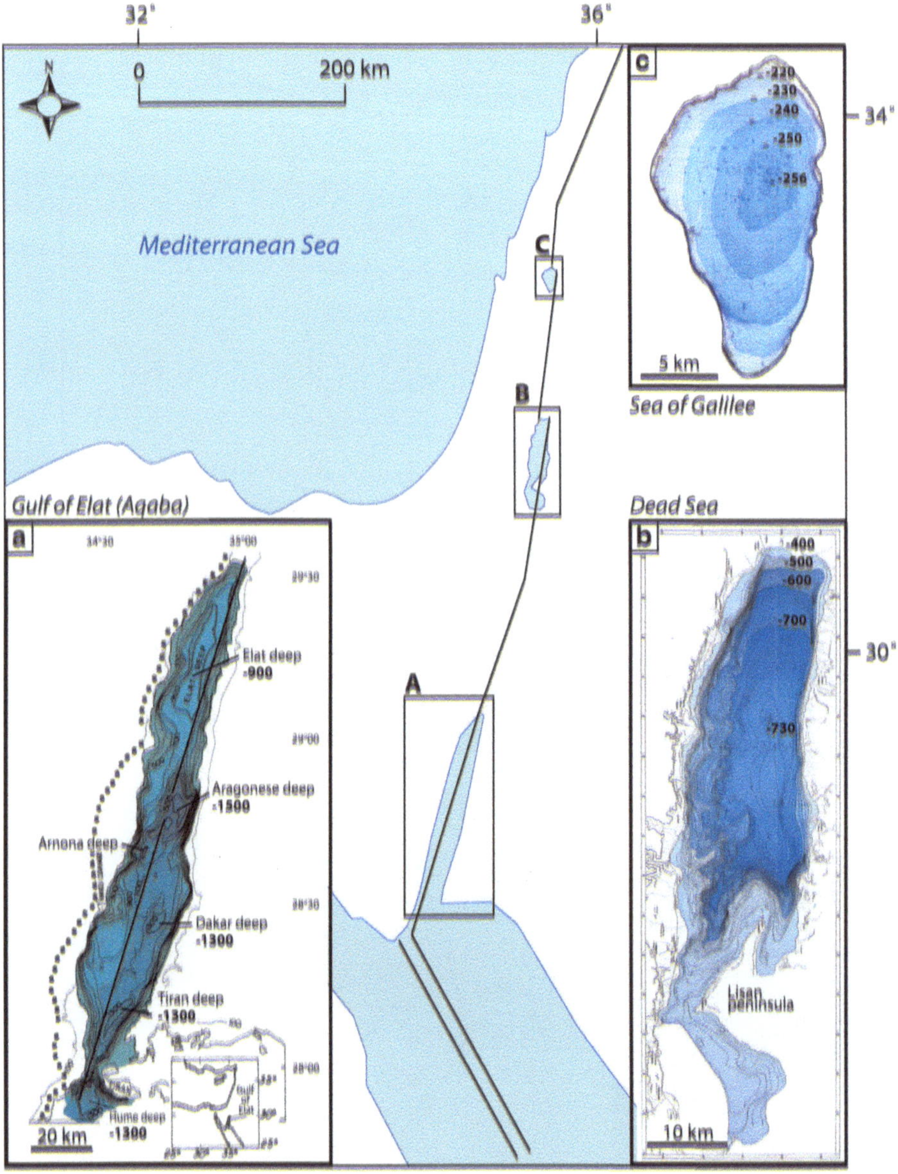

Fig. 1.2 Bathymetry of: (**a**) Gulf of Elat (Aqaba) (Modified after Ben Avraham (1987), Hall and Ben-Avraham (1978)). Location of the geophysical profiles along the axis of the basin (Fig. 1.7) marked by a *solid line*; location of the seismic refraction profiles IS and IN (Fig. 1.6) marked by heavy *dots*. (**b**) Dead Sea (Modified after Hall 1979). (**c**) Sea of Galilee (Modified after Ben-Avraham et al. 1990)

The Gulf of Elat (Aqaba) is about 180 km long and 20 km wide. It is the largest depression along the fault and is comprised of three elongated, en-echelon, actively subsiding pull-apart basins, which strike N20°–25°E (Ben-Avraham et al. 1979; Ben-Avraham 1985; Ben-Avraham and Garfunkel 1986). Bathymetric undulations within the basins form a series of distinct deeps: the Tiran and Dakar deeps in the southern basin, the Arnona and Aragonese deeps in the central basin, and the Elat Deep in the northern basin. The Hume Deep lies to the south of the Straits of Tiran and connects the Gulf to the Red Sea. The northern basin is the largest and shallowest with average water depth of ~900 m. The central basin is the deepest with average water depth of ~1,500 m and the southern basin has an average water depth of ~1,100 m.

The Dead Sea basin is the lowest continental depression (about 424 m below mean sea level) and is one of the largest pull-apart basins on Earth. It is about 150 km long, 15–17 km wide, and extends from the northern Arava Valley to Jericho (Neev and Emery 1967; Freund et al. 1970; Garfunkel 1981, 1997; ten Brink et al. 1993; Garfunkel and Ben-Avraham 1996; Lazar et al. 2006). It is divided into two main basins, a northern basin and a southern basin, separated by the Lisan Peninsula. The northern basin is occupied by a terminal lake known as the Dead Sea. The average water depth of the deep lake floor is ~300 m (Neev and Hall 1979). The southern basin was once covered by a very shallow lake (part of the Dead Sea), is now exposed and is used by the Dead Sea Works (Israel) and by the Arab Potash Company (Jordan) as artificial evaporation ponds.

The Kinneret-Bet Shean Basin is located north of the Dead Sea. The Sea of Galilee (Lake Kinneret) is located at the northern section of the basin. It is a fresh-water lake ~12 km at its widest point and ~20 km long. Its surface is approximately 210 m below mean sea level and has a maximum water depth of 46 m (Ben-Avraham et al. 1990). North of it the rather small and exposed Hula Basin is located. It is ~22 km long, ~7 km wide, and lays about 800 m below its surrounding mountains.

1.3 Architecture of Sedimentary Fill

Large sedimentary sequences were accumulated within the basins. Their shape varies along the length of the basins. In most cases the sedimentary fill is asymmetric in cross section, but occasionally it is symmetrical.

1.3.1 Gulf of Elat (Aqaba)

Within the narrow sea arm of the Gulf of Elat (Aqaba) three large sedimentary basins are expressed in the sea floor bathymetry. The basins are mostly asymmetrical in cross section. Ben-Avraham (1985) has suggested that the asymmetry of the strike-slip basins within the gulf of Elat is probably due to the fact that the basins are bounded only on one side by a strand of the Dead Sea transform fault. Thus, only one of the longitudinal faults is characterized by a strike-slip component, while the other longitudinal fault is predominantly normal. The basin is asymmetric toward the transform fault. This idea is supported by geological and geophysical evidence such as the offset of certain sedimentary sequences and magnetic anomalies (Ben-Avraham 1985). The asymmetry within the basins suggests that in the southern basin and the southern part of the central portion of the gulf the active strike-slip motion is taking place primarily on the western boundary faults of the basins, while in the northern portion of the gulf it is taking place primarily along the eastern boundary fault (Fig. 1.3). The only place where an en-echelon arrangement of strands of the DSF exists is in the central portion of the Gulf where the configuration of the basin is symmetric. Interesting, this is also where the sedimentary sequences are folded, suggesting that the area is under compression.

1.3.2 Dead Sea

The Dead Sea basin is divided into two deep basins, the northern and the southern basins. A remarkable feature of the sedimentary sequences in the northern portion of the southern basin and the southern portion of the northern basin is their symmetry (ten Brink and Ben-Avraham 1989; Al-Zoubi et al. 2002). Deformation in the Dead Sea basin takes place mainly along the transverse and longitudinal faults, while the basin fill is relatively undeformed and hardly tilted (Fig. 1.4). The basin has widened by the collapse and tilting of arcuate block from the western margin but the original shape is a full-graben (ten Brink and Ben-Avraham 1989). They further noticed that the basin is divided into 20–30 km long segments by transverse faults. The most impressive transverse fault is the Amazyahu fault, dividing the southern basin into two segments. Advanced processing of seismic reflection profiles in the area indicated that the Amazyahu fault, as well as the Boqeq fault that forms the northern boundary of the southern basin, are in fact normal faults. These faults cut through the sedimentary section down to the crystalline basement, similar to the eastern and western boundary faults of the basin (Ginzburg et al. 2006).

At the northern edge of the northern basin the Jericho sub-basin, that extends from the lake northward onto land, is asymmetric towards the Jericho fault, which is the active strand of the DSF in this area (Lazar et al. 2006). The Jericho sub-basin is much shallower than the sub-basins in the central part of the Dead Sea basin where they are symmetrical.

1.3.3 Sea of Galilee

The Sea of Galilee (Lake Kinneret) basin forms the northern portion of the Kinneret – Bet Shean basin. The Kinneret basin itself is divided into two distinct parts. The southern part, which includes the southern lake and the Kinarot Valley, located

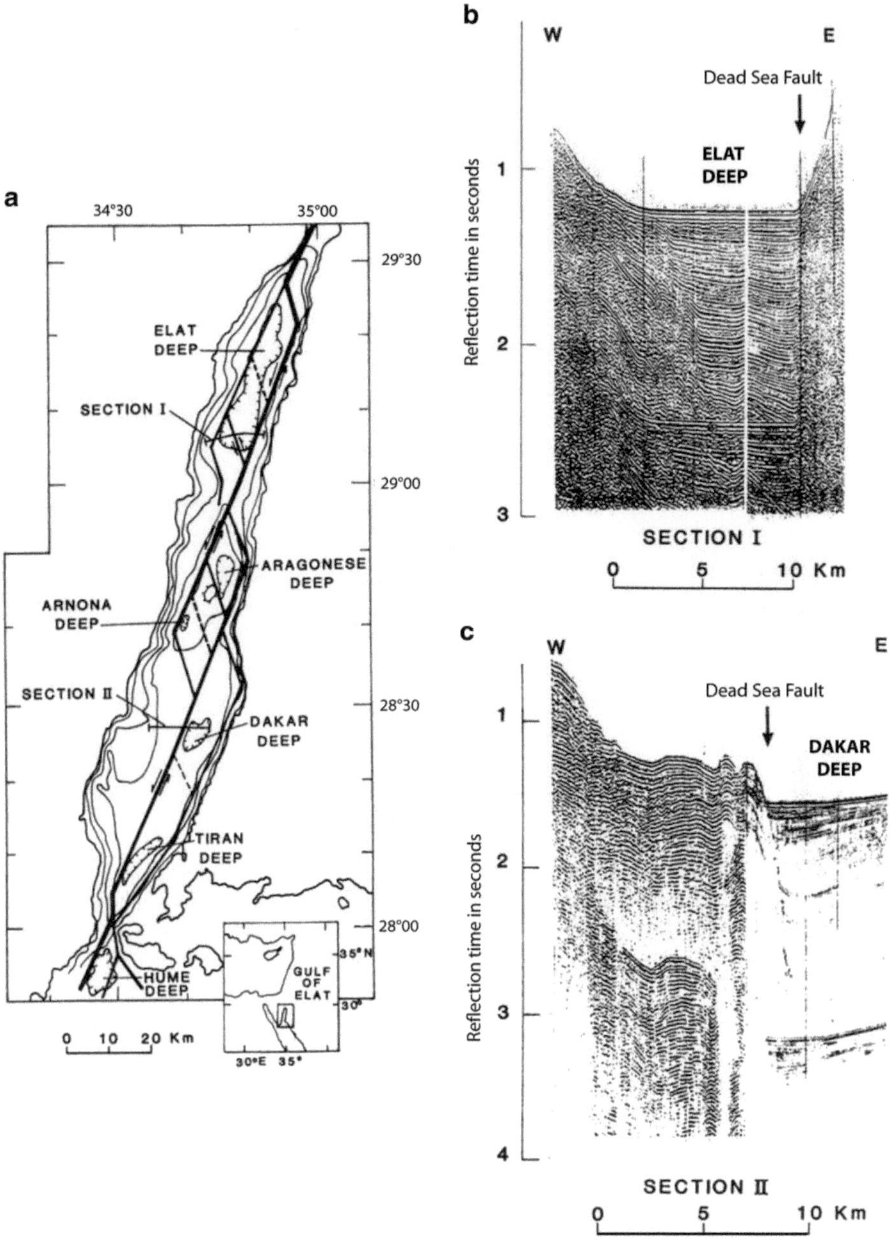

Fig. 1.3 (**a**) Structure of the Gulf of Elat. Segments of the Dead Sea Fault are shown with *bold lines* and *arrows*, other faults in the gulf are shown with *thin lines*. The strike-slip faults, both longitudinal and transverse, also have a normal component (After Ben-Avraham 1985; Ben-Avraham and Tibor 1993). (**b**) and (**c**) Continuous seismic profiles across the northern and southern basins of the Gulf of Elat (Modified after Ben-Avraham 1992) (Note the asymmetry of the basin fill to the east (Elat Deep) and to the west (Dakar Deep)). Profiles were scanned from analog seismic data acquired in the 1970s

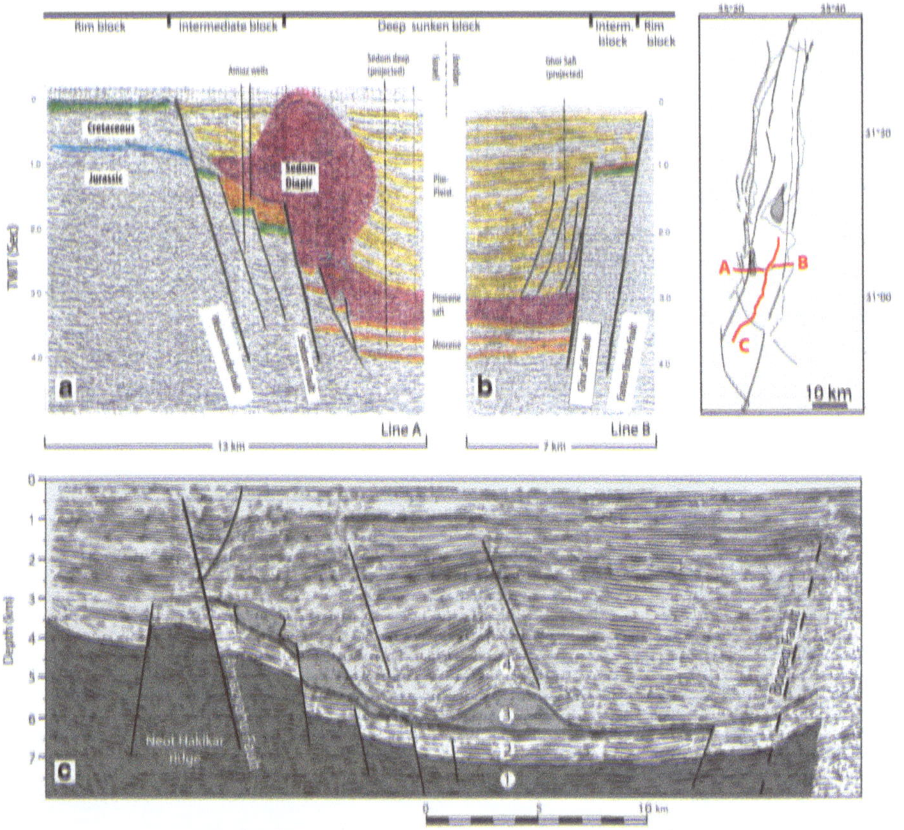

Fig. 1.4 Seismic profiles across and along the southern basin of the Dead Sea. Interpreted E-W seismic lines (**a**) and (**b**) show the main structural elements on both sides of the southern Dead Sea basin: Rim Blocks, Intermediate Blocks, Deep Sunken Block and the faults that separate them (Note that the Sedom Deep-1 is projected into line (**a**) to a position that fit the stratigraphy of the deep block). Color code: *yellow*, Pleistocene; *purple*, Pliocene salt; *orange*, Miocene; *green*, Cretaceous; *blue*, Jurassic (Modified after Al-Zoubi et al. 2002). (**c**) Interpreted pre-stack depth migrated seismic profile. Layer 1 is the pre-fill Cretaceous basement, layer 2 comprises the Hazeva Formation, layer 3 is the Pliocene evaporite series, and layer 4 comprises the Plio-Pleistocene sediments of the basin fill (Modified after Ginzburg et al. 2006). Profile locations are marked on map (Modified after Ginzburg et al. 2006) at *top right-hand* side

farther south, is narrow, deep and symmetrical, whereas the northern part is wider, shallower and asymmetrical to the east (Fig. 1.5). In the south, two strands of the DSF border the basin, while in the north only the eastern strand exists (Ben-Avraham et al. 1996; Reznikov et al. 2004). This situation is similar to the Gulf of Elat, where the central basin, bordered by two strands of the DSF, is symmetric and the northern basin, which is bordered by only one strand of the DSF, is asymmetric to the east. Another interesting similarity to the Gulf of Elat is observed in the narrowest part of the Sea of Galilee, where two strands of the DSF border the basin, and the sedimentary

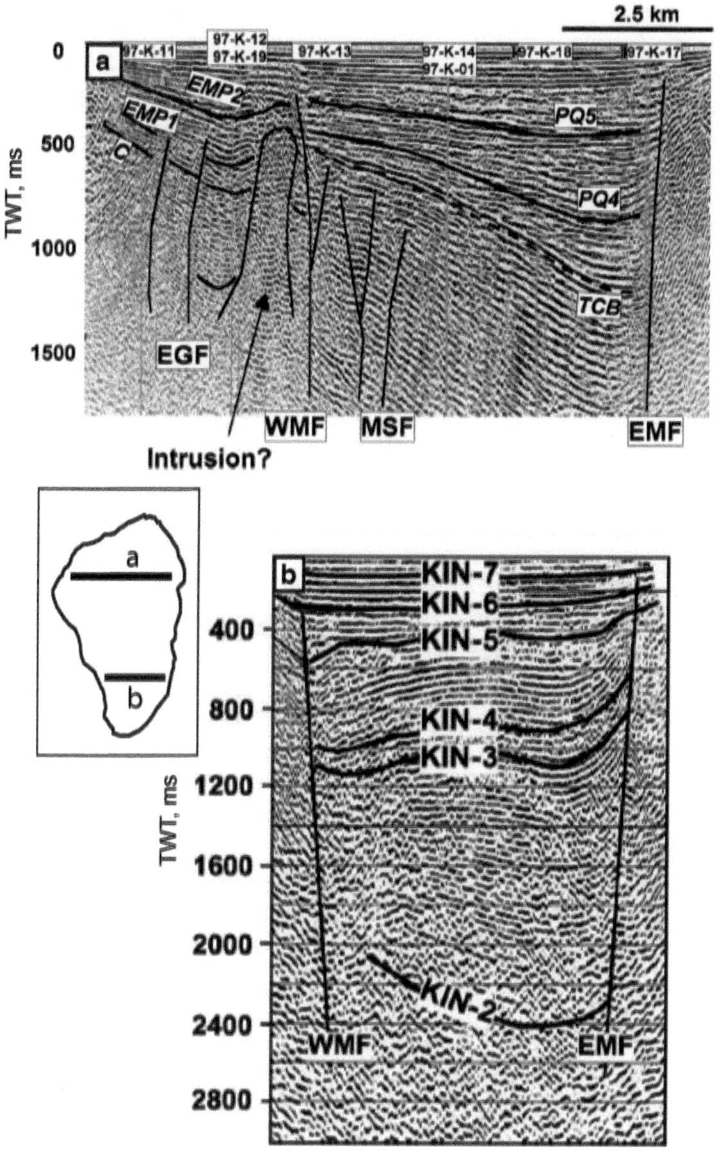

Fig. 1.5 Seismic profiles across the Sea of Galilee. (**a**) Migrated seismic line across the northern part of the lake: *EMF* eastern marginal fault; *MSF* median step fault, which separates the western part of the basin from the deep eastern part; *WMF* proposed continuation of the west marginal fault, which delimits the basin in the west; *TCB* top cover basalt. A possible intrusion of material into the overlying sediments is marked on the east (Modified after Reznikov et al. 2004) (Note the asymmetry to the east of the basin fill). (**b**) Seismic reflection profile across the southern part of the lake: *WMF* western marginal fault, *EMF* eastern marginal fault (Modified after Hurwitz et al. 2002) (Note the symmetry of the basin fill). Profile locations are marked on the insert map

sequences are folded (Hurwitz et al. 2002) indicating that the area is under compression. Reznikov et al. (2004) have suggested that the deepest sub-basin is bordered by deep normal transverse faults, similar to the situation in the deepest sub-basin of the Dead Sea. A complex zone of deformations delimits the northern sub-basin to the west.

1.4 Crustal Structure

1.4.1 Seismic Refraction

During the 1970s a series of seismic refraction profiles were obtained along the DSF and on its margins (Ginzburg et al. 1979a, b, 1981). These data show that the DSF is associated with a crustal structure which is slightly different than its surroundings. The seismic data obtained in the experiment were first evaluated using first-arrival information (Ginzburg et al. 1979a). The high quality of the data made possible an evaluation of later arrival information resulting in a detailed velocity structure of the crust underneath the rift and the adjacent areas (Ginzburg et al. 1979b). An interesting observation is the existence of a 5 km thick transition zone along the DSF within the lower crust, above the crust-mantle boundary, in which the velocity increases rapidly and smoothly from about 6.72 km s^{-1} to upper mantle velocity (~7.9 km s^{-1}). The profiles outside the rift valley show a different velocity structure in the crust. Here, no transition zone was found to exist in the base of the lower crust (Ginzburg et al. 1979b).

A revised interpretation of the seismic refraction data was presented by Ginzburg et al. (1981). The structural models of the rift profiles were plotted on a longitudinal section from the Sea of Galilee to the Red Sea (Fig. 1.6). On the same section the relative Bouguer anomaly is shown. As can be seen the regional gravity gradient is conformable with crustal thinning from Elat northward and particularly southward in the direction of the Red Sea. The crust thins considerably to the south along the Gulf of Elat. The observed thinning is of some 8 km, down to a thickness of only 27 km, some 80 km north of the Red Sea. The gravity observations indicate a continuing thinning to the south beyond the point where seismic refraction control is available. This is compatible with available data on the structure of the Red Sea (Ben-Avraham 1987; Lazar et al. 2012). Important point stated by Ginzburg et al. (1981) is that the rift and its immediate surroundings are underlain by a thinner than usual crust.

In the last few years several seismic refraction-reflection studies were carried out across the southern DSF. They were focused on the DSF segment between the Dead Sea and the Red Sea. These include several interdisciplinary efforts, including the DESERT project (DEad SEa Rift Transect), the DESIRE project (Dead Sea Integrated REsearch project), which are presented by Weber et al. (2014, this volume) and the USGS profile (ten Brink et al. 2006). The DESERT profile cut across the central Arava Valley, while the DESIRE and USGS profiles across the southern Dead Sea basin.

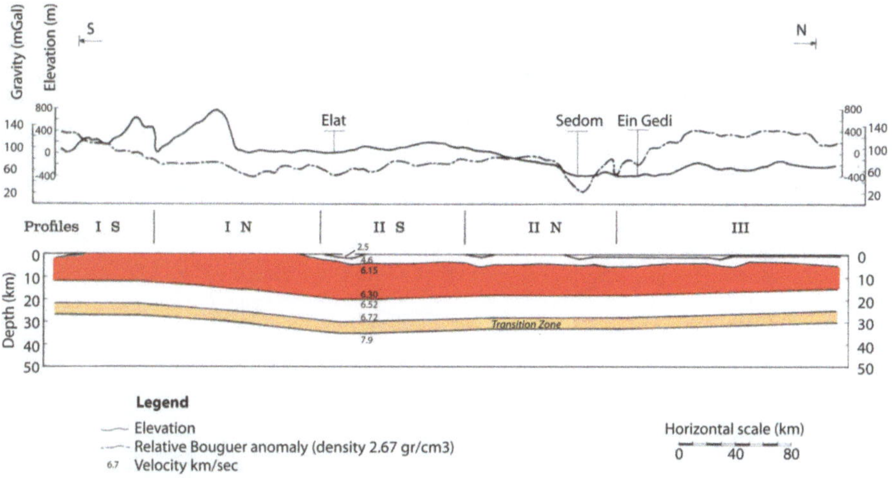

Fig. 1.6 Crustal model along the DSF from the Sea of Galilee (N) to the Red Sea (S), calculated from seismic refraction profiles. Coinciding elevation (*solid line*) and relative Bouguer gravity anomaly (*dashed line*) are also plotted (Modified after Ginzburg et al. 1981)

The three seismic profiles crossing the DSF from the Mediterranean to the Jordanian highlands demonstrate an increase in Moho depth from about 25 km to about 35 km, with only a small asymmetric topography of the Moho beneath the DSF at the Arava Valley. In the southern Dead Sea basin proper the imaging of the Moho is complicated by its relatively narrow width while crossing it in EW direction. Nevertheless, Weber et al. (2014, this volume) and ten Brink et al. (2006) concluded that the Dead Sea basin is a mostly upper crustal feature with a decoupling zone at about 20 km depth.

1.4.1.1 Gulf of Elat (Aqaba)

As a part of the seismic refraction survey of Ginzburg et al. (1979a, b), a profile was obtained on land along the western shore of the Gulf of Elat. In the northern part of the Gulf, the depth to Moho is about 32 km. From the middle part of the Gulf, the Moho shallows quite sharply until the southern edge where it is about 8 km shallower than in the north. The uplift of the Moho is also reflected in the gravity, magnetic and heat flow data (Fig. 1.7). The seismic refraction data where used along with the gravity and heat flow data to model the structure of the crust in the Gulf of Elat.

1.4.1.2 Dead Sea

A seismic refraction experiment was conducted along a north–south profile in the northern basin of the Dead Sea (Ginzburg and Ben-Avraham 1997). It involved the use of 9 OBS (Ocean Bottom Seismometers) and 11 portable seismic land

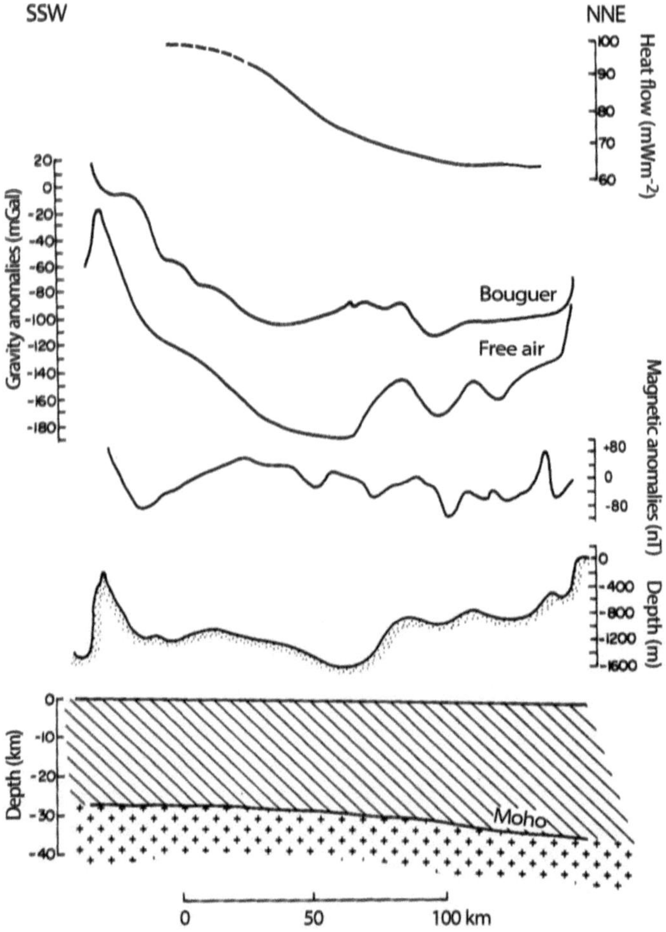

Fig. 1.7 Geophysical profiles along the Gulf of Elat (see Fig. 1.2 for location). *Top to bottom*: heat flow, Bouguer gravity anomaly, free air gravity anomaly, magnetic anomaly, bathymetry and depth to Moho on the western margin (After Ben-Avraham 1987)

stations which were placed north and south of the lake. The interpretation of the results indicates that the basement lies at 6–8 km under the northern basin. South of the basin, a major fault affecting the basement was detected with apparent sense of motion to the south. This major faulting which downthrows the basement some 4–5 km to the south, forms the northern boundary of the southern basin (Fig. 1.8). The basement at the northern part of the southern basin, away from the boundary fault, was found to be at about 14 km depth, making this area one of the deepest pull-apart basins in Earth.

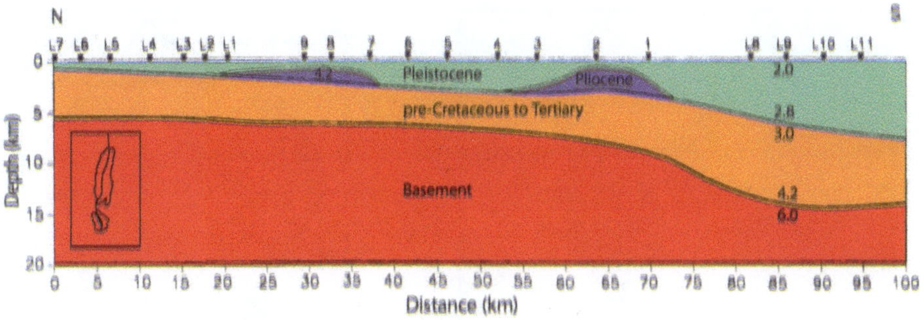

Fig. 1.8 Velocity-depth section along the Dead Sea. The section is calculated from N-S trending seismic refraction profile. The 2.0 km/s velocity represents the Pleistocene fill of the basin. The 4.2 km/s velocity is associated with Pliocene evaporites. The 6.0 km/s velocity represents the top of the crystalline basement, while the overlaying 3.0–3.8 km/s is associated with the Tertiary to pre-Cretaceous sediments (Modified after Ginzburg and Ben-Avraham 1997)

1.4.1.3 Sea of Galilee

Two shallow seismic refraction lines were obtained across the Sea of Galilee in the north and south (Ben-Avraham et al. 1981). Seismic explosives were placed in shot-holes at either end of each line below the water table. The signals were picked up by hydrophones and were transmitted to the shore based recording stations by Telseis[R] transmitters. Although penetration was limited to about 600 m, the two reversed refraction profiles show large differences in the velocity structure between the various parts of the lake. The southern part is more homogeneous than the northern one. In the north, the northwest section is quite different than the northeast section. It seems thus that the lake is divided into at least three separate regions with different sedimentary sequences. Drilling south of the lake and potential field studies (see below) have provided valuable information on the deep structure of the basin.

1.4.2 Gravity and Magnetics

1.4.2.1 Gulf of Elat (Aqaba)

Overall, the northern half of the Gulf of Elat has a larger amount of magnetic anomalies than the southern half, although on land over eastern Sinai the area off the southern half has a much more disturbed magnetic field than the area off the northern half of the Gulf (Ben-Avraham 1985). In the southern third of the gulf the magnetic field is rather smooth (Fig. 1.9) and its shape almost follows the bathymetric contours. Ben-Avraham (1987) proposed that crustal spreading processes propagate from the Red Sea into the southern gulf and caused a remagnetization of the crust in this area.

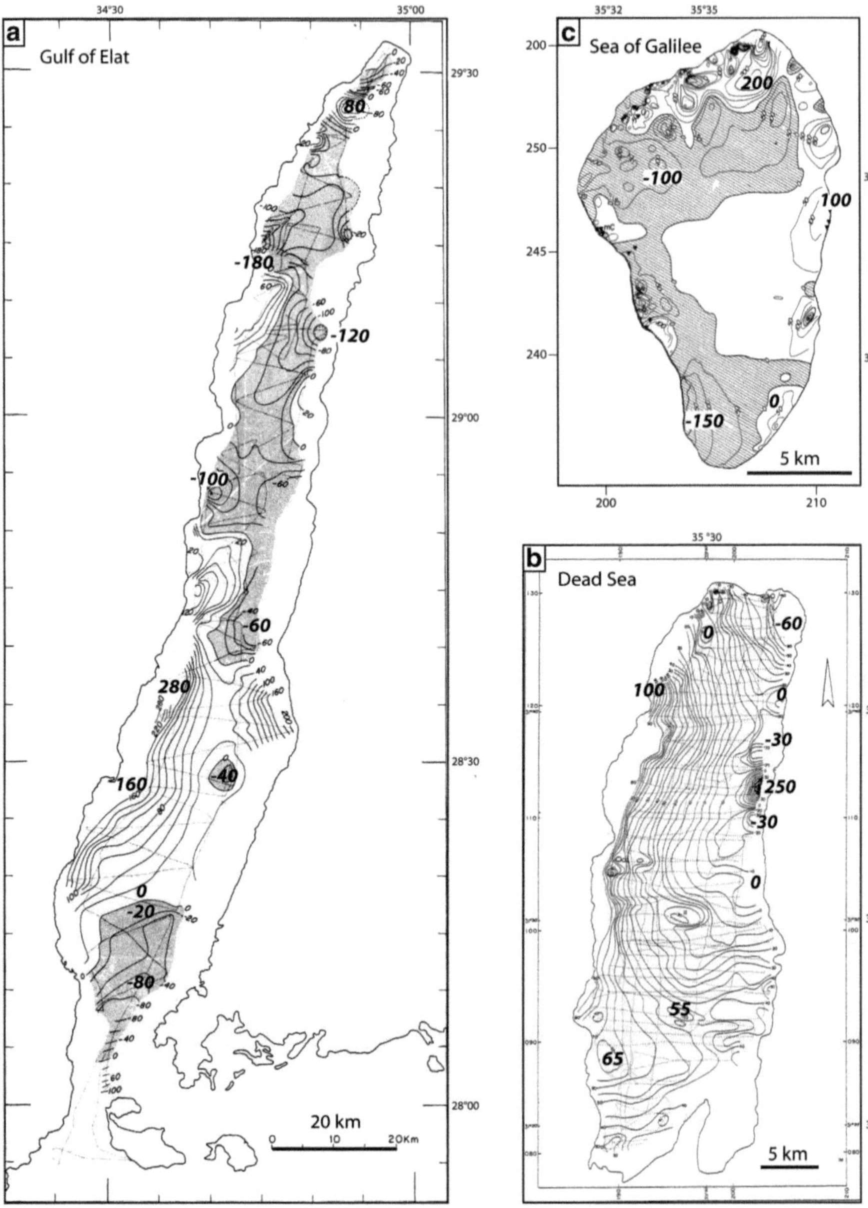

Fig. 1.9 Magnetic anomaly maps. (**a**) Gulf of Elat; contour intervals 10 nT (Modified after Ben-Avraham 1985, 1987). (**b**) Dead Sea; contour intervals 5 nT; *survey track lines* indicated by *dots* (Modified after Frieslander and Ben-Avraham 1989). (**c**) Sea of Galilee. Hatched areas mark negative anomalies; contour intervals 50 nT (Modified after Ben-Avraham et al. 1980; Ginzburg and Ben-Avraham 1986)

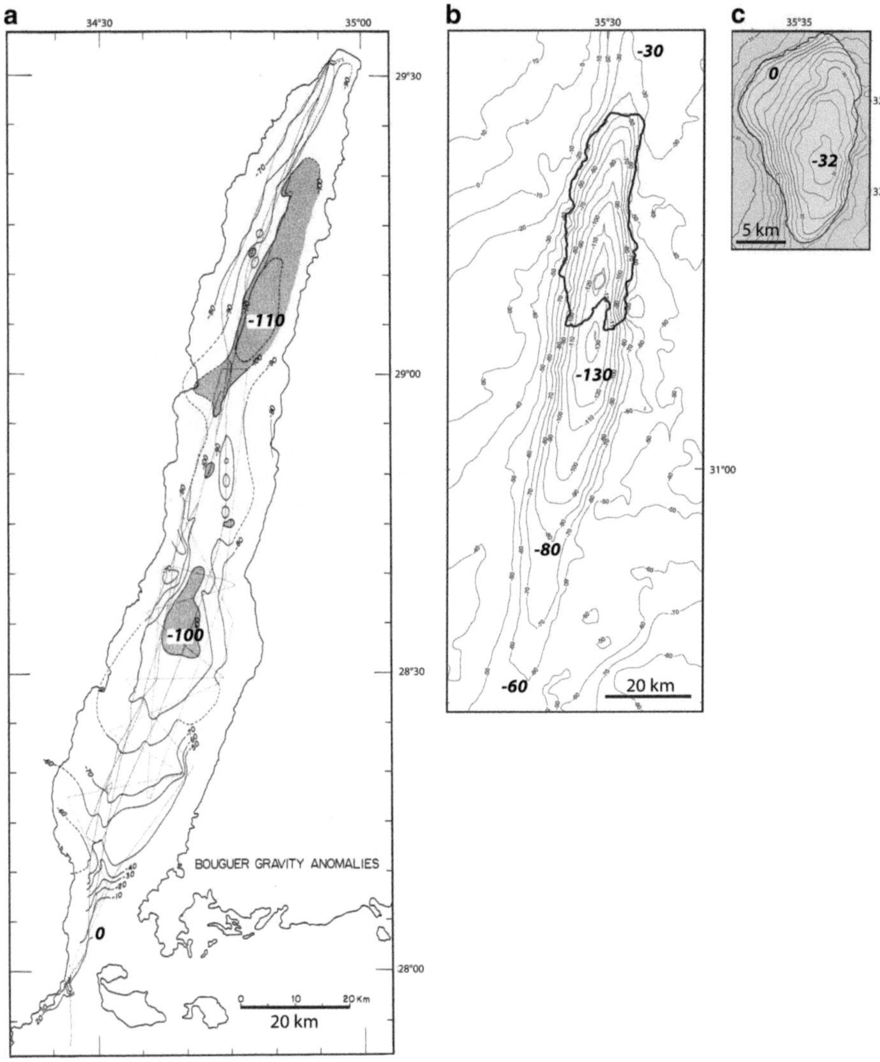

Fig. 1.10 Bouguer gravity anomaly maps. (**a**) Gulf of Elat; density: $\rho = 2{,}670$ kg m^{-3} (Modified after Ben-Avraham 1985). (**b**) Dead Sea; cropped and modified after ten-Brink et al. (2001); density $\rho = 2{,}670$ kg m^{-3}. (**c**) Sea of Galilee; densities: $\rho = 2{,}150$ kg m^{-3} below 150 mbsl; $\rho = 2{,}550$ kg m^{-3} above 150 mbsl (Modified after Ben-Avraham et al. 1996)

The free air gravity anomaly map shows two large elongated lows, the southern one reaches −200 mGal and the northern one −170 mGal. The Bouguer anomaly map (Fig. 1.10) also shows two lows, the northern one reaches −110 mGal and the southern one −100 mGal. At similar water depths in the Red Sea, the free air anomalies are around 0 mGal and the Bouguer anomalies are around +150 mGal. This indicates a large mass deficiency under the Gulf of Elat as well as under the basinal areas north of it, as is shown below.

1.4.2.2 Dead Sea

The magnetic field over the Dead Sea is quite smooth (Frieslander and Ben-Avraham 1989). Most of the magnetic anomalies have long wavelengths and low amplitudes (Fig. 1.9). However, a few short wavelength anomalies exist, some of which have high amplitudes. A comparison between the magnetic anomaly map of the Dead Sea and aeromagnetic maps over the land areas to the west and east, shows that the magnetic anomalies extend uninterrupted from the land area to the west into the basin. In the east the magnetic contours are discontinuous across the eastern margin of the basin. This suggests throughout most of development of the basin, the eastern bounding fault acted as a strike-slip fault.

Gravity studies indicate that the Dead Sea basin becomes shallower and narrower to the north and south (Fig. 1.10). Based on the gravity data ten Brink et al. (1993) suggested that the basin sags toward its deepest part in the center. These data also show faulted blocks, several kilometers in width, along the western side of the basin, while the eastern side is being occupied by a wide graben. These blocks were interpreted as evidence for passive collapse into the deepening graben (ten Brink et al. 1993). Gravity also suggests that the Moho is not significantly elevated under the basin, with deformation being limited to the crust. The gravity data indicate the division of the basin into sub-basins along its length.

1.4.2.3 Sea of Galilee

Several magnetic surveys have been conducted over the Sea of Galilee and its vicinity (Ben-Avraham et al. 1980; Ginzburg and Ben-Avraham 1986). The data of a recent new magnetic survey, which was carried out using a denser grid and an advanced system, are now being analyzed (Rofe et al. 2014). The magnetic map (Fig. 1.9) shows considerable anomalies near the margins of the lake, especially the northern margin, which are probably caused by fault-controlled occurrences of volcanic rocks, while the central portion of the lake is magnetically quite.

Gravity data (Fig. 1.10) indicate that the Sea of Galilee can be divided into two distinct sections (Ben-Avraham et al. 1996). The largest gravity anomaly is located over the southern section which is narrow and bordered by two N-S-trending faults (Fig. 1.5). This structure is thus the deepest part of the basin. It lies south of the bathymetric low. Gravity data suggest that the southern sub-basin may extend southward into the Kinnarot Valley south of the lake. The southern sub-basin is probably a pull-apart, which formed as a result of the transform motion along the main faults.

The wider northern section of the basin was probably formed as a result of interactions between the main N-S and secondary NW-SE- to E-W-trending fault systems (Ben-Avraham et al. 1981), i.e., rotational opening and transverse normal faulting. The northeastern part of the basin is occupied by an asymmetrical (to the east) half-graben structure (Fig. 1.5) which differs from the deep asymmetrical sub-basin of the southern section (Ben-Avraham et al. 1996; Hurwitz et al. 2002;

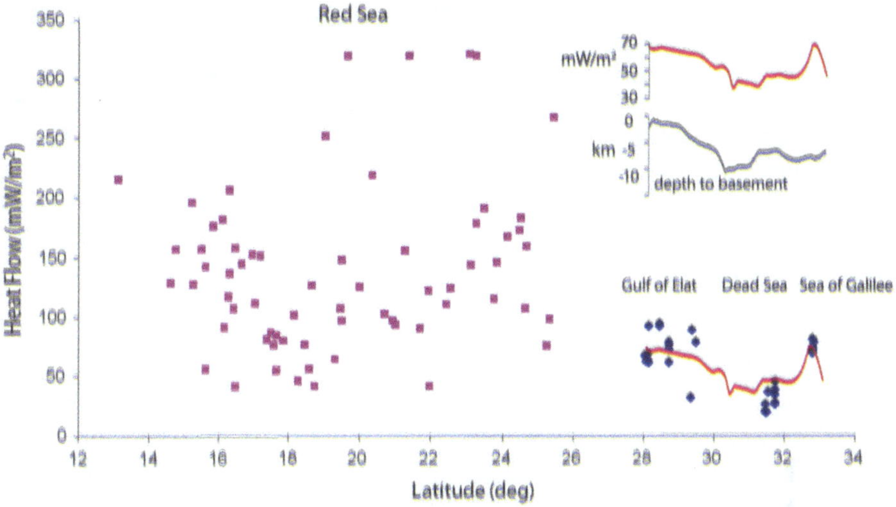

Fig. 1.11 Heat flow profile along the Red Sea (*pink rectangles*; After Ben-Avraham and Von Herzen 1987), Gulf of Elat, Dead Sea and Sea of Galilee (*blue diamonds* After Ben-Avraham et al. 1978) Heat flow profile along the Dead Sea Fault (*red line*; Modified after Shalev et al. 2012) superimposed on data from Ben-Avraham et al. (1978) on *bottom right*, and compared with depth to basement on *top right* (Modified after Shalev et al. 2012)

Reznikov et al. 2004). Although bathymetrically this is the deepest part of the lake and thus, probably the most actively subsiding area, gravity data suggest that the basin here is shallower than in the southern section.

1.4.3 Heat Flow

The average measured heat flow in the northern Dead Sea Basin is 38 mW/m^2 (Ben-Avraham et al. 1978; Ben-Avraham 1997) and it is 42 mW/m^2 (Eckstein and Simmonsi 1977) west of the basin. These values are very similar to the uniform heat flow measured in the eastern Mediterranean (Erickson 1970). A recent re-evaluation of the heat flow data (Shalev et al. 2007) confirmed this low value. A heat flow profile from the Sea of Galilee to the Red Sea (Fig. 1.11) show relatively high values in the Sea of Galilee area dropping to anomalously low values in the Dead Sea area and then increasing again in the southern part of the Gulf of Elat (Fig. 1.7) towards the Red Sea which is characterized by high and variable values.

A new comprehensive study of the surface heat flow distribution in Israel was derived from well measurements spread all over the country (Shalev et al. 2011). This study shows that the surface heat flow in Israel is indeed low over most of the country, only 40–45 mW/m^2 on the average. The Dead Sea basin, though, is not exceptional in this respect. Recent update to heat flow data estimated from xenoliths (Shalev et al. 2012, Fig. 1.11) suggests that the temperature gradient at the Dead Sea basin is relatively low, resulting in low heat flux (<40 mW/m^2) and a relatively deep seismicity extending to lower crustal depths, in agreement with earthquake depths (>25–30 km) (Aldersons et al. 2003). Higher heat fluxes were estimated at the Sea of Galilee (70 mW/m^2) and at the Gulf of Elat (65 mW/m^2). A comparison of earlier (Ben-Avraham et al. 1978; Ben-Avraham and Von Herzen 1987) and recent (Shalev et al. 2012) heat flow data is presented in Fig. 1.11, demonstrating a consistency of heat flow results among these works.

1.4.4 Seismicity

The DSF is associated with deep crustal seismicity in the area from the northern Gulf of Elat to the Sea of Galilee (Aldersons et al. 2003; Shamir 2006; Aldersons and Ben-Avraham 2014, this volume). The deepest seismicity, just above the Moho, occurs under the Dead Sea basin.

Aldersons et al. (2003) studied the seismicity along the DSF between Elat-Aqaba and the Sea of Galilee for the period 1984–1997. Out of 2,283 earthquakes routinely recorded by the network of the Geophysical Institute of Israel (GII) and expanded with stations from the Jordan Seismological Observatory (JSO), the location and depth of 410 earthquakes were recomputed. Within the Dead Sea basin itself, 42 well-constrained microearthquakes were relocated from seismograms repicked by hand. 60 % of these earthquakes nucleated at depths between 20 and 32 km and more than 40 % occurred below the depth of peak seismicity situated at 20 km. With the Moho at 32 km, the upper mantle appeared to be aseismic during the 14-year data period.

The depth distribution of background seismicity, in agreement with an average heat flow of 40–45 mW/m^2, suggests a seismogenic thickness T_S of 28.4 ± 2.2 km in the Dead Sea area (Aldersons and Ben-Avraham 2014, this volume). This value is much higher than previously assumed.

1.5 Discussion

The DSF is associated with several of the deepest pull-apart basins on Earth. It is also associated with lower crustal seismicity and low heat flow. The formation of these basins must be linked somehow with the processes that have led to the formation of the DSF itself. The origin of the DSF was the topic of several studies. Earlier models (Steckler and ten Brink 1986) have suggested that the rifting activity along the DSF propagated from the northern Red Sea in the south to the north. Indeed, at the southern part of the DSF, within the Gulf of Elat, geophysical data indicate that seafloor spreading processes propagate from the Red Sea northward (Ben-Avraham 1987). More recent studies, however, have suggested that stresses generated at the collision belt along the Taurus Mountains are responsible for the formation of the DSF. A simulation of faulting processes along the northern DSF and the Levant margin (Ben-Avraham and Lyakhovsky 1992) suggested that the formation of the DSF could be explained as a result of simultaneous propagation from both north and south. Recently Ben-Avraham and Schubert (2006) and Ben-Avraham et al. (2010) have suggested that the two tips of the propagating cracks from north and south met in the area where the Dead Sea basin was developed.

1.5.1 Geometry of Deep Basins

An en-echelon arrangement of the main faults with a full graben configuration of the basin in-between is quite rare along the DSF. In most other cases the basins are asymmetrical (Figs. 1.3 and 1.5) with the main strand of the transform occurring on one side, while the other is bounded mainly by normal faults (Ben-Avraham 1992; Ben-Avraham and Zoback 1992). This situation also characterizes basins along other continental transforms. For example, basins along the El Pilar fault in Venezuela, such as the Cariaco basin (Schubert 1984), are strongly asymmetric.

A notable exception is in the Dead Sea basin. Here, strands of the DSF are located on the east and west (ten Brink and Ben-Avraham 1989; Al-Zoubi et al. 2002; Larsen et al. 2002). These are the Sedom Fault on the west and Ghor Safi Fault on the east. Fault plane solutions indicate that both faults are strike–slip in this area (van Eck and Hofstetter 1990). As a result, the basin here is symmetric with an original shape of a full-graben. The basin has widened with time by the collapse and tilting of blocks from the eastern and western margins. The only other two places where such a situation occurs along the DSF are the Aragonese Deep at the central part of the Gulf of Elat (Ben-Avraham 1985) and the southern Sea of Galilee at the central part of the Kinneret–Bet Shean basin (Ben-Avraham et al. 1996; Fig. 1.12).

The transverse faults bordering the basins on their north and south sides and within the basins are less clear. Very few of them are well imaged by seismic reflection profiles. One of the best studied transverse faults is the Amazyahu fault on the southern margin of a deep segment within the southern sub-basin of the Dead Sea. The trace of this fault is expressed on the surface as an easily recognized escarpment in the southern sub-basin of the Dead Sea. Several surveys of seismic reflection lines were shot across this fault over the last 35 years. Earlier profiles led to new models about the nature of transverse faulting in the Dead Sea basin. These models describe it as a major listric fault which reaches the basement. These old profiles were in the time domain.

The results of processing recent seismic reflection data across the Amazyahu fault using advanced techniques shed new light on the nature of the fault (Ginzburg et al. 2006). The pre-stack depth migration gives a much better definition of the faults and better continuity of the pre-fill reflections, thus enabling the study of the basement faulting and its role in the internal structure of the basin. It indicates that the Amazyahu fault, which was previously interpreted as a listric fault, is a deep basement fault. Ginzburg et al. (2006) also noted that the Boqeq Fault, at the northern margin of the southern sub-basin of the Dead Sea, cuts through the deeper part of the young fill and the Cretaceous and older beds, probably to the crystalline basement. It appears thus, that at least in several cases the transverse faults are deep normal faults cutting deep into the base of the basins.

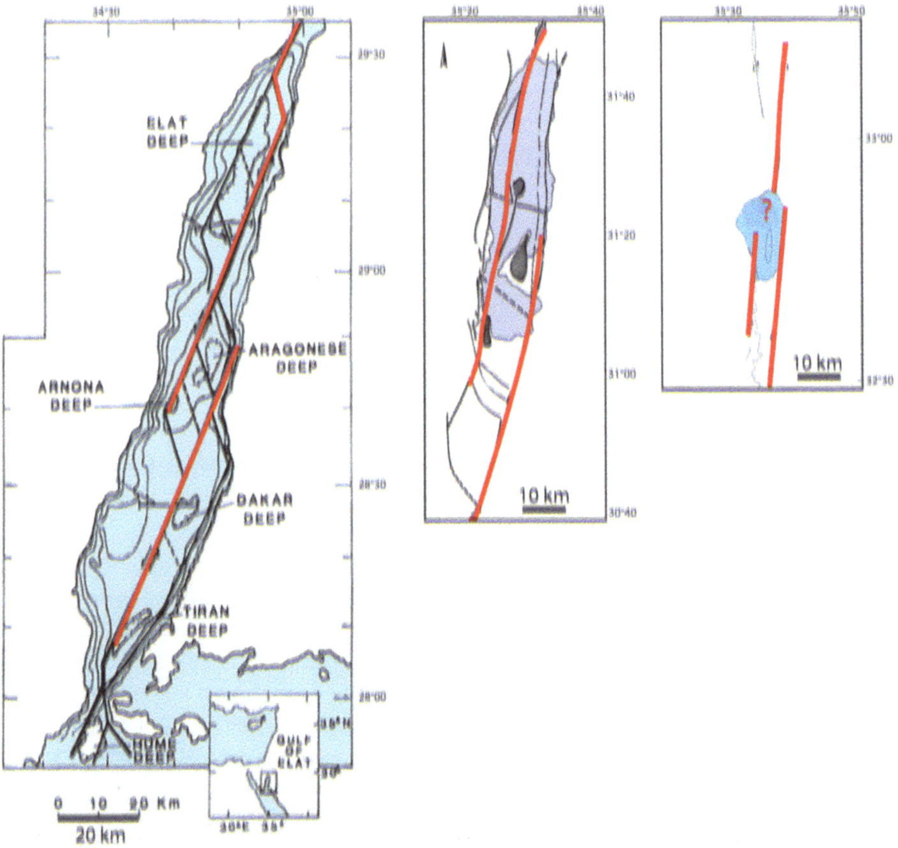

Fig. 1.12 En-Echelon arrangement of the strands of the Dead Sea Fault. *Left*: Gulf of Elat (Modified after Ben-Avraham 1985; Ben-Avraham and Tibor 1993). *Middle*: Dead Sea (Modified after Ben-Avraham 1997); Hatched areas are salt bodies. *Right*: Sea of Galilee (Modified after Ben-Avraham et al. 1996; Reznikov et al. 2004; Hurwitz et al. 2002)

1.5.2 Depth of Basins

The basins along the southern DSF vary in size and depth. The deepest segments are the ones where en-echelon arrangement of the strands of the DSF exists. A particular deep segment is the northern part of the southern basin of the Dead Sea graben. Here, seismic refraction and reflection profiles and gravity data indicate that the basin is 12–15 km deep with a ratio of width/depth of less than one. This unusual situation is unique to the entire length of the DSF. In the southern Sea of Galilee the ratio of width/depth is also less than 1 (Fig. 1.13). In this area the basin is 6–8 km deep. In the Gulf of Elat the basins reach a depth of 6 km under relatively deep water column.

The geometry of the southern Dead Sea basin is anomalous compared to the rest of the DSF. Here, the northern segment is bounded by deep vertical faults on all sides.

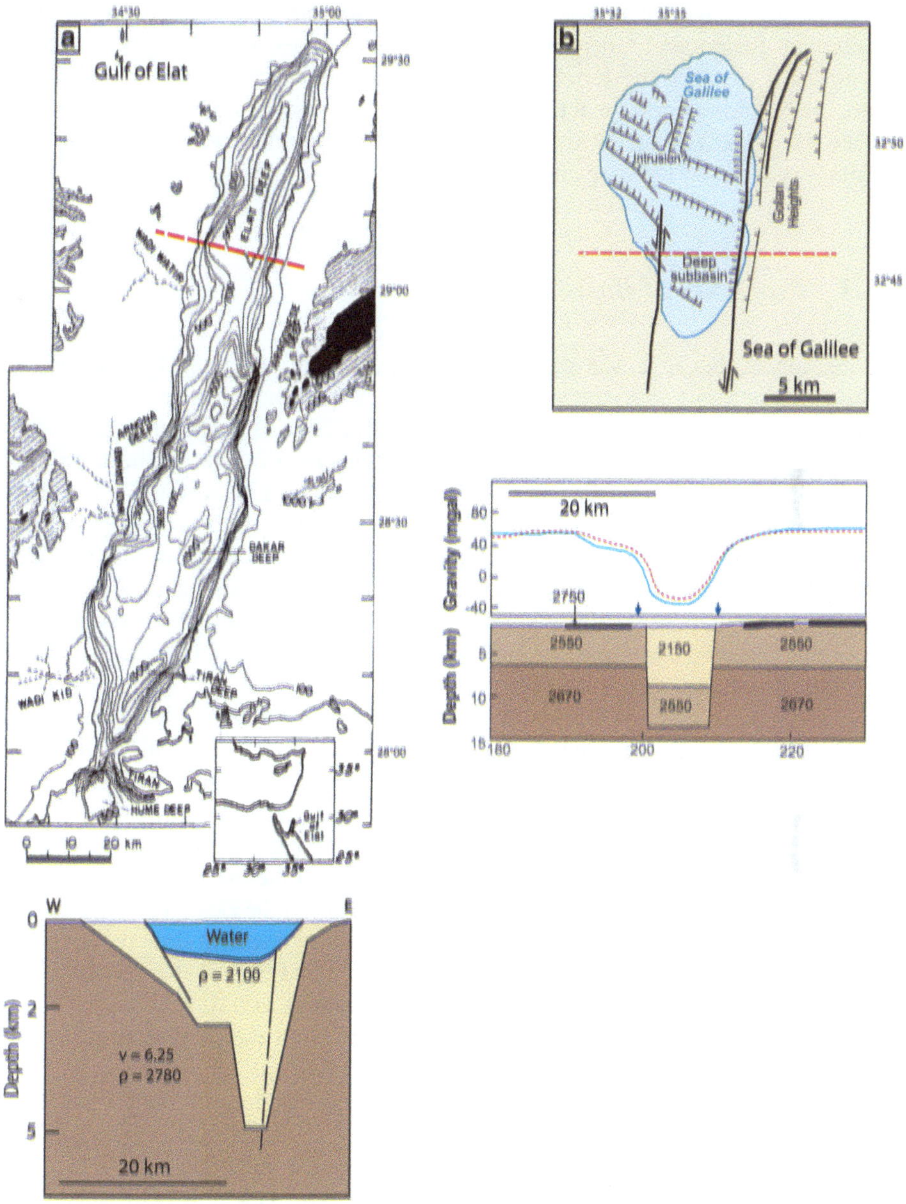

Fig. 1.13 Crustal structure models of the Gulf of Elat and the Sea of Galilee based on gravity data. (**a**) Elat Deep. Faults within the basin are based on continuous seismic profiles. The model was tied to refraction data along the western coast of the Gulf of Elat (Ginzburg et al. 1979a, b, 1981); densities (ρ) in kg m^{-3}; seismic velocities (v) in km s^{-1}. See location on map at *top* (Modified after Ben-Avraham 1985). (**b**) Sea of Galilee. Observed (*dashed line*) free air gravity data compared with calculated gravity (*solid lines*) from 2-D density-depth models presented below; densities (ρ) in kg m^{-3}; *black layers* represent basalt flows; *arrows* mark the boundaries of the lake. See location on map at *top* (Modified after Ben-Avraham et al. 1996, 2008) (Note the different depth scale of each model; horizontal scales are similar). A crustal structure model for the Dead Sea is presented in Fig. 1.14

These are the transverse Boqeq and Amazyahu faults in south and north, and the longitudinal Sedom and Ghor Safi faults in west and east, which are the southern and northern edges of the Jericho and Arava faults, respectively (Fig. 1.14). As a result, the basin is symmetrical both in east–west and north–south directions. At the surface the basin is about 30 km long and about 18 km wide; however, at the sub-bottom it is about 20 km long and 13 km wide.

1.5.3 Processes of Basin Formation

As mentioned above, it was suggested that the formation of the DSF could be explained as a result of simultaneous propagation from the north and south (Ben-Avraham and Lyakhovsky 1992; Lyakhovsky et al. 1994). This concept was used in an attempt to explain the formation of the deep basin in the northern part of the southern basin of the Dead Sea by Ben-Avraham and Schubert (2006). They have suggested that it was formed as a "drop down" basin during early stages of the evolution of the DSF when the tips of propagating faults, one from the collision front in the north and one from the Red Sea in the south met in this area. The fault tips overlapped and curved towards each other, isolating a block of crust and lithosphere that dropped into the mantle. The low heat flow and deep seismicity in the Dead Sea area support the brittle behavior of the lower crust.

In contrast, Sobolev et al. (2005) and Petrunin and Sobolev (2006) presented results of a three-dimensional thermo-mechanical model of a pull-apart basin formed at an overstepping of an active continental transform fault such as the Dead Sea basin. They adopted the classical scheme of pull-apart basin formation and demonstrated that the major parameter controlling basin length, thickness of sediments and deformation pattern beneath the basin is the thickness of the brittle layer. However, the average measured surface heat flow at the Dead Sea basin is less than 40 mW/m^2 (Ben-Avraham et al. 1978), well below the heat flow required for their models. The low surface heat flow values along the DSF are also supported by the depth distribution of local seismicity (Aldersons et al. 2003).

Ben-Avraham et al. (2010) simulated the formation of the deep segment of the Dead Sea basin and other deep basins along the DSF and demonstrated that the "drop down" mechanism of the Dead Sea basin formation suggested by Ben-Avraham and Schubert (2006) is possible. Density heterogeneities formed in the crust or upper mantle during a previous stage of regional magmatism, drop into the upper mantle when strike-slip faults are created and detach them from the surrounding lithosphere. The simulations indicate that the resulting basin is rhomb-shaped and that with time it grows by the addition of distinct segments to its edges. The proposed mechanism could account for the formation and evolution of large sedimentary basins along other strike-slip fault systems, such as the San Andreas Fault and other continental transform faults.

This mechanism could also account for the fact that the large basins along the DSF do not reduce in their width with time. Measurements of the geometric complexity of faults (Wesnousky 1988, 1994; Stirling et al. 1996) demonstrate that the density

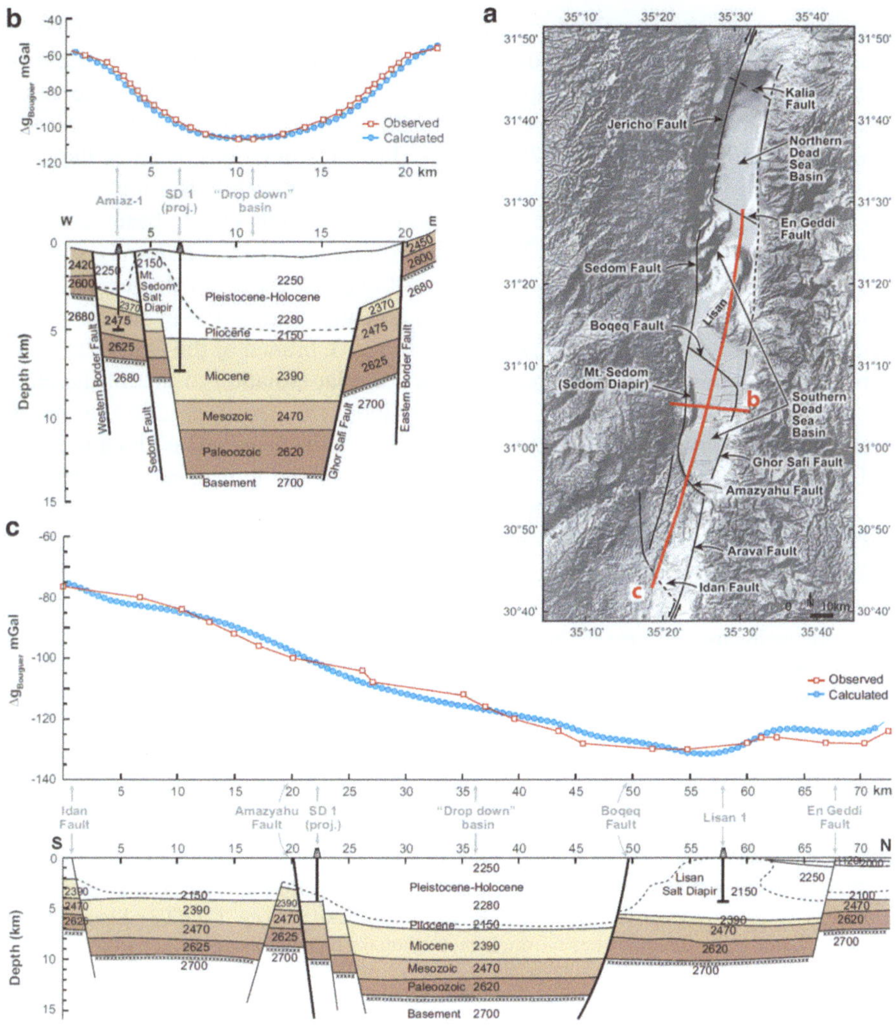

Fig. 1.14 Structural models of the Dead Sea basin. (**a**) A Digital Terrain Map (DTM) image showing the main tectonic elements of the Dead Sea basin. Faults are marked as *black lines* (After Ben-Avraham 1997). The Dead Sea basin is divided into two sub-basins, which are separated by the Lisan Peninsula, which is a large buried salt diapir. The two basins are thought to be divided by a large oblique normal fault, the Boqeq Fault. The two main strands of the DSF in this area are the Jericho fault, which borders the northern sub-basin on the west, and the Arava Fault, which borders the southern sub-basin on the east. Geological profiles shown in (**b**) and (**c**) are indicated by *red lines* (background DTM by Hall 1996). (**b**) E-W geological cross section showing the deep sub-basin ("Drop down basin") located in the northern part of the southern Dead Sea basin. The section is based on seismic reflection, seismic refraction, and borehole data (SD 1, Sedom Deep-1 borehole). The deep sub-basin is bordered by deep vertical faults; the Sedom Fault in the west and the Ghor Safi Fault (which does not extend to the surface) in the east. (**c**) N-S geological cross section based on prestack depth migration seismic reflection profiles, as well as on the seismic refraction, gravity, and borehole data. The section shows that the area of the deep sub-basin is the deepest part ("Drop down basin") of the Dead Sea basin. It is bordered by deep vertical faults, the Boqeq Fault in the north, and the Amazyahu Fault in the south. Densities are given in kg m^{-3}. Gravity models are shown on top (Modified after Ben-Avraham and Schubert (2006) and Ben-Avraham et al. (2008))

of step offsets in the examined fault traces decreases as a function of cumulative slip, showing an evolution with continuing deformation from a disordered network of linked fault segments to simpler dominant localized fault zones. The same phenomena have been demonstrated in analog modeling (e.g. Rahe et al. 1998; Wu et al. 2009; Smit et al. 2008a, b, 2010). Indeed several small basins along the DSF show localization of faulting. An interesting example is the Hula Basin north of the Sea of Galilee. While the surface expression is of a rhomb shaped basin, a recent study (Schattner and Weinberger 2008) showed that the basin is relatively shallow and the main strand of the DSF in this area traverses the basin diagonally from southeast to northwest. Nemer et al. (2008) recently studied the evolution of the Yammouneh sedimentary basin along the main active branch of the DSF within the Lebanese restraining bend and showed a very similar situation to that of the Hula Basin. In the south, Ehrhardt et al. (2005) and Tibor et al. (2010) described a similar situation in the northern Gulf of Elat. How then were the deep basins along the DSF and other shear zones formed, when the tendency of the fault zone is to become smooth and localized? The "drop down" mechanism might possibly be in part responsible for this. The deep seismicity and low heat flow values suggest that the deformation might be brittle in the lower crust in these areas.

Acknowledgements The help of Mor Kanari in preparing the figures as well as with other technical aspects of the paper is greatly appreciated. I am very grateful to him for his effort. Uri Schattner helped with earlier versions of Figs. 1.11 and 1.12.

References

Aldersons F, Ben-Avraham Z (2014) The seismogenic thickness in the Dead Sea area. In: Garfunkel Z, Ben-Avraham Z, Kagan E (eds) The Dead Sea Transform. Springer, Heidelberg, pp 53–89

Aldersons F, Ben-Avraham Z, Hofstetter A, Kissling E, Al-Yazjeen T (2003) Lower-crustal strength under the Dead Sea basin from local earthquake data and rheological modeling. Earth Planet Sci Lett 214:129–142. doi:10.1016/s0012-821x(03)00381-9

Al-Zoubi A, Shulman H, Ben-Avraham Z (2002) Seismic reflection profiles across the southern Dead Sea basin. Tectonophysics 346:61–69. doi:10.1016/s0040-1951(01)00228-1

Ben-Avraham Z (1985) Structural Framework of the Gulf of Elat (Aqaba), Northern Red Sea. J Geophys Res 90:703–726. doi:10.1029/JB090iB01p00703

Ben-Avraham Z (1987) Rift propagation along the southern Dead Sea rift (Gulf of Elat). Tectonophysics 143:193–200. doi:10.1016/0040-1951(87)90088-6

Ben-Avraham Z (1992) Development of asymmetric basins along continental transform faults. Tectonophysics 215:209–220. doi:10.1016/0040-1951(92)90082-h

Ben-Avraham Z (1997) Geophysical framework of the Dead Sea: structure and tectonics. In: Niemi TM, Ben-Avraham Z, Gat JR (eds) The Dead Sea: the lake and its setting, Oxford monographs on geology and geophysics. Oxford University Press, New York, pp 22–35

Ben-Avraham Z, Garfunkel Z (1986) Character of transverse faults in the Elat Pull-Apart Basin. Tectonics 5:1161–1169. doi:10.1029/TC005i007p01161

Ben-Avraham Z, Lyakhovsky V (1992) Faulting processes along the northern Dead Sea transform and the Levant margin. Geology 20:1139–1142. doi:10.1130/0091-7613(1992)020<1139:fpatnd>2.3.co;2

Ben-Avraham Z, Schubert G (2006) Deep "drop down" basin in the southern Dead Sea. Earth Planet Sci Lett 251:254–263. doi:10.1016/j.epsl.2006.09.008

Ben-Avraham Z, Tibor G (1993) The northern edge of the Gulf of Elat. Tectonophysics 226:319–331. doi:10.1016/0040-1951(93)90125-4

Ben-Avraham Z, Von Herzen RP (1987) Heat flow and continental breakup: the Gulf of Elat (Aqaba). J Geophys Res 92:1407–1416. doi:10.1029/JB092iB02p01407

Ben-Avraham Z, Zoback MD (1992) Transform-normal extension and asymmetric basins: an alternative to pull-apart models. Geology 20:423–426. doi:10.1130/0091-7613(1992)020<0423:tne aab>2.3.co;2

Ben-Avraham Z, Hänel R, Villinger H (1978) Heat flow through the Dead Sea rift. Mar Geol 28:253–269. doi:10.1016/0025-3227(78)90021-x

Ben-Avraham Z, Almagor G, Garfunkel Z (1979) Sediments and structure of the Gulf of Elat (Aqaba)–Northern Red Sea. Sediment Geol 23:239–267. doi:10.1016/0037-0738(79)90016-2

Ben-Avraham Z, Shoham Y, Klein E, Michelson H, Serruya C (1980) Magnetic survey of Lake Kinneret-central Jordan Valley, Israel. Mar Geophys Res 4:257–276. doi:10.1007/bf00369102

Ben-Avraham Z, Ginzburg A, Yuval Z (1981) Seismic reflection and refraction investigations of Lake Kinneret – central Jordan Valley, Israel. Tectonophysics 80:165–181. doi:10.1016/0040-1951(81)90148-7

Ben-Avraham Z, Amit G, Golan A, Begin ZB (1990) The bathymetry of Lake Kinneret and its structural significance. Isr J Earth Sci 39:77–83

Ben-Avraham Z, ten Brink U, Bell R, Reznikov M (1996) Gravity field over the Sea of Galilee: evidence for a composite basin along a transform fault. J Geophys Res 101:533–544. doi:10.1029/95jb03043

Ben-Avraham Z, Garfunkel Z, Lazar M (2008) Geology and evolution of the Southern Dead Sea Fault with emphasis on subsurface structure. Annu Rev Earth Planet Sci 36:357–387. doi:10.1146/annurev.earth.36.031207.124201

Ben-Avraham Z, Lyakhovsky V, Schubert G (2010) Drop-down formation of deep basins along the Dead Sea and other strike-slip fault systems. Geophys J Int 181:185–197. doi:10.1111/j.1365-246X.2010.04525.x

Eckstein Y, Simmonsi G (1977) Measurement and interpretation of terrestrial heat flow in Israel. Geothermics 6:117–142. doi:10.1016/0375-6505(77)90023-2

Ehrhardt A, Hubscher C, Ben-Avraham Z, Gajewski D (2005) Seismic study of pull-apart-induced sedimentation and deformation in the Northern Gulf of Aqaba (Elat). Tectonophysics 396:59–79

Erickson AJ (1970) The measurement and interpretation of heat flow in the Mediterranean and Black Sea, Thesis, Massachusetts Institute of Technology (unpublished)

Freund R, Garfunke Z, Zak I, Goldberg M, Weissbrod T, Derin B (1970) Shear along Dead-Sea Rift. Philos Trans R Soc Lond Ser A Math Phys Sci 267:107–130

Frieslander U, Ben-Avraham Z (1989) Magnetic field over the Dead Sea and vicinity. Mar Pet Geol 6:148–160. doi:10.1016/0264-8172(89)90018-4

Garfunkel Z (1981) Internal structure of the dead-sea leaky transform (rift) in relation to plate kinematics. Tectonophysics 80:81–108

Garfunkel Z (1989) Tectonic setting of Phanerozoic magmatism in Israel. Isr J Earth Sci 38:51–74

Garfunkel Z (1997) The history and formation of the Dead Sea basin. In: Niemi TM, Ben-Avraham Z, Gat JR (eds) The Dead Sea: the lake and its setting, Oxford monographs on geology and geophysics. Oxford University Press, New York, pp 36–56

Garfunkel Z, Ben-Avraham Z (1996) The structure of the Dead Sea basin. Tectonophysics 266:155–176. doi:10.1016/s0040-1951(96)00188-6

Garfunkel Z, Ben-Avraham Z (2001) Basins along the Dead Sea Transform. In: Ziegler PA, Cavazza W, Robertson AHF, Crasquin-Soleau S (eds) Peri-Tethyan rift/wrench basins and passive margins, vol 186, Mémoires du Muséum national d'histoire naturelle. Publications scientifiques du Muséum, Paris, pp 607–627

Ginzburg A, Ben-Avraham Z (1986) Structure of the Sea of Galilee graben, Israel, from magnetic measurements. Tectonophysics 126:153–164. doi:10.1016/0040-1951(86)90225-8

Ginzburg A, Ben-Avraham Z (1997) A seismic refraction study of the north basin of the Dead Sea, Israel. Geophys Res Lett 24:2063–2066. doi:10.1029/97gl01884

Ginzburg A, Makris J, Fuchs K, Prodehl C, Kaminski W, Amitai U (1979a) A seismic study of the crust and upper mantle of the Jordan-Dead Sea rift and their transition toward the Mediterranean Sea. J Geophys Res 84:1569–1582. doi:10.1029/JB084iB04p01569

Ginzburg A, Makris J, Fuchs K, Perathoner B, Prodehl C (1979b) Detailed structure of the crust and upper mantle along the Jordan-Dead Sea rift. J Geophys Res 84:5605–5612. doi:10.1029/JB084iB10p05605

Ginzburg A, Makris J, Fuchs K, Prodehl C (1981) The structure of the crust and upper mantle in the Dead Sea rift. Tectonophysics 80:109–119. doi:10.1016/0040-1951(81)90144-x

Ginzburg A, Reshef M, Ben-Avraham Z, Schattner U (2006) The style of transverse faulting in the Dead Sea basin from seismic reflection data: the Amazyahu fault. Isr J Earth Sci 55:129–139

Hall JK (1979) Bathymetric chart of the Dead Sea 1:100,000. Geological Survey of Israel, Jerusalem

Hall JK (1996) Digital topography and bathymetry of the area of the Dead Sea depression. Tectonophysics 266:177–185. doi:10.1016/s0040-1951(96)00189-8

Hall JK, Ben-Avraham Z (1978) New bathymetric map of the Gulf of Elat (Aqaba). Tenth international on congress on sedimentology, Abstract, vol 1. p 285

Hurwitz S, Garfunkel Z, Ben-Gai Y, Reznikov M, Rotstein Y, Gvirtzman H (2002) The tectonic framework of a complex pull-apart basin: seismic reflection observations in the Sea of Galilee, Dead Sea transform. Tectonophysics 359:289–306. doi:10.1016/s0040-1951(02)00516-4

Larsen BD, Ben-Avraham Z, Shulman H (2002) Fault and salt tectonics in the southern Dead Sea basin. Tectonophysics 346:71–90. doi:10.1016/s0040-1951(01)00229-3

Lazar M, Ben-Avraham Z, Schattner U (2006) Formation of sequential basins along a strike-slip fault-Geophysical observations from the Dead Sea basin. Tectonophysics 421:53–69. doi:10.1016/j.tecto.2006.04.007

Lazar M, Ben-Avraham Z, Garfunkel Z (2012) The Red Sea – New insights from recent geophysical studies and the connection to the Dead Sea Fault. J Afr Earth Sci 68:96–110. doi:http://dx.doi.org/10.1016/j.jafrearsci.2012.04.001

Lyakhovsky V, Ben-Avraham Z, Achmon M (1994) The origin of the Dead Sea rift. Tectonophysics 240:29–43. doi:10.1016/0040-1951(94)90262-3

Neev D, Emery KO (1967) The Dead Sea: depositional processes and environments of evaporites, Israel geological survey bulletin 41. Jerusalem, 147p

Neev D, Hall JK (1979) Geophysical investigations in the Dead Sea. Sediment Geol 23:209–238. doi:10.1016/0037-0738(79)90015-0

Nemer T, Gomez F, Al Haddad S, Tabet C (2008) Coseismic growth of sedimentary basins along the Yammouneh strike-slip fault (Lebanon). Geophys J Int 175:1023–1039. doi:10.1111/j.1365-246X.2008.03889.x

Petrunin A, Sobolev SV (2006) What controls thickness of sediments and lithospheric deformation at a pull-apart basin? Geology 34:389–392. doi:10.1130/g22158.1

Picard L (1987) The Elat (Aqaba)-Dead Sea-Jordan subgraben system. Tectonophysics 141: 23–32. doi:10.1016/0040-1951(87)90172-7

Rahe B, Ferrill DA, Morris AP (1998) Physical analog modeling of pull-apart basin evolution. Tectonophysics 285:21–40. doi:10.1016/s0040-1951(97)00193-5

Reznikov M, Ben-Avraham Z, Garfunkel Z, Gvirtzman H, Rotstein Y (2004) Structural and stratigraphic framework of Lake Kinneret. Isr J Earth Sci 53:131–149

Rofe M, Ben-Avraham Z, Tibor G (2014) Detection of sub-bottom basalt flows in Lake Kinneret using high resolution magnetic survey. (In preparation)

Schattner U, Weinberger R (2008) A mid-Pleistocene deformation transition in the Hula basin, northern Israel: implications for the tectonic evolution of the Dead Sea Fault. Geochem Geophys Geosyst 9:Q07009. doi:10.1029/2007gc001937

Schubert C (1984) Basin formation along the Bocono-Moron-El Pilar fault system, Venezuela. J Geophys Res 89:5711–5718. doi:10.1029/JB089iB07p05711

Shalev E, Lyakhovsky V, Yechieli Y (2007) Is advective heat transport significant at the Dead Sea basin? Geofluids 7:292–300. doi:10.1111/j.1468-8123.2007.00190.x

Shalev E, Lyakhovsky V, Weinstein Y, Ben-Avraham Z (2011) The thermal structure of Israel. Solid Earth Discuss 3:431–452. doi:10.5194/sed-3-431-2011

Shalev E, Lyakhovsky V, Weinstein Y, Ben-Avraham Z (2012) The thermal structure of Israel and the Dead Sea Fault. Tectonophysics. doi:http://dx.doi.org/10.1016/j.tecto.2012.09.011

Shamir G (2006) The active structure of the Dead Sea depression. In: Enzel Y, Agnon A, Stein M (eds) New frontiers in Dead Sea paleoenvironmental research, vol 401, Geological society of America special papers. Geological Society of America, Boulder, pp 15–32. doi:10.1130/2006.2401(02)

Smit J, Brun JP, Cloetingh S, Ben-Avraham Z (2008a) Pull-apart basin formation and development in narrow transform zones with application to the Dead Sea basin. Tectonics 27:n/a–n/a. doi:10.1029/2007tc002119

Smit J, Brun JP, Fort X, Cloetingh S, Ben-Avraham Z (2008b) Salt tectonics in pull-apart basins with application to the Dead Sea basin. Tectonophysics 449:1–16. doi:http://dx.doi.org/10.1016/j.tecto.2007.12.004

Smit J, Brun JP, Cloetingh S, Ben-Avraham Z (2010) The rift-like structure and asymmetry of the Dead Sea Fault. Earth Planet Sci Lett 290:74–82. doi:http://dx.doi.org/10.1016/j.epsl.2009.11.060

Sobolev SV, Petrunin A, Garfunkel Z, Babeyko AY (2005) Thermo-mechanical model of the Dead Sea Transform. Earth Planet Sci Lett 238:78–95. doi:10.1016/j.epsl.2005.06.058

Steckler MS, ten Brink US (1986) Lithospheric strength variations as a control on new plate boundaries: examples from the northern Red Sea region. Earth Planet Sci Lett 79:120–132. doi:10.1016/0012-821x(86)90045-2

Stirling MW, Wesnousky SG, Shimazaki K (1996) Fault trace complexity, cumulative slip, and the shape of the magnitude-frequency distribution for strike-slip faults: a global survey. Geophys J Int 124:833–868. doi:10.1111/j.1365-246X.1996.tb05641.x

ten Brink US, Ben-Avraham Z (1989) The anatomy of a pull-apart basin: Seismic reflection observations of the Dead Sea basin. Tectonics 8:333–350. doi:10.1029/TC008i002p00333

ten Brink US, Ben-Avraham Z, Bell RE, Hassouneh M, Coleman DF, Andreasen G, Tibor G, Coakley B (1993) Structure of the Dead Sea pull-apart basin from gravity analyses. J Geophys Res 98:21877–21894. doi:10.1029/93jb02025

ten Brink US, Al-Zoubi A, Rybakov M (2001) Bouguer gravity anomaly map of the Dead Sea Fault system, Jordan and Israel: contour interval 2 mGal. USGS open-file report 01–216

ten Brink US, Al-Zoubi AS, Flores CH, Rotstein Y, Qabbani I, Harder SH, Keller GR (2006) Seismic imaging of deep low-velocity zone beneath the Dead Sea basin and transform fault: implications for strain localization and crustal rigidity. Geophys Res Lett 33:L24314. doi:10.1029/2006gl027890

Tibor G, Niemi TM, Ben-Avraham Z, Al-Zoubi A, Sade RA, Hall JK, Hartman G, Akawi E, Abueladas A, Al-Ruzouq R (2010) Active tectonic morphology and submarine deformation of the northern Gulf of Eilat/Aqaba from analyses of multibeam data. Geo-Mar Lett 30:561–573. doi:10.1007/s00367-010-0194-y

van Eck T, Hofstetter A (1990) Fault geometry and spatial clustering of microearthquakes along the Dead Sea-Jordan rift fault zone. Tectonophysics 180(1):15–27. doi:10.1016/0040-1951(90)90368-I, ISSN 0040–1951

Wdowinski S, Zilberman E (1997) Systematic analyses of the large-scale topography and structure across the Dead Sea rift. Tectonics 16:409–424. doi:10.1029/97tc00814

Weber M, Abu-Ayyash K, Ben-Avraham Z, Choi S, Darwish J, El-Kelani R, Garfunkel Z, Götze H-J, Hofstetter R, Koulakov I, Laske G, Mechie J, Meyer U, Mohsen A, Petrunin A, Rioseco EM, Ryberg T, Rümpker G, Sobolev SV (2014) Geophysical studies of the lithosphere along the Dead Sea Transform. In: Garfunkel Z, Ben-Avraham Z (eds) The Dead Sea Transform. Springer, Heidelberg, pp 29–52

Wesnousky SG (1988) Seismological and structural evolution of strike-slip faults. Nature 335:340–343

Wesnousky SG (1994) The Gutenberg-Richter or characteristic earthquake distribution, which is it? Bull Seismol Soc Am 84:1940–1959

Wu JE, McClay K, Whitehouse P, Dooley T (2009) 4D analogue modelling of transtensional pull-apart basins. Mar Pet Geol 26:1608–1623. doi:10.1016/j.marpetgeo.2008.06.007

Chapter 2
Geophysical Studies of the Lithosphere Along the Dead Sea Transform

Michael Weber, Khalil Abu-Ayyash, Zvi Ben-Avraham, Sungchan Choi, Jaser Darwish, Radwan El-Kelani, Zvi Garfunkel, Hans-Jürgen Götze, Abraham Hofstetter, Ivan Koulakov, Gabi Laske, James Mechie, Uwe Meyer, Ayman Mohsen, Alexei Petrunin, Ernesto Meneses Rioseco, Trond Ryberg, Georg Rümpker, Stephan V. Sobolev, and DESERT & DESIRE Groups

Abstract In this chapter we report on the deep structure of the Dead Sea Transform (DST) as derived from geophysical observations and numerical modelling, calibrated by geological and geodynamic evidence.

We use seismics, seismology and gravity to study the crust and lithosphere of the Dead Sea Transform (DST) system. These observations are integrated with 3D thermo-mechanical modelling of the evolution of the DST through time to understand the deeper structure of the DST. The three seismic profiles crossing the DST

M. Weber (✉) • J. Mechie • A. Mohsen • A. Petrunin • T. Ryberg • S.V. Sobolev
Deutsches GeoForschungsZentrum, GFZ, Telegrafenberg, Potsdam 14473, Germany
e-mail: mhw@gfz-potsdam.de; jimmy@gfz-potsdam.de; ayman@gfz-potsdam.de; Petrunin.alexey@gfz-potsdam.de; trond@gfz-potsdam.de; stephan@gfz-potsdam.de

K. Abu-Ayyash • J. Darwish
Natural Resources Authority, P.O. Box 7, Amman 11118, Jordan
e-mail: Khalil.abuayyash@gmail.com; jaser@nra.gov.jo

Z. Ben-Avraham
Department of Geophysical, Atmospheric and Planetary Sciences, Tel Aviv University, Ramat Aviv, Tel Aviv 69978, Israel

Charney School of Marine Sciences, University of Haifa, Mt. Carmel, Haifa 3190501, Israel
e-mail: zviba@tau.ac.il

S. Choi • H.-J. Götze
Christian-Albrechts-Universitaet zu Kiel, Christian-Albrechts-Platz 4, 24118 Kiel, Germany
e-mail: choi@geophysik.uni-kiel.de; hajo@geophysik.uni-kiel.de

R. El-Kelani
An-Najah National University, P.O. Box 7, Nablus, Palestine
e-mail: radwan@najah.edu

Z. Garfunkel
The Fredy and Nadine Herrmann Institute of Earth Sciences, The Hebrew University of Jerusalem, Edmond J. Safra campus, Givat Ram, Jerusalem 91904, Israel
e-mail: Zvi.garfunkel@huji.ac.il

Z. Garfunkel et al. (eds.), *Dead Sea Transform Fault System: Reviews*, Modern Approaches in Solid Earth Sciences 6, DOI 10.1007/978-94-017-8872-4_2,

from the Mediterranean in the West to the Jordan highlands in the East show an increase in Moho depth from about 25 km to about 35 km; with only minor topography. This depth increase of about 10 km of the Moho from West to East is also found in tomographic images using regional and teleseismic events, which shows additionally a N – S trending thickening of the crust under the Arava/Araba Fault (AF). In the Dead Sea Basin (DSB) proper the imaging of the Moho is complicated by the presence of the Lisan Salt dome. From these results and other evidence we conclude that the Dead Sea basin is a mostly upper crustal feature with a decoupling zone at about 20 km depth. Using SKS waves we find below the Moho under the DST a narrow, ca. 20 km wide, vertical decoupling zone reaching into the mantle, representing the boundary layer between the African and Arabian plates. This observation agrees with the results from the study of surface waves that also show a region of reduced S-velocities under the DST, reaching down into the lithosphere. Whereas the lithosphere thins gradually east of the DST from N to S from ca. 80 to ca. 67 km, below about 120 km depth little structure can be observed in tomographic images.

The abovementioned observational constraints can all be fitted with the classical pull-apart model, if the lithosphere was thermally eroded to 80 km thickness about 20 Ma ago, combined with weak rheologies for crust and upper mantle. The most likely explanation of the features described is thus a thinning of the lithosphere around the DST in the Late Cenozoic, likely following by rifting and spreading of the Red Sea.

A. Hofstetter
Geophysical Institute of Israel, Lod 71100, Israel

Ecole et Observatoire des Sciences de la Terre,
University of Strasbourg, Strasbourg 67000, France
e-mail: ramih@gii.co.il

I. Koulakov
Institute of Petroleum Geology and Geophysics,
3, Akademika Koptyuga Prosp., Novosibirsk 630090, Russia
e-mail: koulakovIY@ipgg.nsc.ru

G. Laske
University of California, San Diego, 9500 Gilman Drive, La Jolla, CA 92093-0225, USA
e-mail: glaske@ucsd.edu

U. Meyer
Bundesanstalt fuer Geowissenschaften und Rohstoffe,
Stilleweg 2, 30655 Hannover, Germany
e-mail: Uwe.meyer@bgr.de

E.M. Rioseco
Leibniz-Institut fuer Angewandte Geophysik, Stilleweg 2, 30655 Hannover, Germany
e-mail: Ernesto.menesesrioseco@liag-hannover.de

G. Rümpker
Goethe-Universitaet, Grueneburgplatz 1, 60323 Frankfurt am Main, Germany
e-mail: rumpker@geophysik.uni-frankfurt.de

DESERT & DESIRE Groups
Deutsches GeoForschungsZentrum, GFZ, Telegrafenberg, Postsdam 14473, Germany

Keywords Geophysics • Modelling • Moho • Arava/Araba fault • Lithosphere • Upper mantle

2.1 Introduction

The Dead Sea Transform (DST) represents a key site to study structure and evolution of large plate boundaries. The DST is a left-lateral fault zone containing a series of deep pull-apart basins, most prominently the Dead Sea Basin (DSB), and extends from the Red Sea spreading centre in the South to the Zagros zone of plate convergence in the North. The present-day relative motion along the DST is ~5 mm/year (Klinger et al. 2000a; Wdowinsky et al. 2004; Bartov and Sagy 2004; Mahmoud et al. 2005; Le Beon et al. 2008; Reilinger et al. 2006). For a review of its development see Garfunkel (1988, 2014, this volume) and Ben-Avraham et al. (2008) and references therein.

The DST is seismically active with catastrophic earthquakes with magnitudes up to 7 (Garfunkel et al. 1981; Ambraseys et al. 1994; Amiran et al. 1994; Klinger et al. 2000b; Ken-Tor et al. 2001; Migowski et al. 2004; Agnon et al. 2006; Hofstetter et al. 2003; Grünthal et al. 2009). Its ongoing micro-seismic activity has been documented by van Eck and Hofstetter (1989), Amrat et al. (2001), Salamon et al. (2003), Aldersons et al. (2003), Alderson and Ben-Avraham (2014, this volume), Grünthal et al. (2009), and Hofstetter et al. (2014, this volume). Since the establishment of seismic networks in Jordan and Israel in about 1980, seismicity can be monitored down to a magnitude of 2–3. For details see Hofstetter et al. (2014, this volume).

The study presented here focuses on the DST segment between the Dead Sea and the Red Sea (Fig. 2.1). It integrates results from several interdisciplinary efforts, especially the DESERT project (DEad SEa Rift Transect) and the DESIRE project (Dead Sea Integrated REsearch project). Here we focus on the results of lower crustal and upper mantle structure, whereas the shallow structure of the DST is presented by Ben-Avraham (2014, this volume). The general layout of the DESERT and DESIRE experiments is given in Fig. 2.1. A description of the main results of DESERT, including near surface studies and a comparison of the DST to other large faults such as the San Andreas Fault, can be found in Weber et al. (2009, 2010). A recent joint interpretation of DESERT and DESIRE results for crustal structure can be found in Weber et al. (2009/2010).

In the next section we summarize the results of seismic, seismological and gravity experiments for the DSB and the Arava/Araba Fault (AF) for deep crustal and upper mantle structure. These results are then integrated with findings from thermo-mechanical modelling, followed by the summary.

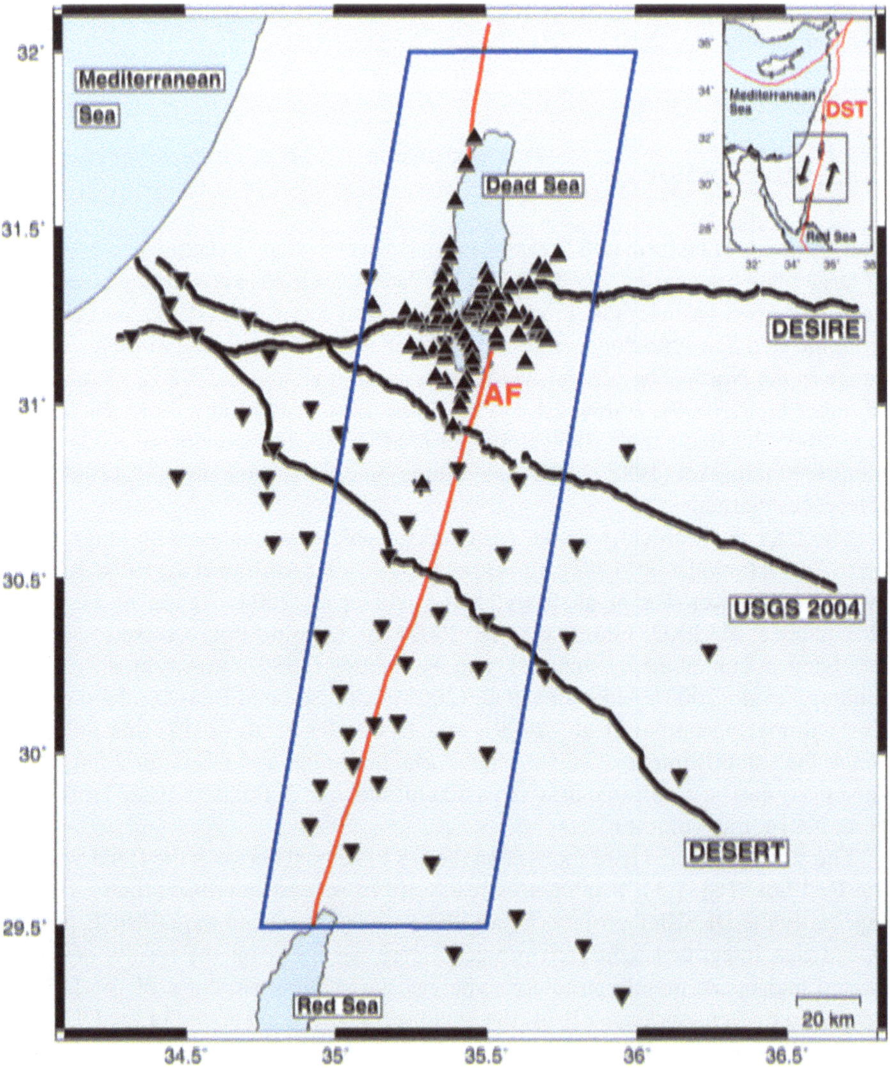

Fig. 2.1 Setting of the Dead Sea Transform (DST) and location of geophysical experiments. The *red lines* indicate the main branches of the DST, e.g. the Arava/Araba Fault (AF). Wide-angle Reflection/Refraction (WRR) profiles of DESERT, USGS2004 and DESIRE are given in *black*, from N to S. DESIRE – 235 km long, 11 shots, 600 receivers (Mechie et al. 2009), USGS2004 – 250 km long, 5 shots, 334 receivers (ten Brink et al. 2006) and DESERT – 260 km long, 13 shots, 99 receivers; (DESERT Group 2004; Mechie et al. 2005). Passive long-term deployments (1–1.5 years) of seismological stations of DESERT (48 stations) and DESIRE (80 stations) are shown as inverted and *upright triangles*, respectively. The *blue box* shows the area of the helicopter gravity survey of DESIRE (Götze et al. 2010a, b). (**Insert**) Main faults of the Levant, see e.g. Garfunkel (1981), Ben-Avraham (1985). The *left lateral* displacement of 107 km at the DST is indicated by *black arrows*

2.2 Deep Crustal Structure

Figure 2.2 gives the P-velocity models derived along three wide-angle reflection/refraction (WRR) profiles (from South to North, DESERT – DESERT Group 2004; USGS 2004 – ten Brink et al. 2006; DESIRE – Mechie et al. 2009).

The DESERT experiments were performed in 2000, for parameters see the caption of Fig. 2.1. The DESERT profile shows a Moho with a steady increase in depth from about 26 km at the Mediterranean (NW) to about 39 km under Jordan (SE) (Fig. 2.2c) with only a small but visible, asymmetric topography of the Moho under the AF. In the NW the Moho depth at the coast is in agreement with the value determined by Makris et al. (1983). Towards the SE the Moho has been continued to dip to the SE as also indicated by El-Isa et al. (1987a, b), although the Moho depths in El-Isa et al. (1987a, b) are a few kilometers less than those shown here.

In 2004 ten Brink et al. (2006) shot the USGS experiment located between the DESERT profile and the Dead Sea (Fig. 2.1). They found a low-velocity zone, the signature of the southern end of the DSB, down to the mid-crust. Similar to DESERT the Moho depth increases smoothly from West to East from about 25 km at the Mediterranean to about 35 km under Jordan (Fig. 2.2b). No significant topography of the Moho was derived in their model.

The third WRR experiment across the DST was performed in 2006 through the DSB proper (Mechie et al. 2009 and Fig. 2.1). The model of the P-wave velocity structure in Fig. 2.2a shows a depth of the seismic basement, defined here as the layer with P-wave velocities of at least 5.6 km s^{-1}, beneath the southern DSB of about 11 km, while an earlier experiment (Ginzburg and Ben-Avraham 1997) suggested that the seismic basement reaches a maximum depth of about 14 km, about 10 km south of the DESIRE profile. In contrast to this strong topography in the upper crust the interfaces below about 20 km depth, including the top of the lower crust and the Moho, probably show less than 3 km variation in depth beneath the profile as it crosses the southern DSB. The models of DESIRE and DESERT were forced to agree at their crossing point near the Mediterranean (Fig. 2.1) and thus they are consistent there. No such boundary condition was imposed for the USGS profile, and thus the misfit in Moho depth between the USGS and DESIRE profiles at their crossing point (at about profile kilometer 70 and 60 for DESIRE and USGS, respectively) is 6.5 km. Since a model for the DESIRE profile with a similarly shaped Moho as in the USGS profile provides a significantly worse fit to the data, we think the model presented in Fig. 2.2a is appropriate.

An alternative method to determine the depth of the Moho is the receiver function method (RFM). It uses P-to-S converted phases which are produced as teleseismic P waves cross a seismic discontinuity. Detailed descriptions of the RFM have been reported in a number of papers (e.g. Vinnik 1977; Kind et al. 1995). Figure 2.3 gives the results of two studies using the stations from DESERT and DESIRE (Fig. 2.1 left and right), respectively (Mohsen et al. 2005, 2011).

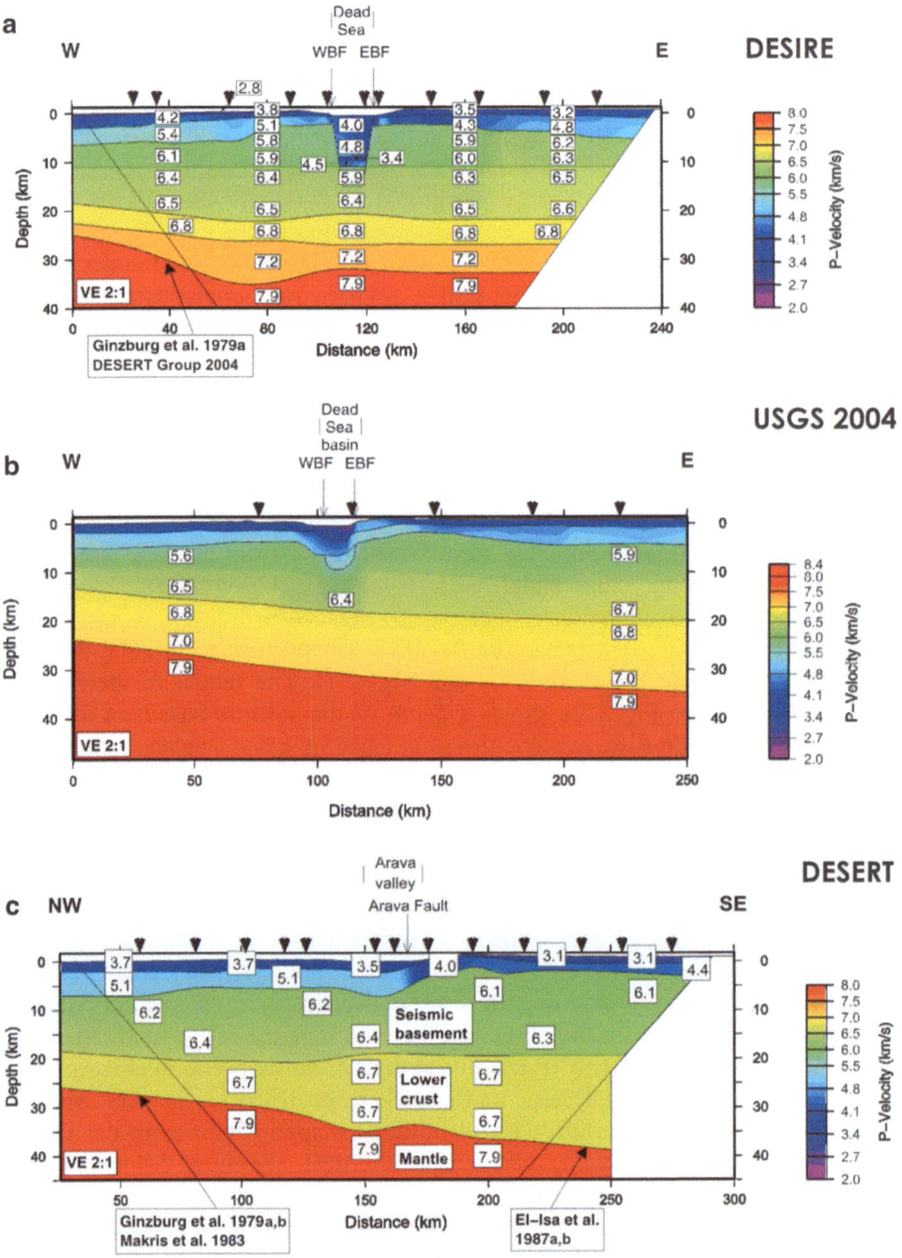

Fig. 2.2 Crustal structure under the 235, 250 and 260 km long WRR profiles of DESIRE, USGS2004 and DESERT (*black lines* in Fig. 2.1) from the Mediterranean coast (W) to the Jordan highlands (E), respectively. (**a**) DESIRE: P-wave velocity model (Modified from Mechie et al. 2009). Velocities are in km/s. *Triangles* at the *top* represent the shot-points. Only the region within the *diagonal lines* is resolved in this study. To the W the boundaries and P-wave velocities are based on previous work by Ginzburg et al. (1979a) and DESERT Group (2004). WBF = Western

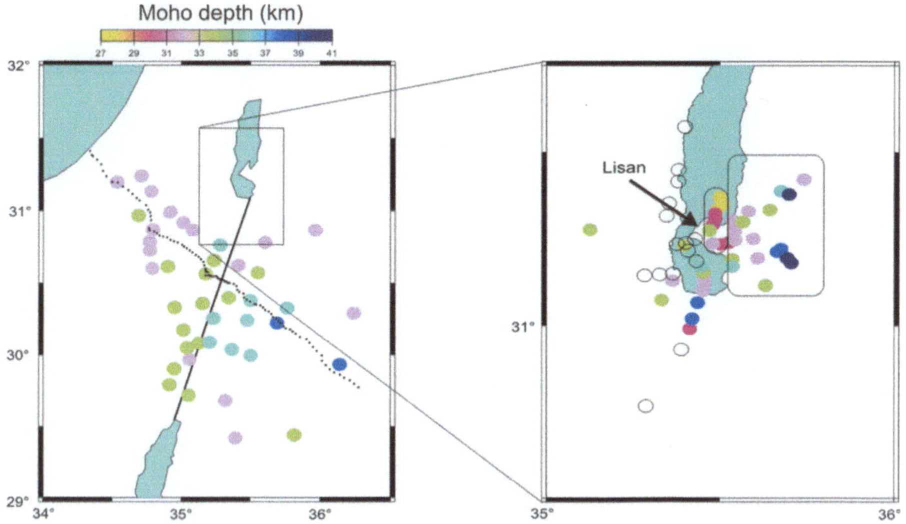

Fig. 2.3 Map of Moho depths derived with the Receiver Function Method (RFM; Mohsen et al. 2005, 2011). The *circles* indicate the conversion points at the Moho and their depths, see colour code at *top*. *White circles* are locations where no Moho depth could be determined. The WRR profile of DESERT is given as *black dots*, see Fig. 2.2 for depth-sections. The **blow-up** shows the results of DESIRE in the vicinity of the Dead Sea in more detail. The smaller cluster (mostly *red* and *yellow*) is located under the Lisan Peninsula; the larger cluster covers the Moho under the eastern shoulder (Modified from Mohsen et al. 2005, 2011)

In the DESERT project (Fig. 2.3, left) the Moho depth increases smoothly across the DST from about 31 km in the NW towards 34 to 38 km in the SE. As shown in Mohsen et al. (2005, Fig. 2.10) the Moho depths of the WRR experiment, from reflection seismics along the DESERT profile (DESERT Group 2004; Fig. 2.11) and from the RFM shown in Fig. 2.3 (left), agree very well. The Moho east of the DST forms a trough-like feature with the deepest parts along the WRR profile of DESERT (dotted line in Fig. 2.3, left). This feature is limited to the DESERT profile, and no Moho deepening north and south of the WRR profile is observed.

Along the DESIRE profile the Moho from the WRR experiment (Fig. 2.2a) and the RFM (Fig. 2.3, right) generally agree better than 3 km (Mohsen et al. 2011, Fig. 2.11). Unfortunately the vertical-incident reflection seismic experiments along the DESIRE profile do not show a clear Moho (Weber et al. 2009/2010, Fig. 2.3). Imaging problems due to the complicated 2-D/3-D structure of the DSB manifest themselves in the 15 stations/piercing points on the western shoulder of the DSB in

---

Fig. 2.2 (continued) Boundary Fault, EBF = Eastern Boundary Fault. Note vertical exaggeration of 2:1 (Modified from Mechie et al. 2009). (**b**) As (a) but for USGS2004 (Modified from ten Brink et al. 2006). (**c**) As (a) but for DESERT. To the NW the boundaries and P-wave velocities are based on previous work by Ginzburg et al. (1979a, b) and Makris et al. (1983) while to the SE the boundaries and the P-wave velocities are based on El-Isa et al. (1987a, b) (Modified from DESERT Group (2004) and Mechie et al. (2005))

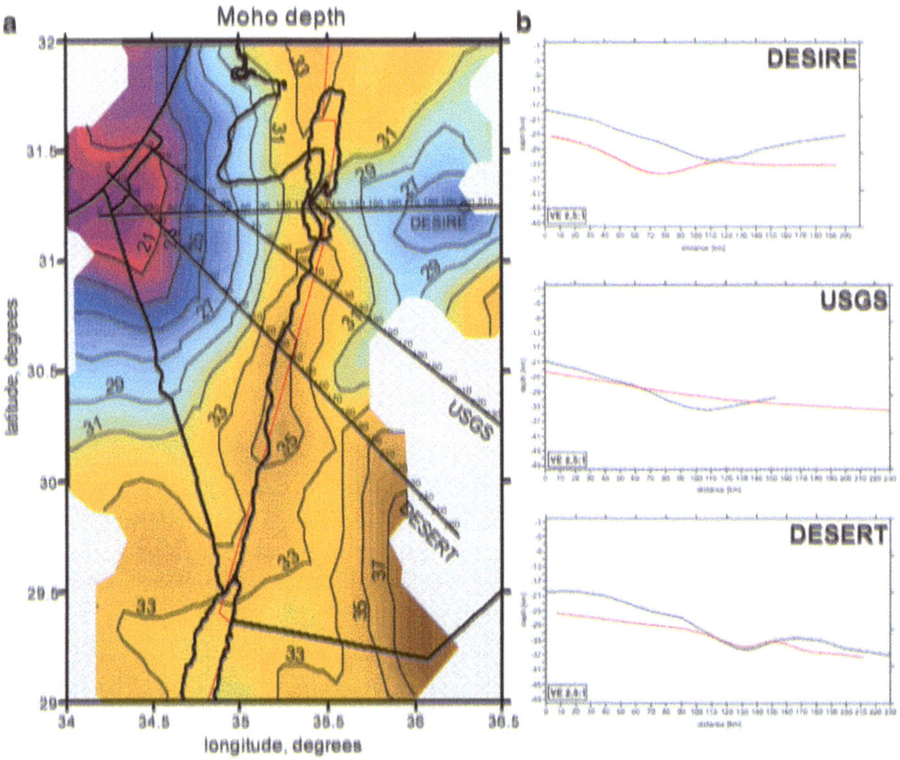

Fig. 2.4 (**a**) Map of Moho depths derived from tomographic inversion using local and regional earthquakes and local and regional stations (Koulakov and Sobolev 2006). Isolines of the Moho depths in kilometres are shown and marked. The WRR profiles of DESIRE, USGS and DESERT are shown with distance along the profiles (see also Fig. 2.1) (Modified from Koulakov and Sobolev 2006). (**b**) *Black lines* – Moho along the three profiles according to the tomographic inversion shown in (**a**). *Red lines* – Moho derived in the WRR experiments (DESIRE, USGS, DESERT from N to S, respectively; see Fig. 2.2) (Note vertical exaggeration of 2.5:1)

Fig. 2.3(right) where no Moho could be determined (open circles). The locations with good data fall into two groups. The larger group east of the DSB shows a general agreement with the results of the WRR experiment in Fig. 2.2a. The smaller cluster in Fig. 2.3(right) is centered on the Lisan peninsula, underlain by a large salt dome. It seems to have a Moho which is shallower by 3–4 km, compared to the Moho under Jordan. Since Fig. 2.2a does not show such a feature we speculate that this discrepancy is due to 2-D/3-D effects, especially on and in the immediate vicinity of the Lisan peninsula.

Using P- and S-wave picks from the ISC catalogue (ca. 250 stations and 3,000 events) in the eastern Mediterranean Koulakov and Sobolev (2006) derived a map of Moho depths for this region (Fig. 2.4a). With this data set a larger spatial coverage than in DESIRE + USGS + DESERT, albeit with lower resolution, could be achieved. The general trend of increased crustal thickness from the Mediterranean towards the East can again be seen. To be able to compare these results with those from the WRR studies Fig. 2.4b shows depth sections along the three WRR profiles. The agreement between the ISC results (black lines in Fig. 2.4b) and the WRR results (red lines in Fig. 2.4b) is usually better then 3–4 km. The biggest discrepancies occur along the DESIRE profile. We attribute these discrepancies to the strong lateral inhomogeneities in the DSB and its immediate vicinity, since, as shown in Weber et al. (2009/2010), on a scale of a few kilometers velocities might change by up to 50 % laterally down to a depth of several kilometers. The robust feature of the tomographic image in Fig. 2.4 is the narrow, ca. 50 km wide band of crustal thickening under the AF between the Red Sea and the Dead Sea. This structure is most likely associated with the minimum of lithospheric strength (Sobolev et al. 2005) and, therefore, explains the occurrence of the Arava/Araba Valley.

Simultaneously with the Moho depth Koulakov and Sobolev (2006) also inverted for the 3-D P- and S-velocity in the crust and upper-most mantle under the DST (Fig. 2.5). The N-S stretching low-velocity anomalies above the Moho (left column in Fig. 2.5) are interpreted as sediments and a zone of fractured and deformed rocks in the middle and lower crust under the Arava/Araba valley between the Dead Sea and the Red Sea. The anomalies in the upper mantle below the Moho (right column in Fig. 2.5) are similarly interpreted as a zone of fractured and deformed rocks due to the transform motion, for more details see also next chapter.

Further constraints on the deep crustal structure come from 3-D density modeling (Götze et al. 2006, 2010a, b; Choi et al. 2011). The 3-D density model is based on a newly compiled Bouguer anomaly map of this area, integrating regional gravity data from Israel and Jordan, local ground-based gravity surveys in DESERT and DESIRE and data from a recently acquired helicopter gravity survey (blue box in Fig. 2.1, Götze et al. 2010a, b; Choi et al. 2011). Figure 2.6 shows a map of the Moho depth using constraints from the three WRR experiments (DESIRE, USGS, DESERT; Fig. 2.2) and the RFM study by Mohsen et al. (2005; Fig. 2.3). Except for the coastal area, where little seismic control is available, the gravity and seismological results usually agree better than 2 km. Again the most obvious feature is the gradual increase in Moho depth from West to East. The most significant deviation from this gradual change in Moho depth from the Mediterranean in the NW to the Arabian platform in the East is the significant increase in crustal thickness at the SE end of the DESERT profile to about 37 km.

Integrating the seismic (WRR profiles), seismological (RFM and regional tomography) and gravity results we conclude that little variation exists in the depth to the Moho or the top of the lower crust except directly under the DST. The main difference in N–S direction is in the thickness of the upper crust. Following ten Brink et al. (2006), Mechie et al. (2009), and Weber et al. (2009) we argue that the Dead Sea pull-apart basin is essentially an upper crustal feature with predominantly N-S upper crustal extension, associated with the left-lateral motion along the DST. This motion is accompanied by side-stepping of the major strike-slip motion from the east side of the valley at the southern end of the basin to the west side of the

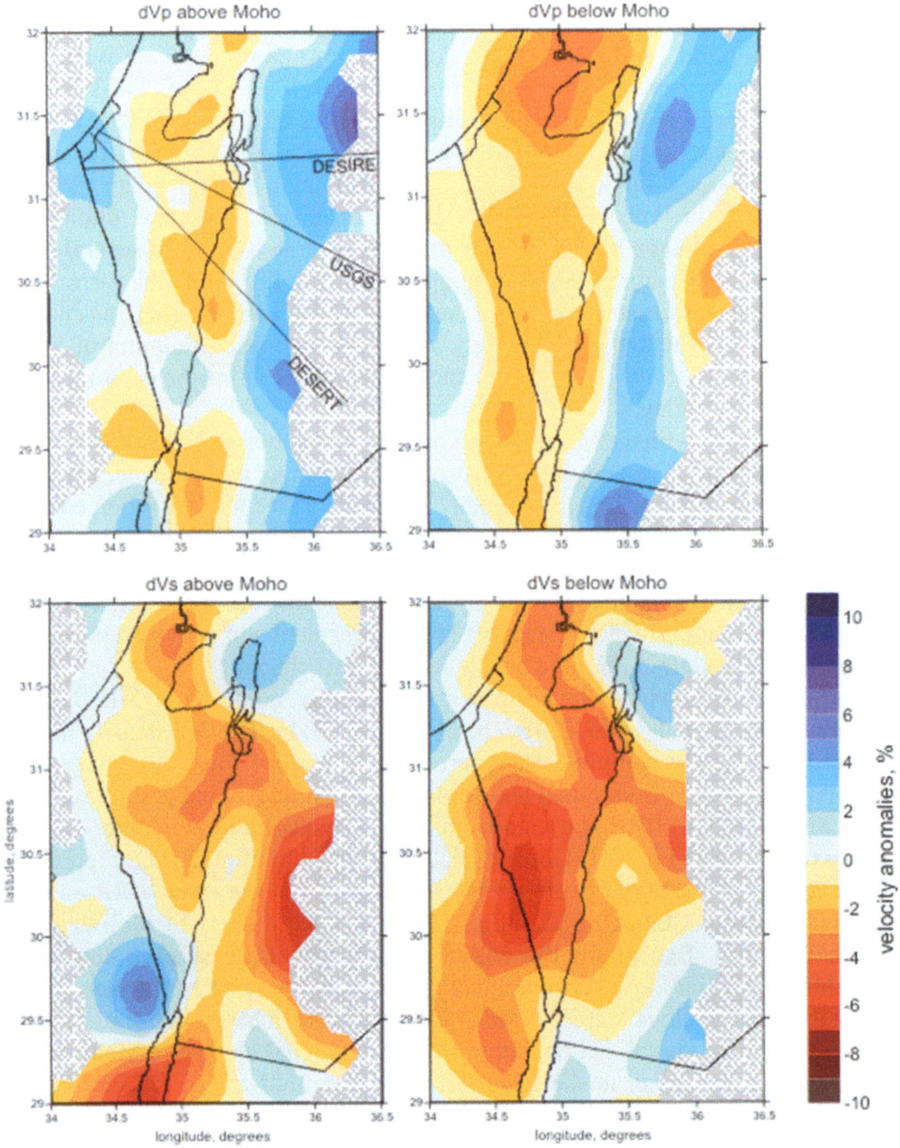

Fig. 2.5 P and S-velocities above and below the Moho derived from tomographic inversion using local and regional earthquakes and stations (Koulakov and Sobolev 2006). The WRR profiles of DESERT, USGS and DESIRE are indicated by *black lines* (Modified from Koulakov and Sobolev 2006)

valley north of the basin. The boundary between the upper and lower crust at about 20 km depth might act as a decoupling zone. Below this boundary the two plates move past each other in essentially a shearing motion.

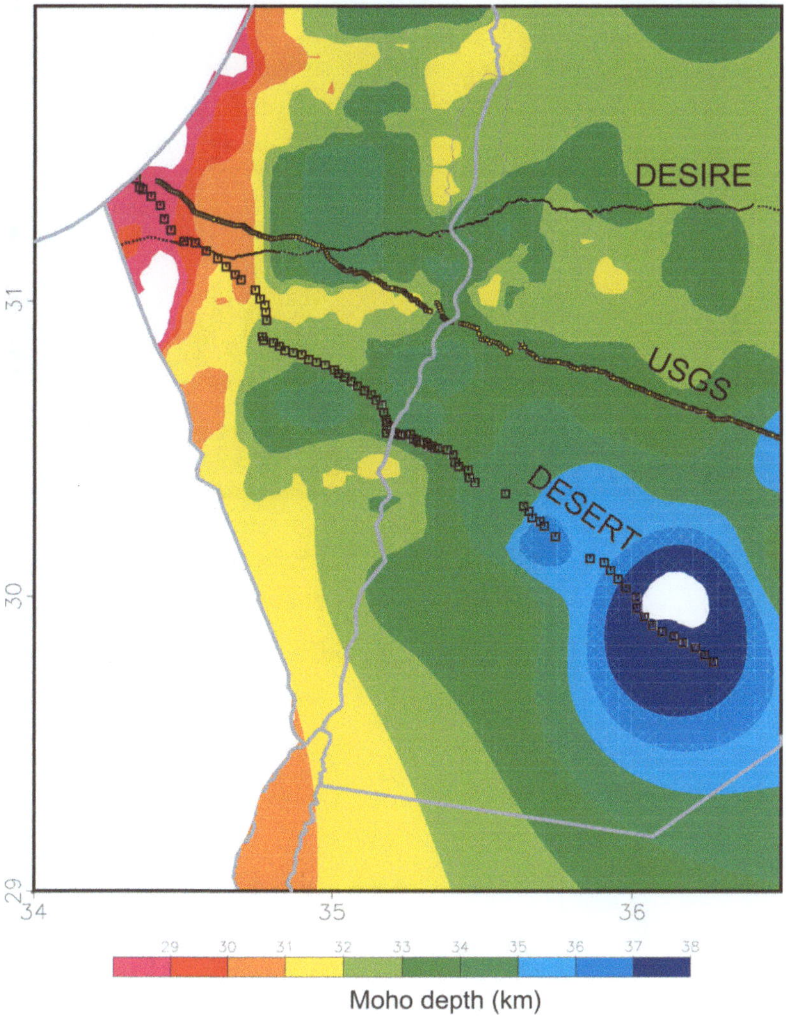

Fig. 2.6 Moho depth map beneath the 3D gravity modelling area (Götze et al. 2006, 2010a, b; Choi et al. 2011). Constraints used in the gravity modelling are seismic velocities and interfaces from the three seismic lines (DESIRE, USGS, DESERT; Fig. 2.2) and the results of the RF study by Mohsen et al. (2005). The three profiles are indicated by three *straight black lines* (Modified from Choi et al. 2011)

2.3 Upper Mantle Structure

Delineating plate boundary deformation in the sub-crustal lithosphere is difficult, especially if these zones of deformation are vertical and narrow. The analysis of shear-wave splitting of teleseismic phases (SKS) provides a tool to infer deformation and crystal alignment in the lithosphere in such situations. Rümpker et al.

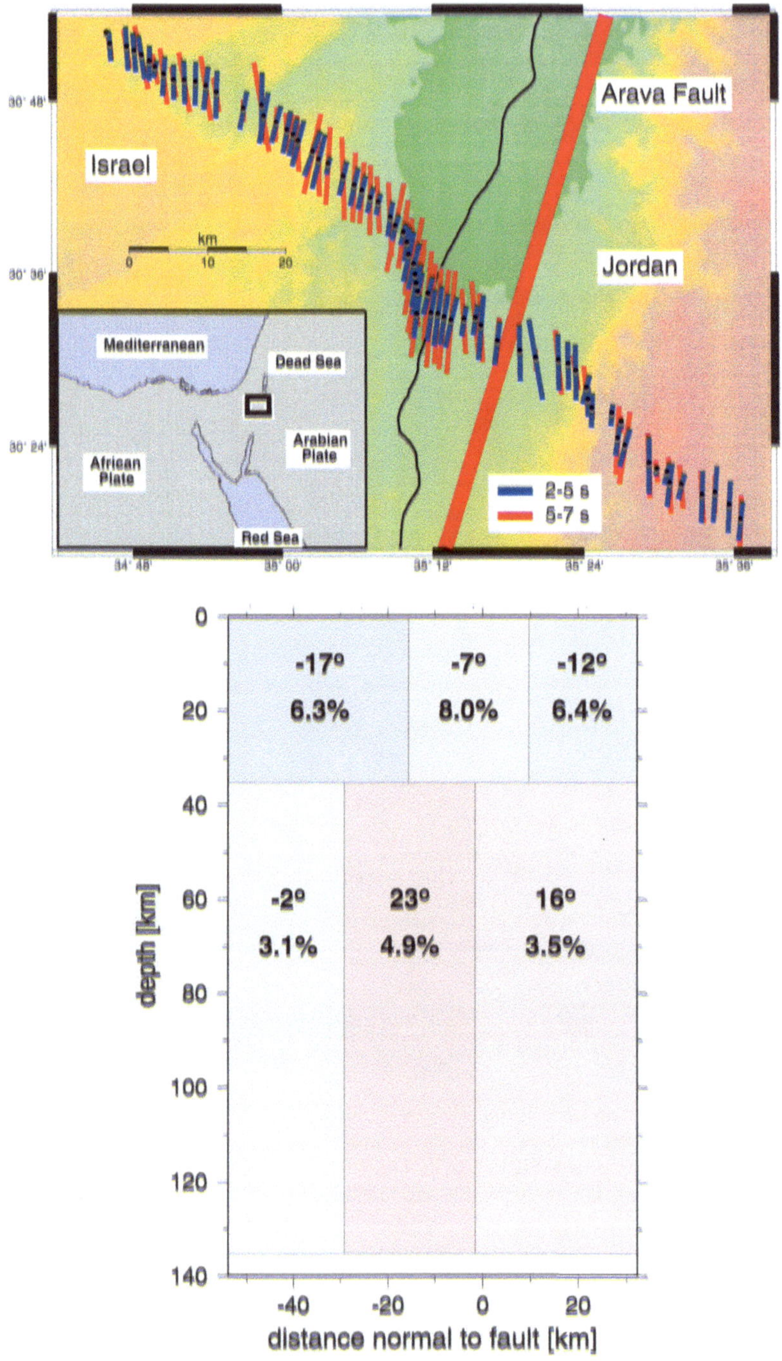

Fig. 2.7 (*Top*) Map with topography and the locations of seismic stations (100 km profile) for which the shear-wave splitting analysis was performed. The AF (*red line*) strikes at approximately N20E. The bars indicate measured SKS splitting parameters for the period range of 2–5 s (*blue*) and 5–7 s (*red*). The orientation corresponds to the polarization direction of the fast shear wave (fast polarization Φ) and the length is proportional to the delay time (δt–1) s. (*Bottom*) Preferred P wave

(2003) and Ryberg et al. (2005) used the recording of a large teleseismic event at a closely spaced profile of DESERT (spacing ≈ 1 km) looking at frequency dependent splitting effects (Fig. 2.7, top). A best fitting model of anisotropic structure in the crust and upper mantle under the DST was determined from these data using local optimization techniques and a global genetic algorithm search. This model is characterized by three crustal blocks (Fig. 2.7, bottom) with the central one containing the present day trace of the DST. Below the Moho a narrow, approximately 20 km wide, vertical decoupling zone in the mantle is found. It is characterized by 20–60 % higher anisotropy and an orientation which is different relative to the neighboring blocks. This zone with a fault-parallel symmetry axis is interpreted as a vertical boundary layer accommodating the transform motion between the Arabian and African plates in the upper mantle.

As part of DESERT Laske et al. (2008) used fundamental mode Rayleigh waves at intermediate periods to show that the sub-crustal S-velocity under the region of the DST is, on average, 5 % lower than in the preliminary reference Earth model (PREM) (Dziewonski and Anderson 1981). This velocity reduction reaches down to at least 200 km (Fig. 2.8). Such a feature is in agreement with larger-scale studies such as Pasyanos and Nyblade (2007), indicating that such an area of reduced velocities in the lithosphere stretches along the DST. As shown in Fig. 2.8 (top row) Laske et al. (2008) found in this region a ca. 80 km thick lid of normal S-velocity (blue colors) to the west of the DST indicative of thermally unaffected crust and upper mantle. The eastern part in contrast is slow relative to PREM. We would also like to point out that the whole region is underlain by an upper mantle, down to 410 km depth, with 3–4 % reduced S-velocities compared to PREM (Mohsen et al. 2006).

Complementary to the study of the upper mantle with surface waves Koulakov et al. (2006) imaged the P-velocity in the upper mantle under the DST using a tomographic inversion of body waves recorded at the DESERT deployment (Fig. 2.1). The low-velocities in the DSB are mapped down to a depth of ca. 40 km (Fig. 2.9). Another large feature in Fig. 2.9 is the high-velocity P-wave anomaly of about +1–2 % in the SE of the area covered, that spans the depth range from approximately 40–120 km depth. A possible explanation for this could be a slightly lower temperature ($\Delta T <$ ca. 100 °C) and/or a preexisting lithospheric compositional anomaly in this region. Below about 120 km depth very little structure and anomalies can be detected in the tomographic images, indicating that inhomogeneities and effects of the plate-boundary DST are mostly limited to the top 100 km. This holds especially for the lithosphere-asthenosphere boundary that shows very little topography.

Fig. 2.7 (continued) velocity anisotropy model for crust and uppermost mantle under the DST on the 100 km long profile perpendicular to the AF (at 0 km) (Rümpker et al. 2003; Ryberg et al. 2005). The azimuth of the symmetry axis and the magnitude of the anisotropy are given in each block. The model is characterized by a central upper mantle zone of increased anisotropy, which is differently oriented with respect to its neighbors. No vertical exaggeration (Modified from Rümpker et al. (2003) and Ryberg et al. (2005))

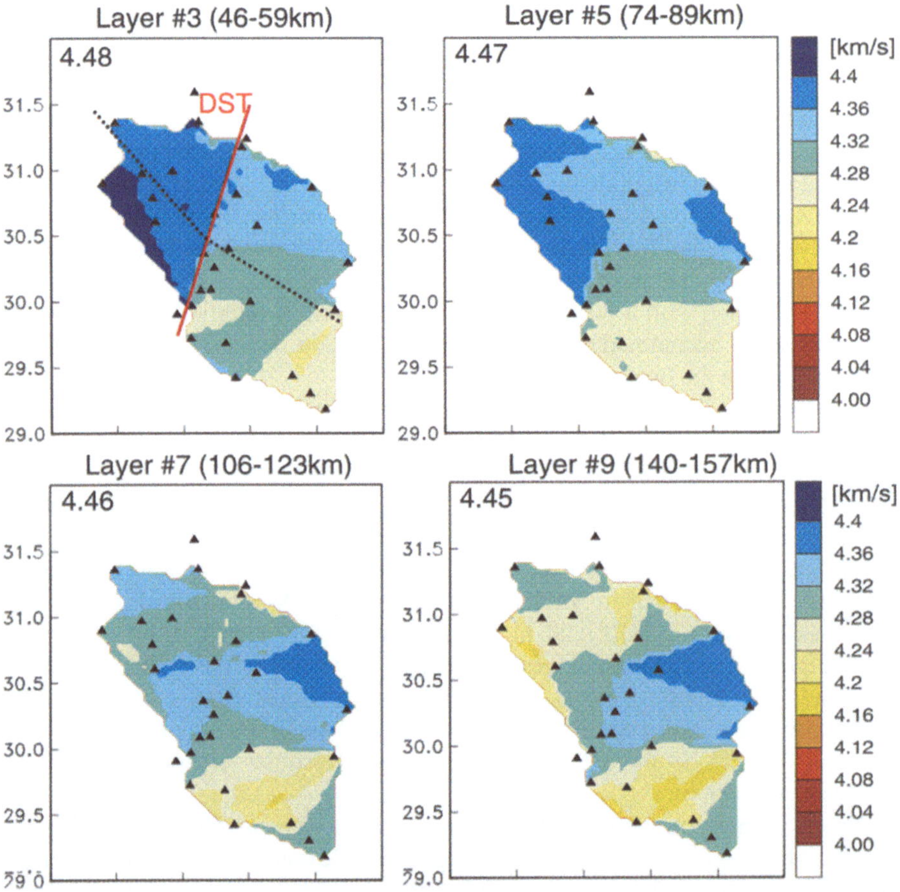

Fig. 2.8 Shear velocity variation in four sub-crustal layers (46–157 km depth) in a velocity model derived from surface waves (Laske et al. 2008). The model is defined on a 0.05 deg grid and then smoothed. The number in the upper left-hand corner in each panel denotes average velocities in km/s in isotropic PREM (Dziewonski and Anderson 1981). The DST is given in *red*, and the DESERT profile as a *dotted black line* for reference, respectively (Modified from Laske et al. 2008)

This observation is confirmed by Mohsen et al. (2006, 2011) who used S-to-P converted waves to delineate the lithosphere-asthenosphere boundary (LAB). The results for the DESERT deployment are given in Fig. 2.10(left) and those of DESIRE in Fig. 2.10 (right) respectively. Very gradual thinning of the lithosphere on the eastern side of the DST from ca. 80 km in the north of the Dead Sea to about 67 km at the Red Sea can be observed. On the western side of the DST only very few piercing points are available, since most of the events are located in the NE quadrant, but the few data indicate a thin LAB of about 67 km. The dense

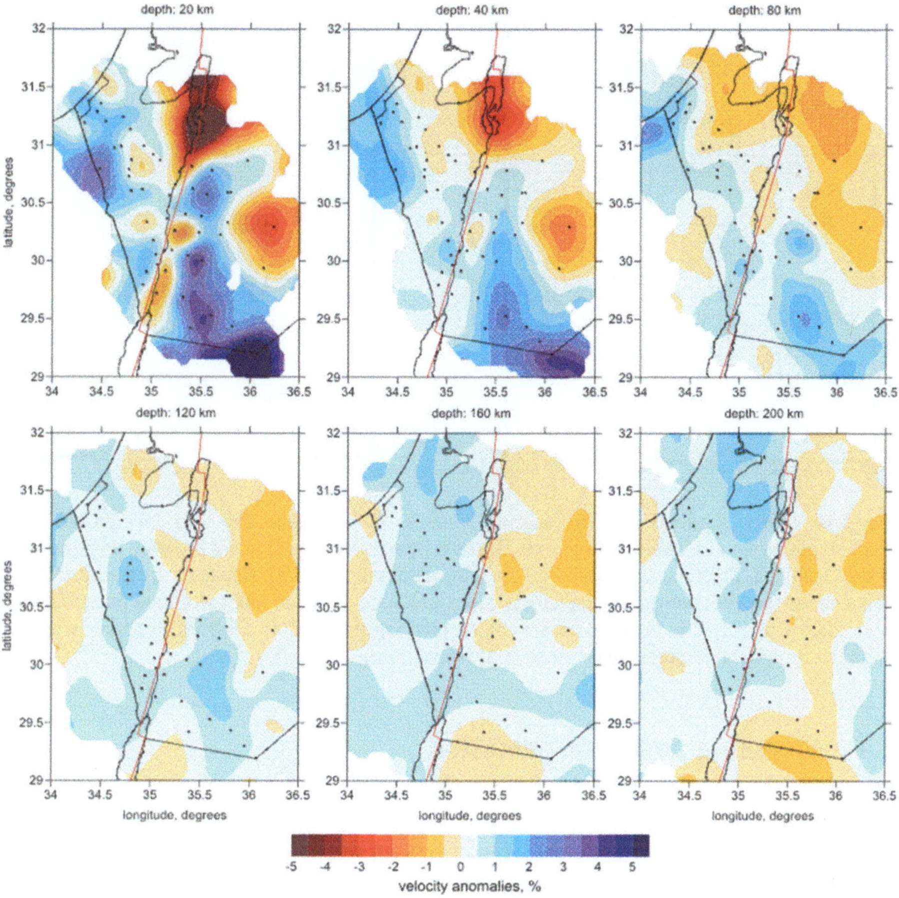

Fig. 2.9 P-velocity anomalies, see color bar at *bottom*, in horizontal sections from 20 to 200 km depth. The stations from DESERT used are indicated by *small black dots* (Modified from Koulakov et al. 2006)

deployment of DESIRE allows testing the values in block 5 of Fig. 2.10 (left) since more than 350 piercing points cover the insert (Fig. 2.10, right). The LAB value of 75 ± 5 km from DESERT agrees well with the values of 73 km derived with DESERT data alone.

The observations presented in Fig. 2.10 support the conclusions of previous studies based on xenolith data, heat flow observations, regional uplift history and geodynamic modeling (Förster et al. 2004, 2007; Steinitz and Bartov 1991; Sobolev et al. 2005; Petrunin and Sobolev 2006) that postulated that the lithosphere around the DST has been significantly thinned in the Late Cenozoic, likely following rifting and spreading of the Red Sea.

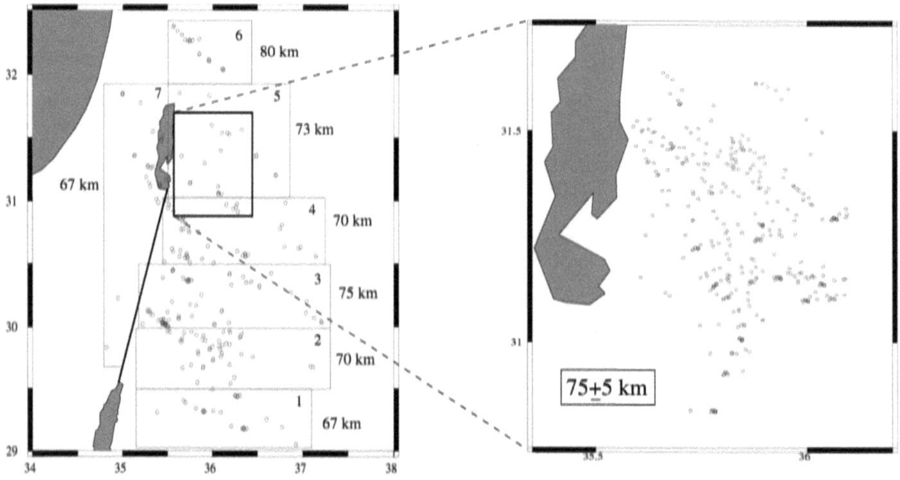

Fig. 2.10 (*Left*) Distribution of *Sp* piercing points at 80 km depth. The area covered has been divided into seven non-overlapping boxes denoted by numbers (1–7) and a blow-up area just east of the Dead Sea. Each box (*the blow-up*) contains more than 20 (350) stacked traces, respectively. The average thickness of the lithosphere is given in km beside each box (Modified from Mohsen et al. 2006). (*Right*) The blow-up shows the *Sp* piercing points from DESIRE together with the average lithosphere thickness and one standard deviation **δ**

2.4 Modeling and Interpretation

The more than 10 km thick Neogene Dead Sea Basin (DSB) has long been considered as the classical pull-apart basin (Garfunkel and Ben-Avraham 1996, and references therein). However, recently Ben-Avraham and Schubert (2006) and Ben-Avraham et al. (2010) dismissed a pull-apart origin for the deepest, southern-most part of the DSB. One major argument was the apparently anomalously cold crust of the DSB, based on surface heat flow of below 50 mW/m^2, and seismicity penetrating to the lower crust. Indeed, numerical thermo-mechanical models (Petrunin and Sobolev 2006, 2008) suggest that the formation of a pull-apart basin is most likely impossible in a lithosphere with steady state surface heat flow below 50 mW/m^2, because in such a lithosphere the entire crust is very strong and is mechanically attached to the even stronger upper mantle. Petrunin et al. (2012) by means of 3D thermo-mechanical modeling show however, that the observed surface heat flow, seismic structure and distribution of seismicity in the lithosphere below the DSB can be reconciled with the pull-apart origin of the DSB. The results of this study are (1) a more realistic estimate of surface heat flow, unaffected by fluid-flow at the flanks of the DSB, of 50–55 mW/m^2, (2) a transient thermal structure of the lithosphere affected by lithospheric thinning, that has likely happened in the region at about 20 Ma, and (3) a rheologically weak upper crust and uppermost mantle beneath the DSB.

2.4.1 Apparently Contradicting Observations

Estimates of the present-day heat flow in the Dead Sea Transform area are controversial. Heat flow lower than 50 mW/m^2 is reported for Israel close to the Mediterranean Sea (Eckstein and Simmons 1978; Eckstein 1979). Measurements at the Dead Sea (Ben-Avraham et al. 1978) also indicate extremely low heat flow (about 30 mW/m^2) that could however be explained by active groundwater circulation (Shalev et al. 2007). Heat flow higher than 60 mW/m^2 is reported at the flanks of the DSB (also likely affected by the groundwater circulation, Shalev et al. 2007) and in Jordan closer to the Red Sea (Förster et al. 2007). Reasonable estimates for the surface heat flow at the location of the DSB (unaffected by the fluid-flow) are therefore likely to be around 50–55 mW/m^2.

This heat flow of 50–55 mW/m^2, if at steady state, corresponds to a thickness of the lithosphere of more than 130 km (e.g. Förster et al. 2007). However, receiver function studies (Mohsen et al. 2006) demonstrate, that present day lithospheric thickness in the area is much less, i.e. between 60 and 80 km.

The DSB region is characterized by unusual seismicity pattern. Aldersons et al. (2003) report that 60 % of the micro-earthquakes are located at depths of 20–32 km (lower crust) and the peak seismicity is situated just below the upper-lower crustal boundary (20 km). Double difference relative relocation suggests, that the seismic activity at the DSB is mostly confined to the upper crust (0–20 km), and that it disappears in the lower crust at depth of 25–26 km (Shamir 2006). In a more recent study Braeuer et al. (2012) also find, that the seismicity is concentrated in the upper crust, between 7 and 17 km depth, a result similar to Alderson and Ben-Avraham (2014). The seismicity in the lower crust may indicate a low temperature there, which correlates with the low values of the surface heat flow of about 50–55 mW/m^2. However, both these studies report an absence of earthquakes beneath the Moho, which is consistent with an uppermost mantle temperature of more than 600 deg C (Hirth 2002) which would, in turn, correspond to a steady-state surface heat flow of more than 60 mW/m^2.

2.4.2 Model Reconciling Contradictory Observations

We have designed a 3D thermo-mechanical model addressing the apparently contradicting observations mentioned above. The model includes a three-layer crust and two-layer lithospheric mantle with visco-elasto-plastic rheology (Fig. 2.11). Rheological models are taken from laboratory data and thermal properties according to the measurements on the rocks from the region (Förster et al. 2007). Brittle failure is simulated by a Mohr-Coulomb friction rheology with strain softening (see Sobolev et al. 2005, for details). We also introduce weak seeds of faults in the upper crust where the friction coefficient is lowered to 0.1. The model setup, similar to the previous models by Petrunin and Sobolev (2006, 2008), is shown in Fig. 2.11a. As in previous models we assume that strike-slip motion with a velocity of 6 mm/year started at 17 Ma forcing a pull-apart basin to develop between the nuclei of two pre-defined faults. The key new feature of the model is

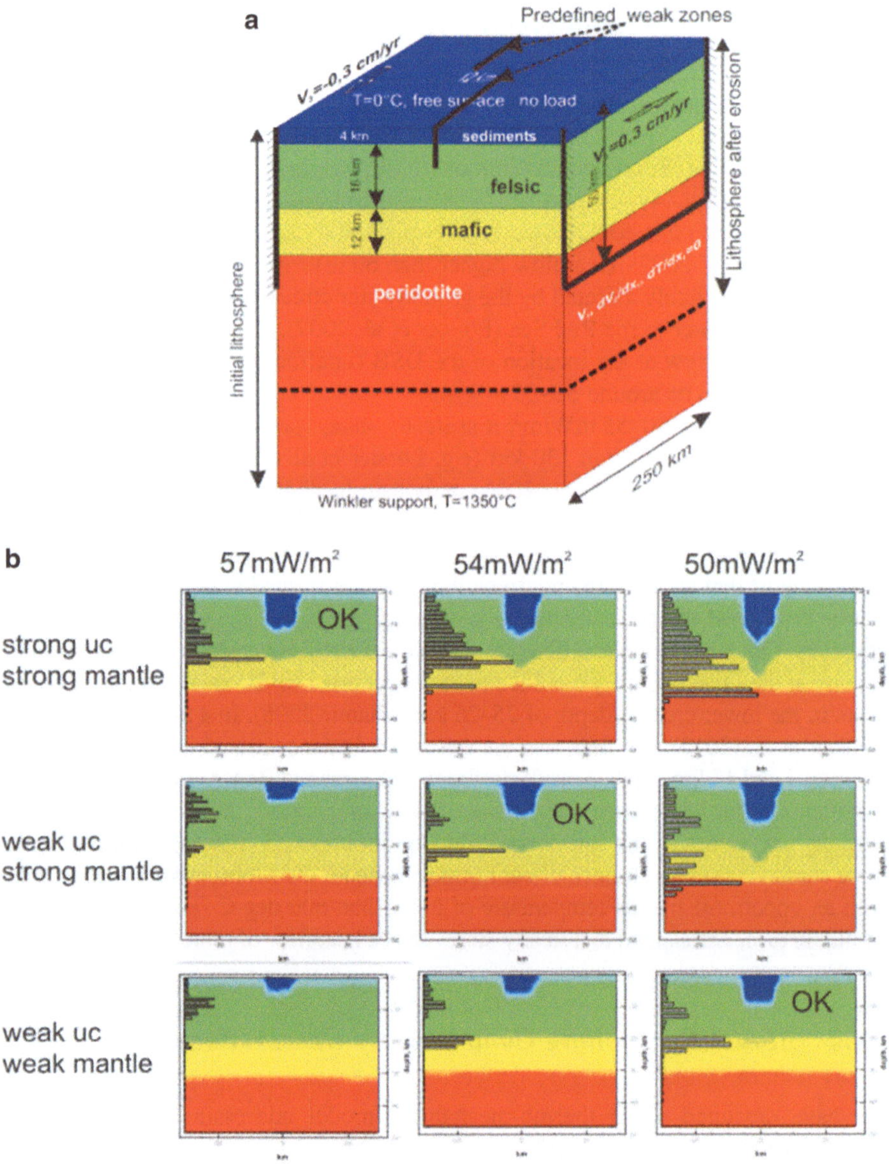

Fig. 2.11 Results of the numerical modelling of a pull-apart basin. (**a**) Setup of the thermo-mechanical model. It is assumed that initially thick (120–180 km) lithosphere has been thermally eroded to 80 km at 20 Ma. Strike-slip motion with velocity of 6 mm/year has started at 17 Ma forcing a pull-apart basin to develop between the nuclei of two pre-defined faults. Model features are similar to models in Petrunin and Sobolev (2006, 2008). (**b**) Cross-section through the middle of the basin after 105 km of strike-slip motion for models with different initial lithospheric thickness and different rheology of the upper crust and upper mantle. *Bars* indicate normalized mechanical energy dissipation in brittle regime, here used as a proxy for the seismic energy release. *Numbers* indicate predicted present-day surface heat flow at the flanks of the basin. *uc* upper crust (Modified from Petrunin et al. 2012)

the assumption that the initially thick lithosphere beneath the future DSB was thermally eroded to about 80 km at 20 Ma.

We have carried out a set of numerical experiments with various initial thicknesses of lithosphere from 120 to 180 km and different rheological models of the upper crust and upper mantle. The aim of the models was to find whether rheological models exist supported by experimental data that allow fitting the following set of observations for the DSB, namely

1. A present day surface heat flow at the DSB flanks of 50–55 mW/m^2,
2. A present-day lithospheric thickness of 80 km,
3. A thickness of the Cenozoic sediments in the basin of 8–12 km,
4. No prominent topography of the Moho and intra-crustal boundary and
5. No earthquakes in the mantle.

As a proxy for the seismic activity we use energy dissipation in the brittle (Mohr-Coulomb friction plastic) deformation mode. For the mantle we also assume that deformation is seismic if it is brittle and temperature is below 600 °C.

The modeling results are presented in Fig. 2.11b, where the lithospheric structure at the central cross-section of the model after 105 km of strike-slip motion is plotted for different initial lithospheric thickness and different rheology of the upper crust and upper mantle. Models which fit the observations are marked with "OK". The rheological models with the highest viscosities in both the upper crust (Gleason and Tullis 1995) and upper mantle (Hirth and Kohlstedt 2003), shown in Fig. 2.11b in the upper row are not consistent with the heat flow constraint, but would fulfill all others constraints, if present day surface heat flow was slightly higher than observed (57 mW/m^2). However, if a lower-viscosity rheological model for the upper crust (Ranalli 1995) is used (Fig. 2.11b, middle row), then the entire set of observations is fulfilled for the surface heat flow of 54 mW/m^2. In the case when additionally the lower Peierls stress rheological model for the upper mantle is employed, according to the recent data by Katayama and Karato (2008), the entire set of observational constraints is fitted, even at surface heat flows as low as 50 mW/m^2 (Fig. 2.11b, lower row).

From our numerical experiments, we thus conclude, that the entire set of observational constraints for the DSB including surface heat flow, magnitude of late Cenozoic subsidence, lithospheric thickness and crustal structure as well as seismicity pattern can be fitted with the classical pull-apart model assuming that the lithosphere has been thermally eroded to 80 km at about 20 Ma and the upper crust and uppermost mantle have relatively weak rheologies, in agreement with experimentally determined values for quartz (upper crust) and olivine (mantle).

2.5 Summary

Based on a multi-disciplinary study, involving seismics, seismology and gravity, we find the following main characteristics for the deep crustal structure of the DST: (1) an increase in Moho thickness from ca. 25 km at the Mediterranean in the West to ca. 35 km in the Jordan Highlands in the East, (2) a strong local inhomogeneity overlying this general trend in the Dead Sea Basin, and (3) a North–south trending crustal thickening in the middle of the Arava/Araba valley. In the upper mantle under the DST we observe (1) an area of reduced S-velocities in the lithosphere along the DST, (2) a narrow, ca 20 km wide, decoupling zone between the Arabian and African plate characterized by fault parallel anisotropy and (3) that below ca. 120 km depth no significant inhomogeneities can be detected. Integrating these results with xenolith data, heat flow studies, the lack of earthquakes in the mantle and 3D thermo-mechanical modeling shows that a classical pull-apart model triggered by the thinning of the lithosphere around the DST to about 80 km in the Late Cenozoic is able to fit the observables and explain the evolution of the DST.

Acknowledgements The work presented here was supported by the Deutsche Forschungsgemeinschaft (DFG) and the Deutsches GeoForschungs-Zentrum Potsdam (GFZ). We thank the editors Z. Garfunkel, Z. Ben-Avraham and E. Kagan, and W. Mooney for their comments to this manuscript. We are grateful to U. ten Brink for providing the model of his studies of the DST and A. Siebert for her help in drafting many of the figures, most of them created with the GMT software (Wessel and Smith 1998). We thank the National Ministry of Infrastructure of Israel, the Natural Resources Authority (NRA) of Jordan, the German embassies in Tel Aviv and Amman and the An-Najah National University in Nablus, Palestine, for their support during the field work. We thank our contractors the Geophysical Institute of Israel, the Site Group (Jordan) and the Chemical and Mining Industries (Jordan) for their excellent work under difficult logistic conditions. The instruments for the field work were provided by the Geophysical Instrument Pool of the GFZ. For more information on DESERT and DESIRE, respectively, see:

http://www.gfz-potsdam.de/portal/gfz/Struktur/Departments/Department+2/sec22/projects/past_projects/DESERT

http://www.gfz-potsdam.de/portal/gfz/Struktur/Departments/Department+2/sec22/projects/current_projects/DESIRE

References

Agnon A, Marco S, Migowski C (2006) Intraclast breccia layers in laminated sequences: recorders of paleo-earthquakes, New Frontiers In: Enzel Y, Agnon A, Stein M (eds) Paleoenvironmental research. Geol Soc Amer Spec Paper 401, pp 195–214. doi: 10.1130/2006.2401(13)

Aldersons F, Ben-Avraham Z (2014) The seismogenic thickness along the Dead Sea Transform. In: Garfunkel Z, Ben-Avraham Z, Kagan E (eds) The Dead Sea Transform. Springer, Dordrecht, pp 53–89

Aldersons F, Ben-Avraham Z, Hofstetter A, Kissling E, Al-Yazjeen T (2003) Lower-crustal strength under the Dead Sea basin from local earthquake data and rheological modelling. Earth Planet Sci Lett 214:129–142

Ambraseys NN, Melville CP, Adams RD (1994) The seismicity of Egypt, Arabia and the Red Sea: a historical review. Cambridge University Press, Cambridge, p 201

Amiran DHK, Arieh E, Turcotte T (1994) Earthquakes in Israel and adjacent areas, macroscopic observations since 100 B.C.E. Isr Explor J 44:260–305

Amrat AQ, Feldman L, Shapira A (2001) The unified earthquake catalog of the Dead Sea Region. http://seis.gii.co.il/heb/hazards/docs/Catalog-rprt.pdf

Bartov Y, Sagy H (2004) Late Pleistocene extension and strike slip in the Dead Sea Basin. Geol Mag 141:565–572

Ben-Avraham Z (1985) Structural framework of the Gulf of Elat (Aqaba), northern Red Sea. J Geophys Res 90:703–726
Ben-Avraham Z (2014) Geophysical studies of the crustal structure along the Dead Sea Transform. In: Garfunkel Z, Ben-Avraham Z, Kagan E (eds) The Dead Sea Transform. Springer, Dordrecht, pp 1–27
Ben-Avraham Z, Schubert G (2006) Deep "drop down" basin in the southern Dead Sea: earth planet. Sci Lett 251:254–263
Ben-Avraham Z, Hänel R, Villinger H (1978) Heat flow through the Dead Sea rift. Mar Geol 28:253–269
Ben-Avraham Z, Garfunkel Z, Lazar M (2008) Geology and evolution of the Southern Dead Sea fault with emphasis on subsurface structure. Annu Rev Earth Planet Sci 36:357–387. doi:10.1146/annurev.earth.36.031207.124201
Ben-Avraham Z, Lyakhovsky V, Schubert G (2010) Drop-down formation of deep basins along the Dead Sea and other strike-slip fault systems. Geophys J Int 181:185–197
Braeuer B, Asch G, Hofstetter R, Haberland C, Jaser D, El-Kelani R, Weber M (2012) Microseismicity distribution in the southern Dead Sea area and its implications on the structure of the basin. Geophys J Int 188(3):873–878. doi:10.1111/j.1365-246X.2011.05318.x
Choi S, Götze HJ, Meyer U, DESIRE Group (2011) 3D density modelling of underground structures and spatial distribution of salt diapirism in the Dead Sea Basin. Geophys J Int 184:1131–1146. doi:10.1111/j.1365-246X.2011.04939.x
DESERT Group (2004) The crustal structure of the Dead Sea Transform. Geophys J Int 156:655–681
Dziewonski AM, Anderson DL (1981) Preliminary reference earth model. Phys Earth Planet Inter 25:297–356
Eckstein Y (1979) Heat flow and hydrologic cycle: examples from Israel. In: Cermak V, Rybach L (eds) Terrestrial heat flow in Central Europe. Springer, Heidelberg, pp 88–97
Eckstein Y, Simmons G (1978) Measurements and interpretation of terrestrial heat flow in Israel. Geothermics 6:117–142
El-Isa Z, Mechie J, Prodehl C, Makris J, Rihm R (1987a) A crustal structure study of Jordan derived from seismic refraction data. Tectonophysics 138:235–253
El-Isa Z, Mechie J, Prodehl C (1987b) Shear velocity structure of Jordan from explosion seismic data. Geophys J R Astron Soc 90:265–281
Förster HJ, Oberhänsli R, Sobolev SV, Förster A, Stromeyer D (2004) Lithosphere composition and thermal regime across the Dead Sea Transform in Israel and Jordan. Eos Trans AGU 85(17), Jt Assem Suppl, Abstract T11A-05
Förster AH, Förster HJ, Masarweh R, Masri A, Tarawneh K, DESERT Group (2007) The terrestrial heat flow of the Arabian shield in Jordan. Int J Earth Sci 30:271–284. doi:10.1016/j.jseaes.2006.09.002
Garfunkel Z (1981) Internal structure of the Dead Sea leaky transform (rift) in relation to plate kinematics. Tectonophysics 80:81–108
Garfunkel Z (1988) The pre-Quaternary geology of Israel. In: Yom-Tov Y, Ternov E (eds) The zoogeography of Israel. W. Junk, Dordrecht, pp 7–34
Garfunkel Z (2014) Overview of geological features of the Dead Sea Transform. In: Garfunkel Z, Ben-Avraham Z, Kagan E (eds) The Dead Sea Transform. Springer, Dordrecht, pp 109–150
Garfunkel Z, Ben-Avraham Z (1996) The structure of the Dead Sea basin. Tectonophysics 266:155–176
Ginzburg A, Ben-Avraham Z (1997) A seismic refraction study of the north basin of the Dead Sea, Israel. Geophys Res Lett 24:2063–2066
Garfunkel Z, Zak I, Freund R (1981) Active faulting in the Dead Sea rift. Tectonophysics 80:1–26
Ginzburg A, Makris J, Fuchs K, Prodehl C, Kaminski W, Amitai U (1979a) A seismic study of the crust and upper mantle of the Jordan-Dead Sea rift and their transition toward the Mediterranean Sea. J Geophys Res 84:1569–1582

Ginzburg A, Makris J, Fuchs K, Perathoner B, Prodehl C (1979b) Detailed structure of the crust and upper mantle along the Jordan-Dead Sea rift. J Geophys Res 84:5605–5612

Gleason GC, Tullis J (1995) A flow law for dislocation creep of quartz aggregates determined with the molten salt cell. Tectonophysics 247:1–23. doi:10.1016/0040-1951(95)00011-B

Götze HJ, El-Kelani R, Schmidt S, Rybakov M, Förster HJ, Ebbing J, DESERT Group (2006) Integrated 3-D density modelling and segmentation of the Dead Sea Transform. Int J Earth Sci 289–302. doi:10.1007/s00531-006-0095-5

Götze HJ, Meyer U, Choi S (2010a) Helicopter gravity survey in the Holy Land. Eos Trans AGU 91(12):109. doi:10.1029/2010EO120002

Götze HJ, Meyer U, Choi S, Bartov Y, Ben-Avraham Z, Jasser D, Abu Ayyash K, El-Kelani R (2010b) Helicopter-borne gravity survey in the Dead Sea area. Eos Trans AGU Electronic supplement 91(12):109, 2010. doi:10.1029/2010EO120002

Grünthal G, Hakimhashemi A, Schelle H, Bosse C, Wahlström R (2009) Study of the long-term temporal behavior of the seismicity of the Dead Sea fault zone and its implication for time-dependent seismic hazard assessments. Scientific technical report STR 09/09, GFZ German Research Centre for Geosciences, Potsdam

Hirth G (2002) Laboratory constraints on the rheology of the upper mantle. Rev Mineral Geochem 51(1):97–120. doi:10.2138/gsrmg.51.1.97

Hirth G, Kohlstedt D (2003) Rheology of the upper mantle and the mantle wedge: a view from the experimentalists. In: Eiler J (ed) Inside the subduction factory, Geophysical monograph series, 138. AGU, Washington, DC, pp 83–105

Hofstetter A et al (2014) Instrumental data on the seismic activity along the Dead Sea Transform. In: Garfunkel Z, Ben-Avraham Z, Kagan E (eds) The Dead Sea Transform. Springer, Dordrecht, pp 263–278

Hofstetter A, Thio HK, Shamir G (2003) Source mechanism of the 22/11/1995 Gulf of Aqaba earthquake and its aftershock sequence. J Seismol 7:99–114

Katayama I, Karato SI (2008) Low-temperature, high-stress deformation of olivine under water-saturated conditions. Phys Earth Planet Inter 168:125–133

Ken-Tor R, Agnon A, Enzel Y, Stein M, Marco S, Negendank JFW (2001) High-resolution geological record of historic earthquakes in the Dead Sea basin. J Geophys Res 106:2221–2234

Kind R, Kosarev GL, Petersen NV (1995) Receiver functions at the stations of the German Regional Seismic Network (GRSN). Geophys J Int 121:191–202

Klinger Y, Avouac J, Dorbath L, Karaki NA, Tisnerat N (2000a) Seismic behavior of the Dead Sea fault along Araba valley, Jordan. Geophys J Int 142:769–782

Klinger Y, Avouac J, Karaki NA, Dorbath L, Bourles D, Reyss JL (2000b) Slip rate on the Dead Sea Transform fault in northern Araba valley (Jordan). Geophys J Int 142:755–768

Koulakov I, Sobolev SV (2006) Moho depth and three-dimensional P and S structure of the crust and uppermost mantle in the Eastern Mediterranean and Middle East derived from tomographic inversion of local ISC data. Geophys J Int 218–235. doi:10.1111/j.1365-246X.2005.02791

Koulakov I, Sobolev SV, Weber M, Oreshin S, Wylegalla K, Hofstetter R (2006) Teleseismic tomography reveals no signature of the Dead Sea Transform in the upper mantle structure, Earth Planet. Sci Lett 252:189–200. doi:10.1016/j.epsl.2006.09.039

Laske G, Weber M, DESERT Group (2008) Lithosphere structure across the Dead Sea Transform as constrained by Rayleigh waves observed during the DESERT experiment. Geophys J Int 173:593–610. doi:10.1111/j.1365- 246X.2008.03749.x

Le Beon M, Klinger Y, Amrat AQ, Agnon A, Dorbath L, Baer G, Ruegg JC, Charade O, Mayyas O (2008) Slip rate and locking-depth from GPS profiles across the southern Dead Sea Transform. J Geophys Res 113:B11403. doi:10.1029/2007JB005280

Mahmoud S, Reilinger R, McClusky S, Vernant P, Tealeb A (2005) GPS evidence for northward motion of the Sinai Block: implications for E. Mediterranean tectonics. Earth Planet Sci Lett 238:217–224

Makris J, Ben-Avraham Z, Behle A, Ginzburg A, Giese P, Steinmetz L, Whitmarsch RB, Eleftheriou S (1983) Seismic refraction profiles between Cyprus and Israel and their interpretation. Geophys J R Astron Soc 75:575–591

Mechie J, Abu-Ayyash K, Ben-Avraham Z, El-Kelani R, Mohsen A, Rümpker G, Saul S, Weber M (2005) Crustal shear velocity structure across the Dead Sea Transform from two-dimensional modelling of DESERT project explosion seismic data. Geophys J Int 160:910–924. doi:10.1111/j.1365-246X.2005.02526

Mechie J, Abu-Ayyash K, Ben-Avraham Z, El-Kelani R, Qabbani I, Weber M, DESIRE Group (2009) Crustal structure of the southern Dead Sea basin derived from project DESIRE wide-angle seismic data. Geophys J Int 178:457–478. doi:10.1111/j.1365-246X.2009.04161

Migowski C, Agnon A, Bookman R, Negendank JFW, Stein M (2004) Recurrence pattern of Holocene earthquakes along the Dead Sea Transform revealed by varve-counting and radiocarbon dating of lacustine sediments. Earth Planet Sci Lett 222:301–314

Mohsen A, Hofstetter R, Bock G, Kind R, Weber M, Wylegalla K, Rümpker G, DESERT Group (2005) A receiver function study across the Dead Sea Transform. Geophys J Int 160:948–960. doi:10.1111/j.1365-246X.2005.02534

Mohsen A, Kind R, Sobolev S, Weber M, DESERT Group (2006) Thickness of the lithosphere east of the Dead Sea Transform. Geophys J Int 167:845–852

Mohsen A, Asch G, Mechie J, Kind R, Weber M, Stiller M, Hofstetter R, Abu-Ayyash K, DESIRE Group (2011) Crustal structure of the Dead Sea Basin (DSB) from a receiver function analysis. Geophys J Int 184:463–476

Pasyanos M, Nyblade AA (2007) A top to bottom lithospheric study of Africa and Arabia. Tectonophysics 444:27–44. doi:10.1016/j.tecto.2007.07.008

Petrunin AG, Sobolev SV (2006) What controls thickness of sediments and lithospheric deformation at a pull-apart basin? Geology 34:389–392. doi:10.1130/G22158.1 DOI:10.1130/G22158.1#_blank

Petrunin AG, Sobolev SV (2008) Three-dimensional numerical models of the evolution of pull-apart basins. Phys Earth Planet Inter 171(1–4):387–399. doi:10.1016/j.pepi.2008.08.017 DOI:10.1016/j.pepi.2008.08.017#_blank

Petrunin AG, Meneses Rioseco E, Sobolev SV, Weber M (2012) Thermomechanical model reconciles contradictory geophysical observations for the Dead Sea Basin. Geochem Geophys Geosys 13(4). doi:10.1029/2011GC003929

Ranalli G (1995) Rheology of the earth, 2nd edn. Chapman & Hall, London, p 413

Reilinger R, 25 co-authors (2006) GPS constraints on continental deformation in the Africa-Arabia-Eurasia continental collision zone and implications for the dynamics of plate interaction. J Geophys Res 111:B0541. doi:10.1029/2005JB004051

Rümpker G, Ryberg T, Bock G, DESERT Seismology Group (2003) Boundary-layer mantle flow under the Dead Sea Transform fault inferred from seismic anisotropy. Nature 425:497–501

Ryberg T, Rümpker G, Haberland C, Stromeyer D, Weber M (2005) Simultaneous inversion of shear waves splitting observations from seismic arrays. J Geophys Res 110:03301. doi:10.1029/2004JB003303

Salamon A, Hofstetter A, Garfunkel Z, Ron H (2003) Seismotectonics of the Sinai subplate – the eastern Mediterranean region. Geophys J Int 155:149–173. doi:10.1046/j.1365-246X.2003.02017.x

Shalev E, Lyakhovsky V, Yechieli Y (2007) Is advective heat transport significant at the Dead Sea basin? Geofluids 7:292–300

Shamir G (2006) The active structure of the Dead Sea Depression. Geol Soc Am Spec Pap 401:15–32

Sobolev SV, Petrunin A, Garfunkel Z, Babeyko AY, DESERT Group (2005) Thermo-mechanical model of the Dead Sea Transform. Earth Planet Sci Lett 238:78–95. doi:10.1016/j.epsl.2005.06.058

Steinitz G, Bartov Y (1991) The Miocene-Pliocene history of the Dead Sea segment of the rift in light of K-Ar ages of basalt. Isr J Earth Sci 40:199–208

ten Brink US, Al-Zoubi AS, Flores CH, Rotstein Y, Qabbani I, Harder SH, Keller GR (2006) Seismic imaging of deep low-velocity zone beneath the Dead Sea basin and transform fault: implications for strain localization and crustal rigidity. Geophys Res Lett 33:L24314. doi:10.1029/2006GL027890

van Eck T, Hofstetter A (1989) Micro-earthquake activity in Dead Sea region. Geophys J Int 99:605–620
Vinnik LP (1977) Detection of waves converted from P to SV in the mantle. Phys Earth Planet Inter 15:39–45
Wdowinsky S, Bock Y, Baer G, Prawirodirdjo L, Bechor N, Naaman S, Knafo R, Forrai Y, Melzer Y (2004) GPS measurements of current crustal movements along the Dead *Sea fault*. J Geophys Res 109:B05403. doi:10.1029/2003JB002640
Weber M, 68 coauthors (2009) Anatomy of the Dead Sea Transform from lithospheric to microscopic scale. Rev Geophys 47:RG2002. doi:10.1029/2008RG000264
Weber M, 68 coauthors (2010) Correction to "Anatomy of the Dead Sea Transform from lithospheric to microscopic scale". Rev Geophys 48:RG1003
Weber M, Alasonati-Tašárová A, Abu-Ayyash K, Ben-Avraham Z, Choi S, Darwish J, El-Kelani R, Garfunkel Z, Götze HJ, Grünthal G, Hofstetter A, Kesten D, Mechie J, Meyer U, Mohsen A, Paschke M, Petrunin A, Ryberg T, Sobolev SV, Stiller M, DESERT and DESIRE Groups (2009/2010) Results of geophysical studies across the Dead Sea Transform: the Arava/Araba valley and the Dead Sea Basin. Isr J Earth Sci 58:147–161. doi:10.1560/IJES.58.3-4.147
Wessel P, Smith WHF (1998) New, improved version of the Generic Mapping Tool released. EOS Trans Am Geophys Union 79:579

Chapter 3
The Seismogenic Thickness in the Dead Sea Area

Freddy Aldersons and Zvi Ben-Avraham

Abstract Four independent distributions of seismicity in the Dead Sea area underline the occurrence of lower-crustal seismic activity down-to at least 27 km and possibly as deep as 33.6 km. From these distributions, the seismogenic thickness is estimated to be 28.4 ± 2.2 km. The existence of a seismogenic zone extending deep into the lower crust is consistent with an average heat flow of only 40–45 mWm^{-2} over most regions of Israel, and around 40 mWm^{-2} in the Dead Sea area in particular. The seismogenic thickness in the Dead Sea area is thus nearly twice the average seismogenic thickness of 15 km observed in southern California. The fact that some seismic activity occurs down-to the Moho in the Dead Sea area suggests that the state of fully plastic deformation is probably not reached in the crust under the seismogenic zone.

The ISC – GEM (Storchak et al. 2013) relocation of the M_W 6.3 earthquake of 11 July 1927 from regional and teleseismic instrumental data resulted in a well-constrained epicenter located in the Jordan Valley, not far from the epicenter reported in the 1927 bulletin of the ISS. Since the causative fault of this earthquake is likely to be the Dead Sea transform, we propose a preferred epicenter at 31.92°N–35.56°E. The focal depth determined instrumentally by the ISC – GEM relocation is 15 ± 6 km, and we found an average macroseismic depth of 21.5 ± 2.5 km. Our results as a whole underline also the seismogenic importance of the transition between the upper and the lower crust in the Dead Sea area for moderate and probably also for large earthquakes.

F. Aldersons (✉)
Seismological Consultant, 29, ch. du pré des fourneaux, 1293 Bellevue, GE, Switzerland
e-mail: fred@faldersons.net

Z. Ben-Avraham
Department of Geophysical, Atmospheric and Planetary Sciences, Tel Aviv University, Ramat Aviv, Tel Aviv 69978, Israel

Charney School of Marine Sciences, University of Haifa, Mt. Carmel, Haifa 3190501, Israel
e-mail: zviba@tau.ac.il

Z. Garfunkel et al. (eds.), *Dead Sea Transform Fault System: Reviews*, Modern Approaches in Solid Earth Sciences 6, DOI 10.1007/978-94-017-8872-4_3,

Keywords Seismogenic thickness • Dead Sea • Seismicity • Depth • Rheology • Heat flow • Brittle • Ductile • 1927 • 2004 • Earthquake • Macroseismic • Isoseismal • Seismic moment

3.1 Introduction

The purpose of this chapter is to determine the seismogenic thickness or maximum possible depth extent of rupture during moderate to large earthquakes in the Dead Sea area. The seismogenic thickness is important to assess the downdip width of faulting and hence to improve the estimate of realistic rupture area for a given fault or faulting system. Whether the seismogenic thickness represents or not a significant part of the crustal thickness has also direct implications regarding the prevailing mode of deformation under the seismogenic zone. This in turn can provide precious constraints on thermomechanical models in a given region. The question is particularly relevant to the Dead Sea area where for a long time, it had been accepted that the seismicity was entirely confined to the upper crust (Ben-Menahem et al. 1976).

We summarize the results of four independent studies of seismicity showing each on its own that the ultimate decline in seismicity in the Dead Sea area occurs in fact deep into the lower crust. We then use these depth distributions to estimate the seismogenic thickness according to criteria defined in southern California by Nazareth and Hauksson (2004). We also relocate the M_W 5.3 earthquake of 11 February 2004 from the seismograms recorded by the GII and by the JSO.

Since Zohar and Marco (2012) published the macroseismic intensities determined by Avni (1999) for the M_W 6.3 earthquake of July 1927, we contour by kriging several sets of intensities and determine the macroseismic depth of the earthquake according to Medvedev's (1962) and Shebalin's (1973) methods. Thanks to the recent ISC – GEM instrumental relocation (Storchak et al. 2013), there is no doubt left today that the epicenter of the 1927 earthquake is well located in the Jordan Valley, not far from the original ISS location.

Finally, we conclude that the existence of a seismogenic zone extending deep into the lower crust is also quite consistent with the average heat flow over most regions of Israel, and in the Dead Sea area in particular.

3.2 From Microearthquakes to Seismogenic Thickness

3.2.1 Instrumental Seismicity

Aldersons et al. (2003) studied the seismicity along the Dead Sea Transform (DST) between Aqaba-Elat and the Sea of Galilee for the period 1984–1997. Out of 2,283 earthquakes, routinely recorded by the network of the Geophysical Institute of Israel (GII) expanded with stations from the Jordan Seismological Observatory (JSO),

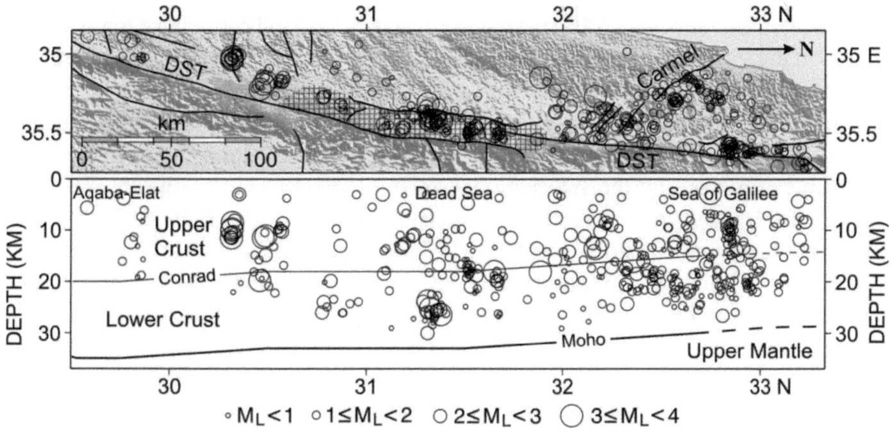

Fig. 3.1 Depth section of well-constrained seismicity (410 earthquakes, 1984–1997) along the Dead Sea transform (DST) from Aqaba-Elat to the Sea of Galilee. The square grid fill defines the Dead Sea Basin on the map. Conrad and Moho discontinuities from Ginzburg et al. (1981) (From Aldersons et al. 2003)

a first selection resulted in 653 earthquakes with at least 8 bulletin P-readings and an azimuthal gap smaller than 180°. To better exclude earthquakes with poorly constrained depths, the epicentral distance to the nearest station was required to be less than about twice the depth estimate according to an initial relocation. Four hundred and ten earthquakes qualified and are displayed in Fig. 3.1.

In order to make sure that the abundant lower-crustal seismicity observed between the Dead Sea basin and the Sea of Galilee was not an artifact produced by bulletin phase picking inconsistencies, Aldersons et al. (2003) gathered a data set of seismograms from the Dead Sea basin and carefully repicked the seismograms by hand. With the same quality criteria as those defined above, 42 well-constrained earthquakes ($M_L \leq 3.2$) were selected. Among those, 60 % nucleated at depths between 20 and 32 km and more than 40 % occurred below the depth of peak seismicity situated at 20 km. With the Moho at 32 km, the upper mantle appeared to be aseismic during the 14-year data period. A relocation involving the simultaneous use of three velocity models revealed that this apparently anomalous distribution of focal depths was also not an artifact created by strong lateral velocity variations (Aldersons et al. 2003). An upper bound depth uncertainty of ±5 km was estimated below 20 km, but for most earthquakes the depth uncertainty should not exceed ±2 km.

Several more recent publications provide well-constrained focal depths of earthquakes in the Dead Sea area. We briefly summarize below the general purpose and methodology of studies for which we could obtain sufficient details.

In a study of the active structure of the Dead Sea basin, Shamir (2006) used the double-difference method (Waldhauser and Ellsworth 2000; Waldhauser 2001) to relocate 176 earthquakes ($2.5 \leq M_L \leq 5.2$) that occurred during the period 1984–2005. Shamir used bulletin phase data augmented by P and S first-arrivals picked by hand. Waveform cross-correlation was not used.

Koulakov and Sobolev (2006) performed a large-scale local earthquake tomography study of the Eastern Mediterranean. The purpose of their work was to create a regional map of the Moho depth and 3-D velocity models of the crust and uppermost mantle, more detailed than existing models. They performed an iterative simultaneous inversion for source parameters, P- and S-velocities, and Moho depth using local traveltimes reported by the International Seismological Center between 1964 and 2001 (I.S.C. 2001). About 3,000 well-constrained events (with at least 25 phase picks and azimuthal gap not greater than 180°) within a radius of 6° around the Dead Sea Basin qualified for the tomography. Hypocentral locations were determined in three steps. In the first step, initial absolute locations were calculated according to a 1-D velocity model corrected for a variable Moho depth. In the second step, a relative relocation was performed according to the double-difference method (Waldhauser and Ellsworth 2000; Waldhauser 2001). Hypocentral locations were computed simultaneously with the determination of velocity anomalies and the Moho depth during the final tomographic inversions.

Braeuer et al. (2012b) performed a high-resolution local earthquake tomography study of the southern Dead Sea basin within the framework of the DESIRE project (DEad Sea Integrated REsearch project). The purpose of their work was to draw a detailed image of the distributions of the P-wave velocity and the V_P/V_S ratio, and to gain a better understanding of the deeper structure of the basin. During the 18-months period between October 2006 and March 2008, a dense array of 65 3-C stations recorded 530 well-locatable earthquakes with coda magnitudes M_D ranging between −0.5 and 4.5, but with most magnitudes confined between 0.0 and 1.0 and a completeness magnitude (Rydelek and Sacks 1989) of only 0.5, a value much lower than the completeness magnitude of about 2.0 for the GII catalogue since 1984 (Shapira and Hofstetter 2002; Begin and Steinitz 2005). The study started (Braeuer et al. 2012a) by the determination of 1-D velocity models (V_P, V_P/V_S and V_S) and station corrections with Velest (Kissling et al. 1994; Kissling 1995). Tomographic inversions for hypocentral parameters and velocities (V_P and V_P/V_S) were subsequently performed with Simulps (Thurber 1984) following a gradual approach from 1-D to 2-D and from 2-D to 3-D (Braeuer et al. 2012b).

Finally, seismograms from the Geophysical Institute of Israel (GII) and from the Jordanian Seismological Observatory (JSO), already used in part by Aldersons et al. (2003), are further analyzed here. The complete dataset (1986–2001), principally derived from 1-C vertical seismometers, was repicked for P-wave arrival times and highly-consistent observation weights by the automatic picking system MPX. MannekenPix or MPX (Aldersons 2004) is an advanced automatic phase picking system satisfying the stringent quality requirements of high-quality studies. In particular, it tackles in a most advanced way the delicate issue of picking uncertainty (Küperkoch et al. 2012). MPX was successfully applied on a significant scale during a local earthquake tomography study of Italy (Di Stefano et al. 2006, 2009), and it has also been used more recently in several other projects involving large datasets (Li et al. 2008; Diehl et al. 2009; Valoroso et al. 2013).

As a general background to the study of the local seismicity in the Dead Sea region, it is important to know that gathering a high-quality dataset leading to truly well-constrained earthquake locations in the Dead Sea basin and along the DST is not as trivial as it might seem. A first problem results from the fact that along the rather linear border between Israel and Jordan, the two national networks (GII and JSO) operate nearly completely separately. Several stations close to, and on both sides of the border, were well directly recorded by the neighbor network during the year 1995 and afterwards. But this is apparently no longer true. As a consequence, most earthquakes close to the DST and in particular those in the Dead Sea basin, are recorded as separate events by each network, typically with high azimuthal gaps around 180°. In order to make matters worse, a very unstable time signal was used by the JSO until at least the year 2010, with severe clock errors commonly reaching several minutes. When common stations (signals recorded by both networks at a given station) are available for an event, very accurate corrections (± 5–10 ms) can usually be determined by cross-correlation even when clocks differ by several minutes. For most events, however, only very few stations from the JSO located close to the border and directly recorded by the GII can sometimes be used. As a result, the maximum epicentral distance for GII seismograms in our dataset is 200–300 km, but typically less than 120 km for most JSO seismograms. For some events, however, our data integrate a complete set of stations from the GII and from the JSO.

Our dataset includes 188 earthquakes (1986–2001) with coda duration magnitudes ranging between 0.3 and 3.5 according to the GII. Only events with at least 8 acceptable to good P-phases determined by MPX and an azimuthal gap smaller than 180° were selected according to a preliminary relocation. We also relocated the M_W 5.3 (Harvard CMT) earthquake of 11 February 2004, for which very different epicentral locations have been computed and depth estimates range between 15 and nearly 29 km. A well-constrained focal depth for this earthquake is highly relevant to the determination of the seismogenic thickness because this event is already a moderate earthquake ($5 \leq M < 7$; Hagiwara 1964; Lee and Stewart 1981) and as such, it does not belong to the seismic background of micro-earthquakes. The crust in the Dead Sea area is then certainly seismogenic down-to at least the true focal depth of this earthquake, or any larger one like the M_W 6.3 earthquake of July 1927 that we treat separately due to the fact that its study involves both instrumental and macroseismic data.

In order to determine hypocentral locations and related uncertainties as a function of our approximate knowledge of the seismic velocity field, the final location for each of the 189 earthquakes was determined statistically from perturbations applied on a 1-D velocity model computed with Velest and called here the 'Dead Sea 2013' model. In other words, we did not relocate the dataset using a single 1-D velocity model with a classical uncertainty ellipsoid as measure of location quality. Instead, we relocated each earthquake according to a great number of different 1-D velocity models and derived the final location and uncertainties for each event from the distribution of individual results.

The Dead Sea 2013 P-Wave velocity model was derived with Velest from the best events of our dataset. The selection criteria were: at least 10 acceptable to good P-phases with at least 4 JSO phases, and an azimuthal gap not exceeding 150°. Fifty-two events qualified and they represent a dataset of 1,038 P-phases, or 20 P-phases per

Table 3.1 Velocity model Dead Sea 2013 (in *bold*) and perturbations

Layer	V_p (km/s)	Perturbation (V_p, %)	Bottom depth (km)	Perturbation (depth, %)
1	6,000	−4.0	7.000	−6.7
	6,120	−2.0	7.250	−3.3
	6,240	**−0.0**	**7.500**	**+0.0**
			7.750	+3.3
			8.000	+6.7
2			10.000	−9.1
	6,240	−2.0	10.500	−4.5
	6,370	**−0.0**	**11.000**	**+0.0**
	6,530	+2.5	11.500	+4.5
			12.000	+9.1
3			31.500	−1.6
	6,530	**+0.0**	**32.000**	**+0.0**
	6,750	+3.4	32.500	+1.6
			33.000	+3.1
4	7,910	−1.5	Half-space	
	7,950	−1.0		
	7,990	−0.5		
	8,030	**−0.0**		

event on the average. The M_W 5.3 earthquake of February 2004 is one of the selected events. Due to the smallness of this dataset, we did not compute station corrections. Based on previous results (Aldersons et al. 2003), we explored models with 4 layers and with 5 layers. Models with 4 layers displayed usually a lower global misfit than models with 5 layers, and model Dead Sea 2013, the one with the lowest global misfit, is also a model with only 4 layers (Table 3.1).

Model Dead Sea 2013 is certainly a good model, but it is not because it was determined by Velest that it is an optimal minimum 1-D model. We will only be able to derive a true minimum 1-D model for the Dead Sea region as a whole when several hundreds of high-quality events will be available with a sufficiently balanced number of GII and JSO stations. This is especially difficult to do for distant stations of the JSO due to the clock problems affecting the vast majority of seismograms from the JSO and preventing far too often an accurate merging between GII and JSO seismograms.

From perturbations defined around model Dead Sea 2013 in Table 3.1, 7,200 models were applied on the full dataset of 189 earthquakes. The global average RMS misfit on all the earthquakes relocated with the 7,200 models ranges, however, only between 0.27 and 0.31 s. Out of the 7,200 models, we selected a broad pool of the best models, those for which the global average RMS misfit was not higher than 0.28 s. Three thousand two hundred and ninety models qualified and, for each earthquake, we selected as final hypocentral location the median value of the 3,290 hypocenters relocated with Velest in single-earthquake mode. Similarly, we defined hypocentral uncertainty limits at the 5th and 95th percentiles (90 % confidence) of the 3,290 locations selected for each earthquake.

Table 3.2 Selected parameters for the M_W 5.3 earthquake of 11 February 2004, 08 h 15 m, according to various sources

Source	O.T. (s)	JSO corr.	Lat N (deg)	Lon E (deg)	Depth (km)	Gap (deg)	RMS (s)	Nr phases
ISC (2011)	02.63	+6.000 s	31.711	35.452	**26.2** ± 4.0 **25** (pP)	20	1.08	751
EMSC	04.00	0 s	31.688	35.544	**28.6**	122	0.98	435
JSO	(14 m) 57.70	0 s	31.694	35.581	**21.8**	167	0.20	12
EAT et al. (2006)	N/A	N/A	31.755	35.667	**18**±5	72	N/A	42
HOF (2008)	03.30	+6.000 s	31.700	35.557	**17** ± 0.7	~75	0.41	50
ELN et al. (2009)	02.10	N/A	31.666	35.678	**15** ± 0.2	70	0.60	96
This study (GII+JSO)	03.87	+6.125 s	31.691 ± 0.2 km	35.542 ± 0.2 km	−1.4 **21.7** +1.1	75	0.35	39

Depth values in *bold*. JSO Corr. is the time correction applied to JSO phases

Before focusing on the distribution of depths from our data and from the other datasets, we examine the relocation results for the M_W 5.3 earthquake of 11 February 2004. Selected parameters determined by the most important sources are gathered in Table 3.2 and corresponding epicenters are plotted in Fig. 3.2.

According to Fig. 3.2, there are four distinct locations for the M_W 5.3 earthquake of 11 February 2004. First, a cluster of epicenters is located in or at the edge of the Dead Sea basin. Solutions from the GII, the EMSC and from this study locate the epicenter in the Dead Sea basin, while the JSO locates the epicenter close to the Eastern Dead Sea fault. The solution of the ISC is located further west beyond the Jericho Fault. Al-Tarazi et al. (2006) relocated this earthquake and its aftershocks far outside the basin NE of the main cluster. Finally, Abou Elenean et al. (2009) relocated the earthquake at about the same longitude as Al-Tarazi et al. (2006) but further south.

In order to possibly explain why there is such a wide range of epicenters for an M_W 5.3 earthquake that should be very-well located and constrained, we examine first the origin times as determined by the GII and by the JSO (Table 3.2, second column). Assuming that the clock of the GII is accurate, the origin time of the earthquake is 08 h 15 m 03.30 s (Hofstetter et al. 2008). The JSO origin time, on the other hand, is 08 h 14 m 57.70 s (JSO time reported by the ISC), or 5.60 s earlier than the GII origin time. We can thus reasonably suspect that the JSO clock was late by a value roughly equal to the difference in origin times, or about 6 s. During the merging of JSO seismograms with GII seismograms in our relocation, only station AQBJ (Aqaba) from the JSO was available as an active common station recorded by both the JSO and the GII. It was thus possible to determine that JSO phase arrival times should be corrected by +6.125 s to match the GII clock according to station AQBJ. Table 3.2 shows that the ISC corrected the JSO arrival times by

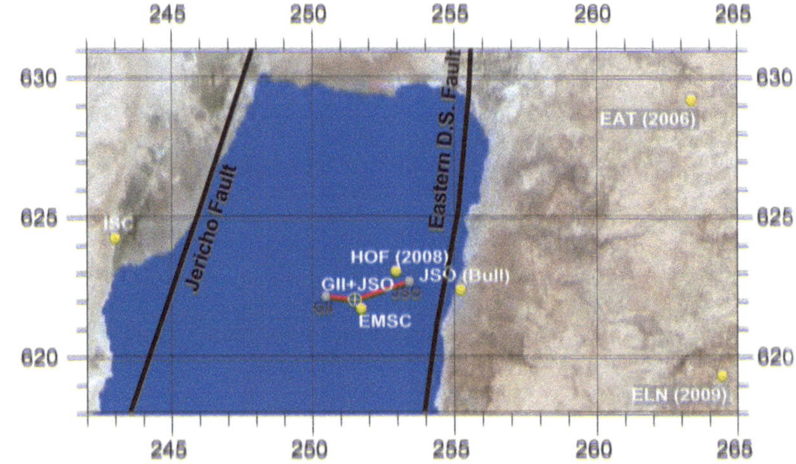

Fig. 3.2 M_W 5.3 earthquake of 11 February 2004, 08 h 15 m. Epicenters according to ISC, EMSC, JSO Bulletin, HOF: Hofstetter et al. (2008), EAT: Al-Tarazi et al. (2006), ELN: Abou Elenean et al. (2009), GII+JSO: this study (separate locations from GII seismograms alone and from JSO seismograms alone in *dark grey*, connected to final location by *red lines*). Israel Transverse Mercator (ITM) grid coordinates (km)

+6.000 s, a value also applied by the GII to JSO arrival times according to details available from the ISC. The JSO located the earthquake only from its own stations and reported uncorrected times. The EMSC also did not correct the original JSO times but used more than 400 phase readings from local to teleseismic distances. Regarding Al-Tarazi et al. (2006) and Abou Elenean et al. (2009), it is not known whether they corrected the JSO arrival times or not.

Our relocated epicenter (Fig. 3.2) is only 0.4 km away from the EMSC epicenter, and it is also located about 2 km SW of the epicenter determined by Hofstetter et al. (2008). In our location procedure detailed earlier, when only GII seismograms are used, the epicenter shifts by about 1 km west, and by about 2 km toward the JSO Bulletin location when only JSO seismograms are used (Fig. 3.2). This allows to infer that the JSO Bulletin epicenter located slightly further east at the edge of the basin can be explained by the absence of GII phases in the JSO location procedure for this event. It is very difficult, however, to understand why the Al-Tarazi et al. (2006) and Abou Elenean et al. (2009) solutions are located so far away from the cluster of earthquakes in the Dead Sea basin where the agreement and the quality of epicentral locations are quite good.

Figure 3.3 displays the hypocentral depth of the M_W 5.3 earthquake of 11 February 2004 according to the sources listed in Table 3.2. Despite the considerable scatter of values, our well-constrained depth of 21.7 km, virtually identical to the 21.75 km median value derived from all the estimates, suggests a focal depth in the lower crust albeit near its top.

The complete distribution of depths for the 188 earthquakes of our MPX dataset and for the M_W 5.3 earthquake of 2004 is displayed in Fig. 3.4. The average depth

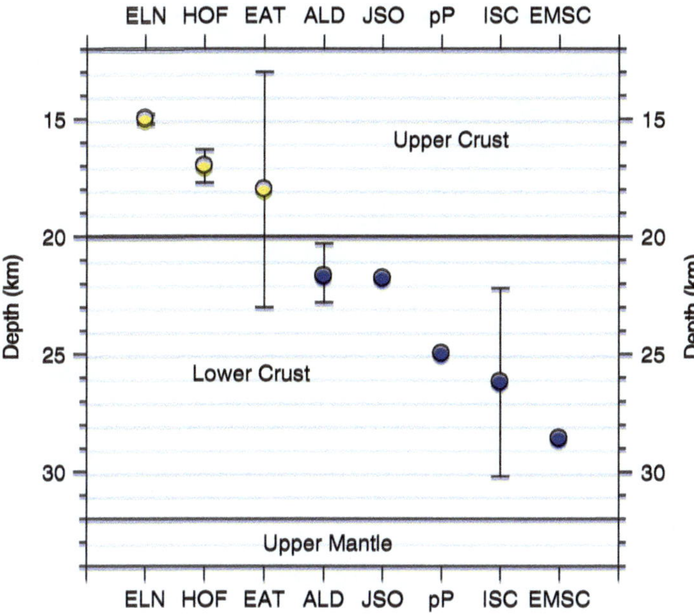

Fig. 3.3 M_W 5.3 earthquake of 11 February 2004, 08 h 15 m. Depths and uncertainties according to ELN: Abou Elenean et al. (2009), HOF: Hofstetter et al. (2008), EAT: Al-Tarazi et al. (2006), ALD: this study, JSO Bulletin, pP (ISC), ISC and EMSC. Upper-crustal depths in *yellow*, lower-crustal depths in *dark blue*

uncertainty is ±1.5 km in the upper crust with higher uncertainties close to the surface. In the lower crust, the average depth uncertainty is only ±1.0 km.

Earthquakes from the MPX dataset (Figs. 3.4 and 3.5a) are distributed over the whole crust from 0.0 to 31.4 km with the largest peak of seismicity at 12–14 km, and a smaller one at 18–20 km right above the transition between the upper crust and the lower crust. A region of increased activity occurs also between 8 and 12 km in the upper crust, and another one in the lower crust between 24 and 28 km, depth below which the number of earthquakes rapidly decreases. The M_W 5.3 earthquake of 11 February 2004, relocated in the lower crust at 21.7 km, occurred slightly below the peak of 18–20 km, a depth at which the seismic activity is only average.

Depths from Shamir (2006; Fig. 3.5b) range from 3 to only 27 km and are bimodal with a broad peak between 7 and 11 km in the upper crust, and another one between 17 and 19 km that seems to correspond to the 18–20 km peak in the MPX dataset. The seismicity is largely upper crustal, with a flatter envelope in the upper crust than in the MPX dataset. The lower crust is also seismogenic but the decline in seismicity starts immediately at the transition between the upper crust and the lower crust at 20 km depth.

The seismic activity in the Braeuer et al. dataset (2012b; Fig. 3.5c) extends between 2.8 and 28.9 km with a strong peak at 14–16 km, where nearly 25 % of the

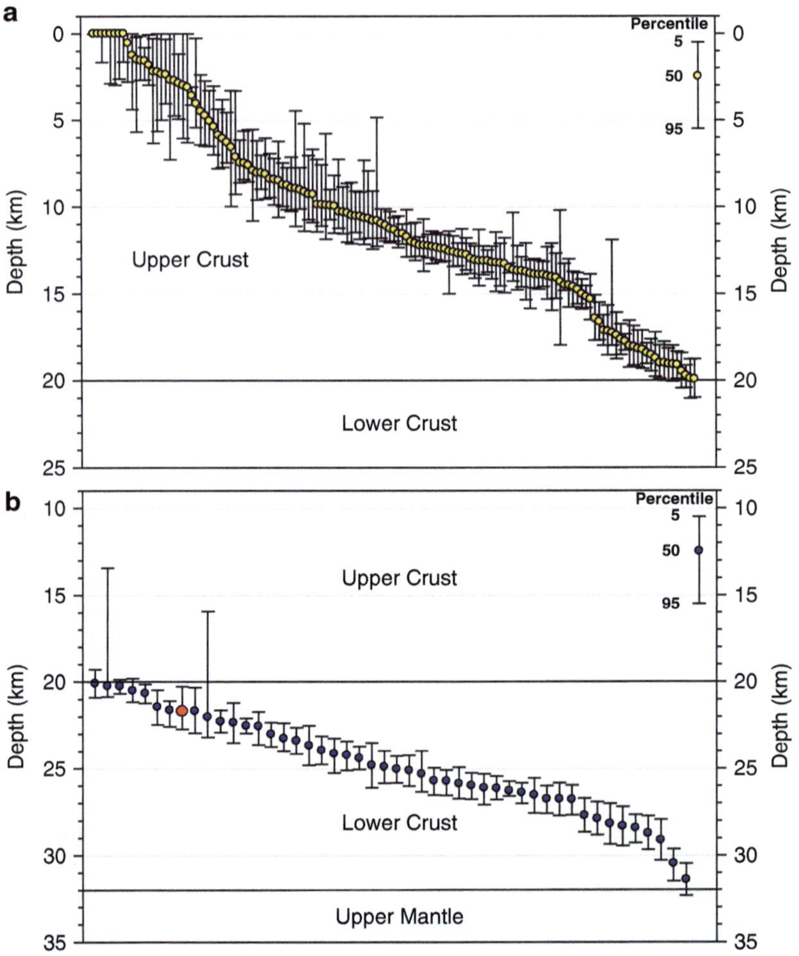

Fig. 3.4 Depth distribution for the 188 earthquakes (1986–2001) of the MPX dataset in the Dead Sea basin ($0.3 \leq M_L \leq 3.5$), and depth of the M_W 5.3 earthquake of 11 February 2004. (**a**) Upper-crustal depths and uncertainties (90 % confidence). (**b**) Lower-crustal depths and uncertainties (90 % confidence). M_W 5.3 earthquake of February 2004 in *red*

micro-earthquakes occurred. In addition, two larger earthquakes of coda magnitude 4.3 and 4.5 respectively also nucleated in this narrow range of depths. A second peak occurs between 8 and 10 km and it correlates with a similar peak in the MPX dataset (Fig. 3.5a). In fact, the whole range between 8 and 16 km displays a good similarity over the two datasets, even if the largest peak from Braeuer et al. (2012b) occurs at 14–16 km or 2 km deeper than in the MPX dataset. The top of the lower crust seems to be nearly depleted in seismic activity in Braeuer et al. (2012b), but an increase of activity is then observed between 24 and 28 km as in the MPX dataset. The seismic activity declines then rapidly and stops nearly at 29 km depth.

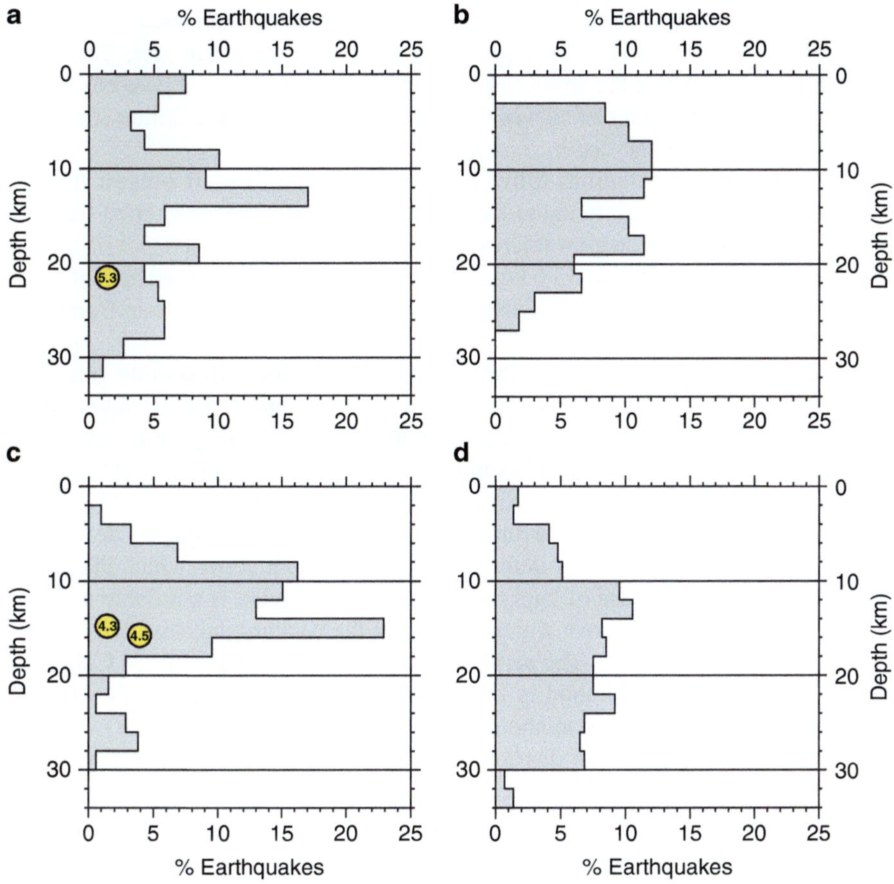

Fig. 3.5 Focal depths of well-constrained seismicity in the Dead Sea basin (30°25′N < Lat < 32°12′N, 34°52′E < Lon < 35°40′E). Braeuer et al. (2012b) in the southern Dead Sea basin (31°00′N < Lat < 31°28′N, 35°18′E < Lon < 35°39′E). (**a**) Depth distribution of 188 earthquakes ($0.3 \leq M_L \leq 3.5$) from the MPX dataset (This study). M_W 5.3 earthquake of 11 February 2004 in *yellow*. (**b**) Depth distribution of 176 earthquakes ($2.5 \leq M_L \leq 5.2$) (Adapted from Shamir 2006). (**c**) Depth distribution of 524 earthquakes ($-0.5 \leq M_L \leq 2.68$) (Adapted from Braeuer et al. 2012b). Depth of M_D 4.3 and 4.5 earthquakes in *yellow*. (**d**) Depth distribution of 294 earthquakes (Adapted from Koulakov and Sobolev 2006)

Finally, depths from Koulakov and Sobolev (2006; Fig. 3.5d) range from 0.0 to 33.6 km with a nearly flat envelope between 10 and 30 km. The seismicity is slightly higher in the 12–14 km range and in the 22–24 km range. The lower crust appears to be highly seismogenic in this dataset, especially toward the DST. A few earthquakes were also located below 32 km and could have nucleated in the upper mantle. The authors, however, consider these rare events as an exception or as belonging to the lower crust rather than a distinct population of earthquakes occurring in the uppermost mantle. They suggest also a simple explanation for the deepening of the seismicity they observe toward the DST. First, the failure stress in the brittle deformation regime does not depend on the strain rate but increases with depth. Second, the flow stress in the ductile deformation regime increases with the strain rate. Therefore, closer to the DST, where the strain rate is expected to be higher compared to further away, the increased strain rate moves the intersection between the failure stress and the flow stress to a greater depth.

The four depth distributions differ in their details but they all suggest that the ultimate decline in seismicity in the Dead Sea area starts in the lower crust between 20 and 30 km, with the deepest micro-earthquakes nucleating at a depth of at least 27 km and possibly as deep as 33.6 km. Similar results are also reported by Hofstetter et al. (2012) in their study of the crustal structure of the Dead Sea basin from local earthquake tomography.

It has been suggested already a while ago (Sibson 1982; Magistrale and Zhou 1996) that in southern California, the depth extent of rupture during major earthquakes can be reliably determined from the distribution of background seismicity. More recently, Nazareth and Hauksson (2004) studied a dataset of about 250,000 earthquakes ranging in magnitude from 0.1 to 7.3. The authors assumed that the maximum depth extent of rupture during moderate to large earthquakes defines the seismogenic thickness T_S. They determined that the depth above which 99.9 % of the seismic moment release of background seismicity occurs is a reliable estimate of T_S, and that the depth above which 98.3 % of the background seismicity occurs is also a good estimate of T_S. The seismogenic thickness for the southern California crust determined by this method is 15 km on the average but it is highly variable, ranging from less than 10 km to about 25 km (Nazareth and Hauksson 2004).

In order to derive the depth distribution of the seismic moment M_0 and to determine T_S from the 99.9 % depth criterion (Nazareth and Hauksson 2004), we used Bakun's (1984) relation (Eq. 3.1) defined in central California to compute M_0 from the magnitude of microearthquakes:

$$M_0 = 10^{(1.2M+17)} 1.0 \leq M \leq 3.5 \tag{3.1}$$

where M is either M_D or M_L. Among the four datasets, we determined the release of seismic moment only for the Braeuer et al. (2012b) dataset and for the MPX dataset (this study) due to the lack of magnitude details for the two other datasets. For all datasets, we also derived T_S directly from the depth distribution or 98.3 % depth criterion (Nazareth and Hauksson 2004). Results are provided in Table 3.3. The seismogenic thickness T_S derived from the background seismicity (both criteria, all datasets) ranges in the Dead Sea area from 25.1 to 30.9 km, with an average value of 28.4 km ± 2.2 km at the 95.4 % confidence level according to a *t*-test. It is also interesting to know that in the Braeuer et al. (2012b) dataset of micro-earthquakes ($-0.5 \leq M_D \leq 2.68$), the deepest earthquake occurred at a depth of 28.9 km with a coda magnitude M_D of 2.54 at the 99.2th percentile in magnitude. This single and deepest earthquake accounts for more than 5 % of the total seismic moment released by the micro-earthquakes of Braeuer et al. (2012b). Finally, Table 3.3 shows that all datasets agree about the fact that the seismic activity is greater in the upper crust than in the lower crust. With the exception of the dataset from Shamir (2006),

Table 3.3 Time span of each dataset, Relative Frequency of the seismicity in the upper crust and lower crust, and Seismogenic Thickness T_S according to two criteria

Source	Time span (years)	U.C. (%)	L.C. (%)	T_S (99.9 % M_0) (km)	T_S (98.3 % N) (km)
Shamir (2006)	22	85	15	N/A	25.1
Braeuer et al. (2012b)	1.5	91	9	28.9	27.0
MPX dataset (this study)	16	75	25	29.7	28.7
Koulakov and Sobolev (2006)	38	61	39	N/A	30.9

UC upper crust, *LC* lower crust, M_0 seismic moment, *N* number of earthquakes. Results sorted by increasing Seismogenic Thickness T_S

Table 3.3 suggests also that datasets with a longer time span seem to report an increased relative frequency of seismicity in the lower crust compared to shorter datasets.

3.2.2 *The Earthquake of 11 July 1927*

The M_W 6.3 earthquake that occurred on 11th July 1927 in the Dead Sea area is also very important for the determination of the seismogenic thickness because it is the largest earthquake in the Dead Sea area recorded instrumentally and because unfortunately it was destructive, for which widespread macroseismic observations were also reported. Until now, however, studies have dealt more specifically either with the instrumental data (Ben-Menahem et al. 1976; Shapira et al. 1993) or with the macroseismic effects caused by the earthquake (Vered and Striem 1977; Avni et al. 2002; Zohar and Marco 2012). Various epicenters have well been determined but the instrumental and macroseismic results available did not allow to single out a unique location among those that have been proposed so far.

Figure 3.6 displays the location of instrumental epicenters and macroseismic epicentral regions for the 1927 earthquake according to various sources. The ISS (1927) instrumental epicenter was not computed, however, from the arrival times reported in the ISS (1927) bulletin despite the publication of about 120 residual times. The bulletin states instead that the ISS epicenter was given by the observatory of Ksara in Lebanon. Ksara is located only 205 km away (NNE 10°) from the ISS epicenter. In 1927, the observatory operated two Mainka horizontal pendula. The accuracy of the time signal of Ksara was usually excellent, regular corrections being made by meridian transit observations (Wood 1921). Due to the strength of the 1927 earthquake occurring only about 200 km from Ksara, the NS pendulum stopped to work less than 5 s after the P onset time and the pen of the EW component was quickly destroyed, preventing any possible timing of the S arrival (Berloty 1927). The analysis of the two truncated seismograms revealed however that the azimuth of the epicenter was about SSW 185° from Ksara, on a great circle running through Es-Salt in Transjordan. What became later on the ISS epicenter (Table 3.4; ISS 1927) according to Ksara, was in fact a preliminary location determined by the

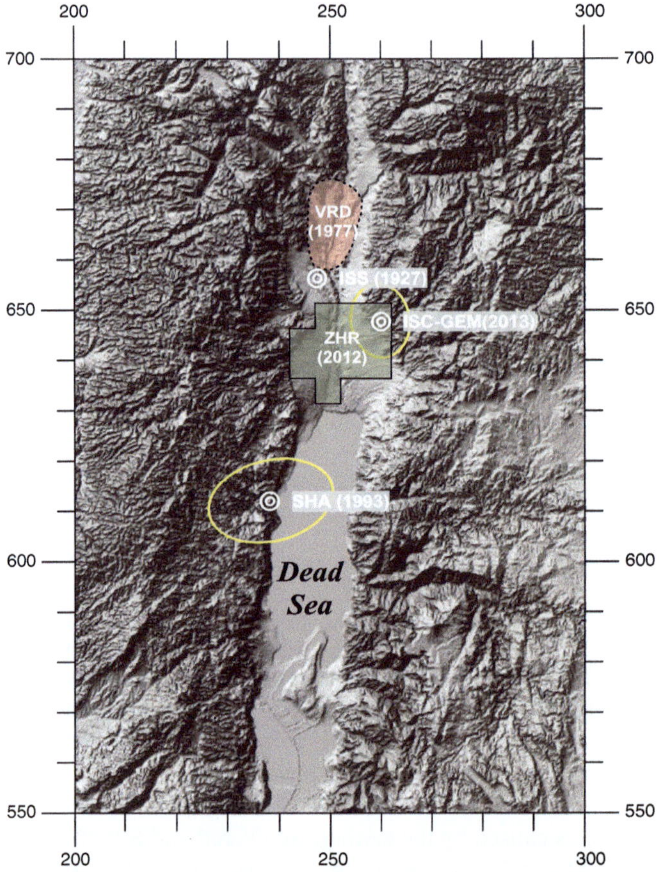

Fig. 3.6 M_W 6.3 earthquake of 11 July 1927: instrumental epicenters and macroseismic epicentral regions. ISS (1927): instrumental; Vered and Striem (1977): macroseismic (*red*); Shapira et al. (1993): instrumental; Zohar and Marco (2012): macroseismic (*green*); ISC-GEM: instrumental (Storchak et al. 2013). Uncertainty ellipses in *yellow*. Israel Transverse Mercator (ITM) grid coordinates (km)

Institut de Physique du Globe (IPG) of Strasbourg (Berloty 1927) derived from several arrival times. The final and published IPG epicenter is however different and located further north at 32.5 ° N – 35.6 ° E (Rothé 1928).

Gutenberg and Richter (1954) repeat the Ksara – ISS (1927) location, but they provide a depth estimate of 35 km and an mb magnitude of 6.25. Ben-Menahem et al. (1976) did not relocate the epicenter of the earthquake as they also repeat the ISS location (Table 3.4), but they determined a well-constrained M_S magnitude of 6.19 ± 0.05. They estimated the hypocentral depth to be 5–7 km according to spectral amplitude of surface waves, a modeling technique by which they also determined complete source parameters according to some assumptions (Table 3.5). From Table 3.5, the motion was determined by Ben-Menahem et al. (1976) as sinistral strike-slip with a small normal component on a nearly vertical and N10° E fault

Table 3.4 Earthquake of 11 July 1927: main parameters according to various sources of instrumental results

Source	Time	Epicenter	Depth	Mag	N def [a]	RMS	Gap
ISS (1927)	13 h 03 m 55 s	32.0 N 35.5 E (KSA – IPG)	N/A	N/A	N/A	N/A	N/A
GUT (1954)	13 h 04 m 07 s	32.0 N 35.5 E (ISS)	35 km	mb 6.25	N/A	N/A	N/A
ABM (1976)	13 h 04 m 07 s	32.0 N 35.5 E (ISS)	5–7 km	M_S 6.19 ± 0.05	N/A	N/A	N/A
SHA (1993)	13 h 04 m 06 s	31.60 N 35.40 E Smax [b] ~ 12 km Smin [c] ~ 8 km Strike [d] ~ 70°	Normal	N/A	38 phs 30 sta	2.9 s 76 ms/phs	139°
ISC-GEM	13 h 04 m 10 s	31.922 N 35.633 E Smax [b] 7.2 km Smin [c] 5.7 km Strike [d] 175°	15 ± 6 km	M_S 6.15 M_W 6.29 ± 0.21	110 phs 65 sta	3.8 s 35 ms/phs	86°

ISS International Seismological Summary, *KSA* Ksara station in Lebanon (33°49.38″N 35°53.40″E +920 m), *IPG* Institut de Physique du Globe of Strasbourg, *GUT* Gutenberg and Richter (1954), *ABM* Ben-Menahem et al. (1976), *SHA* Shapira et al. (1993), ISC-GEM Storchak et al. (2013)
[a]Number of distinct phases with location weight > 0, and number of stations
[b]Uncertainty Semi-Major axis length
[c]Uncertainty Semi-Minor axis length
[d]Semi-Major axis azimuth (clockwise from north)

Table 3.5 Earthquake of 11 July 1927. Source parameters from spectral amplitudes of surface waves (Ben-Menahem et al. 1976)

Strike azimuth	Slip angle	Dip angle	Potency[a] ($cm*km^2$)	Length	Slip	Rupture velocity	Rupture propagation
N 10° E	350°	85°	18,000	45 km	0.4±0.1 m	2.2 km/s	N to S

[a]Potency $P_0 = A\ D$ with A the fault area and D the average slip on the fault (Ben-Menahem and Singh 2000). Seimic moment $M_0 = \mu\ P_0$ with μ the rigidity

segment 45 km long. The vertical extent of the rupture was estimated at about 10 km. A pronounced directivity effect on the spectra suggests that the epicenter was moving towards the south and that the rupture started thus close to the northern tip of the causative fault segment.

About 12 km north from the ISS (1927) instrumental epicenter endorsed by Ben-Menahem et al. (1976) and close to Damiya bridge, Braslavski had reported in 1938 the damming of Jordan River for 21 h by a massive landslide triggered by the 1927 earthquake. This is close to where Vered and Striem (1977) located the macroseismic epicentral region of the earthquake (Fig. 3.6), with an intensity value of IX on the Modified Mercalli scale (Wood and Neumann 1931). Vered and Striem (1977) also determined the depth of the earthquake according to Medvedev's (1962)

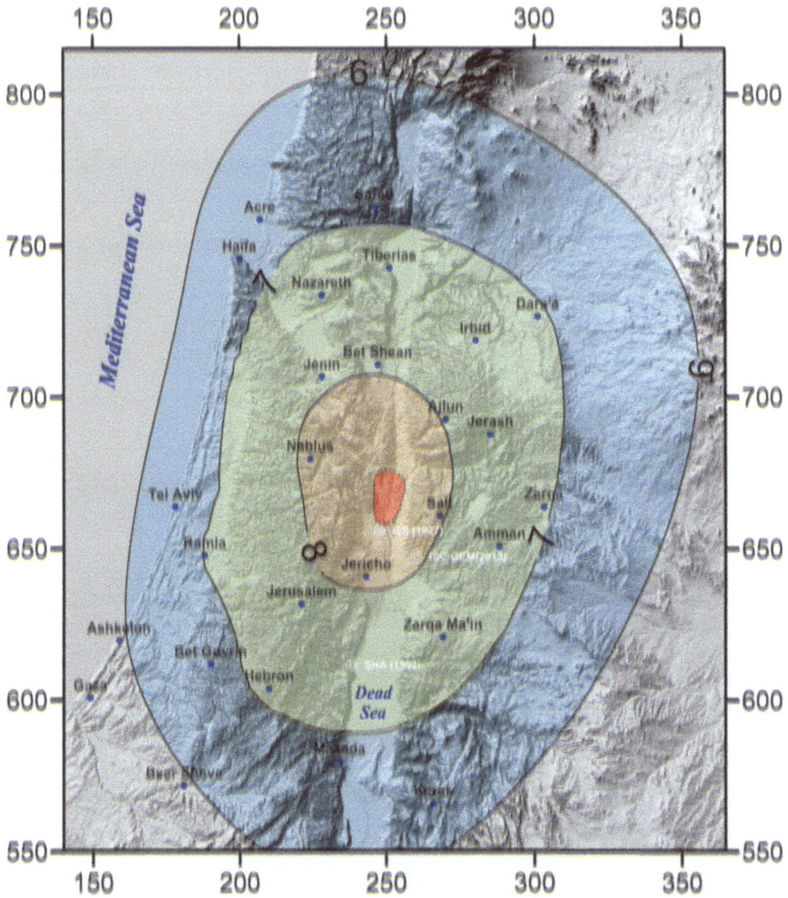

Fig. 3.7 Isoseismal map (MMI) of the earthquake of 11 July 1927 according to Vered and Striem (1977). ISS (1927): instrumental epicenter from ISS. SHA (1993): instrumental epicenter from Shapira et al. (1993). ISC-GEM: instrumental epicenter from ISC-GEM (Storchak et al. 2013). The area in red (MMI = IX) defines the macroseismic epicentral region according to the authors. Israel Transverse Mercator (ITM) grid coordinates (km). (Redrawn from original publication)

methods. Due to the ellipticity of the isoseismal curves (Fig. 3.7), the authors estimated 16 km as lower bound for the source depth and 28 km depth as upper bound, or an average depth of 22 ± 6 km.

Shapira et al. (1993) used the IASP91 (Kennett and Engdahl 1991) velocity model to relocate the 1927 instrumental epicenter according to 28 P-arrivals and 10 S-arrivals from stations within a radius of 40° from the Dead Sea region. Their instrumental epicenter is located on the western shore of the Dead Sea basin, 45 km SSW of the initial ISS (1927) location (Fig. 3.6). An average uncertainty radius of about 10 km was determined (Fig. 3.6 and Table 3.4) for the new location. The depth value is not provided. Instead, the authors report a "normal" depth, probably meaning a crustal depth. After the determination of the new epicenter, Avni (1999) and Avni et al. (2002) argued that the damming of Jordan River reported by Braslavski (1938) never occurred and that it derived exclusively from Garstang (1931), the only source to have reported such a phenomenon that he did not witnessed himself. Furthermore, Avni et al. (2002) suggested also that the new instrumental epicenter location was more consistent with the set of macroseismic intensities reported by Avni (1999). They favored thus another causative fault segment located further south compared to the results of Ben-Menahem et al. (1976). As a whole, it seems that the new instrumental location determined by Shapira et al. (1993) and apparently further supported by macroseismic evidence (Avni 1999; Avni et al. 2002), has been little challenged. Abou Karaki (1999), however, questioned the reliability of the relocated epicenter, and argued that the new location was far less convincing than the original ISS (1927) location.

Zohar and Marco (2012) re-estimated the epicenter of the 1927 earthquake from macroseismic intensities reported by Avni (1999). They proceeded in two main steps. First, they corrected the initial intensities according to local site-attributes. Out of 133 sites, 111 or 83 % of them were corrected. Second, in 5 km by 5 km cells, they correlated the corrected intensities with a logarithmic function of the hypocentral distance to each site and used Pearson's correlation coefficient to estimate the quality of fit for each cell. An area of 300 km^2, in which the best correlations are found (Fig. 3.6), indicates the most appropriate location of the epicentral region according to the authors. The centroid of this area coincides with a location on the Dead Sea transform. The authors consider, however, that their macroseismic epicentral region is about equally consistent with the instrumental epicenter of Ben-Menahem et al. (1976) and with the instrumental epicenter of Shapira et al. (1993).

Early 2013, the ISC – GEM Global Instrumental Earthquake Catalogue (1900–2009) was released (Bondár et al. 2012; Di Giacomo et al. 2012; Storchak et al. 2013). The new catalogue is the result of a special effort to improve and extend existing bulletin data. The project was funded by the GEM (Global Earthquake Model) Foundation. It was led by the ISC and performed by a team of international experts following recommendations from a panel of IASPEI observers. The relocation procedure benefitted from improved depths determined by the EHB technique (Engdahl et al. 1998), and more accurate epicenters due to the development of a new location algorithm (Bondár and Storchak 2011) at the ISC. The catalogue also provides for all the earthquakes their moment magnitude M_W with an uncertainty estimate. The epicenter of the 1927 earthquake according to the ISC – GEM catalogue is located only 15 km SE of the old ISS (1927) epicenter (Table 3.4 and Fig. 3.6), with an average epicentral uncertainty radius of about 6.5 km. The computed depth is 15 ± 6 km. The moment magnitude M_W has been determined as 6.29 ± 0.21 and derives apparently from native surface-wave magnitudes M_S converted to M_W by regression. According to the parameters reported in Table 3.4, the ISC-GEM relocation (Storchak et al. 2013) of the 1927 earthquake is much better constrained than the Shapira et al. (1993) relocation. The ISC – GEM results are based on more than twice the number of stations and nearly three times the number of active (non-zero weight) phases. The average residual time per phase is less than half the value from Shapira et al. (1993) and the azimuthal gap is reduced by nearly 40 % due to the greater number of stations, distant up to about 80° from the epicentral region. In addition, the location procedure at the ISC has been significantly improved and refined during the 20 years separating the two relocations. Consequently, there is little doubt that the epicentral region of the 1927 earthquake is well located somewhere in the Jordan Valley as initially determined already more than 80 years ago by the IPG of Strasbourg (Berloty 1927; Rothé 1928), and as argued by Abou Karaki (1999).

But if no damming of Jordan River occurred in 1927 as demonstrated by Avni (1999) and by Avni et al. (2002), the epicentral region determined by Vered and Striem (1977) around Damiya bridge (Fig. 3.6) and related depth estimates might then be invalidated. Since Zohar and Marco (2012) published macroseismic intensities as determined by Avni (1999) as well as their own site-dependent corrected values, we drew several isoseismal maps from which we derived depth estimates according to methods from Medvedev (1962) and from Shebalin (1973).

These methods are based on relations between the seismic source depth and the areal extent of isoseismal contour lines. Medvedev (1962) developed two approximative methods. The first method is described by Eq. 3.2, in which m is the serial number of an isoseismal curve of intensity I, H_m is the corresponding source depth estimate, Δ_m is the equivalent radius of a circle with same area as the isoseismal, and n is a coefficient of amplitude attenuation with distance.

$$H_m = \frac{\Delta_m}{\sqrt{2^{\frac{2(m-0.5)}{n}} - 1}} \tag{3.2}$$

The serial number m is given by

$$m = I_0 - I + 1, \tag{3.3}$$

where I_0 is the highest intensity value of the isoseismals, for which m = 1. Typical values for the attenuation coefficient n are n = 2 or n = 1.5. If the equivalent radius Δ_m is given in km, the depth H_m is also expressed in km. Equation 3.2 allows to determine a depth value from any isoseismal contour. The attenuation coefficient n is usually chosen as the value minimizing the dispersion of results. In agreement with Vered and Striem (1977), n = 1.5 appears to be the best value for the 1927 earthquake. Following the observation that the second and third isoseismals are usually the most reliable ones, Medvedev (1962) developed a second method to determine the depth, given by Eq. 3.4:

$$H = \beta\sqrt{S_2 + S_3} \tag{3.4}$$

In Eq. 3.4, H is the source depth in km, S_2 and S_3 are the areas bounded by the second (m = 2) and third (m = 3) isoseismals expressed in thousands of km^2, and β is a coefficient depending on the coefficient of attenuation n. For n = 1.5, β = 5.2. For n = 2.0, β = 7.0.

Shebalin (1973), in an attempt to determine several earthquake parameters from anomalies of intensity near the source, provides Eq. 3.5, in which m is the serial number of an isoseismal curve of intensity I, H_m is the corresponding source depth, Δ_m is the radius of the circle with same area as the isoseismal of serial number m, and n is the coefficient of amplitude attenuation with distance:

$$H_m = \frac{D_m}{\sqrt{10^{\frac{2m}{n}} - 1}} \tag{3.5}$$

The serial number m is given by

$$m = I_0 - I, \tag{3.6}$$

where I_0 is the highest intensity value of the isoseismals. Equation 3.5 requires the intensity I to be lower than I_0. The attenuation coefficient n in Eq. 3.5 varies between 3.0 and 4.5 for crustal earthquakes (Shebalin et al. 1974). We have selected a value of n = 3.5 at which the dispersion of depths from Eq. 3.5 is minimal for all datasets except the one from Vered and Striem (1977), for which n = 4 minimizes the dispersion.

In order to estimate the source depth of the 1927 earthquake according to Medvedev's and Shebalin's methods, we contoured by kriging several sets of MSK (Medvedev and Sponheuer 1969) macroseismic intensities determined by Avni (1999), and published by Zohar and Marco (2012) who also introduced their own variants. Although several spatial interpolation techniques can be used, kriging (Krige 1951; Matheron 1963) is one of the most robust of them, producing curves usually as smooth as allowed by the data. The principal benefit of drawing isoseismals by kriging is that it avoids the subjectivity usually introduced when contouring is done by hand (Ambraseys and Douglas 2004).

According to tests on the sets of intensity data, we found that a linear variogram model with lags up to about 100 km provided adequate results. Anisotropy did not appear to be an issue and was thus not modeled. Ordinary (no trend) point (no averaging in blocks) kriging was used to interpolate the observations and produce the isoseismals.

The depth estimates according to various intensity determinations are detailed in Table 3.6. The magnitudes in Table 3.6 derive from Eq. 3.7 below defined by Ambraseys (2006) for the region of the Dead Sea transform:

$$M_S = -0.138 + 0.554\ I + 0.0033\ R_I + 1.54\ \mathrm{Log}\ R_I \tag{3.7}$$

where M_S is the surface wave magnitude, I is the MSK intensity and R_I is the equivalent average radius of the isoseismal of intensity I.

The most consistent results are those of Vered and Striem (1977). In particular, the M_S magnitudes determined by Eq. 3.7 for isoseismals VIII and VII are remarkably similar (6.72 and 6.77), even if they are too high compared to the instrumental M_S value of 6.15 determined by the ISC – GEM (Storchak et al. 2013).

Table 3.6 Depth estimates according to Medvedev's methods and Shebalin's method. MSK Intensities from Avni (1999): *MD* (Mode), *MN* (Mean), *MX* (Max); MSK Intensities corrected by Zohar and Marco (2012): *MD-C1* (Mode Corrected 1); MMI Intensities from Vered and Striem (1977): VRD. I_0: highest Intensity isoseismal. m-MED and m-SHB: Isoseismal Index number according to Medvedev and Shebalin respectively. M_S: Surface Wave Magnitude according to Eq. 3.7 from Ambraseys (2006). h: source depth (km) according to Medvedev's methods 1 (MED- 1) and 2 (MED-2), and Shebalin (SHB). Valid depths in **bold**. Cells with inconsistent values have been *greyed* out

Data	I_0	I	m MED	m SHB	Area (km^2)	Radius (km)	M_S AMB	h MED-1	h MED-2	h SHB
MD	IX	VIII	2	1	137	7	5.60	3.8		4.0
		VII	3	2	14,133	67	6.80	**22.2**		**18.7**
			2 + 3		14,270				**19.6**	
MN	VIII	VIII	1	0	1,463	22	6.40	28.2		
		VII	2	1	12,148	62	6.70	35.9		37.7
MX	IX	VIII	2	1	656	14	6.10	8.3		8.7
		VII	3	2	15,214	70	6.80	**23.1**		**19.4**
			2 + 3		15,870				**20.7**	
MD -C1	VIII	VIII	1	0	40	3.6	5.20	4.7		
		VII	2	1	3,369	33	6.20	**18.9**		**19.8**
		VI	3	2	20,976	82	6.40	**27.1**		**22.8**
			2 + 3		24,345				**25.7**	
VRD	IX	VIII	2	1	3,200	32	6.72	**18.4**		**21.7**
		VII	3	2	14,000	67	6.77	**22.2**		**22.3**
			2 + 3		17,200				**21.6**	

The fact that Vered and Striem used the MMI scale instead of the MSK scale is not an issue. Similar magnitude overestimations occur also for most MSK intensities reported in Table 3.6. But as can be seen from Fig. 3.7, the isoseismals from Vered and Striem are so concentric and so homothetic that they seem to have been idealized rather than just smoothed. A recurrent problem with all the other intensity variants except Mean (MN) intensities (Avni 1999; Fig. 3.8), is that the isoseismal VIII areas (Table 3.6) appear to be systematically underestimated as can be seen from the too low M_S magnitudes predicted by Eq. 3.7. The underestimation seems to be particularly significant for the Mode intensities corrected (MD–C1) by Zohar and Marco (2012), for which the kriging procedure did not produce any isoseismal VIII contour at all. The tiny area of 40 km^2 reported in Table 3.6 was drawn by hand (Fig. 3.17 in Appendix), in an attempt to determine an epicentral region from the highest intensities. In fact, besides the isoseismal VIII of Vered and Striem, there is no other dataset resulting in consistent depths for that level of intensity. The area of isoseismal VIII in Mean (MN) intensities (Avni 1999; Fig. 3.8) appears to be realistic, but unlike the data of Vered and Striem, there is no intensity IX value in this dataset. Consequently, all depths derived from Mean (MN) intensities (Avni 1999) appear to be inconsistent. If no intensity IX had been present in the data of Vered and Striem, the depth estimate from isoseismal VIII according to Medvedev's first method would not

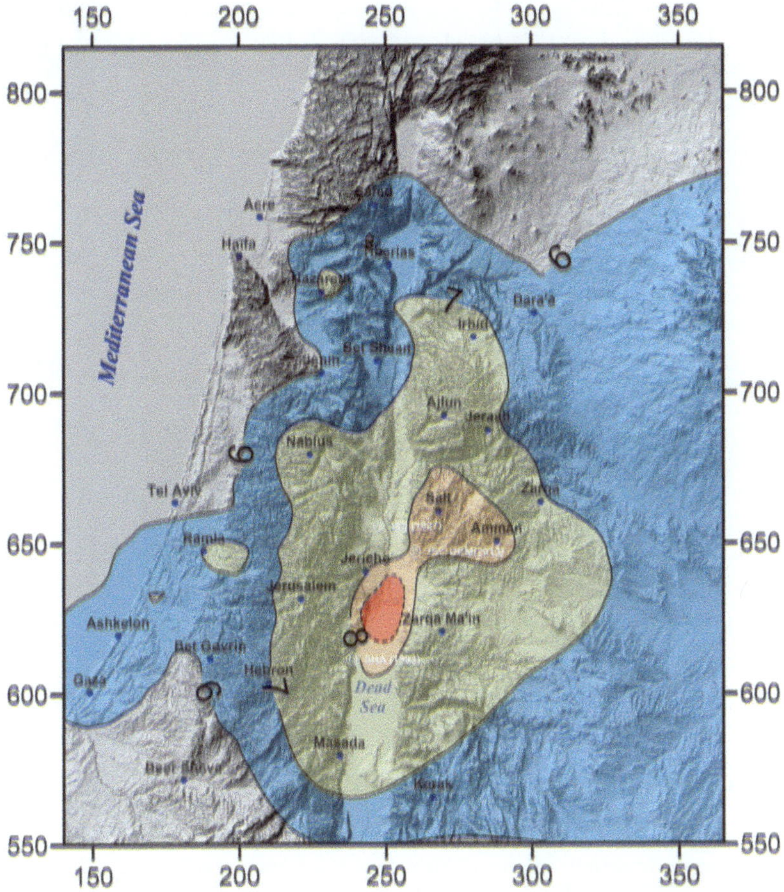

Fig. 3.8 Isoseismal map (MSK) of the earthquake of 11 July 1927 produced by kriging of Mean Intensities determined by Avni (1999), and published by Zohar and Marco (2012). ISS (1927): instrumental epicenter from ISS. SHA (1993): instrumental epicenter from Shapira et al. (1993). ISC-GEM: instrumental epicenter from ISC-GEM (Storchak et al. 2013). The area in red defines the macroseismic epicentral region (drawn by hand) according to this dataset. Israel Transverse Mercator (ITM) grid coordinates (km)

have been 18.4 km but 41.6 km instead, a value also inconsistent with the other estimates. Isoseismal areas considered to be inconsistent were not included to derive the source depth. According to 16 valid estimates (Table 3.6), we found a macroseismic source depth average of 21.5 ± 2.5 (STD) km.

If the DST is well the causative fault of the 1927 earthquake as suggested by source parameters determined by Ben-Menahem et al. (1976; Table 3.5), we can also assume that the true epicenter is located not far from the ISC – GEM instrumental epicenter (Storchak et al. 2013). We propose thus as preferred epicenter 31.92 ° N – 35.56 ° E (Fig. 3.9) with a latitude uncertainty of at least ± 7 km taken

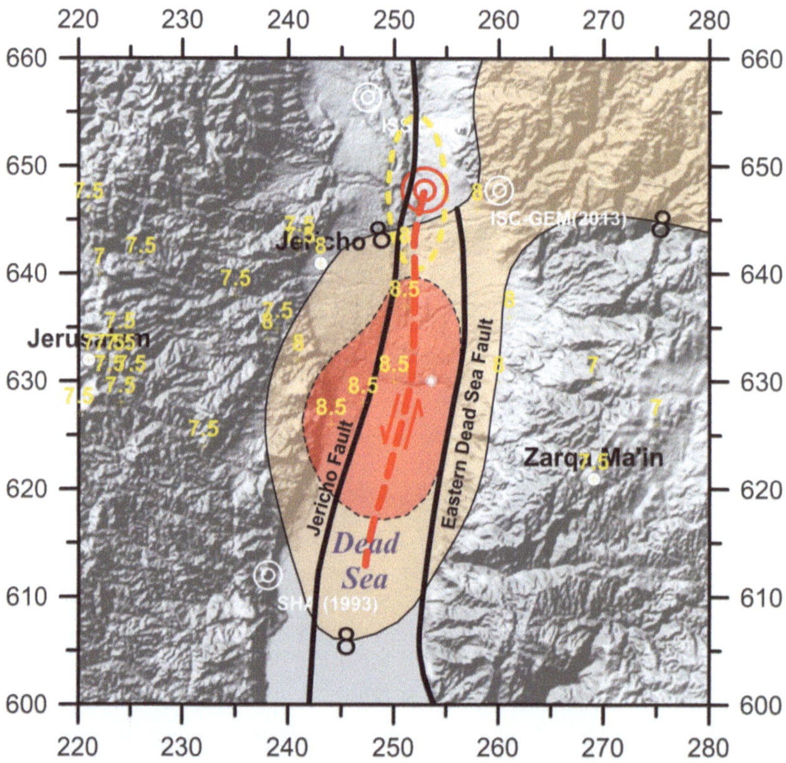

Fig. 3.9 Preferred epicenter, and tentative causative fault (Version 1) of the M_W 6.3 earthquake of 11 July 1927. Epicenter at red concentric circles. Epicentral location uncertainty as *dashed yellow ellipse*. Subsurface causative fault trace as *red dashed line* (36 km long). Isoseismal intensity MSK=VIII curve produced by kriging of Mean Intensities determined by Avni (1999), and published by Zohar and Marco (2012). Mean Intensity (MSK) values determined by Avni (1999) in *yellow*. Macroseismic epicentral region, according to this dataset, drawn by hand in *red*. ISS (1927): instrumental epicenter from ISS. SHA (1993): instrumental epicenter from Shapira et al. (1993). ISC-GEM: instrumental epicenter from Storchak et al. (2013). *White asterisk* on the northern shore of the Dead Sea: alledged location of photograph in Fig. 3.11. Israel Transverse Mercator (ITM) grid coordinates (km)

from the ISC – GEM relocation. The longitude uncertainty should be much smaller than the latitude uncertainty if the DST is indeed the causative fault.

Regarding the causative fault segment, the prime candidate is the Jericho Fault but we cannot exclude the Eastern Dead Sea Fault. Furthermore, as suggested by Lazar (2004) and supported by results from Hofstetter et al. (2012), there might well be only a single fault, the Dead Sea Fault, acting near the central axis of the basin from about 15–20 km depth. In that case, if the hypocenter was located around 20 km depth in agreement with the macroseismic data, the causative fault could be this unique Dead Sea Fault and not any of the two border faults observed at the surface. If the hypocenter was shallower, as determined instrumentally, then the motion should have been initiated along one of the two border faults and the source would have moved deeper during the history of the earthquake.

In Fig. 3.9, our preferred epicenter is plotted on a detailed view of Mean Intensity values determined by Avni (1999) and the isoseismal VIII contoured by kriging. Although the isoseismal VIII extends surprisingly beyond Salt and Amman to the NE (Fig. 3.8), it does not even reach our preferred epicenter in the Jordan Valley (Fig. 3.9). This is apparently due to the fact that the closest point located further north in the Jordan Valley is about 30 km away from our preferred epicenter, where an intensity of only VII was estimated. In Fig. 3.9, we have also drawn by hand a possible macroseismic epicentral zone encompassing the sites where the highest intensity values have been observed (see also Fig. 3.11). The fact that the macroseismic epicentral region does not include the instrumental epicenter is not necessarily an anomaly. Bakun (2006) observed that the location of the macroseismic epicentral zone corresponds more to the location of the moment centroid than to the instrumental epicenter where the rupture starts. If Ben-Menahem et al. (1976) are right about the unilateral propagation of the source towards the south, it is logical to assume as they did, that the epicenter was located near the northern tip of the causative fault segment. In that case, the moment centroid and the macroseismic epicenter could be located at some distance south from the epicenter. The relative location of the epicentral macroseismic region with respect to our preferred epicenter (Fig. 3.9) seems to agree with this possibility. The predominance of unilateral propagation in $M_W > 6.0$ earthquakes, especially for strike-slip faulting, has been suggested by McGuire et al. (2002) and by Mai et al. (2005). Unilateral propagation might even be enhanced at plate boundaries such as the DST due to the contrast of elastic properties, at least over some depth range, between the two sides of the transform (McGuire et al. 2002).

It is very difficult, however, to estimate the length of the causative fault segment. Ben-Menahem et al. (1976) suggested a fault segment of 45 km long derived from directivity but this value seems very high for an M_W 6.3 earthquake, a magnitude value for which the average subsurface length of a strike-slip rupture is slightly under 22 km (Wells and Coppersmith 1994). Tentatively, we have drawn in Fig. 3.9 a subsurface fault segment of about 36 km long. But as we have seen before, the M_W 5.3 earthquake of February 2004 occurred in the northern part of the Dead Sea. According to focal mechanism solutions for the mainshock and the distribution of aftershocks, Hofstetter et al. (2008) determined that the causative fault was a dextral strike-slip fault oriented WNW – ESE. According to the location of the aftershocks (Hofstetter et al. 2008) that occurred within 48 h of the mainshock, we inferred a length of about 12 km for the causative fault. From relative relocations by Al-Tarazi et al. (2006) and by Hofstetter et al. (2008), the depth distribution of aftershocks seems to have been largely confined within only about 7 km from the depth of the mainshock. If we transpose this result to our relocated depth of 21.7 ± 1.4 km for the mainshock, the range of depths between 18 and 25 km contributed highly then to the moment release for the 2004 earthquake. With a macroseismic depth average of 21.5 ± 2.5 km, the M_W 6.3 earthquake of 1927 apparently also generated

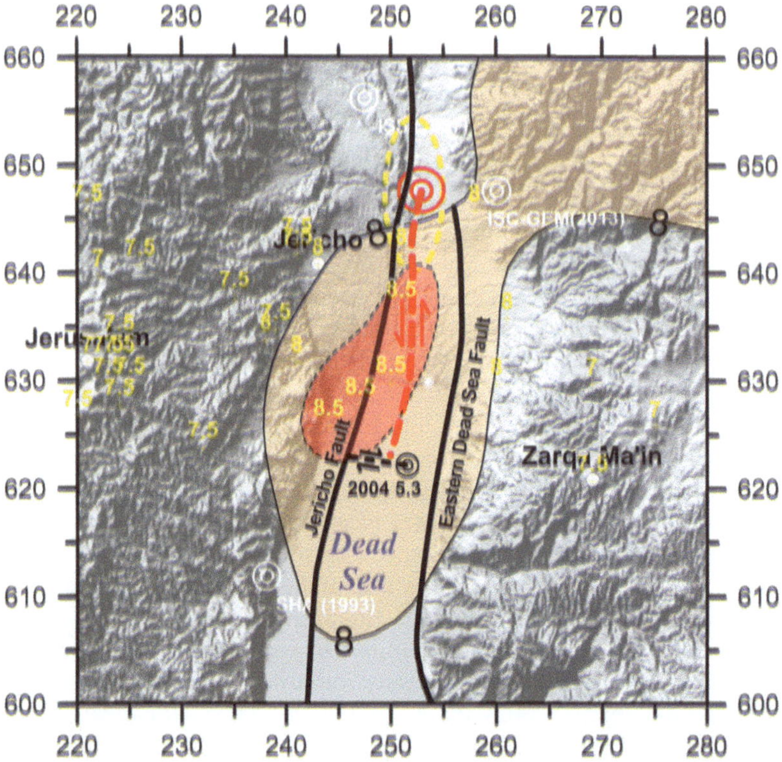

Fig. 3.10 Preferred epicenter, and tentative causative fault (Version 2) of the M_W 6.3 earthquake of 11 July 1927. Epicenter of the 1927 earthquake at *red concentric circles*. Epicentral location uncertainty as *dashed yellow ellipse*. Subsurface causative fault trace as *red dashed line* (25 km long). Epicenter of the M_W 5.3 earthquake of 11 February 2004 at *black concentric circles*, and subsurface fault trace as *black dashed line*. 1927 Isoseismal intensity MSK=VIII curve produced by kriging of Mean Intensities determined by Avni (1999), and published by Zohar and Marco (2012). Mean Intensity (MSK) values determined by Avni (1999) in *yellow*. Macroseismic epicentral region, according to this dataset, drawn by hand in *red*. ISS (1927): instrumental epicenter from ISS. SHA (1993): instrumental epicenter from Shapira et al. (1993). ISC-GEM: instrumental epicenter from Storchak et al. (2013). *White asterisk* on the northern shore of the Dead Sea: alleged location of photograph in Fig. 3.11. Israel Transverse Mercator (ITM) grid coordinates (km)

a great part of its seismic moment at the focal depth of the 2004 earthquake. Furthermore, the earthquake of 1927 might have been caused by a segment of the DST intersected and even possibly interrupted by the causative fault of the 2004 earthquake. It is then not impossible that the 1927 causative fault segment was in fact not longer than about 25 km if the causative fault of the 2004 earthquake interrupted it, as displayed in Fig. 3.10. This possibility is especially relevant since both earthquakes generated apparently a considerable part of their seismic moment at very similar depths.

Fig. 3.11 Ground fissures caused by the 1927 earthquake. Possible location close to the Jordan River estuary at the Dead Sea. View probably from south to north, taken in the morning according to the direction of the shade (see Figs. 3.9 or 3.10 for possible location). Estimated intensity MSK=VIII (IX) (Photograph from American Colony, Jerusalem. Matson Photograph Collection)

3.3 Heat Flow and Rheology

Aldersons et al. (2003) calculated a lithospheric strength profile for the Dead Sea Basin. The average measured heat flow in the northern Dead Sea Basin is 38 mWm^{-2} (Ben-Avraham et al. 1978; Ben-Avraham 1997) and it is 42 mWm^{-2} (Eckstein and Simmons 1978) west of the basin. These values are very similar to the uniform heat flow measured in the eastern Mediterranean (Erickson 1970). Consequently, the surface heat flow of 40 ± 2 mWm^{-2} used to compute the equilibrium geotherm appeared to be well constrained. A recent re-evaluation of the heat flow data (Shalev et al. 2007) confirmed this low value. A strain rate of 2×10^{-15} s^{-1} was used in the model. This value was derived from a relative plate motion of 5 mm/year, a value consistent with recent GPS estimates taken both from the Wadi Araba Fault and from the Jordan Valley Fault (Le Béon et al. 2008). Figure 3.12 displays the results. A narrow brittle to ductile transition occurs in the crust around 380° C at 31 km depth. In the upper mantle, the brittle to ductile transition occurs in the model at 44 km depth and at 490° C.

A study of the surface heat flow in southeastern Jordan (Förster et al. 2007) estimated the average value to be around 60 mWm^{-2}. This result is further supported by a set of samples originating from the upper crust down-to the lithospheric mantle under Jordan (Förster et al. 2010). However, the wells where these measurements were made, are located approximately 100 km east of the DST and more than 180 km south of the Dead Sea, and the authors acknowledge the importance of this

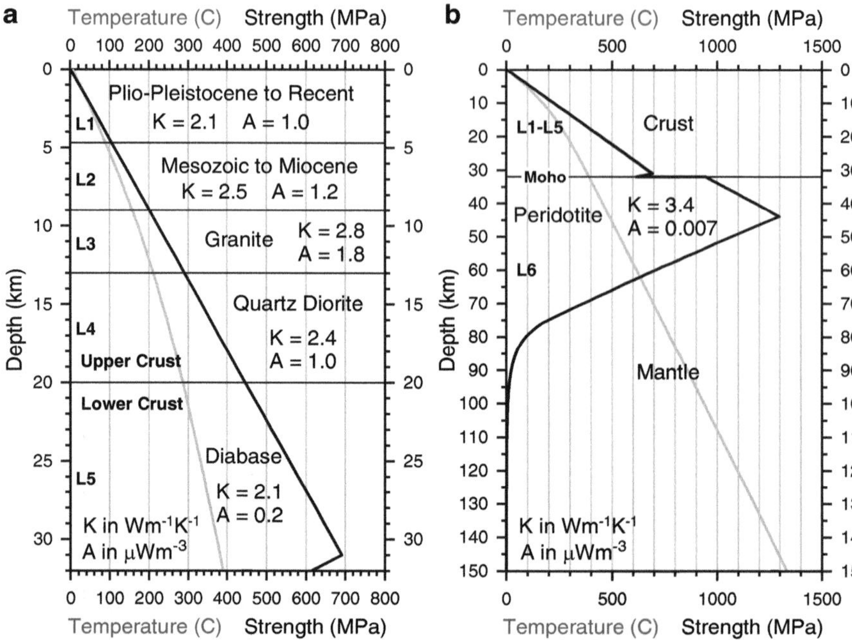

Fig. 3.12 Rheology of the Dead Sea Basin. (**a**) Crustal geotherm (*grey*) and strength profile (*black*). (**b**) Lithospheric geotherm (*grey*) and strength profile (*black*). Surface heat flow of 40 mWm^{-2}. K is the thermal conductivity and A is the radiogenic heat production rate (From Aldersons et al. 2003)

fact. They notice, however, that their value of 60 mWm^{-2} is identical to the heat flow constrained by thermomechanical modeling (Petrunin and Sobolev 2006) to be the pre-Miocene heat flow for the Dead Sea basin, instead of the measured value of 40 mWm^{-2} used by Aldersons et al. (2003). We agree that a direct match between the two values is possible but we do not think that it is the most likely scenario. If the 60 mWm^{-2} value of Förster et al. (2007) is projected perpendicularly along the DST (Fig. 3.13, pt. 4), it fits quite well with the decay of extreme heat flow observed in the Northern Red Sea axis and associated with the opening of the Red Sea. We think that this interpretation is more realistic than a direct match with a hypothetical heat flow estimate derived from numerical modeling, in conflict with the measured values, and located about 200 km away from the Dead Sea. Our preferred scenario is also in agreement with an extensive study of the surface heat flow distribution in Israel derived from well measurements spread all over the country (Shalev et al. 2012). This study shows that the surface heat flow in Israel is indeed low over most of the country, only 40–45 mWm^{-2} on the average, and around 40 mWm^{-2} in the Dead Sea (Fig. 3.14, top) in particular.

Figure 3.14 (from Shalev et al. 2012) shows also that the isotherm of 350 ° C, between the Sea of Galilee and Aqaba-Elat, is deepest in the Dead Sea area where it reaches 25–30 km depth. In agreement with earthquakes located by the EMSC

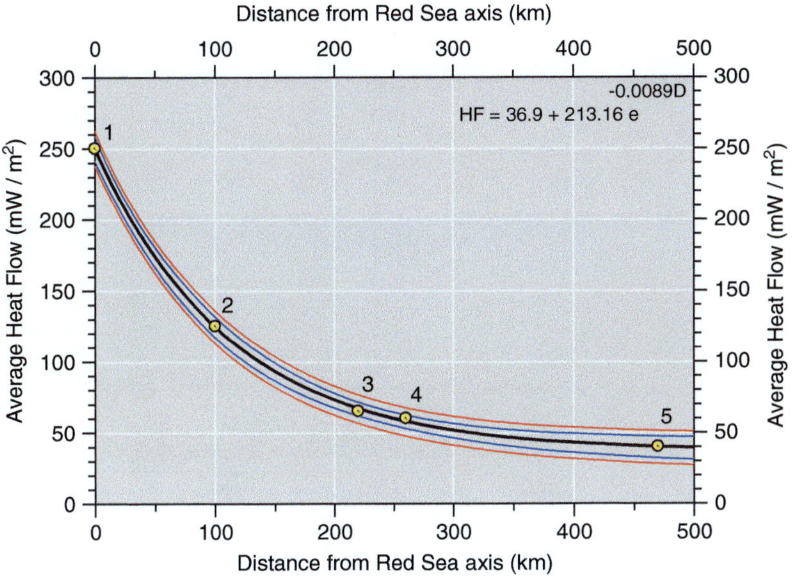

Fig. 3.13 Surface Heat Flow from the Red Sea axis to the Dead Sea basin. *1*: Martinez and Cochran (1989) and Makris et al. (1991). *2*: Martinez and Cochran (1989) and Makris et al. (1991). *3*: Ben-Avraham and Von Herzen (1987). *4*: Förster et al. (2007). *5*: Ben-Avraham et al. (1978), Eckstein and Simmons (1978), and Shalev et al. (2007). 95 % confidence and prediction bands in *red* and *blue* respectively

(Godey et al. 2006), Aldersons et al. (2003) and Shamir (2006), Shalev et al. (2012) determined 26–27 km as the seismogenic thickness in the Dead Sea area (Fig. 3.14). Towards the Sea of Galilee, the heat flow increase is slow but south-east of the Sea of Galilee the heat flow reaches about 90 mWm^{-2}. This anomaly is probably due to ground water flow along faults or from recent magmatic activity (Roded 2012). Towards Aqaba-Elat, the heat flow increase is steadier than towards the Sea of Galilee, and heat flow values above 60 mWm^{-2}, similar to those found by Förster et al. (2007, 2010) in SE Jordan, are observed near Aqaba-Elat. Shalev et al. (2012) agree with Aldersons et al. (2003) who suggested that the higher heat flow in southern Israel is probably reflecting the opening of the Red Sea and related processes.

3.4 Discussion and Concluding Remarks

We focused on depth distributions of seismicity from four independent studies, with some overlap in the datasets, but with extensive differences in their respective methodology. The results of the four studies unmistakably underline the occurrence, in the Dead Sea area, of lower-crustal microearthquakes down-to at least 27 km and

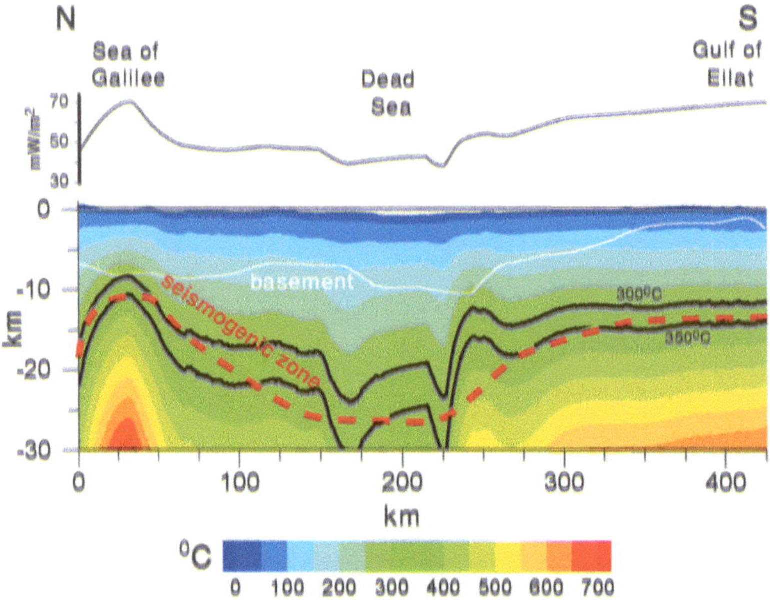

Fig. 3.14 Temperature distribution along the Dead Sea Fault from the Sea of Galilee to the Gulf of Aqaba-Elat. Temperatures derived from the surface heat flow (*top*). Seismogenic zone (*red dashed line*) derived from the EMSC 2011 catalogue (Godey et al. 2006) for the entire profile, Aldersons et al. (2003) and Shamir (2006) for the Dead Sea area, and Navon (2011) for the Sea of Galilee area (From Shalev et al. 2012)

possibly as deep as 33.6 km. In addition, a moderate earthquake of magnitude M_W 5.3 nucleated in February 2004 in the Dead Sea Basin at a depth of 21.7 ± 1.4 km according to our relocation of the mainshock. From relative relocations of aftershocks by Al-Tarazi et al. (2006) and by Hofstetter et al. (2008), we believe that the range of depths between 18 and 25 km highly contributed to the seismic moment release for this earthquake.

The seismogenic thickness T_S, or maximum possible depth extent of rupture during moderate to large earthquakes, is estimated in the Dead Sea area to be 28.4 ± 2.2 km following two criteria defined by Nazareth and Hauksson (2004) to determine T_S from the depth distribution of microearthquakes. The existence of a seismogenic zone extending deep into the lower crust is also consistent with an average heat flow of only 40–45 mWm^{-2} over most regions of Israel, and around 40 mWm^{-2} in the Dead Sea area in particular (Shalev et al. 2007, 2012). From their isotherm of 350 ° C and the depth distribution of earthquakes according to various sources, Shalev et al. (2012) determined a seismogenic thickness of 26–27 km in the Dead Sea area. With all the evidence available today, it is thus no longer possible to argue that the seismic activity gradually decreases below 15 km depth and completely stops around 20 km as suggested by Ben-Menahem et al. (1976) from a small and poorly-constrained sample of earthquakes.

The seismogenic thickness in the Dead Sea area is nearly twice the average seismogenic thickness of 15 km observed in southern California. It is important to know that for a seismogenic thickness of about 28 km, large earthquakes can nucleate at or close to 28 km depth. Sibson (1982) suggested a significant tendency of large earthquakes to nucleate in fact near the base of the seismogenic zone and to propagate upwards (Sibson 1982; Das and Scholz 1983). The strength of this tendency appears today to be somewhat weakened after the occurrence of a number of large and shallow earthquakes in California, but the possibility of large earthquakes to nucleate very deep in the seismogenic zone remains unaffected by the existence of large and shallower earthquakes as well. Latter earthquakes tend to demonstrate that the depth distribution of large earthquakes is perhaps not so different as previously thought from the rather unfocused depth distribution of microearthquakes.

The recent ISC – GEM relocation (Storchak et al. 2013) of the M_W 6.3 ± 0.2 earthquake of 11 July 1927 from regional and teleseismic data results in a well-constrained instrumental location in the Jordan Valley, not far from the initial epicenter determined by the Institut de Physique du Globe of Strasbourg (Berloty 1927) and reported in the ISS bulletin of 1927. Since the DST appears to be the causative fault of this earthquake as suggested by source parameters determined by Ben-Menahem et al. (1976), we propose a preferred epicenter at 31.92 ° N – 35.56°E. The focal depth determined instrumentally by the ISC – GEM procedure (Storchak et al. 2013) is 15 ± 6 km, and we found an average macroseismic depth of 21.5 ± 2.5 km. The depth value derived from macroseismic data, however, should not necessarily be equal to the focal depth determined instrumentally. The instrumental focal depth represents the depth of the source at the beginning of the fault rupture, but the depth derived from isoseismals in the far-field represents more some average depth of the source during its complete history. Since seismic slip is usually not homogeneously distributed across the whole rupture plane but is concentrated at asperities (Lay et al. 1982) where it can be much higher than average, it is possible that the depth derived from macroseismic data in the far-field corresponds more to the average depth of the asperities than to the average depth of rupture on the fault plane. We could adopt a unique depth of about 20 km that would be consistent both with the instrumental depth and with the macroseismic depth, taking into account their respective uncertainties. In that case, the 1927 earthquake would have nucleated and propagated close to the transition between the upper and the lower crust. It is also possible that the earthquake nucleated around 15 km as determined instrumentally, or shallower, and that the source deepened with time. Since the latter scenario is more complex and not required by the data taking into account the uncertainties on the instrumental and macroseismic depths, we prefer the simpler solution of a unique depth around 20 km. The M_W 5.3 earthquake of February 2004 nucleated at a similar depth (21.7 ± 1.4 km), with the depth range between 18 and 25 km contributing highly to the moment release. It is thus possible that the larger M_W 6.3 earthquake of July 1927 ruptured the seismogenic zone even deeper than 25 km, perhaps down-to its limit around 28 km. It is not possible, however, to have any certainty regarding this possibility due to the lack of reliable constraints on fault length, width and slip for the earthquake of 1927. Among parameters determined by Ben-Menahem et al. (1976), their fault length of 45 km appears to be unrealistically high, the vertical extent of 10 km results from a focal depth estimate of 5–7 km and the hypothesis, invalidated today, that the seismicity in the Dead Sea area quickly decreases below 15 km depth. Finally, the slip value of 0.4 m results from the need to end up with values of potency and seismic moment consistent with the observed magnitude.

If the Dead Sea basin is indeed confined to the upper crust as suggested by wide-angle seismic data from the DESIRE project (Mechie et al. 2009), then the M_W 6.3 earthquake of 1927 and the M_W 5.3 earthquake of 2004 generated a great part of their seismic moment near the base of the Dead Sea basin (18–20 km), and immediately below the Dead Sea basin (20–25 km) in the lower crust. Our results for the M_W 6.3 earthquake of 1927 and for the M_W 5.3 earthquake of 2004 underline the seismogenic importance of the transition between the upper and the lower crust for moderate and probably also for large earthquakes, in agreement with Mechie et al. (2009) who suggested that this transition might even act as a decoupling zone at the base of the Dead Sea basin.

Finally, with a seismogenic zone nearly as thick as the entire crust, it is unlikely that a fully plastic mode of deformation could prevail under the seismogenic zone in the crust because the brittle to plastic transition extends over a range of pressure and temperature in which the mode of deformation gradually changes from fully brittle to fully plastic (Scholz 2002). The fact that some microearthquakes nucleate down-to the depth of the Moho in the Dead Sea area according to this study and from Koulakov and Sobolev (2006), suggests that the state of fully plastic deformation is probably not reached in the crust under the seismogenic zone. In the Dead Sea area, brittle deformation appears thus to be the dominant deformation mechanism prevailing throughout the crust, a constraint already integrated in the drop-down model of Ben-Avraham and Schubert (2006; Ben-Avraham et al. 2010; Ben-Avraham 2014, this volume).

Acknowledgements We greatly thank I. Koulakov and B. Braeuer for providing the details of their distributions of seismicity in the Dead Sea region. We thank R. Hofstetter, D. Zakosky and L. Feldman for providing the GII seismograms of the M_W 5.3 earthquake of 2004. We thank T. Al-Yazjeen, W. Olimat and B. Al-biss for providing the JSO seismograms of the M_W 5.3 earthquake of 2004. Many thanks to D. Storchak, I. Bondár and D. Di Giacomo from the ISC for providing the details of the ISC-GEM relocation for the M_W 6.3 earthquake of 1927. Thanks also to O. Heidbach and an anonymous reviewer for their criticism, and advice to improve the manuscript.

Appendix

M_W 6.3 Earthquake of 11 July 1927: Additional isoseismal maps (Figs. 3.15, 3.16, and 3.17)

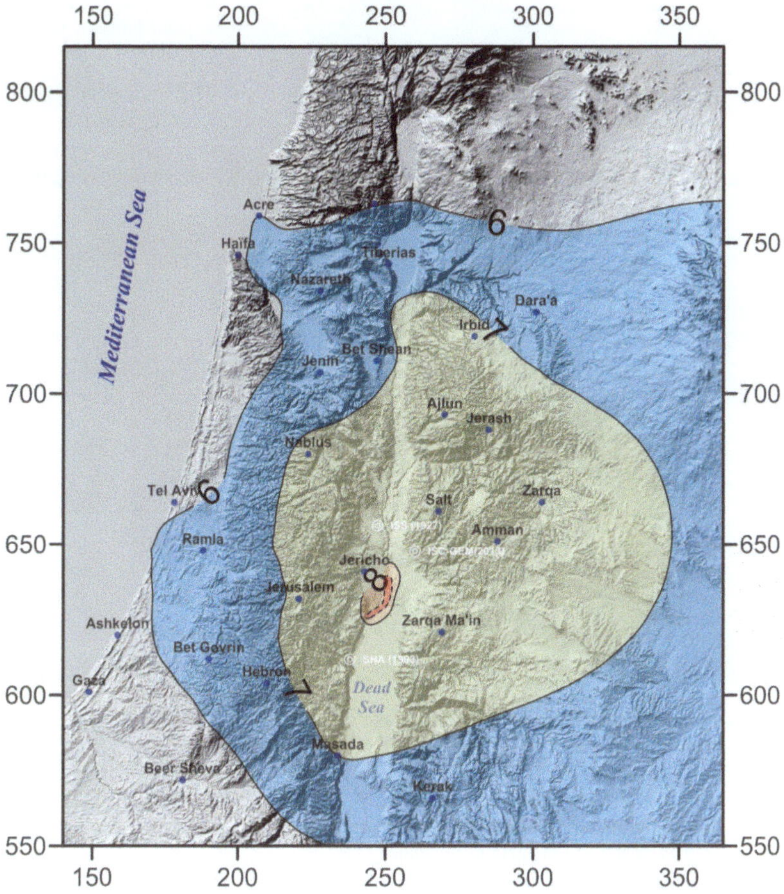

Fig. 3.15 Isoseismal map (MSK) of the earthquake of 11 July 1927 produced by kriging of Mode Intensities determined by Avni (1999), and published by Zohar and Marco (2012). ISS (1927): instrumental epicenter from ISS. SHA (1993): instrumental epicenter from Shapira et al. (1993). ISC-GEM: instrumental epicenter from ISC-GEM (Storchak et al. 2013). The area in *red* (MSK = IX) defines the macroseismic epicentral region (drawn by hand) according to this dataset. Israel Transverse Mercator (ITM) grid coordinates (km)

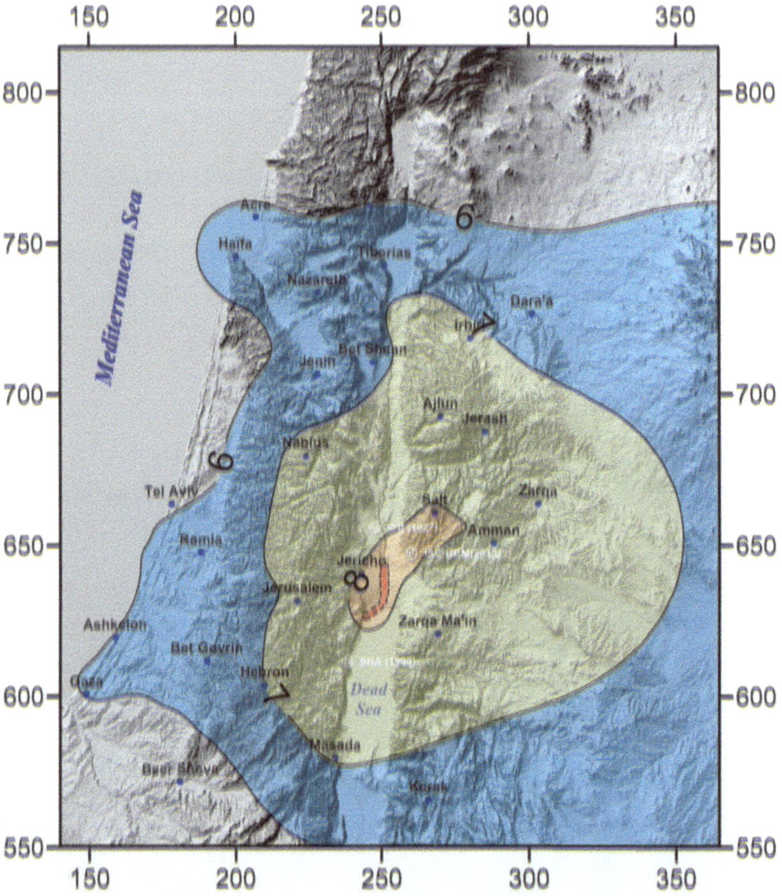

Fig. 3.16 Isoseismal map (MSK) of the earthquake of 11 July 1927 produced by kriging of Max Intensities determined by Avni (1999), and published by Zohar and Marco (2012). ISS (1927): instrumental epicenter from ISS. SHA (1993): instrumental epicenter from Shapira et al. (1993). ISC-GEM: instrumental epicenter from ISC-GEM (Storchak et al. 2013). The area in red (MSK = IX) defines the macroseismic epicentral region (drawn by hand) according to this dataset. Israel Transverse Mercator (ITM) grid coordinates (km)

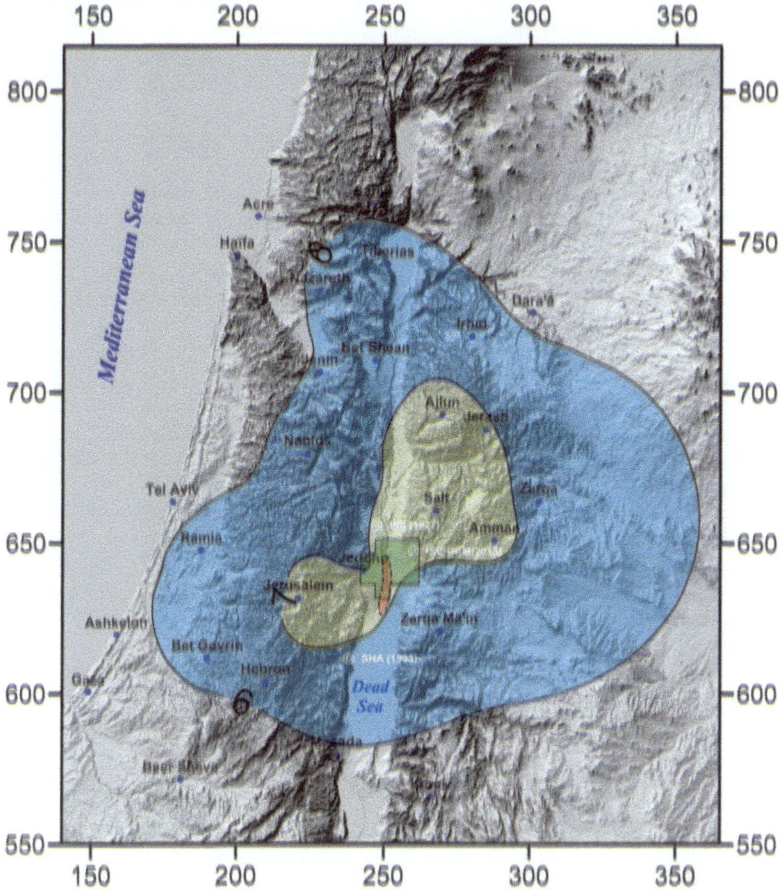

Fig. 3.17 Isoseismal map (MSK) of the earthquake of 11 July 1927 produced by kriging of Modal Intensities determined by Avni (1999) and Corrected for site attributes (± 1 unit MSK) by Zohar and Marco (2012). ISS (1927): instrumental epicenter from ISS. SHA (1993): instrumental epicenter from Shapira et al. (1993). ISC-GEM: instrumental epicenter from ISC-GEM (Storchak et al. 2013). The area in red (MSK = VIII) defines the macroseismic epicentral region (drawn by hand) according to this dataset. The region in *green* defines the macroseismic epicentral region according to the regressions of Zohar and Marco (2012). Israel Transverse Mercator (ITM) grid coordinates (km)

References

Abou Elenean K, Aldamegh K, Zharan H, Hussein H (2009) Regional waveform inversion of 2004 February 11 and 2007 February 09 Dead Sea earthquakes. Geophys J Int 176:185–199

Abou Karaki N (1999) Location of the earthquake of Palestine 11-7-1927: a critical assessment, vol 8. Abhath al-Yarmouk, Basic sciences and engineering series, 1. pp 9–34 (In Arabic, English Abstract)

Aldersons F (2004) Toward a three-dimensional crustal structure of the Dead Sea region from local earthquake tomography. PhD thesis, Tel-Aviv University, p 123

Aldersons F, Ben-Avraham Z, Hofstetter A, Kissling E, Al-Yazjeen T (2003) Lower-crustal strength under the Dead Sea Basin from local earthquake data and rheological modeling. Earth Planet Sci Lett 214:129–142

Al-Tarazi E, Sandvol E, Gomez F (2006) The February 11, 2004 Dead Sea earthquake ML = 5.2 in Jordan and its tectonic implication. Tectonophysics 422:149–158

Ambraseys N (2006) Comparison of frequency of occurrence of earthquakes with slip rates from long-term seismicity data: the cases of Gulf of Corinth, Sea of Marmara and Dead Sea Fault Zone. Geophys J Int 165(2):516–526

Ambraseys N, Douglas J (2004) Magnitude calibration of north Indian earthquakes. Geophys J Int 159(1):165–206

Avni R (1999) The 1927 Jericho earthquake – comprehensive macroseismic analysis based on contemporary sources. PhD thesis, Ben-Gurion University of the Negev, vol 1, 64 p, vol 2, p 139 (in Hebrew)

Avni R, Bowman D, Shapira A, Nur A (2002) Erroneous interpretation of historical documents related to the epicenter of the 1927 Jericho earthquake in the Holy Land. J Seismol 6:469–476

Bakun WH (1984) Seismic moments, local magnitudes, and coda-duration magnitudes for earthquakes in central California. Bull Seismol Soc Am 74(2):439–458

Bakun WH (2006) Estimating locations and magnitudes of earthquakes in southern California from modified Mercalli intensities. Bull Seismol Soc Am 96(4A):1278–1295

Begin ZB, Steinitz G (2005) Temporal and spatial variations of microearthquake activity along the Dead Sea Fault, 1984–2002. Isr J Earth Sci 54:1–14

Ben-Avraham Z (1997) Geophysical framework of the Dead Sea: structure and tectonics. In: Niemi TM, Ben-Avraham Z, Gat JR (eds) The Dead Sea: the Lake and its settings, vol 36. Oxford University Press, New York, pp 22–35

Ben-Avraham Z (2014) Chapter1: Geophysical studies of the crustal structure along the Southern Dead sea fault. In: Garfunkel Z, Ben-Avraham Z, Kagan E (eds) Dead Sea Transform Fault System: Reviews. Springer, Dordrecht, pp 1–27

Ben-Avraham Z, Schubert G (2006) Deep "drop down" basin in the southern Dead Sea. Earth Planet Sci Lett 251:254–263

Ben-Avraham Z, Von Herzen RP (1987) Heat flow and continental breakup: the Gulf of Elat (Aqaba). J Geophys Res 92:1407–1416

Ben-Avraham Z, Hänel R, Villinger H (1978) Heat flow through the Dead Sea rift. Mar Geol 28:253–269

Ben-Avraham Z, Lyakhovsky V, Schubert G (2010) Drop-down formation of deep basins along the Dead Sea and other strike-slip fault systems. Geophys J Int 181:185–197

Ben-Menahem A, Singh SJ (2000) Seismic waves and sources, 2nd edn. Dover Publications, Mineola, p 1102

Ben-Menahem A, Nur A, Vered M (1976) Tectonics, seismicity and structure of the Afro-Eurasian junction – The breaking of an incoherent plate. Phys Earth Planet Inter 12:1–50

Berloty SJ (1927) Annales de l'Observatoire de Ksara (Liban), Observations (Section Séismologique) – notice spéciale sur le tremblement de terre de Palestine, 11 juillet 1927, 62–95

Bondár I, Storchak D (2011) Improved location procedures at the International Seismological Centre. Geophys J Int 186:1220–1244

Bondár I, Engdahl ER, Villaseñor A, Storchak D (2012) ISC-GEM: global instrumental earthquake catalogue (1900–2009): I. Location and seismicity patterns. AGU Fall Meeting, San-Francisco, USA

Braeuer B, Asch G, Hofstetter R, Haberland C, Jaser D, El-Kelani R, Weber M (2012a) Microseismicity distribution in the southern Dead Sea basin and its implications on the structure of the basin. Geophys J Int 188:873–878

Braeuer B, Asch G, Hofstetter R, Haberland C, Jaser D, El-Kelani R, Weber M (2012b) High-resolution local earthquake tomography of the southern Dead Sea area. Geophys J Int 191:881–897

Braslavski J (1938) The earthquake that blocked the Jordan in 1546. Zion 3:323–336 (in Hebrew)

Das S, Scholz CH (1983) Why large earthquakes do not nucleate at shallow depths. Nature 305:621–623

Di Giacomo D, Bondár I, Lee WHK, Engdahl ER, Bormann P, Storchak D (2012) ISC-GEM: global instrumental earthquake catalogue (1900–2009): II. Earthquake magnitudes. AGU Fall Meeting, San-Francisco, USA

Di Stefano R, Aldersons F, Kissling E, Baccheschi P, Chiarabba C, Giardini D (2006) Automatic seismic phase picking and consistent observation error assessment: application to the Italian seismicity. Geophys J Int 165:121–134

Di Stefano R, Kissling E, Chiarabba C, Amato A, Giardini D (2009) Shallow subduction beneath Italy: three-dimensional images of the Adriatic-European-Tyrrhenian lithosphere system based on high-quality P wave arrival times. J Geophys Res 114, B05305. doi:10.1029/2008JB005641

Diehl T, Kissling E, Husen S, Aldersons F (2009) Consistent phase picking for regional tomography models: application to the greater Alpine region. Geophys J Int 176:542–554

Eckstein Y, Simmons G (1978) Measurements and interpretation of terrestrial heat flow in Israel. Geothermics 6:117–142

Engdahl ER, van der Hilst R, Buland R (1998) Global teleseismic earthquake relocation with improved travel times and procedures for depth determination. Bull Seismol Soc Am 88(3):722–743

Erickson AJ (1970) The measurement and interpretation of heat flow in the Mediterranean and Black Sea. PhD thesis, Massachusetts Institute of Technology, p 433

Förster A, Förster H, Masarweh R, Masri A, Tarawneh K (2007) The surface heat flow of the Arabian Shield in Jordan. J Asian Earth Sci 30:271–284

Förster H-J, Förster A, Oberhänsli R, Stromeyer D (2010) Lithospheric composition and thermal structure of the Arabian Shield in Jordan. Tectonophysics 481:29–37

Garstang J (1931) Joshua judges. Constable and Co., London, pp 134–139

Ginzburg A, Makris J, Fuchs K, Prodehl C (1981) The structure of the crust and upper mantle in the Dead Sea rift. Tectonophysics 80:109–119

Godey S, Bossu R, Guilbert J, Mazet-Roux G (2006) The Euro-Mediterranean bulletin: a comprehensive seismological bulletin at regional scale. Seismol Res Lett 77(4):460–474

Gutenberg B, Richter CF (1954) Seismicity of the earth. Princeton University Press, Princeton, p 310

Hagiwara T (1964) Brief description of the project proposed by the earthquake prediction research group of Japan. In: Proceedings of U.S.-Japan conference related earthquake prediction problem, Tokyo, pp 10–12

Hofstetter R, Gitterman Y, Pinsky V, Kraeva N, Feldman L (2008) Seismological observations of the northern Dead Sea basin earthquake on 11 February 2004 and its associated activity. Isr J Earth Sci 57:101–124

Hofstetter R, Dorbath C, Calò M (2012) Crustal structure of the Dead Sea Basin from local earthquake tomography. Geophys J Int 189:554–568

International Seismological Centre (2001) Bulletin disks 1–9 [CD-ROM], International Seismological Centre, Thatcham, UK

International Seismological Centre (2011) On-line Bulletin, http://www.isc.ac.uk. International Seismological Centre, Thatcham, UK

International Seismological Summary (1927) University of Oxford, p 472

Kennett B, Engdahl E (1991) Traveltimes for global earthquake location and phase identification. Geophys J Int 122:429–465

Kissling E (1995) Velest user's guide. Internal report 26, Institute of Geophysics, ETH Zürich, Switzerland

Kissling E, Ellsworth WL, EberhartPhillips D, Kradolfer U (1994) Initial reference models in local earthquake tomography. J Geophys Res 99:19635–19646

Koulakov I, Sobolev S (2006) Moho depth and three-dimensional P and S structure of the crust and uppermost mantle in the Eastern Mediterranean and Middle East derived from tomographic inversion of local ISC data. Geophys J Int 164:218–235

Krige D (1951) A statistical approach to some basic mine valuation problems on the Witwatersrand. J Chem Metall Min Soc S Afr 6:119–139

Küperkoch L, Meier T, Diehl T (2012) Automated event and phase identification. In: Bormann P (ed) New manual of seismological observatory practice 2 (NMSOP-2). Deutsches GeoForschungsZentrum GFZ, Potsdam, pp 1–52. doi:10.2312/GFZ.NMSOP-2_ch16

Lay T, Kanamori H, Ruff L (1982) The asperity model and the nature of large subduction zone earthquakes. Earthq Predict Res 1:3–71

Lazar M (2004) Tectonic processes along the northern edges of pull-apart basins in the Dead Sea rift: a case study from the northern Dead Sea. PhD thesis, Tel-Aviv University, p 156

Le Béon M, Klinger Y, Amrat AQ, Agnon A, Dorbath L, Baer G, Ruegg J-C, Charade O, Mayyas O (2008) Slip rate and locking depth from GPS profiles across the southern Dead Sea Transform. J Geophys Res Solid Earth 113:B11403. doi:10.1029/2007JB005280

Lee W, Stewart S (1981) Principles and applications of microearthquake networks. Academic, New York, p 293

Li C, van der Hilst R, Meltzer A, Engdahl E (2008) Subduction of the Indian lithosphere beneath the Tibetan Plateau and Burma. Earth Planet Sci Lett 274:157–168

Magistrale H, Zhou H (1996) Lithologic control of the depth of earthquakes in southern California. Science 273:639–642

Mai PM, Spudich P, Boatwright P (2005) Hypocenter locations in finite-source rupture models. Bull Seismol Soc Am 95(3):965–980

Makris J, Tsironidis J, Richter H (1991) Heat flow density distribution in the Red Sea. Tectonophysics 198:383–393

Martínez F, Cochran J (1989) Geothermal measurements in the northern Red Sea: implications for lithospheric thermal structure and mode of extension during continental rifting. J Geophys Res 94:12239–12265

Matheron G (1963) Principles of geostatistics. Econ Geol 58(8):1246–1266

McGuire JJ, Zhao L, Jordan TH (2002) Predominance of unilateral rupture for a global catalog of large earthquakes. Bull Seismol Soc Am 92(8):3309–3317

Mechie J, Abu-Ayyash K, Ben-Avraham Z, El-Kelani R, Qabbani I, Weber M (2009) Crustal structure of the southern Dead Sea basin derived from project DESIRE wide-angle seismic data. Geophys J Int 178(1):457–478

Medvedev S (1962) Engineering seismology (English translation from the Russian edition). Israel Program for Scientific Translations, Jerusalem, 1965, p 260

Medvedev SV, Sponheuer W (1969) Scale of seismic intensity. In: Proceedings of fourth World Conference of the Earthquake Engineering. Chilean Association on Seismology and Earthquake Engineering, Santiago, pp 143–153

Navon H (2011) Microseismic characterization of Lake Kinneret basin. M.Sc. thesis, Tel-Aviv University

Nazareth J, Hauksson E (2004) The seismogenic thickness of the southern California crust. Bull Seismol Soc Am 94:940–960

Petrunin AG, Sobolev SV (2006) What controls thickness of sediments and lithospheric deformation at a pull-apart basin? Geology 34:389–392

Roded R (2012) Basal heat flow and hydrothermal regime at the Golan-Ajloun hydrological Basins. M.Sc. thesis, Ben-Gurion University

Rothé E (1928) Annuaire de l'Institut de Physique du Globe 1927, deuxième partie: Séismologie, Bulletin du Bureau Central Séismologique Français, Université de Strasbourg, p 130

Rydelek PA, Sacks IS (1989) Testing the completeness of earthquake catalogues and the hypothesis of self-similarity. Nature 337:251–253

Scholz CH (2002) The mechanics of earthquakes and faulting, 2nd edn. Cambridge University Press, Cambridge, p 471

Shalev E, Lyakhovsky V, Yechieli Y (2007) Is advective heat transport significant at the Dead Sea basin? Geofluids 7:292–300

Shalev E, Lyakhovsky V, Weinstein Y, Ben-Avraham Z (2012) The thermal structure of Israel. Tectonophysics http://dx.doi.org/10.1016/j.tecto.2012.09.011

Shamir G (2006) The active structure of the Dead Sea depression. GSA Spec Pap 401:15–32

Shapira A, Hofstetter A (2002) Seismic parameters of seismogenic zones. Appendix C In: Shapira A (2002) An updated map of peak ground acceleration for Israel Standard 413. Geophysical Institute of Israel report 592/230/02, p 74

Shapira A, Avni R, Nur A (1993) A new estimate for the epicenter of the Jericho earthquake of 11 July 1927. Isr J Earth Sci 42:93–96

Shebalin NV (1973) Macroseismic data as information on source parameters of large earthquakes. Phys Earth Planet Inter 6:316–323

Shebalin NV, Karnik V, Hadzievski D (1974) Balkan region – Catalogue of earthquakes, Atlas of isoseismal maps. UNDP – UNESCO Survey of Seismicity of Balkan Region. Skopje, Yugoslavia, p 366

Sibson R (1982) Fault zone models, heat flow, and the depth distribution of earthquakes in the continental crust of the United States. Bull Seismol Soc Am 72:151–163

Storchak DA, Di Giacomo D, Bondár I, Engdahl ER, Harris J, Lee WHK, Villaseñor A, Bormann P (2013) Public release of the ISC-GEM global instrumental earthquake catalogue (1900–2009). Seism Res Lett 84(5):810–815

Thurber CH (1984) SIMUL3. In: Engdahl ER (ed) Documentation of earthquake algorithms, Report SE-35. World Data Center A for Solid Earth Geophysics, Boulder, pp 15–17

Valoroso L, Chiaraluce L, Piccinini D, Di Stefano R, Schaff D, Waldhauser F (2013) Radiography of a normal fault system by 64,000 high-precision earthquake locations: the 2009 L'Aquila (central Italy) case study. J Geophys Res Solid Earth 118:1156–1176

Vered M, Striem H (1977) A macroseismic study and the implications of structural damage of two recent major earthquakes in the Jordan rift. Bull Seismol Soc Am 67:1607–1613

Waldhauser F (2001) HypoDD: a program to compute double-difference hypocenter locations, US Geological Survey open file report 01–113. U.S. Dept. of the Interior, U.S. Geological Survey, Menlo Park, California

Waldhauser F, Ellsworth W (2000) A double-difference earthquake location algorithm: method and application to the northern Hayward fault, California. Bull Seismol Soc Am 90:1353–1368

Wells D, Coppersmith K (1994) New empirical relationships among magnitude, rupture length, rupture width, rupture area, and surface displacement. Bull Seismol Soc Am 84(4):974–1002

Wood HO (1921) A list of seismologic stations of the world, vol 2, Part 7, 15. The National Research Council of the National Academy of Sciences, pp 397–538

Wood HO, Neumann F (1931) Modified Mercalli intensity scale of 1931. Bull Seismol Soc Am 21(4):277–283

Zohar M, Marco S (2012) Re-estimating the epicenter of the 1927 Jericho earthquake using spatial distribution of intensity data. J Appl Geophys 82:19–29

Chapter 4
The Dead Sea Transform and the Volcanism in Northwestern Arabia

Yishai Weinstein and Zvi Garfunkel

Abstract Volcanism is common along the northern segments of the Dead Sea Transform (DST). In this paper we review its distribution and composition and conclude that this tectono-magmatic association has mainly to do with the magma migration toward the surface and less with magma generation, namely: some volcanic activity concentrated along the DST due to better magma channeling and not due to an enhanced mantle partial melting along this lineament. The volcanism along the DST is clearly part of the western Arabia magmatism, and the early phases of this volcanism probably have to do with Red Sea-related extension during the Early to Middle Miocene. Nevertheless, the DST does play a role in the emplacement of lithospheric mantle domains with different compositions next to each other, which is reflected in the derived lavas.

Keywords Dead Sea Transform • Volcanism • Red Sea • Harrat Ash Shaam • Azraq-Sirhan

4.1 Introduction

Volcanic landscapes are very common along the northern part of the Dead Sea Transform fault (hereafter DST; Figs. 4.1 and 4.2a). This mainly includes the north-western edge of the Harrat Ash Shaam (Early Miocene to Pleistocene, Fig. 4.2a, b),

---

Y. Weinstein
Department of Geography and Environment, Bar-Ilan University, Ramat-Gan 52900, Israel
e-mail: weinsty@biu.ac.il

Z. Garfunkel (✉)
The Fredy and Nadine Herrmann Institute of Earth Sciences, The Hebrew University of Jerusalem, Edmond J. Safra campus, 91904 Givat Ram, Jerusalem, Israel
e-mail: Zvi.garfunkel@huji.ac.il

Z. Garfunkel et al. (eds.), *Dead Sea Transform Fault System: Reviews*,
Modern Approaches in Solid Earth Sciences 6, DOI 10.1007/978-94-017-8872-4_4,

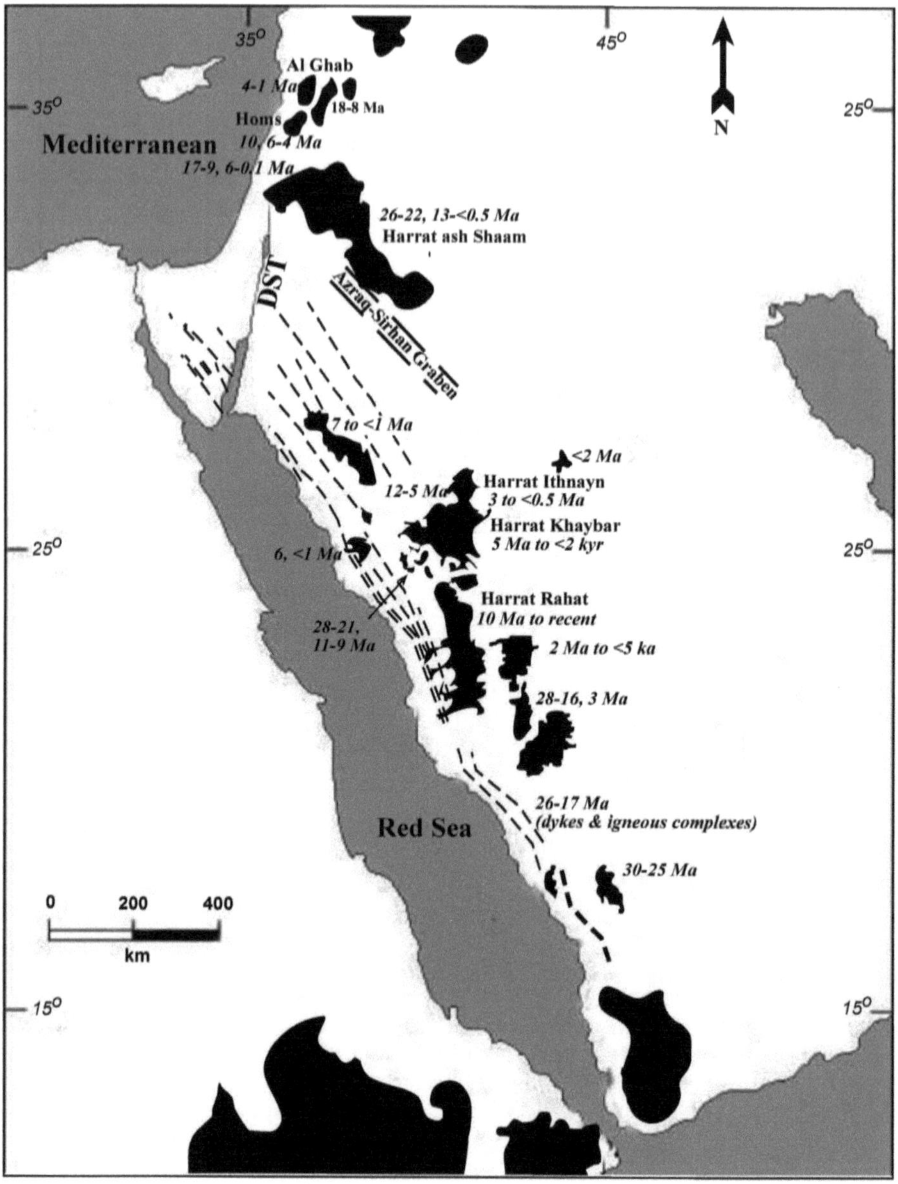

Fig. 4.1 The Arabian volcanic fields (Harrats) with known ages (Data from: Camp and Roobol 1989; Camp et al. 1987, 1991, 1992; Giannérini et al. 1988; Altherr et al. 1990; Coleman 1993; Shaliv 1991; Sebai et al. 1991; Mor 1993; Sharkov et al. 1994, 1998; Heimann et al. 1996; Ilani et al. 2001; Shaanan et al. 2011; Weinstein et al. 2006 & unpublished data; Trifonov et al. 2011). *Dashed lines* parallel to the Red Sea are Early Miocene dykes

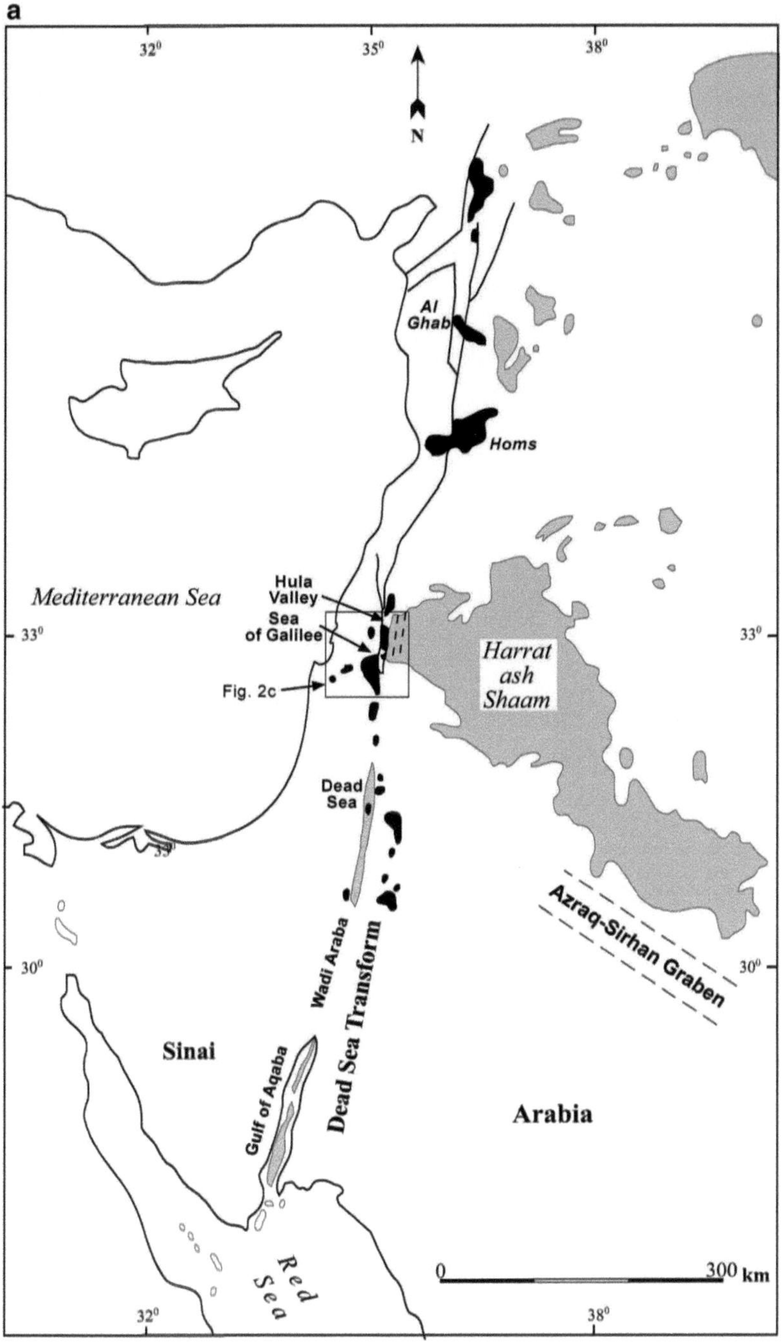

Fig. 4.2 (**a**) Volcanism along the Dead Sea Transform fault (*solid black*). *Gray areas* are for off-transform volcanic fields; (**b**) Map of the volcanic field of Harrat Ash Shaam (from Garfunkel, 1989; Note the ~N10W alignments of cones, in particular close to the transform); (**c**) Map of the northwestern Harrat Ash Shaam (northeastern Israel), where it is bisected by the DST. Note that the thick magmatic section south of the Sea of Galilee (see text) is not shown in this map, since its top is 500 m beneath the surface

b

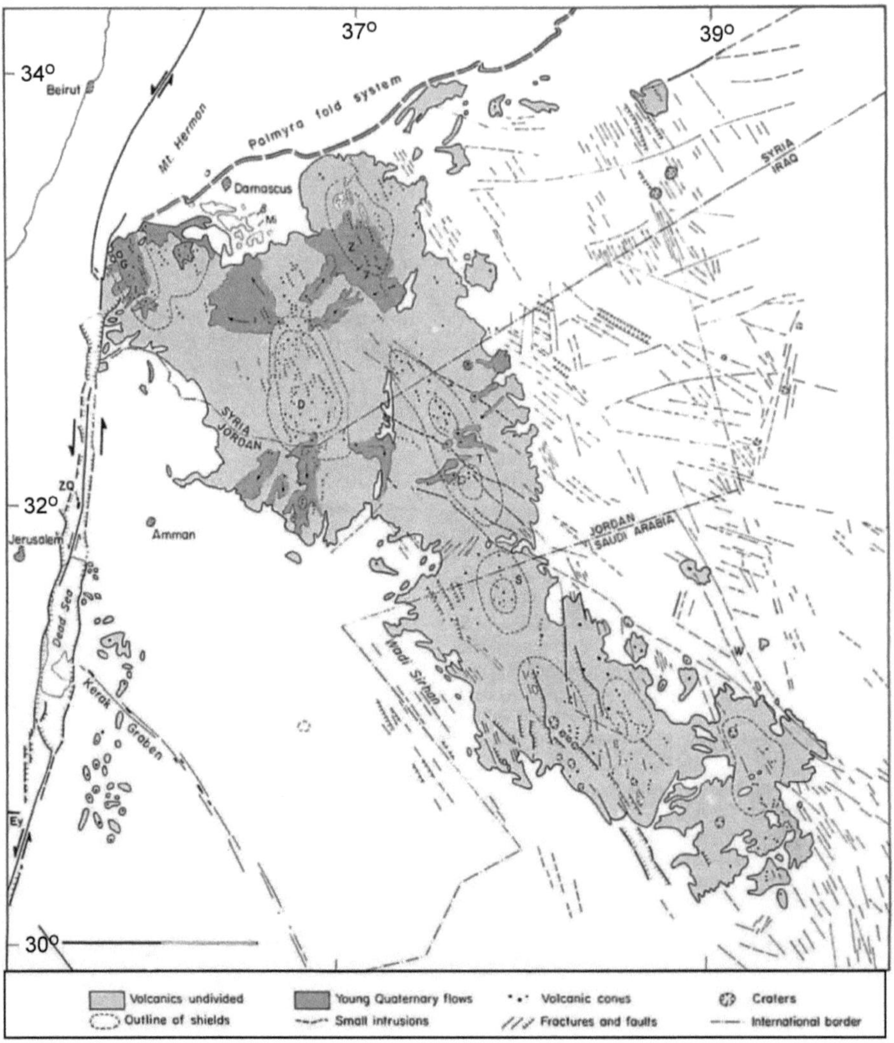

Fig. 4.2 (continued)

which is cut by the DST in the area of the Sea of Galilee and the Hula Valley (Figs. 4.1 and 4.2c), but also Miocene to Pliocene volcanism and dykes along the eastern flanks of the Dead Sea and several small volcanic fields along the northernmost trace of the DST in northern Syria (e.g. Homs and Al Ghab, Ma et al. 2011). The volcanism is alkali-basaltic, similar to that of other intra-plate continental basalts (e.g. Wilson 1993; Shaw et al. 2003; Weinstein et al. 2006).

Intra-plate volcanism is very often associated with rifting (e.g. Sengor and Burke 1978 and reference therein), and the relation between the two is usually

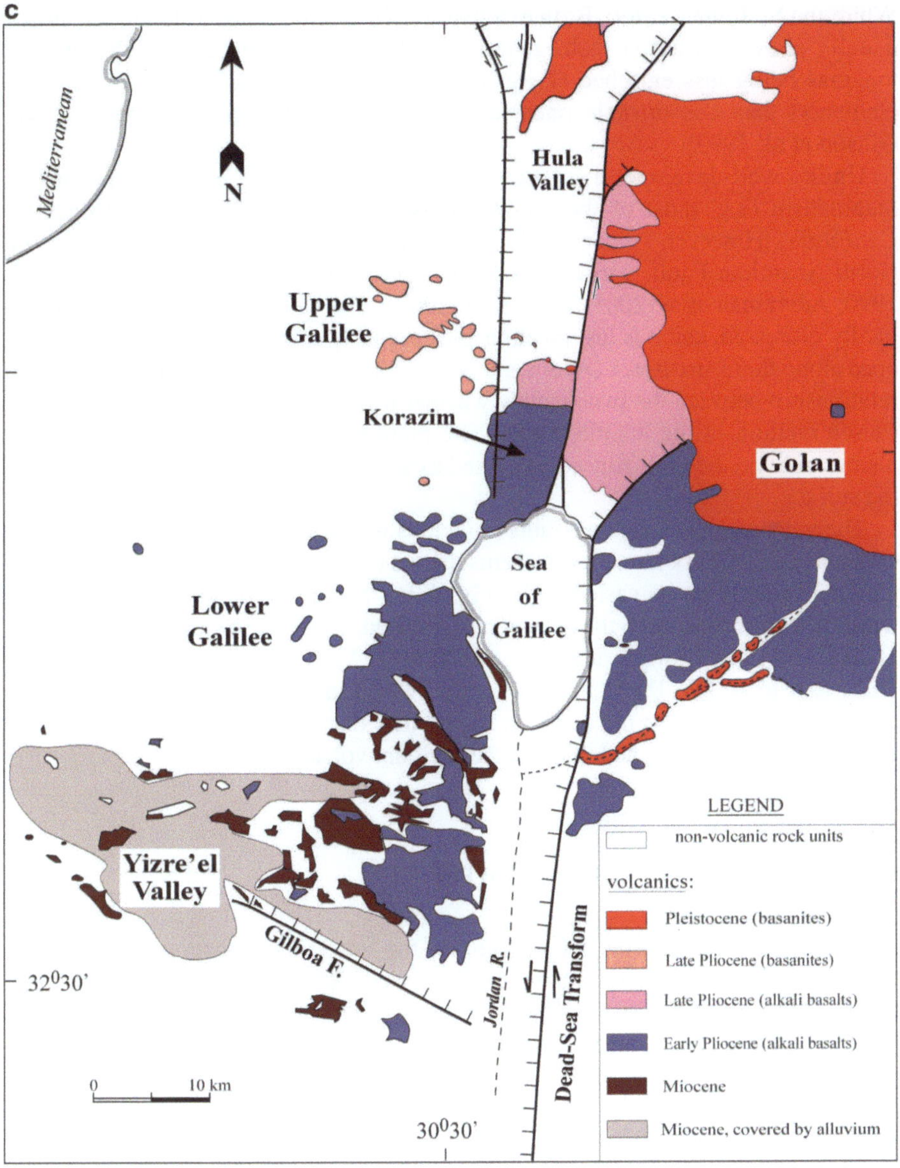

Fig. 4.2 (continued)

interpreted in either the active or the passive way. In the first, both phenomena owe their formation to an ascending mantel plume (hot spot), which induces mantle partial melting due to decompression and causes the rupture of the lithosphere (e.g. Burke and Whiteman 1973). In the passive way, far field stresses cause extension and rupture of the lithosphere, followed by mantle upwelling and partial melting (e.g. Turcotte and Oxburgh 1973; Oxburgh and Turcotte 1974;

White and McKenzie 1989). Both extension-related and plume-related volcanism usually show distinctive petrogenetic/geochemical signature, with the on-axis magmas being less enriched (lower concentrations of alkalis and incompatible elements) than the off-axis ones (e.g. Baker 1987; Camp and Roobol 1992; Gibson et al. 1993).

Unlike well-developed rifts, continental transforms are often devoid of any magmatism (e.g. most of the San Andreas fault system of southern to central California). However, volcanism does occur along some transform faults, e.g. the North Anatolian Fault, particularly in or next to pull-apart basins (Aydin et al. 1990; Adiyaman et al. 2001; Tatar et al. 2007). The cause for this volcanism was hardly discussed, and it is just mentioned as probably related to the extension developed along these structures (Aydin and Nur 1982). It is not clear if there is a genetic relationship between the two, namely whether it is the extension along the fault that caused melting in the mantle, or that the melting occurred due to other mechanisms, and that the fault zones just served as a pathway for the magmas en route to the surface.

The volcanism of western Arabia coincides with the rifting of the Red Sea (e.g. Coleman et al. 1983). However, it seems that its northern parts also partly align with the DST, mainly on its eastern side (Fig. 4.2a). Moreover, volcanic lineaments in western Arabia show both the direction of the Red Sea (~N25W, e.g. the Harrat Ash Shaam, and its northwestern edge in the Yizre'el Valley, Fig. 4.2c) and that of the DST (N-S to N15W, Fig. 4.2b; Garfunkel 1989; Camp and Roobol 1992). In this paper we try to refer to the question of whether there is and what kind of genetic relation exists between the DST and the regional volcanism.

4.2 The Western Arabia 'Harrats'

Basaltic fields cover almost 200,000 km^2 of the western and southwestern Arabian Peninsula (Fig. 4.1). The onset of volcanism was in Yemen, as early as the Oligocene (31 Ma, Menzies et al. 1997), with the eruption of the Ethiopia-Yemen flood basalts (the 'trap series') associated with the emplacement of the Afar plume underneath that region. At ca. 29 Ma, the volcanism changed to a bimodal type, with basalts, rhyolites and ignimbrites erupting simultaneously until 26 Ma (Coleman and McGuire 1988; Menzies et al. 1997). In Saudi Arabia, the first stage of magmatism included plutons and vent lineaments, as well as numerous dykes that stretch along the 1,700 km of the Red Sea eastern coast, associated with the opening of the Red Sea (e.g. Bartov et al. 1980; Baldridge et al. 1991; Sebai et al. 1991; Coleman and McGuire 1988). Nevertheless, most of the volcanism is concentrated off the eastern escarpment of the Red Sea trough, in volcanic fields locally called Harrats, which were active since ca. 30 Ma (ages as old as the Middle Eocene were also reported, Pallister 1987) up to the Pleistocene, and in some cases up to historic times (e.g. the Madinah eruption of 1,256 AD, Camp et al. 1987). Most of the volcanic fields are located within 100–200 km from the Red Sea. However, the largest volcanic field – Harrat Ash Shaam, with an area of 45,000 km^2 – is located 500 km away from the Red-Sea, with a large spatial volcanic gap between it and the close-to-Red Sea Harrats (Fig. 4.1).

Both young and old ages were found from southern Arabia to Syria. However, volcanism at northwestern Arabia (Syria) commenced later than in Saudi Arabia and Yemen (21 and 31–30 Ma, respectively, e.g. Ershov and Nikishin 2004; Krienitz et al. 2009). The western Arabia volcanism was not continuous. Rather, it erupted in two main periods: (1) the Late Oligocene to Early Miocene and (2) the Middle Miocene to Quaternary with a volcanic quiescence of several million years between the two (~20–13 Ma, Camp and Roobol 1992; Ilani et al. 2001). In northwestern Arabia, the volcanism was either continuous during the past 20 Ma (Syria, Krienitz et al. 2009) or with a short volcanic gap at 9–6 Ma (northeastern Israel, Shaliv 1991).

The older magmatic bodies (>20 Ma, dykes, plutons, vent lineaments and volcanic fields) are mostly aligned NW or NNW, parallel to the Red-Sea axis (Fig. 4.1, Coleman and McGuire 1988; Camp and Roobol 1992). In particular, the NW alignment of Harrat Ash Shaam (volcanism commenced at 26 Ma, Ilani et al. 2001) is parallel to the Red-Sea, but also to the strike of the nearby Azraq-Sirhan Graben (Fig. 4.1), which was active (re-activated) since the Oligocene (Segev and Rybakov 2011). Altogether, this suggests a genetic relationship between the early (Oligocene to Middle/Late Miocene) magmatism and the opening of the Red Sea, reflecting a NE-SW regional extension during the Oligocene to Early Miocene. At the northwestern edge of Harrat Ash Shaam (Yizre'el Valley, Fig. 4.2c) and at northern Syria (Homs basalts), NW alignment is also observed in younger volcanism (17–9 and 6 Ma, respectively, Shaliv 1991; Chorowicz et al. 2005). The former (Yizre'el Valley) is probably related to the nearby Gilboa Fault (Fig. 4.2c), which is the northwestern continuance of the Azraq-Sirhan Graben (after restoring for the 105 km sinistral movement along the DST, e.g. Segev and Rybakov 2011) and probably reflects the same regional extension regime as in the early phase Arabian volcanism. The Homs basalts, being significantly younger and located along the northern trace of the DST, could as well be related to an ENE extension along the DST (e.g. Eyal 1996).

It should be noticed that the volcanism is preferably concentrated on the eastern, Arabian shoulder of the Red Sea, with very few exceptions on its western, African side. The reason for that was discussed by several authors (e.g. Almond 1986; Dixon et al. 1989; Camp and Roobol 1992) and is beyond the scope of the current manuscript.

The younger Arabia volcanism (<13 Ma) mainly shows N-S lineaments (e.g. Coleman et al. 1983). This includes large features like the Makkah-Madinah-Nafud line, which includes the Harrats Rahat, Harrat Khaybar and Harrat Ithnayn (Camp and Roobol 1992; Fig. 4.1), as well as numerous lineaments of cinder cones (Garfunkel 1989; Camp and Roobol 1992; Ilani et al. 2001; Mor 1993) within these and other Harrats. Camp and Roobol (1992) suggested that the

Makkah-Madinah-Nafud line is a surface manifestation of a hot asthenospheric lobe that extends northwards from the Afar plume and creates the so-called 'western Arabian swell'. The N-S lineaments of cinder cones, in particular those in the Harrat Ash Shaam are discussed below.

4.3 Volcanism Along the DST

In addition to the Red Sea association, the young volcanism (<6 Ma) often shows close relationship with the DST. This is reflected in the alignment of some of the volcanism along the transform (Fig. 4.2a). It seems that in some of the areas, the regional, western Arabia volcanism is just bisected by the transform (e.g. the Harrat Ash Shaam in the area of northeastern Israel). However, in northern Syria and in the Dead Sea area, some small volcanic fields are clearly aligned along the transform, either within depressions associated with the DST (e.g. Al Ghab, Ma et al. 2011) or on the depression shoulders (e.g. Dead Sea, Barberi et al. 1980; Fediuk and Al Fugha 1999; Fig. 4.2a). In particular at the Dead Sea area, volcanism exists just within 100 km from the DST (solely on its eastern side) and is absent farther away from the transform (Fig. 4.2a). Moreover, in the pull-apart basin of the Sea of Galilee (Fig. 4.2c), where the DST bisects the Harrat Ash Shaam, the (sub-surface) magmatic section is very thick. The cumulative thickness of Early Pliocene basalts and gabbroic intrusives just south of the Sea of Galilee reaches 1,000 m (Marcus and Slager 1985), compared with a few tens of meters on the basin shoulders.

Another DST-associated feature is the common N-S to N15W rows of scoria cones and other lava eruption complexes (Garfunkel 1989), which are sub-parallel to the DST, suggesting possible genetic relationship between the two. This is particularly true for the area close to the DST, e.g. in the Golan, where more than 70 cinder cones are arranged in three rows in a direction of N10W (Lang et al. 1979; Mor 1993). However, as mentioned above, N-S lineaments are also observed in western Arabia, more than 500 km away from the DST, thus inference from these directions on possible DST-volcanism relationship should be carefully treated. Last, although there is no general temporal pattern, there is a N-S age progression in several areas at western and northwestern Arabia (Camp and Roobol 1992; Weinstein et al. 2006; Ma et al. 2011), which according to some authors may be the result of the northward propagation of the DST (Brew et al. 2001a; Ma et al. 2011, and reference therein). On the other hand, it should be noticed that there is hardly any volcanism along the southern 300–400 km of the DST (the Wadi Araba and the Gulf of Aqaba, all the way down to the Red Sea).

Brew et al. (2001a, b) also argued that in Syria, there is a coincidence of volcanic quiescence and a halt in activity of the DST during mid Miocene to Early Pliocene time (14.5–4.5 Ma; Hempton 1987). However, while it is not clear that there was any pause in the DST activity (e.g. Steckler and Ten Brink 1986), updated regional compilations (e.g. Krienitz et al. 2009; Fig. 4.1) also do not support such a long volcanic quiescence and strongly questions this kind of coincidence.

4.4 Composition of the Basalts

The Arabian magmas span a wide range of compositions, from tholeiites to basanites and from basic to acidic (e.g. Camp and Roobol 1992; Baker et al. 1996). The Yemen trap series and the magmas along the Red Sea coast are bimodal, including tholeiites and transitional basalts, as well as rhyolites and ignimbrites (e.g. Coleman et al. 1983; Coleman and McGuire 1988; Chiesa et al. 1989). Farther away from the Red Sea, the Harrat volcanism is more alkalic and enriched with incompatible elements, with compositions of alkali basalts to basanites, and rarely nephelinites (e.g. Coleman et al. 1983; Altherr et al. 1990; Shaw et al. 2003; Weinstein et al. 2006; Ma et al. 2011).

In the Harrats, the basalts of the older volcanic phase are mainly sub-alkali to alkali-basalts. However, in some cases (northern Israel and northern Syria), there is also a trend of enrichment during the Miocene phase, with sub-alkali to alkali basalts erupting during the earlier period (Early Miocene in Syria and the earlier Middle Miocene in Israel), while mainly basanites during the later period (Weinstein 2000; Lustrino and Sharkov 2006). These temporal trends have to be treated cautiously, since in most cases lavas of different age did not erupt exactly at the same location, therefore it could in fact be a spatial pattern (e.g. Weinstein 2000).

In the younger volcanic phase (Late Miocene to recent), there is usually a trend of decreasing silica and enrichment with alkalis and highly incompatible elements from Late Miocene/Early Pliocene to the younger lavas (Late Pliocene or Pleistocene), which are typically basanitic (Camp and Roobol 1992; Shaw et al. 2003; Weinstein et al. 2006; Ma et al. 2011). This trend was interpreted in terms of decreasing degree of melting from the alkali basalts toward the basanites (e.g. Camp and Roobol 1992). However, Weinstein et al. (2006), followed by Ma et al. (2011), showed that the variability in the concentration of highly incompatible elements could not be explained by variable degree of a common source, but rather by source heterogeneity.

No compositional pattern was identified in relation with the DST. At northeastern Israel, where the DST bisects the Harrat Ash Shaam volcanic field, the composition of basalts along the transform (e.g. the Early Pliocene basalts of the Sea of Galilee and Korazim, Fig. 4.2c) is not distinct from basalts of the same age that erupted farther away from the transform (e.g. the southern Golan and Lower Galilee, Figs. 4.2c, 4.3a, and 4.4). Interestingly, a group of Late Pliocene lavas that erupted along the western segment of the DST next to the Hula Valley is more alkalic and enriched than other along transform (Korazim, north of the Sea of Galilee) and off transform Late Pliocene Golan lavas (Figs. 4.3a and 4.4; Weinstein 2012). Similarly, in northern Syria, the composition of basalts along the transform is not distinct from those far away from the transform (Fig. 4.3b; Lustrino and Sharkov 2006). Moreover,

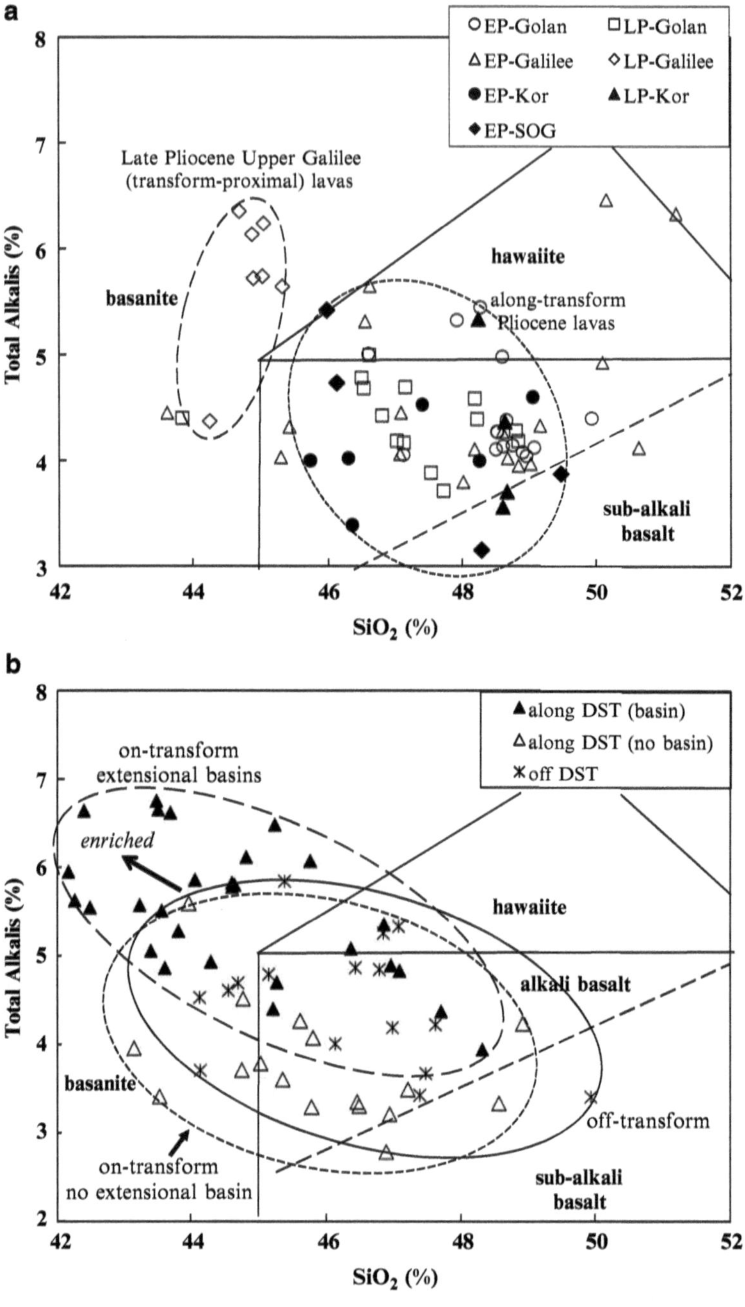

Fig. 4.3 Total Alkalis (Na_2O+K_2O) – Silica (TAS) diagrams of (**a**) northeastern Israel (circum Sea of Galilee) Pliocene basalts (EP stands for Early Pliocene and LP for Late Pliocene; KOR is Korazim, north of the Sea of Galilee, and SOG is the area just south of the Sea of Galilee, see Fig. 4.2c; Data from Weinstein et al. 2006; Weinstein 2012 and unpublished data), (**b**) northern

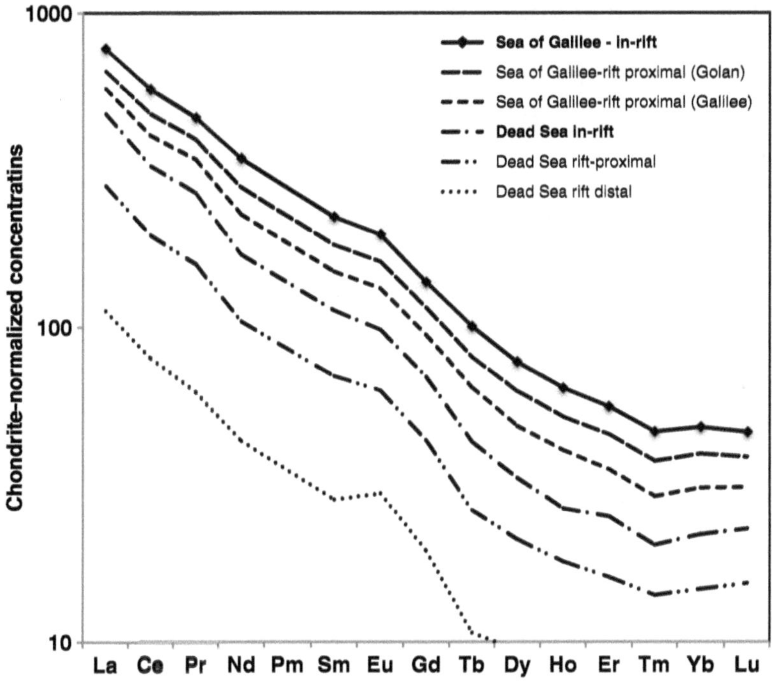

Fig. 4.4 Chondrite-normalized (Boynton 1984) REE diagram, showing (1) average compositions of circum Sea of Galilee Early Pliocene basalts (in-rift – south of the Sea of Galilee, rift-proximal – the southern Golan and Lower Galilee, see Fig. 4.2c for locations), (2) representative compositions of Dead Sea area basalts (in-rift – north of the Dead Sea, rift-proximal – the Dead Sea eastern shoulder, rift-distal – Harrat Ash Shaam (100–150 km to the east), Fediuk and Al Fugha 1999)

basalts associated with an extensional basin (Al Ghab, Figs. 4.1 and 4.2a) are more enriched than other along-transform (e.g. Homs basalts) and off-transform lavas (Fig. 4.3b; Lustrino and sharkov 2006; Ma et al. 2011). Another example is described from the Dead Sea area, where within-rift lavas are more alkali and REE-rich than rift-proximal ones, and both are significantly more enriched than the Harrat Ash Shaam basalts, 100 km to the east. All of this is not easy to explain in conventional terms of extension-related magmatism (e.g. Baker 1987; Gibson et al. 1993), where higher degree of melting, therefore lower enrichment, is expected on-axis.

Fig. 4.3 (continued) Syria Late Miocene to Pleistocene basalts (Al Ghab basin, Homs, Allepo and Euphrates Valley; Lustrino and Sharkov 2006; Ma et al. 2011). In (**a**), along-transform Early Pliocene lavas (south and north of the Sea of Galilee, *dotted circle*) are undistinguished from similar age off-transform Golan and Galilee lavas, while Late Pliocene lavas that erupted along or within a few km from the transform in the Upper Galilee (*dashed circle*) are more enriched than the similar age along-transform Korazim lavas. In (**b**), along-transform lavas are undistinguished or slightly more enriched than the off-transform lavas (*solid circle*)

4.5 The Source and the Cause of Melting

Some authors attributed the western Arabia magmatism to sub-lithospheric/mantle plume sources (e.g. White and McKenzie 1989; Camp and Roobol 1992; Lustrino and Wilson 2007; Krienitz et al. 2009). This concept was challenged in the papers of Altherr et al. (1990) and Stein and Hofmann (1992), who suggested the lithosphere or a fossil plume attached to the base of the lithosphere as the source for all or some of the magmas (Shaw et al. 2003). Follow-up publications showed that the source is hydrated, e.g. amphibole or phlogopite-bearing mantle rocks (e.g. Stein et al. 1997; Weinstein et al. 2006; Ma et al. 2011). This implies solidus temperatures significantly lower than the dry solidus previously used for mantle melting models (e.g. Green 1973; Olafsson and Eggler 1983; Falloon and Green 1990). This further suggests that the mantle source is close to solidus temperatures, therefore more sensitive to thermal perturbations and more available for partial melting. The presence of amphibole also constrains the source depth to less than 90 km (Green 1973; Brey et al. 1983; Wallace and Green 1991), probably within the lithosphere (Shaw et al. 2003; Weinstein et al. 2006).

The next question is what is the driving force for melting? Basically, there are two alternative models. While some authors favor the so-called 'active' melting, which is melting due to an induced thermal perturbation in the mantle, (e.g. mantle plume, Garfunkel 1989; Camp and Roobol 1992; Debayle et al. 2001; Weinstein et al. 2006; Krienitz et al. 2009), others prefer the passive melting model, which involves extension and thinning of the lithosphere (e.g. Bohannon et al. 1989; McGuire and Bohannon 1989; Stein et al. 1993). Camp and Roobol (1992), Weinstein (2000) and Weinstein et al. (2006) suggested that the early stage, Red Sea-associated volcanism (Early to Middle Miocene) was derived by lithosphere extension (passive), while the later (Late Miocene or Pliocene to Pleistocene) occurred due to a thermal perturbation. Weinstein et al. (2006) showed that if melting was of a wet mantle source (e.g. amphibole or CO_2-bearing peridotite), then decompression by itself could hardly produce any melting, thereby it should involve a thermal effect. Also, even though extension was common during the Miocene, field observations provide no evidence for extension during the Plio-Pleistocene, which was a period of extensive volcanism in the area. In some cases, there is even an indication of a syn-volcanic compression regime (e.g. the Hula Valley, Fig. 4.2c, during the past 1 Ma, Schattner and Weinberger 2008).

4.6 Discussion: The Relation Between the DST and the Volcanism

Volcanism is abundant along the northern part of the DST, from the Dead Sea basin northwards. However, the transform fault effect is superimposed on the Red-Sea extension-related regional effect. In particular, the extensive volcanism around the pull-apart basins of the Sea of Galilee and the Hula Valley (Fig. 4.2c) is part of the Ash Shaam volcanic field, which is evidently a Red-Sea or Azraq-Sirhan Graben-related feature as is suggested by its NW direction. In particular, it seems that the earlier phase of volcanism (Early to Middle Miocene) in Harrat Ash Shaam has very little to do with the DST. Early Miocene volcanism erupted at the central and northern parts of the field (e.g. Ilani et al. 2001), away from the DST, while the Middle Miocene volcanism, which is closer to the DST on its west, is concentrated along the NW-trending Gilboa fault (Fig. 4.2c), which is probably a continuation of the Azraq-Sirhan Graben (Weinstein 2000; Segev and Rybakov 2011).

On the other hand, the Late Miocene to Quaternary volcanism shows more affiliation with the DST. This includes a concentration of small volcanic fields along and inside the transform from the Dead Sea northwards (Fig. 4.2a), a very thick Early Pliocene section (~1,000 m) south of the Sea of Galilee (compared with the rift shoulders) and the N-S-aligned structures at the Harrat Ash Shaam (e.g. Garfunkel 1989; Fig. 4.2b). While some of these features (e.g. the off-DST N-S structures) could be related to the regional stress field and are also found far away from the transform (e.g. Camp and Roobol 1992), the DST-aligned volcanism could be petrogenetically-related to the transform, especially when found away from Arabian volcanic centers, e.g. at northern Syria. The latter means that the possibility of partial melting of a mantle source due to limited extension along the DST, in particular along the pull-apart basins (e.g. the Dead Sea area or the Al Ghab basin), should be considered. The way the DST could induce melting, whether in the lithosphere or in the underlying asthenosphere, is by lithosphere thinning and mantle decompression. White and McKenzie (1989) argued that in order for a peridotite (the common upper mantle rock, composed mainly of the mineral olivine) to be partially melted at the upper mantle, an extension factor of 2–5 is needed. Assuming a wet mantle source (amphibole-peridotite or pyroxenite, e.g. Stein et al. 1997; Ma et al. 2011), the extension factor could be significantly reduced. Nevertheless, extension should be manifested regionally or at least close to the transform, which is not the case along the DST. The only subsidence and sedimentary basins associated with the transform during the last few Ma are those within its pull-apart basins, which does not seem to represent a significant extension, such that could cause melting at depth. Therefore, the possibility of a DST-induced partial melting of the mantle is very unlikely. This is supported by the almost absence of volcanism along the southern half of the DST (Wadi Araba and the Gulf of Aqaba) and by the very small volume of volcanism at the Dead Sea basin compared with the fields farther north.

Recent works (e.g. Benallal and Bigoni 2004; Kaus and Podladchikov 2006) have raised the possibility that limited extension could induce local heating within the lithosphere by strain localization and the development of necking instabilities, thereby could possibly cause local partial melting of a wet mantle even under a very limited extension factor (Rosenbaum, pers. Comm. 2011). This possibility should be further investigated in relation with the SDT and other transforms (e.g. the north Anatolian Fault).

One may suggest that if the DST cannot produce melts on its own, it could have enhanced melting in a mantle already affected by the Red Sea or a thermal event (e.g. a hot asthenospheric lobe extending from the Afar plume, Camp and Roobol 1992). However, 'enhanced melting' is usually reflected in the lavas' composition, namely: it should result in along-transform higher degree of melting, thus produce less enriched magmas (lower concentrations of alkalis and incompatible elements, e.g. REE). This is not the case along the DST. The composition of on-transform basalts from the thick section of Early Pliocene basalts and intrusives south of the Sea of Galilee is similar to that of the similar age off-transform southern Golan and Lower Galilee basalts (all are alkali basalts, Weinstein et al. 2006; Fig. 4.3a). Moreover, the on-transform Sea of Galilee basalts are more enriched (in REE) than the off-transform ones (Fig. 4.4), in difference from the above comprehension. Similarly, Late Pliocene alkali-basalts from the Korazim block, north of the Sea of Galilee, have very similar composition to the nearby similar age off-transform Golan basalts (Weinstein 2012). Moreover, Late Pliocene lavas that erupted 1–2 km west of the transform, at the eastern Upper Galilee, are highly-enriched basanites (Fig. 4.3a, Weinstein et al. 2006; Weinstein 2012), compared with the slightly enriched alkali-basaltic composition of the similar age Golan basalts farther away (20 km) from the DST on its east. Similarly, on-transform basalts north of the Dead Sea are more enriched than off-transform and the Harrat Ash Shaam basalts 100 km and farther off-transform (Fediuk and Al Fugha 1999; Fig. 4.4), and the pull-apart Al Ghab basalts are more enriched than the non-extension-related Homs basalts and other off-transform basalts from northern Syria (Lustrino and Sharkov 2006; Ma et al. 2011). Altogether, this suggests that there is no systematic difference in the melting regime (namely, the degree of partial melting) between on-transform and off-transform lavas, which doubt the possibility that the DST has any melting effect.

An alternative DST-volcanism relation was recently suggested by Ma et al. (2011) for the Homs and Al Ghab volcanism of northern Syria (Fig. 4.1). They suggested that the northward migration of volcanism in this area (Late Miocene to Early Pliocene at Homs and Late Pliocene to Pleistocene at Al Ghab) was caused by the northward propagation of the DST, which channeled upwelling asthenosphere at the base of the lithosphere. This, in turn, served as the heat source for a wet lithospheric melting. This should be considered for other areas along the transform.

In a recent paper, Weinstein (2012) presented two examples of magma channeling along the DST north of the Sea of Galilee during the Early and Late Pliocene. Weinstein showed that while allowing magma channeling, the DST also acted as a lithospheric boundary, preventing along-transform magmas from intermixing with off-transform magmas.

Similarly, and in accordance with Garfunkel (1989), we suggest that the DST-associated volcanism is related to crustal/lithospheric drainage of magmas. This means that the DST functions as a plumbing system, allowing easy access of magmas to the surface. This could happen only where magmas are already present at depth (e.g. close to Harrat Ash Shaam), owing their existence to a different thermal/tectonic event, e.g. the Red Sea rifting or a mantle plume. Accordingly, the absence of volcanism along the southern parts of the DST is the result of these areas being far away from both Harrat Ash Shaam and the other Saudi Arabian Harrats.

Finally, the DST does indirectly affect the composition of the magmas by displacing the location of lithospheric magma sources. Weinstein (2012) suggested that the 105 km sinistral movement along the transform has placed different lithospheric domains next to each other, which in certain cases resulted in the eruption of magmas with significantly different chemistry on its both sides. A similar case of a transform acting as a lithospheric boundary was described for the Paleozoic Great Glen Fault (northern Britain, Canning et al. 1998). This aspect should be further examined in other active transform faults.

4.7 Conclusions

1. Volcanism in western and the northwestern Arabia is genetically related to the opening of the Red Sea (Miocene) and possibly to a hot asthenospheric mantle beneath this area (Late Miocene to recent).
2. There is no evidence for any DST-related mantle melting.
3. The DST does function in certain places as a lithospheric channel, which allows magmas an easy access to the surface.
4. The absence of volcanism next to the southern parts of the DST (Wadi Araba and Gulf of Aqaba) is due to its remoteness from the main Arabian volcanic fields.

Acknowledgements We would like to thank M. Wilson and an anonymous reviewer for their very helpful reviews and useful comments on the manuscript. We are also grateful to friends and colleagues, with whom we shared ideas and had discussions regarding the subject of this paper along the years, including: R. Altherr, A. Heimann, S. Hurwitz, D. Mor, O. Navon, R. Weinberger and many others.

References

Adiyaman O, Chorowicz J, Arnaud ON, Gündogdu MN, Gourgaud A (2001) Late Cenozoic tectonics and volcanism along the North Anatolian Fault: new structural and geochemical data. Tectonophysics 338(2):135–165. doi:10.1016/S0040-1951(01)00131-7

Almond DC (1986) Geological evolution of the Afro-Arabian dome. Tectonophysics 131:301–332

Altherr R, Henjes-Kunst F, Baumann A (1990) Asthenosphere versus lithosphere as possible sources for basaltic magmas erupted during formation of the Red Sea: constraints from Sr, Pb and Nd isotopes. Earth Planet Sci Lett 96:269–286

Aydin A, Nur A (1982) Evolution of pull-apart basins and their scale independence. Tectonics 1:91–105

Aydin A, Schultz RA, Campagna D (1990) Fault-normal dilation in pull-apart basins: implications for the relationships between strike-slip faults and volcanic activity. Ann Tecton IV(2):45–52

Baker BH (1987) Outline of the petrology of the Kenya rift alkaline province. In: Fitton JG, Upton BGJ (eds) Alkaline igneous rocks. Geological Society of London Special Publication 30, pp 293–311

Baker JA, Thirlwall MF, Menzies MA (1996) Sr-Nd-Pb isotopic and trace element evidence for crustal contamination of plume-derived flood basalts: Oligocene flood volcanism in western Yemen. Geochim Cosmochim Acta 60(14):2559–2581

Baldridge WS, Eyal Y, Bartov Y, Steinitz G, Eyal M (1991) Miocene magmatism of Sinai related to the opening of the Red Sea. Tectonophysics 197:181–201

Barberi F, Capaldi G, Gasperini P, Marinelli G, Santacroce R, Scandone R, Treuil M, Varet J (1980) Recent basaltic volcanism of Jordan and its implications on the geodynamic history of the

Dead-Sea shear zone. In: Carrelli A (ed) Geodynamic evolution of the Afro-Arabian rift system, vol 47, Atti dei convegni Lincei. Accademia nazionale dei Lincei, Roma, pp 667–683
Bartov Y, Steinitz G, Eyal M, Eyal Y (1980) Sinistral movement along the Gulf of Aqaba – Its age and relation to the opening of the Red Sea. Nature 285:220–222
Benallal A, Bigoni D (2004) Effects of temperature and thermo-mechanical couplings on material instabilities and strain localization of inelastic materials. J Mech Phys Solids 52:725–753. doi:10.1016/S0022-5096(03)00118-2
Bohannon RG, Naeser CW, Schmidt DL, Zimmermann RA (1989) The timing of uplift, volcanism and rifting peripheral to the Red Sea: a case for passive rifting? J Geophys Res 94:1683–1701
Boynton WV (1984) Cosmochemistry of the rare earth elements: meteorite studies. In: Henderson P (ed) Rare earth element geochemistry. Elsevier, Amsterdam, pp 63–114
Brew G, Lupa J, Barazangi M, Sawaf T, Al-Iman A, Zaza T (2001a) Structure and tectonic development of the Ghab basin and the Dead Sea fault system, Syria. J Geol Soc Lond 158:665–674
Brew G, Lupa J, Barazangi M, Al-Maleh AK, Sawaf T (2001b) Tectonic and geologic evolution of Syria. Geoarabia 6(4):573–616
Brey G, Brice WR, Ellis DJ, Green DH, Harris KL, Ryabchikov ID (1983) Pyroxene-carbonate reactions in the upper mantle. Earth Planet Sci Lett 62:63–74
Burke KC, Whiteman AJ (1973) Uplift, rifting and the breakup of Africa. In: Tarling DH, Runcorn SK (eds) Implication of continental drift to earth sciences. Academic, London, pp 735–755
Camp VE, Roobol MJ (1989) The Arabian continental alkali basalt province: Part 1. Evolution of Harrat Rahat, Kingdom of Saudi Arabia. Geol Soc Am Bull 101:71–95
Camp VE, Roobol MJ (1992) Upwelling asthenosphere beneath western Arabia and its regional implications. J Geophys Res 97(B11):15255–15271
Camp VE, Hooper PR, Roobol MJ, White DL (1987) The Madinah eruption, Saudi Arabia: magma mixing and simultaneous extrusion of three basaltic chemical types. Bull Volcanol 49:489–508
Camp VE, Roobol MJ, Hooper PR (1991) The Arabian continental alkali basalt province, Part 2: evolution of Harrats Khaybar, Ithnayn and Kura, Kingdom of Saudi Arabia. Geol Soc Am Bull 103:363–391
Camp VE, Roobol MJ, Hooper PR (1992) The Arabian continental alkali basalt province, Part 3: evolution of Harrat Kishb, Kingdom of Saudi Arabia. Geol Soc Am Bull 104:379–396
Canning JC, Henney PJ, Morrison MA, Van Calsteren PWC, Gaskarth JW, Swarbrick A (1998) The Great Glen Fault: a major vertical lithospheric boundary. J Geol Soc Lond 155:425–428
Chiesa S, Civetta L, De Fino M, La Volpe L, Orsi G (1989) The Yemen trap series: genesis and evolution of a continental flood basalt province. J Volcanol Geotherm Res 36:337–350
Chorowicz J, Dhont D, Ammar O, Rukieh M, Bilal A (2005) Tectonics of the Pliocene Homs basalts (Syria) and implications for the Dead Sea Fault Zone activity. J Geol Soc Lond 162:259–271
Coleman RG (1993) Geologic evolution of the Red Sea, vol 24, Oxford monographs on geology and geophysics. Oxford University Press, New York, p 186
Coleman RG, McGuire AV (1988) Magma systems related to the Red Sea opening. Tectonophysics 150:77–100
Coleman RG, Gregory RT, Brown GF (1983) Cenozoic volcanic rocks of Saudi Arabia. U.S. Geological Survey Open file report USGS-OF-03-93, 82 pp
Debayle E, Lévêque J-J, Cara M (2001) Seismic evidence for a deeply rooted low-velocity anomaly in the upper mantle beneath the northeastern Afro/Arabian continent. Earth Planet Sci Lett 193(423):436
Dixon TH, Ivins ER, Franklin BJ (1989) Topographic and volcanic asymmetry around the Red Sea: constraints on rift models. Tectonics 8:1193–1216
Ershov AV, Nikishin AM (2004) Recent geodynamics of the Caucasus-Arabia-East Africa Region, *Geotectonics* (Engl. Transl.) 38(2):123–136
Eyal Y (1996) Stress field fluctuations along the Dead Sea rift since the middle Miocene. Tectonics 15(1):157–170
Falloon TJ, Green DH (1990) Solidus of carbonated fertile peridotite under fluid saturated conditions. Geology 18:195–199

Fediuk F, Al Fugha H (1999) Dead Sea region: fault-controlled chemistry of Cenozoic volcanics. Geolines 9:29–34

Garfunkel Z (1989) Tectonic setting of Phanerozoic magmatism in Israel. Isr J Earth Sci 38:51–74

Giannérini G, Campredon R, Feraud G, Abou Zakhem B (1988) Deformations intraplaques et volcanisme associe: Exemple de la bordure NW da plaque Arabique au Cenozoique. Bulletin de la Societe Geologique de France (8) t IV, 6:937–947

Gibson SA, Thompson RN, Leat PT, Morrison MA, Hendry GL, Dickin AP, Mitchell JG (1993) Ultrapotassic magmas along the flanks of the Oligo-Miocene Rio Grande rift, U.S.A.: monitors of the zone of lithospheric mantle extension and thinning beneath a continental rift. J Petrol 34:187–228

Green DH (1973) Experimental melting studies on a model upper mantle composition at high pressure under water-saturated and water-undersaturated conditions. Earth Planet Sci Lett 19:37–53

Heimann A, Steinitz G, Mor D, Shaliv G (1996) The Cover Basalt Formation, its age and its regional and tectonic setting: implications from K–Ar and $^{40}Ar/^{39}Ar$ geochronology. Isr J Earth Sci 45:55–71

Hempton M (1987) Constraints on Arabian plate motion and extensional history of the Red Sea. Tectonics 6:687–705

Ilani S, Harlvan Y, Tarawneh K, Rabba I, Weinsberger R, Ibrahim K, Peltz S, Steinitz G (2001) New K–Ar ages of basalts from the Harrat Ash Shaam volcanic field in Jordan: implications for the span and duration of the upper-mantle upwelling beneath the western Arabian plate. Geology 29:171–174

Kaus BJP, Podladchikov YY (2006) Initiation of localized shear zones in viscoelstoplastic rocks. J Geophys Res 111:B04412. doi:10.1029/2005JB003652

Krienitz M-S, Haase KM, Mezger K, van den Bogaard P, Thiemann V, Shaikh-Mashail MA (2009) Tectonic events, continental intraplate volcanism, and mantle plume activity in northern Arabia: constraints from geochemistry and Ar–Ar dating of Syrian lavas. Geochem Geophys Geosyst 10:Q04008. doi:10.1029/2008GC002254

Lang B, Shirav M, Bogosch R (1979) Volcanological aspects of the Har Peres composite volcano, Golan Plateau. Isr J Earth Sci 28:27–32

Lustrino M, Sharkov E (2006) Neogene volcanic activity of western Syria and its relationship with Arabian plate kinematics. J Geodyn 42:115–139

Lustrino M, Wilson M (2007) The circum-Mediterranean anorogenic Cenozoic igneous province. Earth-Sci Rev 81:1–65

Ma GSK, Malpas J, Xenophontos C, Chan GHN (2011) Petrogenesis of latest Miocene-Quaternary continental intraplate volcanism along the Northern Dead Sea Fault System (Al Ghab-Homs Volcanic Field), Western Syria: evidence for lithosphere-asthenosphere interaction. J Petrol 52(2):401–430

Marcus E, Slager J (1985) The sedimentary-magmatic sequence of the Zemah 1 well (Jordan-Dead Sea Rift, Israel) and its emplacement in time and space. Isr J Earth Sci 34:1–10

McGuire AV, Bohannon RG (1989) Timing of mantle upwelling: evidence for a passive origin for the Red Sea rift. J Geophys Res 94:1677–1682

Menzies M, Gallagher K, Yelland A, Hurford AJ (1997) Volcanic and non volcanic rifted margins of the Red Sea and Gulf of Aden: crustal cooling and margin evolution in Yemen. Geochim Cosmochim Acta 61(12):2511–2527

Mor D (1993) A time-table for the Levant volcanic province, according to K–Ar dating in the Golan Heights, Israel. J Afr Earth Sci 16(3):223–234

Olafsson M, Eggler DH (1983) Phase relations of amphibole, amphibole-carbonate and phlogopite-carbonate peridotite: petrologic constraints on the asthenosphere. Earth Planet Sci Lett 64:305–315

Oxburgh ER, Turcotte DL (1974) Membrane tectonics and the East African rift. Earth Planet Sci Lett 22:133–140

Pallister JS (1987) Magmatic history of Red-Sea rifting: perspective from the central Saudi Arabian coastal plain. Geol Soc Am Bull 98:400–417

Schattner U, Weinberger R (2008) A mid-Pleistocene deformation transition in the Hula basin, northern Israel: implications for the tectonic evolution of the Dead Sea Fault. Geochem Geophys Geosyst 9:Q07009. doi:10.1029/2007GC001937

Sebai A, Zumbo V, Feraud G, Bertrand H, Hussain AG, Giannrrini G, Campredon R (1991) $^{40}Ar/^{39}Ar$ dating of alkaline and tholeiitic magmatism of Saudi Arabia related to the early Red Sea rifting. Earth Planet Sci Lett 104:473–487

Segev A, Rybakov M (2011) History of faulting and magmatism in the Galilee (Israel) and across the Levant continental margin inferred from potential field data. J Geodyn 51:264–284

Sengor AMC, Burke K (1978) Relative timing of rifting and volcanism on earth and its tectonic implications. Geophys Res Lett 5:419–421

Shaanan U, Porat N, Navon O, Weinberger R, Calvert A, Weinstein Y (2011) OSL dating of a Pleistocene maar: Birket Ram, the Golan heights. J Volcanol Geotherm Res 201(1–4):397–403. doi:10.1016/j.jvolgeores.2010.06.007

Shaliv G (1991) Stages in the tectonic and volcanic history of the Neogene basin in the Lower Galilee and the valleys. Geological Survey of Israel report GSI/11/91, 94 pp (in Hebrew).

Sharkov EV, Chernyshev IV, Devyatkin EV, Dodonov AE, Ivanenko VV, Karpenko MI, Leonov YG, Novikov VM, Hanna S, Khatib K (1994) Geochronology of late Cenozoic basalts in Western Syria. Petrology 2(4):385–394

Sharkov EV, Chernyshev IV, Devyatkin EV, Dodonov AE, Ivanenko VV, Karpenko MI, Lebedev VA, Novikov VM, Hanna S, Khatib K (1998) New data on the geochronology of Upper Cenozoic plateau basalts from northeastern periphery of the Red Sea rift area (Northern Syria). Dokl Russ Acad Sci Earth Sect 358(1):19–22

Shaw JA, Baker JA, Menzies MA, Thirlwall MF, Ibrahim KM (2003) Petrogenesis of the largest intraplate volcanic field on the Arabian plate (Jordan): a mixed lithosphere – asthenosphere source activated by lithospheric extension. J Petrol 44:1657–1679

Steckler M, Ten Brink U (1986) Lithospheric strength variations as a control on new plate boundaries: examples from the northern Red Sea. Earth Planet Sci Lett 79:120–132

Stein M, Hofmann AW (1992) Fossil plume head beneath the Arabian lithosphere? Earth Planet Sci Lett 114:193–209

Stein M, Garfunkel Z, Jagoutz E (1993) Chronothermometry of peridotitic and pyroxenitic xenoliths: implications for the thermal evolution of the Arabian lithosphere. Geochim Cosmochim Acta 57:1325–1337

Stein M, Navon O, Kessel R (1997) Chromatographic metasomatism of the Arabian-Nubian lithosphere. Earth Plantary Sci Lett 152:75–91

Tatar O, Yurtmen S, Temiz H, Gursoy H, Kocbulut F, Mesci BL, Guezou JC (2007) Intracontinental Quaternary volcanism in the Niksar pull-apart basin, North Anatolian Fault Zone, Turkey. Turk J Earth Sci 16:417–440

Trifonov VG, Dodonov AE, Sharkov EV et al (2011) New data on the late Cenozoic basaltic volcanism in Syria, applied to its origin. J Volcanol Geotherm Res 199:177–192

Turcotte DL, Oxburgh ER (1973) Mid-plate tectonics. Nature 244:337–339

Wallace ME, Green DH (1991) The effect of bulk rock composition on the stability of amphibole in the upper mantle: implications for solidus positions and mantle metasomatism. Mineral Petrol 44:1–19

Weinstein Y (2000) Spatial and temporal geochemical variability in basin-related volcanism, northern Israel. J Afr Earth Sci 30:865–885

Weinstein Y (2012) Transform faults as lithospheric boundaries, an example from the Dead Sea Transform. J Geodyn 54:21–28

Weinstein Y, Navon O, Altherr R, Stein M (2006) The role of lithospheric mantle heterogeneity in the generation of Plio- Pleistocene alkali basaltic suites from NW Harrat Ash Shaam (Israel). J Petrol 47:1017–1050

White R, McKenzie D (1989) Magmatism at rift zones the generation of volcanic continental margins and flood basalts. J Geophys Res 94:7685–7729

Wilson M (1993) Magmatism and the geodynamics of basin formation. Sediment Geol 86:5–29

Chapter 5
Lateral Motion and Deformation Along the Dead Sea Transform

Zvi Garfunkel

Abstract This paper presents an updated summary of the history and shallow structure of the Dead Sea transform (DST) and its plate tectonic context. The DST formed in the Early Miocene as a transform boundary between the Sinai and Arabian plates. The lateral offset was ca. 105 km near the Dead Sea. Since the DST trace is irregular in map view, the lateral motion led to formation of a 10–80 km wide deformation zone along the plate junction. Thus, the structures along the DST can be interpreted within the framework of the Sinai-Arabia plate kinematics. South of ca. lat. 33° transtension, which increased with time, led to variable oblique separation of the plate edges. This produced an almost continuous ca. 5–25 km wide depression whose structure is dominated by a string of pull-apart basins up to 15–20 km wide and up to ca. 12 km deep. The crystalline crust under the largest basins was appreciably thinned and may have been intruded by basalts. The structural pattern changed over time, the present pattern having been mostly established in the second half of the DST history. North of lat. ca. 33°N transpression dominates and the DST flanks are strongly deformed by folding, faulting, and rotation of fault blocks on vertical axes, which together produce shortening perpendicular to the DST and also left lateral shearing of its flanks, qualitatively compatible with the plate kinematic. The deformation can also account for the observed decrease of the lateral offset along the main fault line from ca. 100 km at ca. lat 33°N to 65–70 km at lat. 36°–36.5°N, and also leads to left lateral shearing along the continental margin near the Galilee and farther north. North of lat. 36.5°N the DST now interacts with the Anatolian plate and the Cyprus arc. This resulted from a rearrangement of the plate configuration in that region when westward extrusion of Anatolia began and the East Anatolian Fault formed, but the details of the kinematic changes are incompletely known. This change obscured the structural relations in this region during the early stages of the DST history.

Z. Garfunkel (✉)
The Fredy and Nadine Herrmann Institute of Earth Sciences, The Hebrew University of Jerusalem, Edmond J. Safra campus, 91904 Givat Ram, Jerusalem, Israel
e-mail: zvi.garfunkel@mail.huji.ac.il

Z. Garfunkel et al. (eds.), *Dead Sea Transform Fault System: Reviews*, Modern Approaches in Solid Earth Sciences 6, DOI 10.1007/978-94-017-8872-4_5,

Keywords Dead Sea Transform • Strike slip faults • Plate motions • Pull aparts • Transpression

5.1 Introduction

The Dead Sea transform (DST) (also called Dead Sea or Levant Fault Zone), is one of the new plate boundaries produced by the mid-Cenozoic breakup of the once continuous African-Arabian continent. It forms the NW boundary of the Arabian plate, extending from the Red Sea spreading center to the zone of plate convergence in southern Turkey, a distance of ca. 1,000 km (Fig. 5.1). It is dominated by left lateral motion, ca. 105 km along its southern half. On the west most of the DST is delimited by the Sinai (sub-) plate, but its northernmost part also interacted with the Cyprus arc. The lateral motion along the DST was combined with minor, laterally variable, transtension or transpression which produced a deformed belt along its trace.

The DST attracted geologists already more than 150 years ago, and was recognized as the northern branch of the rift system that extends along East Africa and farther north (Suess 1891; Gregory 1921). At first the DST was interpreted as an extensional rift because parts of it are marked by valleys bordered by normal faults (Gregory 1921; Picard 1943). However, over time it was recognized that left-lateral slip dominates along the DST, so it is a transform plate boundary (Dubertret 1932; Quennell 1959; Freund 1965; Wilson 1965), but the traditional terms "Dead Sea rift" or "Dead Sea fault" are still sometimes used.

The purpose of this work is to present an updated summary of the shallow structure along the DST and how it is related to the lateral motion and to highlight important unsolved issues. Some other important topics are treated in companion papers in this volume.

5.2 Regional Setting

The DST crosses a continental area that was shaped by the Neoproterozoic Pan-African orogeny (Bentor 1985; Garfunkel 1988, 1999; Stern 1994). Later the region became a part of the stable North African-Arabian platform on which an extensive sediment cover accumulated in several depositional phases between Early Cambrian to mid-Cenozoic times (Picard 1943; Bender 1974; Brew et al. 2001a; Garfunkel 1988; Guiraud et al. 2001). Rifting events in latest Paleozoic and early Mesozoic times (ca. 270 to ca. 170 Ma) produced the passive margins of Arabia and NE Africa, including the Levant basin margin and the adjacent east Mediterranean basin (Bein and Gvirtzman 1977; Garfunkel and Derin 1984; Garfunkel 1998; Garfunkel and Ban Avraham 2001; Gardosh et al. 2010). Early Cretaceous intraplate igneous activity and uplifting occurred in the area crossed by the future DST (Garfunkel 1989; Segev 2009), but later there was hardly any igneous activity until

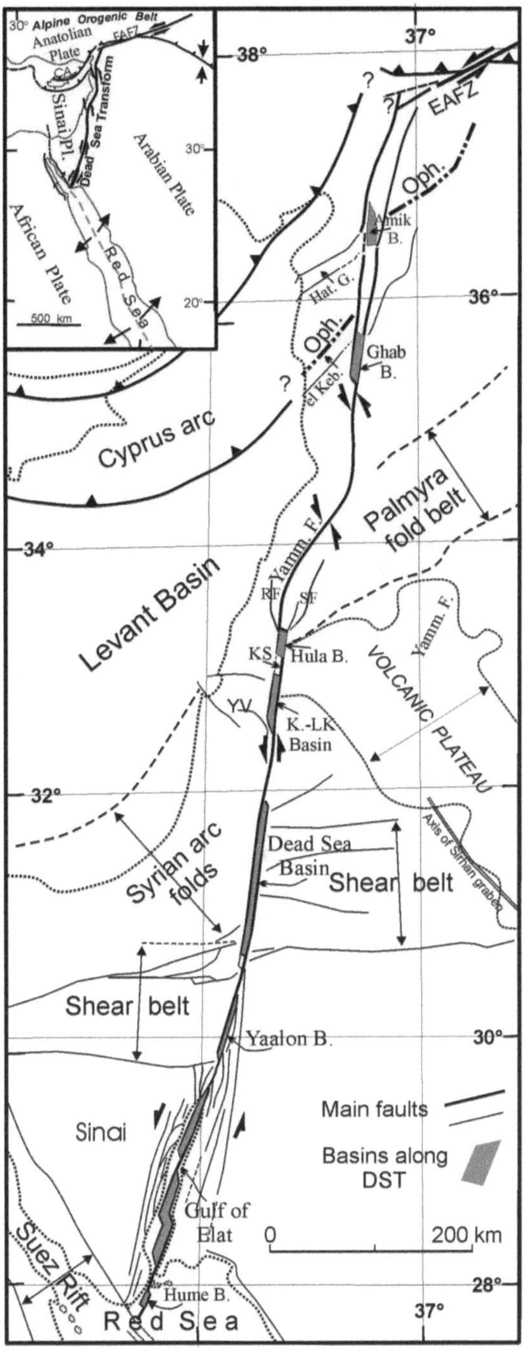

Fig. 5.1 The DST and its regional setting (*inset*). *EAFZ* East Anatolian fault zone, *K.-LK Basin* Kinnarot-Lake Kinneret Basin, *KS* Korazim saddle, *RF* Roum fault, *SF* Serghaya fault, *YV* Yizreel valley. Only main volcanic field is shown

the DST formed. In the Late Cretaceous plate convergence began along the junction of the Arabian platform and the adjacent basins with the Alpine orogenic belt (Şengör and Yilmaz 1981; Le Pichon et al. 1988; Yilmaz 1993) and the Syrian arc compressional structures formed on the platform next to parts of the future DST (Fig. 5.1, Garfunkel 1988; Brew et al. 2001a; Guiraud et al. 2001). This deformation ended when the DST formed. The DST cuts across all the older structures and its formation and development were accompanied by regional-scale uplifting and volcanism.

5.3 The Kinematic Framework

The DST is a prominent geologic discontinuity: the rock sequences and structures facing each other on its two sides are quite different. This is considered to be a result of a left lateral offset of ca. 100 km because its restoration matches all known geologic features across the southern DST, as far north as southernmost Lebanon (Quennell 1959; Freund 1965; Freund et al. 1970; Druckman 1974; Bandel 1981; Bandel and Khouri 1981; Segev 1984; Sneh and Weinberger 2003). The matched features include basement rocks, the facies belts and isopachs of the entire Cambrian to Late Cretaceous sedimentary cover, the lines along which the regional Early Cretaceous erosion truncated older stratigraphic units, as well as the narrow lineaments (belts of faulting and strong folding) of the Central Negev-Sinai shear belt (southern part of the Syrian arc, Fig. 5.1). The latter trend at large angles to the DST, so their matching gives the best estimate of the offset along the DST – ca. 105 km in the Dead Sea area (Fig. 5.1, Quennell 1959; Bartov 1974; Garfunkel 1981). The patterns of magnetic anomalies in Jordan and Israel are also displaced ca. 105 km (Hatcher et al. 1981). Shaliv (1991) noted that the Yizreel Valley could be matched with the Sirhan depression (Fig. 5.1). However, while the latter subsided markedly in the Late Cretaceous (Basha 1982), activity of that age along the Yizreel Valley was not documented. They should have connected beneath the volcanics of the Golan Heights, but such a connecting structure was not observed (Shulman et al. 2004; Meiler et al. 2011), so the significance of this correlation requires further study.

Farther north in Lebanon the DST marks a discontinuity between the mid-Cretaceous carbonate series on its two sides (Saint-Marc 1974; Walley 1998), but an offset marker – the front of ophiolite and related nappes – is known still farther north, near the Syria-Turkey border (Fig. 5.1). These nappes were thrust over the Arabian platform in the Late Cretaceous, and now are intercalated between within the sedimentary section (Ponikarov 1964; Ponikarov et al. 1969; Kopp and Leonov 2000). Their absence in the section 3–10 km SE of their exposures tightly constrains the end-Cretaceous position of their front (it may have been defined by a fault zone: Ponikarov et al. 1969, Figs. 5.1 and 5.6). Matching the ophiolite nappe front across the DST reveals a left lateral offset of 65–70 km (Freund et al. 1970; Al-Maleh et al. 1992; Westaway (2003) favored an offset of 55 km). This is supported by the distribution of Early Miocene marine beds filling a depression SE of the ophiolite nappes (Ponikarov 1964; Ponikarov et al. 1969; Krasheninnikov 2005, see 5.3, 5.5.2.3). An offset of only 20–25 km, as suggested by Trifonov et al. (1991) is unlikely, because when restored this leaves notable bend of the original nappe front where it crossed the future DST. The difference between this estimate and the offset farther south will be discussed below see (5.5.2.4).

The motion along the DST is independently constrained by the regional plate kinematics because it takes up most of the Africa-Arabia plate separation, i.e. the Red Sea opening (Dubertret 1932; Freund 1965, 1970; McKenzie et al. 1970; Le Pichon et al. 1973). Taking into account the relatively small opening of the Suez rift and the stretching of the Red Sea margins, a left-lateral offset of 100–110 km is inferred along the DST, i.e. the Sinai-Arabia plate boundary (Freund 1970; Joffe and Garfunkel 1987). Within error, this is the same as the offset deduced from the geology along the southern half of the DST, though the geometry of the Red Sea opening is deduced from completely independent data. The Sinai-Arabia Euler pole of the motion of these plates, i.e. the DST motion, is estimated to be located in NE Libya or the nearby part of the Mediterranean (Quennell 1959; Freund 1970; Garfunkel 1981). Such a position of the Sinai-Arabia pole implies transtension along the south of the DST and transpression along its northern part (Fig. 5.1). This pole is close to the Arabia-Africa Euler pole (i.e. the Red Sea opening) because these motions are similar. Integration of all these constraints allows to derive well constrained plate kinematic models (e.g. Joffe and Garfunkel 1987; Le Pichon and Gaulier 1988; and see Sect. 5.4).

5.4 The Chronological Framework – The History of Motion

Several observations constrain the history of motion along the DST. It postdates the ca. 20–24 Ma old swarm of Red Sea dikes, as the northern dikes experienced the entire lateral offset (Fig. 5.3, Eyal et al. 1981; Steinitz et al. 1981; Steinitz and Bartov 1991). The ages of the oldest fills of basins along the DST also constrain the time of the beginning of the motion, because the basins resulted from the lateral motion (Garfunkel 1981; Garfunkel and Ben-Avraham 2001, see Sect. 5.5). Drillholes in the Dead Sea basin reached palynologically dated Early Miocene sections (Horowitz 1987) that are considerably thicker than coeval sections (Hazeva Formation) on the western basin flank, indicating that the basin existed already at that time (Garfunkel and Ben-Avraham 1996, 2001). Near Tiberias basalt flows with K-Ar ages of 17 ± 3 Ma to ca. 15.5 Ma occur in the middle of the sediment fill of a small marginal basin (Tiberias sub-basin), while ca. 20 km farther south its entire fill consists of basalts, the oldest ones having K-Ar ages of ca. 15.5 Ma (Shaliv 1991). The nearby 4.25 km deep Zemah 1 well (Z in Fig. 5.5, just south of Lake Kinneret) in the main basin along the DST bottomed in Middle Miocene beds (Horowitz 1987) which are >1 km above the base of the ca. 5 km thick fill (Ben-Avraham et al. 1996). Thus, in this area basins formed already at the beginning of the Middle Miocene or somewhat earlier. Moreover, this part of the DST was a structural boundary already at that time, because in the area that originally was located east of it no basin formed and volcanism was very restricted (Garfunkel 1989; Weinberger et al. 2003), and the early faulting west of the DST did not extend east of it (see below). Evidence from the Sinai triple junction, at the south of the DST, also indicates its early Miocene activity (see Sect. 5.5.1.1).

Along the northern part of the DST marine Early Miocene sediments are preserved in a depression SE of the ophiolite nappes (Ponikarov 1964; Ponikarov et al. 1969; Kopp and Leonov 2000; Krasheninnikov 2005; Hardenberg and Robertson 2007). These works show that west of the DST the Nahr el-Kebir (Latakiye) depression (Figs. 5.1 and 5.6) formed in the Aquitanian, shortly before initiation of the DST in the south, and was accentuated as a graben in the Middle Miocene. In its SW part ca. 1.7 km thick marine sediments accumulated by the end of the Miocene (may be less farther NE). Restoring the DST offset as estimated above aligns these beds with an area east of the DST where Early Miocene sections reach a combined thickness of ca. 0.7 km, which most likely record the eastward extension of the depression west of the DST. Here a Middle Miocene graben is not apparent, but the sediments of this age are folded on NE-SW trending axes, while west of the DST such folds are absent. These features show that here the two sides of DST were deformed differently, which suggests that it was active and formed a structural discontinuity already in the Middle Miocene.

These lines of evidence constrain the beginning of lateral motion along DST to between 20 Ma and ca. 17–16 Ma ago. This is considerably younger than the initial breakaway of Arabia from Africa along the Red Sea-Suez rift line, but is close to the time when faulting along the Suez rift was reduced, seafloor spreading began in the Gulf of Aden, and the Arabia-Africa plate separation accelerated (Garfunkel and Bartov 1977; Bosworth 2005; Garfunkel and Beyth 2006, and references therein). These events are interpreted to show that at that time most of Arabia-Africa motion north of the Red Sea was transferred to the DST. Avni et al. (2012) raised the possibility that earlier faulting without lateral offset occurred along the DST, based on analysis of ancient landscape. Though quite possible, this requires further study.

During the very early stage of the DST lateral motion some right lateral motion may have continued on the Central Negev-Sinai shear belt lineaments (Fig. 5.1, Calvo 2002), but the interaction between these structures and the DST requires further study. However, they need not have obstructed each other, similar to the present situation in southern California where active transverse faults meet the San Andreas Fault (Jennings 1973).

Another constraint on the DST slip history is provided by magnetic anomalies that record seafloor spreading, i.e. Africa-Arabia plate separation, since 5–3 Ma ago in the southern and central Red Sea (Roeser 1975; Chu and Gordon 1998). With plausible Euler poles for the Red Sea opening this translates to 40 ± 2 km of oblique opening near its northern end. Opening of the Suez rift in the last 5 Ma may have taken up ca. 5 km, leaving a lateral slip of about 33–37 km (average rate of 6.5–7 mm/year) along the DST, which is 30–35 % of the total offset. Thus the Miocene motion amounted to ca. 2/3 of the total lateral DST offset.

GPS studies of the present plate motions (Reilinger et al. 2006; Le Beon et al. 2008; ArRajehi et al. 2010; Le Pichon and Kreemer 2010; Al Tarazi et al. 2011; Reilinger and McClusky 2011) found a slip rate of ca. 4.0 to 5.5 mm/year along the DST, compatible with the slip rates indicated by up 50–100 kyr old markers (Garfunkel 2011), but they do not resolve well the much smaller transverse component of motion. These results suggest that the present slip is slower than the average rate in the last 5 Ma. Alchalbi et al. (2010) found a slip rate of 1.8–3.3 mm/year in NW Syria, slower than found in previous studies, but compatible with the evidence that there the total offset was less than farther south.

The constraint from the southern Red Sea provides a datum that can be used for dividing the Sinai-Arabia plate motion into two periods – before and after 5 Ma ago (Garfunkel 1981; Joffe and Garfunkel 1987). This does not imply a discontinuity in motion (as suggested e.g. by Hempton 1987), which is contradicted by the evidence from the Gulf of Aden and the Red Sea showing that plate motions were continuous and quite uniform (Le Pichon and Gaulier 1988; Garfunkel and Beyth 2006 and references therein). Nor does it imply a sudden change of motion at that time or that plate motions were uniform in the two periods. The major young structures along the southern half of the DST (flanked by rigid plates) existed for some time (see Sect. 5.5.1) so they can be used to constrain the geometry of motion during the younger period. Combining their geometry with the above constraint on the amount of motion, allows estimating the entire motion during the last 5 Ma as ca. 1.45° about an Euler pole at 32.8°N, 22.6°E (uncertainty ca. 1°). However, using this pole to reconstruct the total DST motion would produce overlaps of the plate edges, which indicates that the pole position changed. The total pre-5 Ma motion is estimated as ca. 2.7° about a pole located some 4° farther west, which involves a change in the direction of motion of up to ca. 10° along the southern DST (Joffe and Garfunkel 1987). The GPS data seem to imply a continuing eastward shift of the Euler pole and a slight slowing of the slip rate. Reilinger et al. (2006) suggested a present day Euler pole at 32.8°N, 28.4°E (± ca. 3.5 °) for the DST, but this leads (given the Red Sea opening) to problematic inferences regarding the Suez rift (Garfunkel 2011). Westaway (2003) proposed an Euler pole at 31.1°N, 26.7°E, and a motion rate of 0.434°/Ma. This implies a northward increasing slip rate, in conflict with the GPS data (Reilinger et al. 2006; Reilinger and McClusky 2011; Alchalbi et al. 2010), so this model needs revision.

Le Pichon and Kreemer (2010) analyzed the GPS data of the Eastern Mediterranean region. Their results are compatible with the above estimates, though they do not resolve the present Africa-Sinai motion. Most important, their results and those of (Reilinger et al. 2006) reveal an overall rotation of the region surrounding the DST relative to Europe about an Euler pole at the north of the Nile Delta. This profound insight has important implications regarding the regional plate motions, but since this applies to a scale this is outside the scope of the present work.

In summary, the foregoing account shows that the plate kinematic setting of the DST, known is general outline, provides a framework of integrating data from its various parts, but refinement of the present models is still desirable.

5.5 Shallow Structure of the Dead Sea Transform

The DST is not a perfect small circle in map view, so its lateral motion inevitably produces misfits between the edges of the bordering plates, which cause local transtension or transpression (Figs. 5.1 and 5.2, Quennell 1959; Garfunkel 1981). This produces a variety of secondary structures, similar to those found along other major strike slip fault zones (Crowell 1974; Mann et al. 1983; Şengör et al. 1985; Harding et al. 1985). Transtension arises where the major strike slip fault line bends or steps in the same sense as the fault slip (releasing bend). This produces rhombic depressions – pull-apart basins (rhomb grabens) – that grow by becoming longer, so their length cannot be less than the lateral offset during their development (Fig. 5.2a). Thus, when the slip rate is known, the lengths of pull-apart basins put constraints on their ages. Their growth also increases the area enclosed by the bordering plates (Fig. 5.2b). Where major strike slip faults bend or step opposite to the sense of slip (restraining bends), transpression arises, which will produce folds, thrusts, or horsts. Such secondary structures appear all along the DST. The larger secondary structures are well over 10 km long and thus developed over several million years, so the pattern of major faults delimiting them must have persisted over these time intervals. The major faults are actually zones of much fracturing, several hundred meters wide, and along them much shorter structures (1 km to tens of meters long) are developed. Their small size suggests growth in short periods, probably much shorter

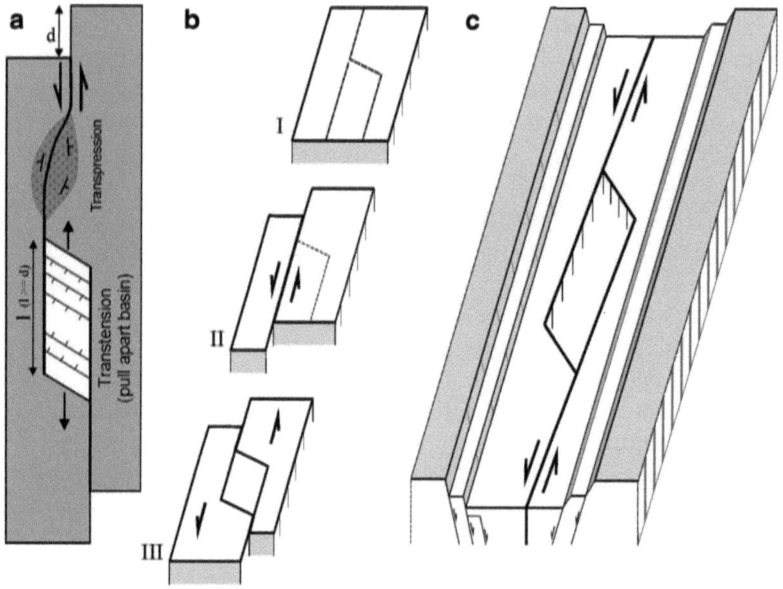

Fig. 5.2 Secondary structures along the DST. (**a**) conditions for transtension and transpression.; (**b**) Increase in are during pull-apart formation. (**c**) structural relations along southern (transtensional) part of DST (See text for discussion)

than 1 Ma, which points at rearrangement of the fractures on which the slip along the fault zones occurred at any given time.

In what follows only the larger secondary structures will be examined. Because of the position Euler pole of the DST, transtension or transpression dominate along its southern and northern parts, respectively (Fig. 5.1), so they will be treated separately.

5.5.1 *Southern Part of the Dead Sea Transform*

This part of the DST, south of Lebanon, is mostly marked by a prominent 5–20 km wide valley (transform valley), mostly with uplifted flanks, in which all or most of the lateral motion takes place. The valley floor is largely covered by very young sediments, so the deeper structure is inferred from drilling and geophysical data. Along the valley the most conspicuous structures are longitudinal faults of two types (Garfunkel 1981; Garfunkel and Ben-Avraham 2001, Fig. 5.2c): (a) normal faults that extend along most of the valley margins, and (b) left stepping strike slip or oblique slip faults within the valley that delimit a string of pull-apart basins. Also present are faults extending across the transform valley. Recent reviews (Ben-Avraham et al. 2008, 2012) summarize many geophysical data along this part of the DST, focusing on the subsurface. Below is an updated review of the structure in the upper crust integrating a wider set of geological and subsurface data and relating it to the lateral offset.

5.5.1.1 The Sinai Triple Junction Area

The southernmost part of the DST is now marked by the ca. 30 km long Hume Deep pull-apart basin (Fig. 5.3a, Ben-Avraham et al. 1979). It is separated from the Gulf of Elat ('Aqaba) by the Tiran strait saddle formed by transpression, east of which reef terraces on the Tiran Island, probably <2 Ma old, were uplifted up to ca. 500 m above sea level (Garfunkel 1981; Goldberg and Beyth 1991). In the SW the NNE-SSW trending DST structures are cut off near the northern extremity of the Red Sea axial depression by the NW-SE trending faults that characterize the entire width of the Red Sea basin and the Suez rift (Cochran 1983, 2005; Garfunkel 1987; Gaulier et al. 1988; Mart and Hall 1984). Moreover, while the Bouguer anomaly over the Hume Deep pull-apart is strongly negative, like over the other pull-apart basins along the DST, a positive Bouguer anomaly characterizes the Red Sea basin (Ben-Avraham et al. 1979; Ben-Avraham 1985; Cochran 2005). This reflects the transition to a crust that was shaped by plate separation rather than in a strike-slip regime.

The early stages of the DST history in this area are recorded by the very early and younger Miocene marine sediments, 0.5–1 km thick known in Midyan and on the southern tip of Sinai (Fig. 5.3a, Dullo et al. 1983; Garfunkel 1987; Cole et al. 1995;

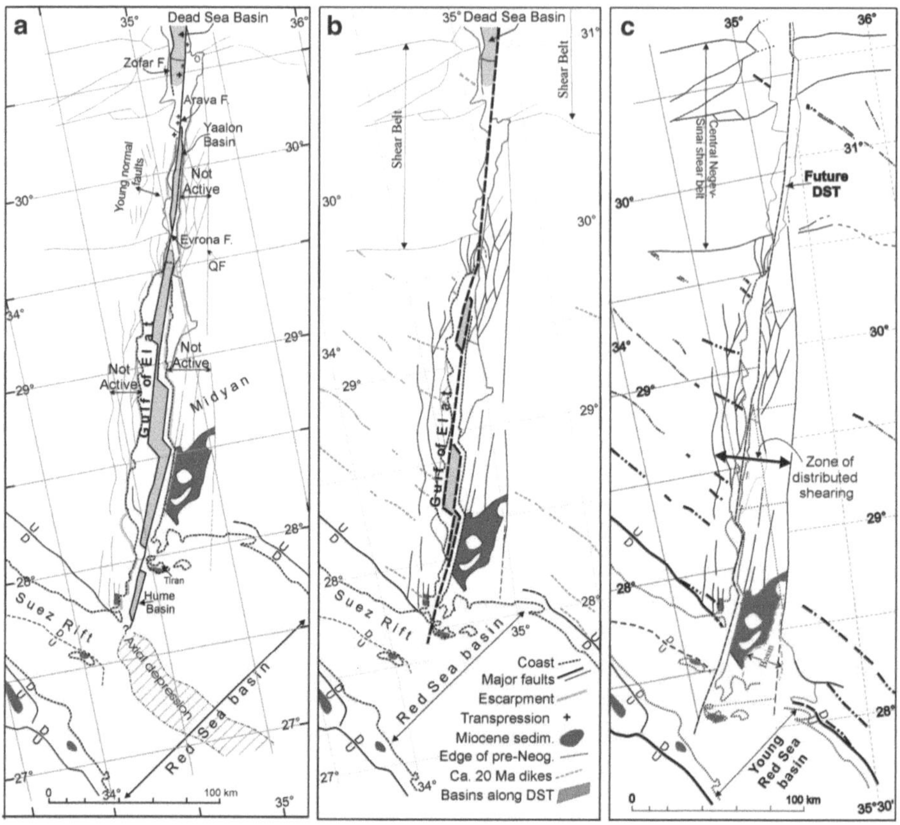

Fig. 5.3 The southern part of the DST. (**a**) present situation. (**b**): reconstruction of 40 km lateral offset (close to end Miocene). (**c**) Reconstruction of the entire transform offset. Abbreviations: *QF* el Quweira faults, *ZF* Zofar fault

Hughes et al. 1999). These beds cover the southern ends of belts of strike-slip faults along the two sides of the Gulf of Elat that were active in the Miocene (further discussed in Sect. 5.5.1.2), and fill a depression in Midyan. Restoration of the lateral motion (Fig. 5.3c) eliminates the Gulf of Elat (between these fault belts) and also aligns the uplifted edges of Sinai and NW Arabia, which shows that before formation of the DST they formed a continuous fault-controlled margin of the very young Red Sea-Suez basin. When the DST formed this margin was breached by a ca. 50 km wide belt of NNE-SSW trending strike slip faults. The above mentioned Miocene beds formed in a marine embayment over the southern part of this belt, so their age constrains the inception of the DST to the Early Miocene. The most subsiding area in Midyan (up to >2 km of sediments) was located 20–30 km east of the break that developed into the Gulf of Elat, but it did not extend much to the north. The later history of these features is treated below (Sect. 5.5.1.2).

5.5.1.2 The Gulf of Elat ('Aqaba) Segment

This segment comprises the ca. 180 km long depression under the Gulf and also the fault belts along its uplifted margins (Fig. 5.3a). The faults in these belts displace left laterally the ca. 20 Ma old Red Sea dikes, proving that they belong to the DST system (Eyal et al. 1981). Lateral offsets of individual faults range from ca. 1 km to >10 km (Eyal et al. 1981); the largest offset, ca. 20 km, probably occurred on the Al-Quweira Fault in SW Jordan (Segev 1984). The total offset reached 40–50 km. Pre-transform sediments and continental Miocene beds are preserved in small pull-aparts along these faults, which proves that faulting occurred before the sediment cover was eroded from the higher standing areas next to these depressions. On the other hand, strike slip faults that extend to the Gulf are truncated by normal faults along its coast, while inland some faults are truncated by an erosion surface or over-stepped by (undated) continental sediments. In Midyan the younger Miocene beds are not visibly displaced by the marginal strike-slip faults. These features indicate that the strike slip faults along the Gulf flanks were active early in the DST history and then the lateral motion was distributed in a wide zone, but later (late in the Miocene?) they became inactive and the lateral motion was concentrated along the depression occupied by the Gulf of Elat ('Aqaba).

The Gulf is delimited on both sides by active normal faults that extend close to its coasts (not shown on Fig. 5.3a for clarity). Within the Gulf three large pull-aparts, well expressed in the bathymetry, are developed between major longitudinal left-stepping strike slip faults (Fig. 5.3a, Ben-Avraham et al. 1979; Ben-Avraham and Garfunkel 1986; Ben-Avraham et al. 2008, 2012). The upper ca. 1 km of the sediment fill, imaged by seismic reflection, records continuing, downward increasing, syn-depositional deformation (tilting, faulting, and in the south also arching), but as the age of the sediments is not constrained, detailed interpretation of the deformation history (e.g. Ehrhardt et al. 2005) are very doubtful. Several narrow ridges in the southern pull-apart may express diapirs of Miocene evaporites (Ben-Avraham et al. 1979) that probably formed next to the Miocene low of Midyan.

Given the lateral offset accommodated in the marginal shear zones, the lateral offset within the Gulf hardly amounted to ca. 2/3 of the total offset. The lengths of the pull-aparts (ca. 40 km, 35 km and 60 km, from north to south) show that they could accommodate only a part of the total DST offset (see Fig. 5.2). This suggests that they formed some time after initiation of the DST, which is supported by reconstruction of part of the motion (Fig. 5.3b). A negative Bouguer anomaly, up to ca. −100 mgal over the pull-aparts (Ben-Avraham 1985), indicates ≥5 km thick sediment fills. However, the negative anomaly extends also over the relatively shallow part of the Gulf west of the southern pull-apart, indicating that a thick sediment fill exists beneath this area as well. This is interpreted as showing that this area was part of a subsiding pull-apart earlier in the evolution of the Gulf, which is supported by the reconstruction of part of the motion (Fig. 5.3b).

Thus, initially the lateral motion along the southernmost part of the DST was distributed over a wide belt of closely spaced strike slip faults that extended in the north to the central Negev-Sinai shear belt (Fig 5.3c). Later, in the late Miocene (?), lateral motion became concentrated in the middle of this belt where the Gulf of Elat depression developed, while the faults along its margins became inactive. As the active faults along the Gulf lead to transtension (Garfunkel 1981), which is not seen along the marginal strike slip fault zones, this history appears to express increasing separation of the flanking plate edges along this part of the DST, which can be related to eastward shifting of the Sinai-Arabia Euler pole (Garfunkel 1981, see Sect. 5.4).

5.5.1.3 The Southern and Central Arava Valley (Wadi Araba) Segment

This segment extends from the Gulf of Elat to the Dead Sea basin (Fig. 5.3a, Garfunkel et al. 1981; Bartov 1994; Ten Brink et al. 1999; Frieslander 2000; Avni et al. 2000; Calvo 2002). In the south, it is separated from the onland northern end of the Gulf of Elat depression by a weakly transpressional structural saddle that is crossed obliquely by the Evrona Fault. North of it the little studied Yaalon basin pull-apart, marked by a ca. 60 km long negative gravity anomaly, is developed between the left stepping Evrona Fault and Arava Fault (Fig. 5.3a). Its southern half, 5–6 km wide, is expressed by a land-locked depression that records continuing subsidence. Farther north the basin narrows to 2–3 km and has no topographic expression, so it is not clear whether it still subsides. The subsurface distribution of Cretaceous beds revealed by drilling in the middle part of the basin indicates a ca. 40 km left lateral offset along its western border fault, so the rest of the lateral motion must have taken place farther east (Bartov 1994).

North of the Yaalon basin exposures of the pre-transform series of the western DST flank extend eastwards as far as the Arava Fault (Frieslander 2000), forming a ca. 20 km long structural saddle between the Yaalon basin and a ca. 8–9 km wide graben farther north that is delimited by the Arava Fault on the east and the Zofar Fault on the west (Fig. 5.3a, Bartov 1994; Bartov et al. 1998; Calvo 2002). This graben is filled by a few kilometers of Miocene continental beds (Hazeva Formation). Seismic data show that the pre-transform series beneath this fill, to a distance of at least 35 km north of the Yaalon basin, is the same as that of the western DST flank, but differs markedly from that exposed a short distance east of the Arava fault (Frieslander 2000; Ryberg et al. 2007). The juxtaposition of different sections across the Arava fault indicates a large lateral offset along this fault.

Originally the graben between the Arava and Zofar Faults extended to the northern Arava. There it continued to subside and became a part of the Dead Sea basin, but its part south of the transverse Buwirida Fault ceased to subside and was tilted westward, probably during a phase of local transpression, exposing ca. 2 km of its Miocene fill (Fig. 5.3, Bartov 1994; Calvo 2002). The age of this event is constrained to the late Miocene or somewhat later, as it postdates the eroded Miocene beds (Hazeva Formation) but predates the hardly disturbed unconformably overlying Pliocene fluviatile beds (Arava Conglomerate; Avni et al. 2000). A few other transpressional structures are developed in this area (Fig. 5.3b, Garfunkel et al.

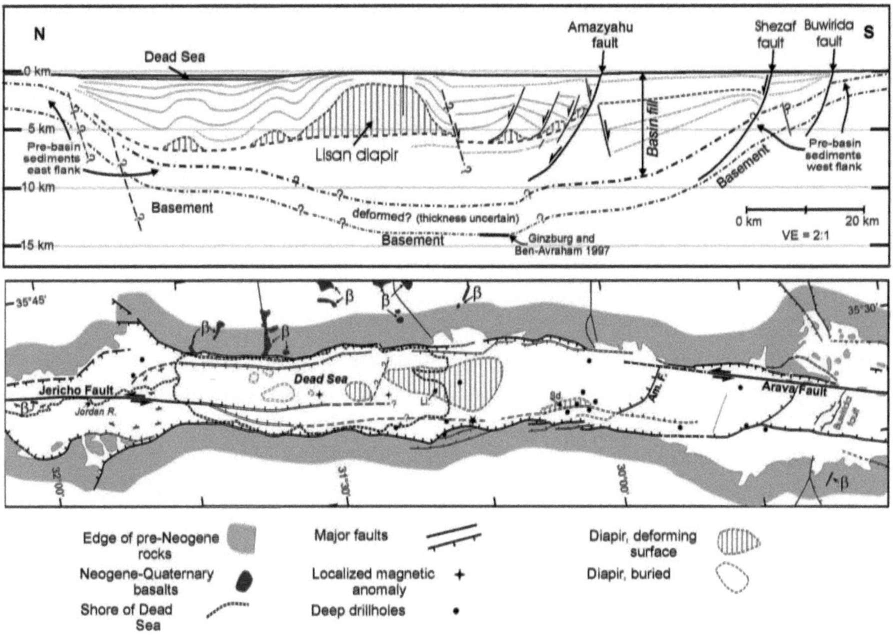

Fig. 5.4 The Dead Sea basin. *Top*: Cross section along basin axis. *Bottom*: main structural features. *Li* Lisan Diapir, *Sd.* Mount Sedom diapir

1981; Galli 1999). A related observation is that in (Early?) Pliocene times a stream originating in the eastern DST flank – the Edom river – flowed across this part of the central Arava Valley and transported characteristic pebbles to the Negev (Ginat and Avni 1994), probably signifying a period of local uplift, due to transpression, of this part of the Arava Valley.

The tectonic situation changed in Late Pliocene-Quaternary times, leading to formation of the present topographic low position of the Arava Valley. This is evidenced by the development at that time of a series of NNE-SSW to N-S striking normal faults (sometimes with a minor left lateral component) in a ca. 30 km wide belt west of the northern Yaalon basin and of the more northern part of the Arava Valley (Fig. 5.3a, Avni et al. 1994). These faults are believed to record a young increase of transtension along the southern part of the DST, resulting from an eastward shift of the Euler pole, as discussed above (Sect. 5.4).

5.5.1.4 The Dead Sea Pull-Apart Basin

The Dead Sea basin, ca. 150 km long, formed between the left stepping left lateral Arava Fault and Jericho Fault (Fig. 5.4, Quennell 1959; Garfunkel 1981; Garfunkel and Ben-Avraham 1996, 2001; Ben-Avraham et al. 2008, 2012). It is topographically well expressed and is also outlined by a conspicuous negative Bouguer anomaly gravity which records the extent of the thick basin fill (Ten Brink et al. 1993). It is the longest pull-apart basin along the DST and the only one that is longer than its total lateral offset. The basin began to subside already in the Early Miocene (>16 Ma), as noted above (see 5.4). Data from outcrops and wells show that the basin fill comprises three divisions: Miocene silici-clastics (Hazeva Formation); latest Miocene-early Pliocene evaporites, mainly halite (Sedom Formation); post-evaporitic lacustrine and fluvial sediments, mainly clastics (Kashai and Crocker 1987; Horowitz 1987; Gardosh et al. 1997; Garfunkel and Ben-Avraham 2001; and references therein).

Available seismic reflection and well data (Neev and Hall 1979; Kashai and Crocker 1987; Ten Brink and Ben-Avraham 1989; Garfunkel and Ben-Avraham 1996; Al-Zoubi and ten-Brink 2002; Larsen et al. 2002; Ginzburg et al. 2006; Ben-Avraham et al. 2008, 2012) allow an updated interpretation of the basin structure (Fig. 5.4). They show that it is dominated by an 8–10 km wide central trough that formed between the extensions of the Arava and Jericho faults, in which the lateral motion takes place. It is separated from the uplifted and little deformed transform shoulders by marginal blocks, wider on the western side. Several transverse faults extend across the basin. The fill of the central trough thickens northward from a few km in the central Arava Valley to ca. 12 km near the transverse Amatzyahu Fault (Khunayzira Fault). Farther north, south of the Lisan diapir, the top of the basement is 13–14 km deep (Ginzburg and Ben-Avraham 1997). These figures are compatible with recent seismic refraction data (Ten Brink et al. 2006; Mechie et al. 2009; ten Brink and Flores 2012). North of the Amatzyahu Fault the base of the evaporite series is at a depth of 3–3.5 km and deepens to 6 km northward (Ginzburg et al. 2006). As here the pre-transform series is at most ca.3 km thick – its thickness on the adjacent western basin margin – the pre-evaporite Hazeva Formation is inferred to be 5–7 km thick. This is much more than in boreholes near the basin margins, implying great early subsidence of the central trough.

The evaporitic series is present only north of the Amatzyahu Fault (Khunayzira Fault) where post-Hazeva subsidence was greatest. It was deposited over a short time interval, perhaps no longer than 2–3 Ma (Zak 1967). The original thickness, probably reaching ca. 2 km, was modified by the growth of the Sedom diapir, Lisan diapir, and smaller disapirs (Fig. 5.4). Superimposed on the northern part of this area is the box-shaped depression now occupied by the waters of the Dead Sea, which defines the extent of the greatest young subsidence that could not yet be filled by sediments.

The available seismic reflection profiles do not reveal much deformation of the basin fill. Broad roll-overs are seen north of the Shezaf Fault and Amatzyahu Fault (Khunayzira Fault) (Fig. 5.4b), suggesting that these faults are listric, but this cannot be resolved from the published seismic data. Smaller scale deformation is not discernible, though considerable deformation is expected beneath the basin (see Sect. 5.5.1.7).

The deep structure of the northern part of the basin, now under the waters of the Dead Sea, is hardly known. A seismic refraction line along its western side records a basement depth of 5.5–7 km (Ginzburg and Ben-Avraham 1997), but this probably represents the marginal blocks where the basement is expected to be 5–6 km deep according to the En Gadi 2 well (Fig. 5.4). The Bouguer anomaly over the central trough is much more negative than along the seismic profile, suggesting a thicker fill and thus a considerably deeper basement. On the other hand, 2–3 km north of the Dead Sea the basement is only ca. 3 km deep and shallows northward according to seismic and borehole data (Lazar et al. 2006; Al-Zoubi et al. 2007). How the basement depth changes it is not known. The data from north of the Dead Sea further show that there the pre-transform sedimentary sequence of the eastern DST flank extends beneath the Jordan Valley as far west as the Jericho Fault. Moreover, farther south the magnetic anomaly of the eastern basin flank extends over the Dead Sea, whereas the anomalies of the western flank are truncated along the marginal step (Frieslander and Ben-Avraham 1989). This shows that the Jericho Fault accommodated the entire lateral transform offset, and also that considerable lateral motion occurred on its southward continuation along the western side of the central trough, probably at least as far south as the site of strong deformation at ca. lat 31°16′ (shown as star in Fig. 5.4, Bartov and Sagy 2004).

The growth of the Dead Sea basin was accompanied by igneous activity on its flanks. Along the eastern margin 9–5 Ma old basalt flows and younger small vents and flows are present (Steinitz and Bartov 1991). West of its southern part a 6.4 Ma old dike and a nearby vent are present. In addition, a short-wavelength magnetic anomalies north of the Lisan diapir (Frieslander and Ben-Avraham 1989) most likely record small igneous bodies inside the basin.

5.5.1.5 The Southern Jordan Valley

This little studied segment (Fig. 5.4) extends to ca. 70 km north of the Dead Sea where it narrows to<4 km at the Bet Shean saddle. Its young fill is much thinner than in the Dead Sea according to gravity anomalies (Ten Brink et al. 1999). The southern part of the Jericho Fault, which extends along the valley floor, records transpression (Garfunkel 1981; Rotstein and Bartov 1989; Gardosh et al. 1990). Ca. 25 km north of the Dead Sea the Zahrat el-Qurein (Grain Sabt) half dome next to the Jericho Fault (Fig. 5.4) exposes ca. 350 m of tilted Miocene (?) coarse clastics (base not seen), unconformably overlain by a basalt flow and by young Quaternary sediments (Bender 1974). The basalt records magma ascent along the DST trace. The margins of the southern Jordan Valley are controlled by normal faults. In the east they trend close to N-S, but in the west their trends are variable, and some branch into the western flank of the valley (Fig. 5.4).

5.5.1.6 The Central and Northern Jordan Valley

This segment extends ca. 90 km north of the Bet Shean saddle to Lebanon. It comprises the Kinnarot-Lake Kinneret (Sea of Galilee) basin (Kinnarot-LK basin) the Korazim saddle, and the Hula basin (Fig. 5.5). It differs from the more southern segments by its more complex structure, by the occurrence of much igneous activity both along the DST and on its flanks, and by the considerable deformation of its western flank.

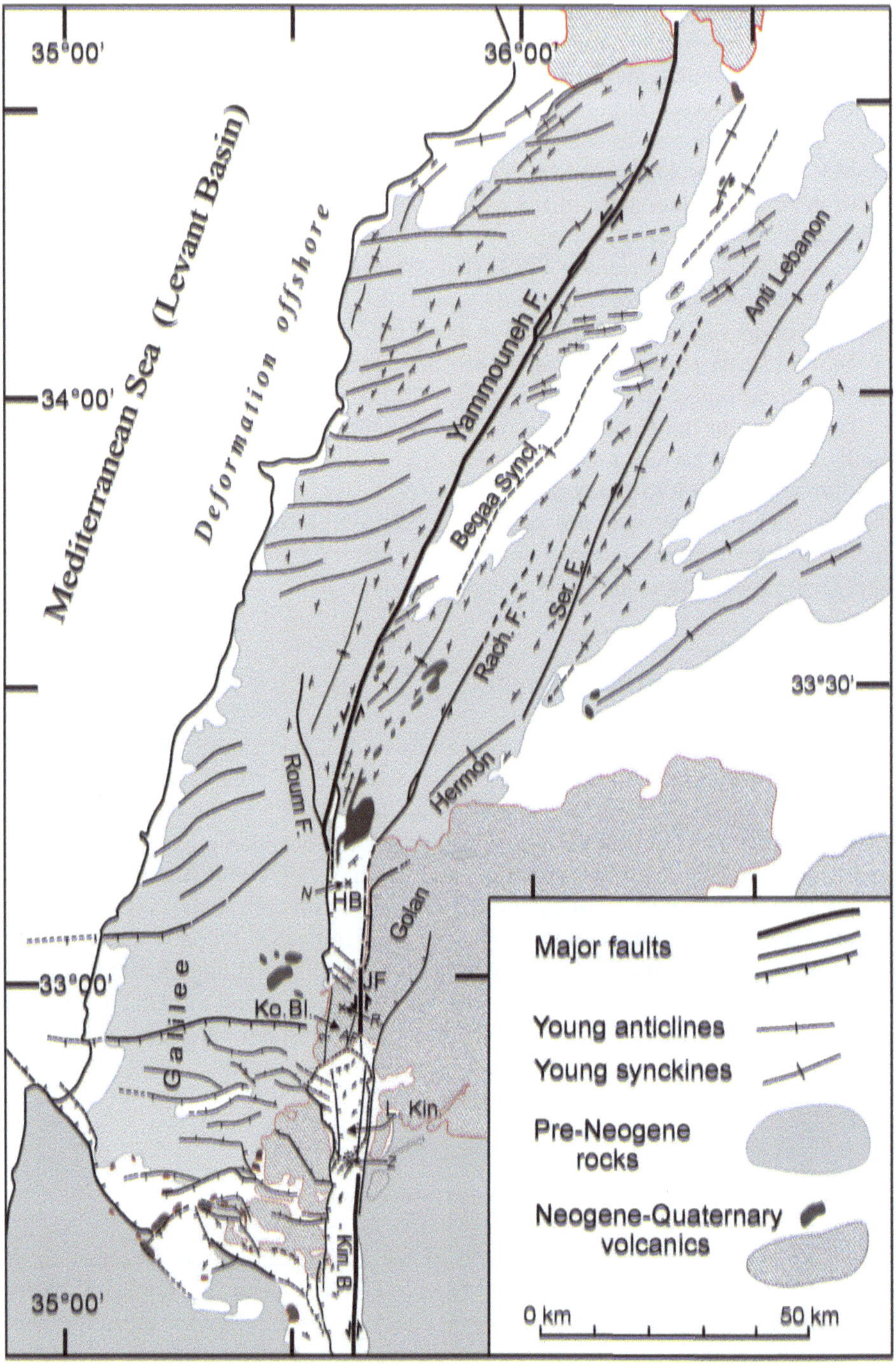

Fig. 5.5 The northern Jordan Valley and Lebanon segments of the DST. *AF* Almagor fault, *HB* Hula Basin, *JF* Jordan Gorge fault, *Ko. Bl.* Korazim block (saddle), *L. Kin.* Lake Kinneret, *Kin. B.* Kinnarot Basin, *R* Rosh Pina drillnole, *Rach. F.* Rachaya fault, *Ser. F.* Serghaya fault

Here the oldest exposed structure related to the DST is seen in a ca. 40 km long and up to 10 km wide low (also called the Tiberias sub-basin) on the western DST flank (Fig. 5.5). Its fill, up to 1 km thick, begins with 17–9 Ma old basalts (Lower Basalt) (see 5.4) interfingering with fluviatile-lacustrine sediments (Hordos Formation), together 400–750 m thick, overlain by continental and some marine sediments and volcanics, with ca. 5.2–4.0 Ma old flows (Cover Basalt) forming the top of the series (Schulman 1962; Shaliv 1991; Heimann et al. 1996).

Nearby the ca. 50 km long Kinnarot-LK basin formed along the main strand of the DST, with a ca. 5 km thick fill south of Lake Kinneret according to gravity data (Ben-Avraham et al. 1996). The 4.25 km deep Zemah 1 well (Fig. 5.5; Marcus and Slager 1986; Mittlefehldt and Slager 1986; Horowitz 1987) bottomed in Middle Miocene conglomerates (with basalt clasts), overlain by sediments that include Late Miocene and early Pliocene evaporites (980 m), mainly halite, which formed when an arm of the Mediterranean reached the DST from the west, and extended southward to the Dead Sea basin. The evaporites are intruded by gabbro sills (total thickness ca. 1.1 km), and are overlain by Pliocene volcanics broadly coeval with the Cover Basalt. The occurrence of igneous rocks, more voluminous than on the basin flanks, indicates preferential magma ascent along the DST. Seismic reflection data and mapping show that after extrusion of the Pliocene basalts the basin fill south of Lake Kinneret was deformed into broad folds with amplitudes reaching 1 km, and the fold crests were much eroded (locally down to the cover basalt), before being covered by a thin veneer of very young sediments. The deformation is interpreted as resulting from a young phase of transpression (Rotstein et al. 1992; Heimann and Braun 2000). In addition, very tightly folded Early Pleistocene beds were revealed in excavations at the Ubeidiya prehistoric site (Picard and Baida 1966), but the young cover obscures the extent of this type of folding.

In contrast, the more northern part of the Kinnarot-LK basin beneath Lake Kinneret subsided up to 2 km and was faulted in post-Cover Basalt times (Fig. 5.5, Ben-Avraham et al. 1996; Hurwitz et al. 2002; Reznikov et al. 2004). The main young structure is a graben, striking N-S, crossed by transverse faults, many of them ending up section. The part of the lake west of the graben is crossed by NW-SE striking faults that extend into the Galilee.

The Korazim saddle north of Lake Kinneret is separated from the basin under the lake by a major transverse fault zone (Fig. 5.5, Hurwitz et al. 2002). This saddle is underlain by a high standing block of pre-transform rocks (revealed by drilling) that is covered by 50–300 m of Neogene sediments overlain by 5.0–3.5 Ma old Cover Basalt, which in the north are covered by still younger flows that erupted through the plateau (Fleischer 1968; Horowitz 1973; Belitzky 1987; Heimann and Ron 1993; Weinstein 2012). Thus here too the DST served as a preferred path for magma ascent. The Korazim saddle is strongly faulted, with the fault blocks having rotated $11 \pm 4°$ ccw on vertical axes according to paleomagnetic data (Heimann and Ron 1993). On the east the Jordan Gorge Fault zone (Fig. 5.5, Garfunkel et al. 1981; Harash and Bar 1988) – the northward extension of the fault on the eastern side of the Kinnarot-LK basin – marks a prominent geologic discontinuity between the Korazim saddle and the Golan Heights (Michelson and Lipson-Benitah 1986).

Its straight trace and the lateral offset of historic and Holocene features (Ellenblum et al. 1998; Marco et al. 2005) show that this fault is the site of the ongoing lateral motion. Seismic reflection data show that it dips westward, indicating local transpression (Rotstein and Bartov 1989). The Almagor Fault west of the Jordan Gorge Fault (Fig. 5.5) is not continuous, so it could not accommodate much lateral motion, but is probably also active. The western boundary of the Korazim saddle is obscured by the Cover Basalt, but a fault with an irregular trace extends along this contact (Belitzky 1987). However, a structural discontinuity between the strongly deformed saddle and the less faulted western transform flank must extend along this boundary. The absence of a through-going fracture suggests there is no young lateral motion along this fault.

The Kinnarot-LK basin is difficult to interpret as a pull-apart, because it is not obviously related to left stepping strike slip faults. A critical point is that the pre-transform section under the Korazim saddle, penetrated by the Rosh Pinna 1 borehole, is offset ca. 50 km left laterally relative to the western DST flank (Freund et al. 1970; Hurwitz et al. 2002). Thus during a substantial part of the DST history lateral motion occurred along the western side of the Korzaim saddle, so originally the Kinnarot-LK basin could have developed as a pull-apart basin between left stepping faults: one west of the Korazim saddle and the other along the east side of the basin, the latter having extended southward to the Bet Shean saddle. At some stage this tectonic setting changed, as now lateral motion is identifiable only along the eastern side of the Korazim saddle. The young transpression south of Lake Kinneret probably formed during the tectonic change. The very young deformation under Lake Kinneret is probably linked to the coeval deformation of the Galilee (Hurwitz et al. 2002, see below).

The Hula basin, north of the Korazim saddle, formed between the left stepping Jordan Gorge fault and Yammouneh Fault, so it is interpreted as a pull-apart basin, (Fig. 5.5; Heimann and Ron 1993; Sneh and Weinberger 2003; Weinberger et al. 2009, 2011). The Notera 3 well (T.D. 2,781 m, Fig. 5.5) in the basin's center crossed a fill of sediments and basalt flows. At ca. 2,340 m a ca. 4.3 Ma old flow overlies an 8.8 ± 0.2 Ma old flow that, in turn, overlies red beds and interbedded basalts (base not reached) (Horowitz and Horowitz 1985; Heimann and Steinitz 1989). This shows that the present Hula basin subsided only since ca. 4.3 Ma ago, after a >4 Ma long period of no subsidence. Such a young age can be expected in view of its small length (ca. 15 km). The pre-ca. 9 Ma series is ca.1 km thick according to gravity data (Rybakov et al. 2003). Fault slivers near the NW corner of the Hula basin expose a strongly folded Late Miocene or older series, >400 m thick, of lacustrine beds and conglomerates built only of Eocene clasts (Kefar Giladi Formation; Glickson 1966; Sneh and Weinberger 2003; Weinberger et al. 2009). The structural setting in which these series formed was obliterated, but it probably differed from the present setting. The younger fill of the Hula basin fill is mostly flat, but in Pleistocene times a diagonal NE-SW striking fault associated with folding formed within the basin, while along its continuation NW of the basin several small uplifted structures formed (Schattner and Weinberger 2008; Heimann et al. 2009; Weinberger et al. 2009, 2011). These authors interpreted these structures as a result of young eastward shifting of the Euler pole, which caused southward expansion of the transpressive segment of the DST.

The area west of the central and northern Jordan Valley, including the Galilee, the Yizreel Valley (Esdarelon Valley), and the Tiberias sub-basin, is crossed by E-W, SSE-NNW, and NW-SE trending normal and oblique slip faults that formed while the DST was active. Here only a few major points can be mentioned. The trends of Middle Miocene faults in the Tiberias sub-basin and the Yizreel Valley form large angles with the DST (Shaliv 1991), but they do not extend east of it, which shows that at that time the DST was already a structural boundary. Paleomagnetic data reveal that the Miocene and Pliocene faulting west of the DST was accompanied by large rotations of the fault blocks about vertical axes, which led to overall E-W shortening and N-S extension of the Galilee (Ron et al. 1984). Since mid-Pliocene times ca. E-W trending normal faults dominated the structure of the southern half of the Galilee which led to N-S extension (Matmon et al. 2003). GPS data were interpreted by Sadeh et al. (2012) as showing that now the northern Galilee moves northward at a rate of 1 mm/year relative to the area south of the Yizreel Valley. Though close to the limit of resolution, this agrees with the structural evidence. The N-S extension (≥5 km) of the western DST flank decreases the lateral transform offset farther north. Some faults extend from the Galilee into Lake Kinneret (Fig. 5.5, Hurwitz et al. 2002). The resulting N-S extension due to these faults may have led to the young deformation and subsidence of Lake Kinneret, which is quite distinct from the transpression north and south of it.

The strong faulting of the western flank of this segment of the DST is considered to express its deformation when it moved laterally along the pronounced bend of the DST trace at the transition between its northern and southern parts. In contrast, only mild faulting affected the Golan Heights east of this segment of the DST (Fig. 5.5, Shulman et al. 2004; Meiler et al. 2011).

5.5.1.7 The Deformation Along the Southern Part of DST

The foregoing summary shows how the shallow structure along the DST was related to the plate kinemtics. It also reveals that most of the time and along most of the southern part of the DST the lateral motion and related deformation took place in an up to a few tens of kilometers wide zone. Only at the central Arava saddle and the Bet Shean saddle was the overall motion close to pure strike slip in narrow zones. Elsewhere the lateral motion took place in wide deformed zone, mainly on left stepping strike slip faults that deviate slightly clockwise from the overall transform. This produced varying amounts of transtension that led to the growth of a string of conspicuous pull-aparts (rhomb grabens) combined with normal faulting along the margins of the transform valley (Fig. 5.2c, Garfunkel 1981). Plate kinematic considerations and the structural history show that the transtension increased with time (see 5.4). This was achieved by structural rearrangements, well seen in the south, while in the Arava Valley and farther north by growth of pull apart basins combined with an increase in the role of the marginal normal faults.

The foregoing discussion focused on the shallow structures, but the features of these structures have also significant implications for processes at deeper level, especially when integrated with geophysical data. This topic is largely outside the scope of the present paper, but since such an analysis is rarely followed, a few points are presented below.

The deformation seen on the surface is expected to extend at least through the entire crust. This is supported by seismic data from the central Arava Valley which show that there the DST extends through the entire crust (Weber et al. 2004, 2009) and also by the difference between coeval volcanics across the DST near the Korazim plateau, which raises the possibility the DST motion affects the magma sources well below the Moho (Weinstein 2012). Here only a few implications of the shallow structures for the deeper deformation and some questions that they raise can be briefly discussed.

The most conspicuous deformation occurs during the growth of the pull-aparts. Most revealing are the data about the Dead Sea basin. As explained above, the southern part of the Dead Sea basin (in the Arava Valley) is floored by the continuation of its western flank, while in its northern part the eastern DST flank extends under the entire basin width (Fig. 5.4). Since the lateral offset along the DST moved these areas apart by ca. 105 km along the basin axis, the floor of the intervening part of the basin must have been stretched and thinned. This is confirmed by seismic and gravity data (ten Brink et al. 2006; Mechie et al. 2009) which show that the Moho under both the basin's center and it's flanks is at about the same depth – ca. 33 km. As the basin fill is up to 12–14 km thick, the crystalline crust under the middle part of the basin is only 22–20 km thick, compared with 30–32 km under the basin flanks. Similar arguments apply also to the deep Gulf of Elat and the Kinnarot-LK basins, where the Moho is not deeper than under their flanks according to gravity data, but there the crustal thinning is probably smaller than under the Dead Sea basin. Stretching of the floors of pull-aparts is expected also in view of the increase in area during their formation (Fig. 5.2b), and it also explains how the lateral slip along the pull-aparts is transferred from their SE to their NW extremities (Garfunkel and Ben-Avraham 1996, their Fig. 6).

Stretching of the crust under the basins should deform the basins' fills. The distribution of earthquake foci under the Dead Sea basin and their source mechanisms (Aldersons et al. 2003; Hofstetter et al. 2007, 2008) show that complex brittle deformation occurs beneath the entire basin width in both the upper and the lower crust, though the deformation pattern is difficult to infer. However, the structures known in the shallow parts of the basin fills (e.g. Gulf of Elat and Dead Sea) – transverse faults and downward increasing dips – probably can not account for the entire basins' growth, so this issue still requires further study.

If the crust beneath the major basins was modified only by stretching, then the basins would be strongly isostatically under-compensated because of the presence of a thick low-density of sediment fill. Exact local compensation of such narrow structures is not expected, but the continuing great subsidence of strongly under-compensated basins and the support of large deviations from isostatic equilibrium are difficult to explain (cf. Ten Brink et al. 2006). However, the occurrence of igneous rocks along the DST (e.g. in the Zemah-1 well, the Zahret el-Qurein dome north of the Dead Sea, and the sources of the magnetic anomalies in its middle) and on its flanks raises the possibility that basalt intrusions in the stretched middle or even shallow crystalline crust contribute to decreasing the mass deficit. Noteworthily, seismic reflection studies (ten Brink et al. 2006; Mechie et al. 2009) show that beneath the middle of the Dead Sea basin material with upper crustal velocities forms a much smaller fraction of the crystalline crust than under the basin flanks. This is difficult to explain by mechanical stretching alone, but can arise if basic intrusions are emplaced into the lower velocity/lower density material of the upper half of the original crust. Underplating by igneous rocks was also inferred in the Imperial Valley, California, and the Baikal rift (Fuis et al. 1984; Nielsen and Thybo 2009).

Thinning of the crust beneath the basins would also influence the thermal regime. The thinned crust under the basins supplies less heat than the intact crust of their flanks, which should be taken into account while interpreting the data. As this is a shallow effect, it will be expressed in the heat flow after a period shorter than the DST history (in contrast to thermal effects originating well below the Moho). The heat flow of 38 mW/m^2 under the northern Dead Sea is similar or lower than that on its western flank (Ben-Avraham et al. 1978; Eckstein 1979; Schütz et al. 2012). In the northern Gulf of Elat the heat flow is 66 ± 30 mW/m^2, similar to 60.3 ± 3.4 mW/m^2 found east of 'Aqaba, and in Lake Kinneret it is 74 ± 4 mW/m^2, higher than in the Galilee (Ben-Avraham et al. 1978; Eckstein 1979; Förster et al. 2007). In these places the thinned crust contributes less to the heat flow at the surface than the intact crust under the transform shoulders. Thus, to account for the observed values, the heat flow from the mantle under the basins may well be higher than under their shoulders. On the other hand the occurrence of earthquakes in the lower crust under the basins (Aldersons et al. 2003) shows that the temperature there is low enough to allow earthquake generation (probably no more than 350–400 °C in the lower crust). More data and modeling of the thermal regime are required to integrate these considerations.

Pull-apart basin formation was also studied by finite element thermo-mechanical models that (Sobolev et al. 2005; Petrunin and Sobolev 2006) and by analogue modeling (e.g. Smit et al. 2010, 2011; Wu et al. 2009 and references therein). Though the models require further refinement to include the effects of lateral structural variations, changes of basin area, igneous activity, and changes of plate motions, they provide very important quantitative insights. However, as they apply to the entire crust and/or lithosphere, their discussion is outside the scope of this paper. Such models should eventually also yield insights into the structural changes that took place along the DST.

In summary, the foregoing considerations reveal how the shallow structures along the DST and their history are related to the motions of the flanking plates, and also show that combining the data regarding the shallow levels with geophysical data from deeper levels can provide insights that are otherwise very difficult to obtain.

5.5.2 *Northern Part of Dead Sea Transform*

The northern part of the DST differs from its southern part by having a sinuous trace and by the significant deformation of its flanks (Fig. 5.1) which is interpreted as expressing transpression. This part of the DST was less studied than the southern part, and its nature was much debated. Here an updated summary of the main available information and unresolved problems is attempted.

5.5.2.1 The Lebanon Restraining Segment

The conspicuous ca. 150 km long Yammouneh Fault extends along the east side of the Lebanon range, and is marked by a ca. 2 km wide damage zone in which much mesoscale fracturing, small scale folding, abundant horizontal slickensides, and small pull-aparts are developed (Heybroek 1942; Hancock and Atiya 1979; Gomez et al. 2006), pointing at the importance of lateral motion. It is believed to take up practically the entire lateral offset of the DST, because the difference between the stratigraphic sections across its southern extremity reveals a lateral offset similar to the offset farther south (Quennell 1959; Freund 1965; Freund et al. 1970; Walley 1983, 1988; Sneh and Weinberger 2003). GPS data (Reilinger et al. 2006; Gomez et al. 2007; Le Beon et al. 2008; Alchalbi et al. 2010) also show that it is the main site of lateral motion in Lebanon. Thus it is not possible to accept interpretations tending to minimize the role of lateral motion on the Yammouneh Fault (e.g. Dubertret 1970).

As the strike of the Yammouneh Fault – 30°–35°NE – deviates considerably clockwise from the ca. N-S strikes of the main DST faults to the south and north, it forms a prominent restraining bend (Quennell 1959; Freund et al. 1970). The resulting transpressive deformation extends to distances of 30–50 km on the two sides of the fault, and is combined with significant vertical motions. This produced the internally deformed, structurally and topographically high, Lebanon and Hemon-Anti Lebanon anticlinal ranges (up to ca. 3 km high) and the synclinal Bekaa Valley syncline between them (Fig. 5.5).

The Bekaa Valley syncline (Fig. 5.5) is a structural low whose axis forms a small angle with the Yammouneh fault. Many smaller scale NE-SW to NNE-SSW trending tight folds (some with overturned flanks), i.e. deviating clockwise from Yammouneh fault, are developed along its flanks (Dubertret 1955; Renouard 1955; Beydoun 1977). This structural pattern points at left-lateral shearing of the entire Bekaa syncline combined with some transverse shortening. The southern part of the syncline is crossed by a longitudinal fault (Hasbaya Fault) whose nature (strike-slip?) is not well constrained; it is sealed by 3.5–2.2 Ma old basalts. Small occurrences of basaltic volcanics occur also farther north along the Bekaa Valley syncline (Dubertret 1955), recording magma ascent close to the DST trace.

The Hermon-Anti Lebanon anticlinal ridge, ca. 30 km wide, comprises several secondary anticlines and is crossed by the Rachaya Fault and Serghaya Fault (Fig. 5.5). The Rachaya Fault branches off the eastern border fault of the Hula basin and extends ca. 40 km through the western flank of the Hermon-Anti Lebanon anticlinal range. The presence of a small pull-apart along its southern part (Heimann et al. 1990) indicates left lateral slip, probably only a few km in view of the limited length of this fault. The Serghaya Fault, ca. 100 km long, extends northward from the SE side of the range across its anticlinal crest, but it becomes indistinct before reaching the Bekaa Valley syncline (Fig. 5.5). Gomez et al. (2001, 2006) identified up to 6 km of left lateral offsets of streams, which indicate continuing lateral motion. Walley (1998) suggested a total offset of ca. 25 km, which appears excessive in view of the limited northward extent of the fault, but this needs further study. In the south this fault extends to a strongly deformed area on the SE side of the Hermon-Anti Lebanon range where Kopp and Leonov (2000) reported thrusting of Jurassic beds over Neogene-Quaternary beds, though they gave no details. The relations of these structures with the Hula basin are hidden by the volcanics of the Golan Heights.

West of the Yammouneh Fault the 30–35 km wide Lebanon range comprises several folds and flexures, mostly sub-parallel to the range, and it is also considerably faulted (Fig. 5.5). The southern part of Lebanon is crossed obliquely by the ca. 35 km long Roum Fault which splays from the Yammouneh Fault and is still active (Nemer and Meghraoui 2006). It displaces left laterally river valleys, but the offset decreases northward (Garfunkel 1981) until there is no visible geologic discontinuity in the pre-DST rocks along its northward prolongation, though much fracturing is recorded (Dubertret 1955; Khair 2001; Griffiths et al. 2000). Thus it appears that the Roum Fault does not extend northward to the Mediterranean coast, so it could not have taken up much of the DST lateral motion (e.g. suggested by Butler et al. 1998). Rather, in view of the very different cross sections of the folds on its two sides it is interpreted as marking a discontinuity in the internal deformation of Lebanon. The area west and SW of the Roum Fault is the northern continuation of the Galilee. It is crossed by NE-SW faults that probably have a right lateral slip component, like the faults with this trend in the adjacent part of the Galilee.

The northern 2/3 of Lebanon forms a major anticlinal arch that is cut by a system of NNE-SSW trending oblique-right lateral faults, offsetting the axis of the arch and the monocline along its western side (Fig. 5.5). As each segment of the arch has a different cross section, it appears that the arch and the faults developed coevally. Some of these faults extend to the coast where they offset Neogene sediments, which shows that they acted coevally with the DST. Because of the lateral DST motion this area must have moved along the bend at the northern end of the Yammouneh fault, so its overall shape must have changed as the motion progressed and it must have been internally deformed, but the mechanism needs further study.

Paleomagnetic data show that the rocks building the Lebanon and Hermon-Anti Lebanon ranges rotated on vertical axes. In northern Lebanon Gregor et al. (1974) found rotations of ca. 60° and ca. 30° ccw (relative to the paleomagnetic poles of Africa) of Jurassic and Early Cretaceous rocks, respectively. Ron (1987) treated these data as a single population and inferred an average rotation of 53° ± 10° ccw, while in southern Mt. Hermon he found that Early Cretaceous rocks rotated 69° ± 13° ccw, and Baer et al. (1998) found rotations of ca. 60° ccw. In contrast, Henry et al. (2010) found an average ccw rotation of 28° ± 6.4° in all exposures of

Aptian and Albian rocks in the Lebanon and Hermon-Anti Lebanon ranges, and in many sites they also identified a Neogene re-magnetization indicating rotations of 11° ccw. Ron et al. (1984) found a ccw rotation of 22° ± 9° in the Galilee next to the border with Lebanon, which very likely applies also to the northward continuation of this domain into Lebanon (SW of the Roum fault). The age of these rotations is not constrained directly, but in analogy with the Galilee (Sect. 5.5.1.) they probably occurred mainly during the activity of the DST. However, Henry et al. (2010), following Gregor et al. (1974), advocated also a pre-Aptian rotation of ca. 30° ccw in northern Lebanon. Clearly, more paleomagnetic data and structural studies are needed.

5.5.2.2 The Syrian Segment

North of Lebanon the DST is also transpressional, but since it trends close to N-S the transverse shortening is much smaller (ca. 1/3) than in Lebanon segment. The two segments are separated by a structural low covered by the ca. 6–4 Ma old Homs (or Shin) basalts (Fig. 5.6, Mouty et al. 1992; Sharkov 2000). West of the DST the volcanics cover an erosion surface that cuts gradually down from Neogene sediments near the coast to Late Jurassic beds next to the DST fault, but ca. 2 km farther east the volcanics cover Middle Cretaceous beds, stratigraphically several hundred meters higher (Ponikarov 1964; Mouty et al. 1992; Chorowicz et al. 2005; Gomez et al. 2006). This shows that here an eroded structural step existed along the trace of the DST before basalt extrusion, indicating that here the DST was active already before the end of the Miocene. This, and the evidence regarding the lateral motion provided by the Miocene sections farther north (see 5.4) show that the Syrian segment of the DST was active considerably earlier than envisaged by Rukieh et al. (2005).

At the transition from Lebanon to the Syrian segment the principal DST fault changes strike quite abruptly, and a ca. 12 km long depression (pull-apart?) is developed (Fig. 5.6). Farther north the principal fault zone is 2–3 km wide and comprises several fractures. Ongoing activity is proven by the offset of a Roman aqueduct and of morphologic features, and also by minor structures (Trifonov et al. 1991; Meghraoui et al. 2003; Chorowicz et al. 2005; Rukieh et al. 2005), confirming that this is where the lateral DST motion now takes place. The shape of the Homs volcanic field strongly suggests that it was offset left laterally, estimated as >20 km by Chorowicz et al. (2005) but considerably less by Rukieh et al. (2005). An exact figure cannot be given because the edges of the volcanic field are much eroded.

Farther north the fault zone bifurcates into two major branches that enclose the ca. 50 km long and ca. 15 km wide Ghab basin pull-apart which has an up to 3.5 km thick young fill (Fig. 5.6, Dubertret 1955; Ponikarov 1964; Kopp and Leonov 2000; Brew et al. 2001a, b; Rukieh et al. 2005). Data from shallow wells show that the basin formed already in the early Pliocene, but the deeper part of its fill, still undated, may well be older. Structural interpretations vary. Brew et al. (2001a, b) inferred that the eastern boundary fault was sub-vertical and the western boundary

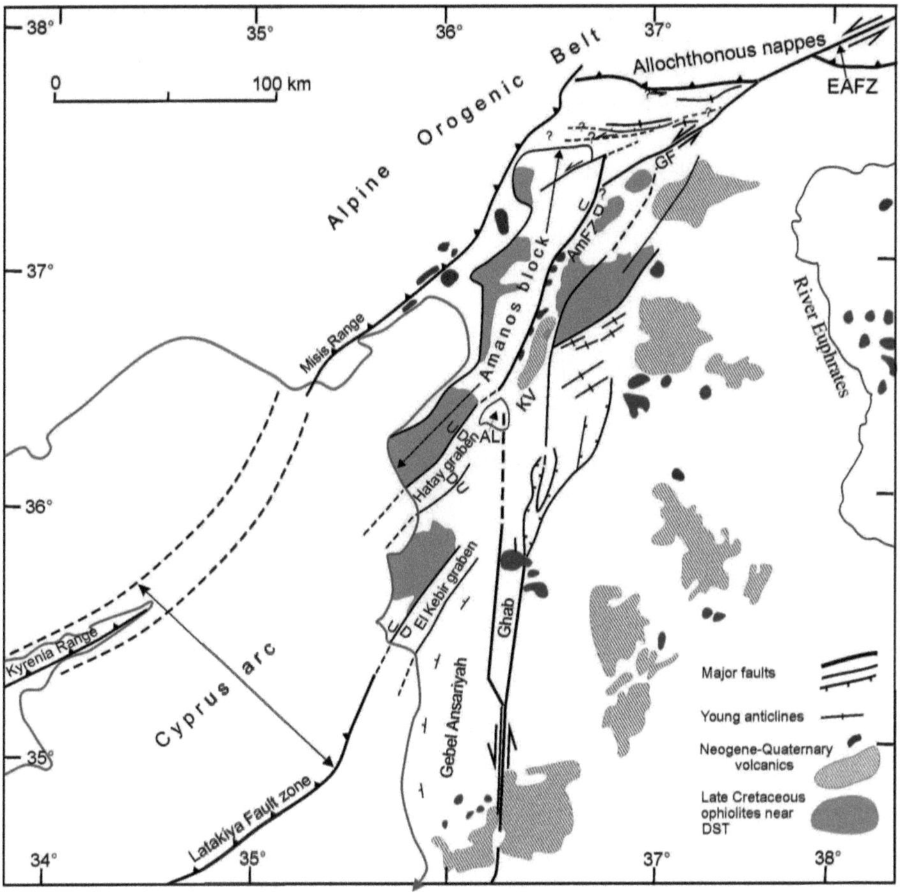

Fig. 5.6 Northern segments of the DST. *AL* Amik Lake, *AmFZ* Amanos fault zone, *EAFZ* East Anatolian fault zone, *GF* Gölbaşi fault, *KV* Karasu Valley

fault was normal, while Kopp and Leonov (2000) reported steep dips and mesoscale structures indicating reverse faulting along both boundary faults, and Zanchi et al. (2002) reported older thrusts and younger E-W extension. Clearly, further structural study is needed.

North of the Ghab basin the principal fault zone comprises a ca. 15 km wide belt of narrow N-S trending horsts and grabens formed by post-Miocene faulting (Fig. 5.6, Ponikarov 1964). Most of the faults in this zone are inactive, as at the northern end of the Ghab basin they are covered by 1–2 Ma old basalts that are not visibly faulted (Ponikarov 1964; Ponikarov et al. 1969; Sharkov 2000), while in the north they do not displace the fill of the Amik basin. East of this belt the possibly active Afrine Fault branches off the Ghab basin, (Fig. 5.6), but its role is insufficiently known. Only the westernmost fault, the prolongation of the western border fault of the Ghab basin, is active and extends northward to the Amik basin (Fig. 5.6, Senel 2002; Karabacak et al. 2010; see 5.5.2.3).

The transform eastern DST flank is little deformed, but its western flank – the uplifted Jebel Ansarieh (An-Nusseriyeh) range (coastal range of Syria) – is much faulted (Ponikarov 1964; Ponikarov et al. 1969). Kopp and Leonov (2000) concluded that the dominant NE-SW striking faults had a right-lateral slip component. They suggested that here blocks rotated ccw on vertical axes, like in N Lebanon, but this has not been investigated. In the north the Nahr al-Kabir (Latakie) graben (see Sect. 5.3) separates the Ansariyeh range from the Bassit block on which Late Cretaceous ophiolites are present (Fig. 5.1, Ponikarov 1964; Rukieh et al. 2005). The motion between these blocks is not known, but there is no obvious evidence that it was significant.

5.5.2.3 The Northern End of the DST

The northernmost, Turkish, segment of the DST is dominated by the Karasu graben, with the Amik basin in its southern part, which are flanked by the high standing Amanos Block in the west and the NE corner of the Arabian plate on the east (Fig. 5.6). These form the direct northward prolongation of the DST structure farther south. Left lateral motion continues northward from the Ghab basin into the Amik basin (Senel 2002; Karabacak et al. 2010), and farther north along the prominent NNE-SSW trending Amanus Fault Zone along the west side of the Karasu graben. Offset lava flows along this fault zone record left-lateral slip at a rate of ca. 2.9 mm/year in the last 1 Ma (Rojay et al. 2001; Seyrek et al. 2007, superseding previous studies). Lateral slip may also occur along the eastern side of the Karasu graben (Perinçek and Çemen 1990; Westaway and Arger 1996), but was not documented in detail. Tatar et al. (2004) inferred Quaternary shearing of the entire Karasu graben based on paleomagneric data, but this is very doubtful – the spread of the data is very large and supporting structural data are lacking.

The Amik basin (Fig. 5.6) is filled with 3–5 km of Neogene-Quaternary beds that overlie Late Cretaceous ophiolite nappes, but farther north the Karasu graben floor exposes the ophiolites and other pre-transform rocks (Coskun 1994; Perinçek and Çemen 1990; Rojay et al. 2001; Senel 2002). This requires a major structural step on the northern side of the Amik basin, but its history was not documented. Young volcanics, often of Quaternary age, extend along the Karasu graben, again demonstrating magma ascent along the DST. In the west the SW-NE trending Hatay graben (Asi graben) branches off the Amik basin (Fig. 5.6). It subsided and was crossed by NE-SW normal faults already in middle and late Miocene times, but the present graben structure was shaped in Pliocene-Quaternary times (Boulton et al. 2006).

The ca. NNE-SSW striking faults delimiting the Karasu graben, which appear to be the direct northward continuation of the DST structures, end abruptly in the north against an E-W trending anticline (near the town Kahraman Maras) (Fig. 5.6, Karig and Kozlu 1990; Yilmaz 1993; Senel 2002). Early plate tectonic models (Freund 1965; Wilson 1965; McKenzie et al. 1970) interpreted this area as a triple junction, called the Maras triple junction, between the Arabia and Sinai plates and the Alpine zone of plate convergence. However, the site of plate convergence – the front of the Alpine allochthonous nappes – is situated 20–25 km north of Maras (Fig. 5.6), and the intervening area exposes Cenozoic sediments typical of the margin of the Arabian platform, which are deformed by N-S compression. Thus the situation is more complex than suggested in these models.

During the early stages of the DST activity, in the Miocene, continuing plate convergence along the Bitlis suture east of it led to emplacement of allochthonous nappes on the northern edge of the Arabian platform, while west of the DST plate convergence took place north of the Eastern Mediterranean basin (Şengör and Yilmaz 1981; Yilmaz 1993; Robertson 2001a, b; Robertson et al. 2004). In (late?) Early Miocene times, close to the time of the DST initiation, the Cyprus arc became the southern part of the orogenic belt and was thrust over the Levant basin (northern part of Sinai plate) along the Latakiya Fault zone (Fig. 5.6, Eaton and Robertson 1993; Robertson 2001b; Vidal et al. 2000; Hall et al. 2005a, b). Thus when the lateral DST motion began in the early Miocene, a triple junction as envisaged in the above models, probably existed along the belt of plate convergence.

However in the Pliocene, 3–4 Ma ago, the geometry of the plate boundaries near the northern end of the DST changed as westward extrusion of Anatolia became important and the left lateral East Anatolian Fault Zone, not yet recognized in the early works, Formed, but a slow change may have begun earlier. It cut obliquely the Alpine nappe front (i.e. the older) and continued into the NE corner of the Arabian platform, becoming the new Arabia-Anatolia plate boundary east of the DST (Fig. 5.6, Şengör et al. 1985; Bozkurt 2001; Hubert-Ferrari et al. 2009). This event was of great significance for the evolution of the entire East Mediterranean region (Le Pichon and Kreemer 2010), but is outside the scope of the present work, so only effects near the DST will be considered here. West of the DST plate convergence continued along the Cyprus arc, so the question arises as to how the new plate boundary east of it continues westward.

The lateral offset along the East Anatolian Fault Zone is estimated as 20–30 km (Şaroĝlu et al. 1992; Westaway and Arger 1996). Offset alluvial fans some distance east of the DST, and GPS data record a young slip rate of 10–11 mm/year (Cetin et al. 2003; Reilinger et al. 2006; Reilinger and McClusky 2011; Le Pichon and Kreemer 2010), compatible with its young age. In the Early Pleistocene and earlier times the Kyrenia-Misis line and perhaps other structures took up some of the convergence (Fig. 5.6, Karig and Kozlu 1990; Hall et al. 2005a, b; Robertson 2001a, b; Robertson et al. 2004), but now the Anatolia-Levant basin convergence is concentrated on the Latakiya Fault Zone according to seismicity and GPS data (Papazachos and Papaioannou 1999; Reilinger et al. 2006). When the East Anatolian Fault Zone formed it must have linked with these structures and a component of lateral slip was most likely added to the motion along them, but how this happened is still not well understood.

Yilmaz (1993 his Fig. 2) and Senel (2002) show that ca. 100 km east of the Karasu graben the East Anatolian Fault Zone splits into several branches that extended north of its presently active trace – one that extends into the northern part of the Amanos Block and offsets its eastern boundary, and another one still farther north (Fig. 5.6). A fault may also extend close to the front of the Alpine nappes, where left lateral motion was reported, though its importance is debated (Karig and Kozlu 1990; Robertson et al. 2004). When these faults were active the motion along the western part of the East Anatolian Fault Zone was partitioned between them, so the link with the structures west of the Amanos Block was apparently through a wide zone, in part extending north of the Amanos Block (Fig. 5.1), but the details still require clarification. As these faults have no clear morphologic expression they are probably inactive at present.

Now the Gölbaşi Fault (GF, Fig. 5.6), marked by a prominent physiographic feature, appears to be the active western strand of the East Anatolian Fault Zone (Perinçek and Çemen 1990; Şaroĝlu et al. 1992; Westaway and Arger 1996). It reaches the Amanus Fault Zone, but an active fault that forms its westward continuation into the Amanos Block was not clearly documented, and the two fault-controlled morphological margins of this block are not visibly offset. Thus it is not easy to accept that the Gölbaşi Fault continues westward across the Amanus Block (as suggested by Westaway 2003; Westaway et al. 2006), but this requires further study. Possibly, the motion along the Gölbaşi Fault is linked to the slip along Latakiye Fault Zone via the Amanus FZ and the Hatay graben (Perinçek and Çemen 1990; Şaroĝlu et al. 1992; Mahmoud et al. 2013).

The stress regime in the area provides additional insights and tends to support this picture. Field studies by Over et al. (2004) reveal that in Recent-Quaternary times a regime of left lateral transtension prevailed along the Hatay graben and the Amanus Fault Zone, compatible with GPS data (Reilinger and McClusky 2011; Mahmoud et al. 2013) and with the normal slip component on the Amanus Fault Zone inferred from field data by Perinçek and Çemen (1990), Rojay et al. (2001), Seyrek et al. (2007) (but not with thrusting on this fault, which Adiyaman and Chorowicz (2002) inferred from remote sensing data). However, if the Amanus Block were rigidly attached to the more southern part of the western DST flank, i.e. to the Sinai plate, then the plate kinematics require transpression in this area (Garfunkel 1981). Thus now this block appears to be decoupled from the Sinai plate and seems to move approximately like the Cyprus arc, which is supported by GPS data (Reilinger and McClusky 2011; Mahmoud et al. 2013). The decoupling probably drives the young activity of the Hatay graben, and possibly also the Nahr el-Kebir graben, though both are relatively old structures. However, this situation is probably young and has not yet produced significant motion relative to the western DST flank, as there is no identifiable geologic discontinuity along these structures. Indeed, the field data (Over et al. 2004) reveal a regime of transpression in Pliocene-Miocene times, indicating considerable coupling of the Amanos block with the western DST flank south of it. However, if the coupling was rigid most of this time, then 20 km or more of transverse shortening across the northernmost DST is expected. Seyrek et al. (2007) concluded that this was the case, but since their plate kinematic model conflicts with the GPS data (see above) and they did not present field data to constrain the amount of E-W shortening, this interpretation requires further study. Thus it seems that the Amanos block tended to behave independently for some time, and recently tends to become attached to the Cyprus arc. So a new Hatay triple junction seems to develop, though it utilizes older structures.

If a significant part of the lateral motion of the East Anatolian FZ was transferred westwards north of the Amanos Block, this would truncate the northern segment of the DST and juxtapose it with a sliver of the Arabian platform that was originally located farther east. Before that the DST was closer to the area (ca. long 36.75°E) where the E-W trending nappes end more western structures farther west strike ca. NE-SW (Fig. 5.6, Yilmaz 1993; Karig and Kozlu 1990; Senel 2002; Robertson et al. 2004). This raises the possibility the shape of the nappe front was influenced to some extent by the DST and not only by the curved shape of the northern boundary of Arabia.

In summary, because of the young changes in the tectonic pattern, the young activity along the Turkish segment of the DST cannot be described in terms of the Sinai-Arabia plate boundary and a simple "Maras triple junction", but incomplete understanding of the new tectonic pattern in this area does not allow reliably restore the older situation. However, since this segment is the direct continuation of the DST structures farther south, and since it is the only major structure that could have accommodated the lateral offset across the north of the Syrian segment, it is believed that it should be considered as part of the DST during the early stages of its development. Additional data are required to resolve these issues.

5.5.2.4 Deformation Along the Northern Part of the DST

The foregoing account shows that deformation along the Lebanon and Syrian segments of the DST is complex, but is considered as a result of transpression because the DST trace deviates clock-wise from the local direction of the relative left lateral plate motion. However, this does not apply to the young history of the Turkish segment because of the young changes in this area (as explained in 5.5.2.3).

Given the orientation of the Yammuneh fault, plate kinematics requires that its flanks should be shortened by up to 40–50 km (Freund et al. 1970; Garfunkel 1981). The actual figure is probably less, because deformation of the adjacent Palmyrides, which began in the Cretaceous but continued during the time of DST activity (Chaimov et al. 1990; Searle 1994; Walley 1998; Kopp and Leonov 2000; Brew et al. 2001a; Rukieh et al. 2005) would bend the DST trace. However, even if a large part of the total ca. 20 km NW-SE shortening of the Palmyrides contributed to bending of this fault, this would amount to <10°, so the remaining transverse shortening of the fault flanks is still expected to be ca. 30 km or more.

The folding of the Yammouneh Fault flanks (Fig. 5.5) indeed records transverse shortening, but the fold geometry shows that it is doubtful that this could have exceeded 5 km. Thus the rest must be explained in another way. The most likely mechanism is rotation on vertical axes of small blocks between sub-parallel strike slip faults, as is revealed by paleomagnetic data. The large ccw rotations (Ron 1987; Henry et al. 2010) provide a very efficient mechanism of large-scale regional deformation (Ron et al. 1984; Garfunkel and Ron 1985). Such a deformation practically preserves surface area, so in contrast with folding and thrusting it does not lead to crustal thickening. Rotations of 25°–30° ccw, as Henry et al. (2010) found in Lebanon, would reduce the E-W dimension of the deformed zone while increasing its N-S length by tens of percent (Freund et al. 1970; Garfunkel and Ron 1985). So

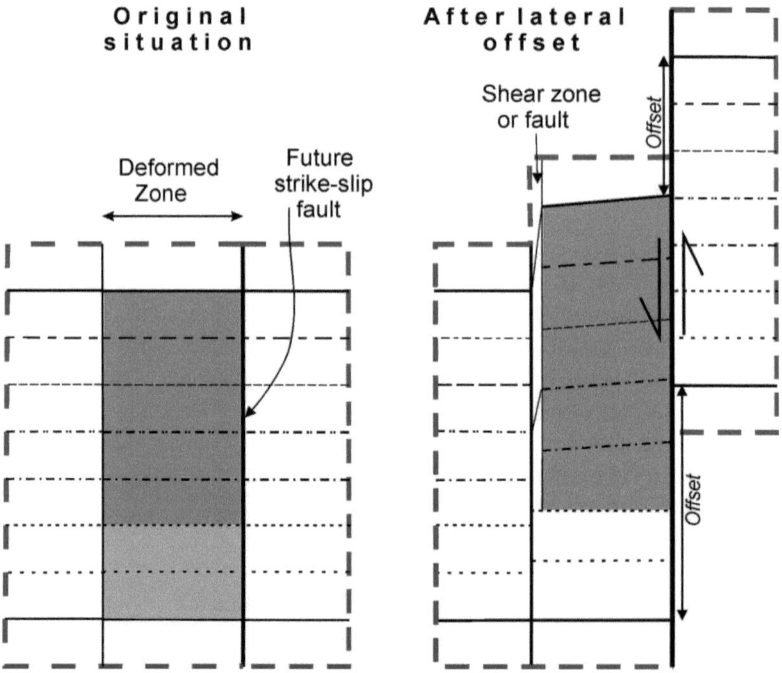

Fig. 5.7 Sketch to show how the amount of lateral offset of marker is affected by deformation of the margins of a strike-slip fault. The *light domain* is affected by fault-parallel extension, while the *darker gray domain* is also sheared. Along the main fault the offset varies, and a fault/shear zone develops along the margin of the deformed domain

qualitatively these results can explain the shortening perpendicular to the DST in Lebanon. The rotation assumed by Ron (1987), based on fewer data and combining measurement from different area, is more difficult to apply, so this matter needs clarification. Along the Syrian segment transpression was much smaller: some folding occurred NE of the Ghab depression, and considerable faulting affected its western flank, and here too rotation of blocks about vertical axes was suggested, but there are no data directly supporting this assumption.

If the left lateral shearing by block rotations indeed results in N-S lengthening of its western flank, then left-lateral shearing must arise along its junction with the adjacent undeformed Levant basin north of the Galilee, as is seen on Fig. 5.7. In this case the lateral offset on the main fault is reduced (see 5.5.1.6) while the total left lateral offset parallel to the DST appears to be preserved (Fig. 5.7). Deformation of the continental margin was indeed observed and was interpreted as related to the deformation onland (Schattner et al. 2006; Carton et al. 2009; Elias et al. 2007), but its magnitude needs further testing.

The considerable deformation of the flanks of the Lebanon and Syrian segments, affecting an up to 65–75 km wide zone at shallow levels, is expected to affect also deeper crustal levels. The occurrence of volcanics close to and along the main trace of the DST from the central Jordan Valley and northward, which shows that the DST was a preferred pathway for magma ascent, raises the possibility that igneous intrusions and resulting heating could have influenced the mechanical properties of the crust under the DST flanks. However, these issues cannot be further explored in the absence of relevant data.

5.6 Discussion

The foregoing summary, though often limited because of insufficient data, highlights some basic questions regarding the plate setting, history, and development of the entire DST, and also stresses major challenges facing future research.

As noted, the DST was superimposed on, and cuts through, older structures. There is no known evidence that any part of it reactivated older structures, though such structures could have influenced the shape of its trace and the details of deformation along it. The bend along the Yammouneh Fault formed where the DST crosses a Permian-Triassic rift underlying the Palmyrides, so it is likely that it was influenced by the older rift. The known data do not resolve differences in the times of initiation of the various parts of the DST before they joined to form a through-going plate boundary. However, if such early lateral offsets were significant (more than a few kilometers) along distinct segments, then next to their terminations visible structures should have formed to accommodate the lateral motions. The only known structures that could have formed (or were accentuated) in this way are the Palmyrides, which could have allowed some early slip on the southern part of the DST. However, since the Palmyrides were deformed also during the DST activity, this could have amounted to only a fraction of the entire ca. 20 km of NW-SE shortening east of the DST trace, which is a small fraction of the total offset. This could delay the northward propagation of the DST by a few Ma only, which is compatible with the lines of evidence mentioned above, that point at its Miocene activity (including its northern pars: see 5.3, 5.5.2.2, and 5.5.2.3). Therefore it is impossible to accept models proposing that the northern part of the DST did not form before the Late Miocene, i.e. when a large lateral offset had already taken place along its southern part (e.g. Kopp and Leonov 2000; Homberg et al. 2010).

It is concluded, therefore, that the entire DST formed over a short, though unknown, period, most likely late in the Early Miocene (17–18 Ma ago). Since then the DST acted as a continuous transform boundary between the Sinai and Arabia plates, except for its Turkish segment (see below). Thus the Sinai-Arabia plate kinematics provides a framework for its interpretation. Since this boundary has an irregular shape in map view, its two sides cannot fit when lateral motion progresses. The result will be either transtension or transpression, depending on the relation between the direction of the relative plate motion and the local trend of the plate boundary. This will lead to different types of deformation along the DST and its flanks that varies along its trace. The foregoing review showed that the kinematics of the Sinai-Arabia plate motion, as constrained by combining data from the DST itself with data from the Red Sea, indeed provide a framework for relating the major structures along the DST to the irregularities of its trace.

In particular, the formation and growth of a string of deep depressions along the southern part of the DST expresses its predominantly transtensional character that is predicted by the plate kinematics. These basins accommodate most of the oblique separation of its flanks which increases as the lateral motion along the DST progresses. In contrast, along the northern part of the DST, where plate kinematics predicts transpression, the deformation is very different and extends some distance into the DST flanks. The expected shortening normal to the DST trace is accommodated by folds with axes sub-parallel to the main DST fault and especially by rotation of fault blocks on vertical axes, which is proven by paleomagnetic data. It turns out that the latter type of deformation leads to a result not appreciated before: it can increase the N-S length of the western DST flank, which in turn will produce left lateral shearing along the junction with the undeformed Levant basin (see 5.5.2.4 and Fig. 5.7). This comes at the expense of the lateral motion along the main DST faults, which may well explain the decrease of the lateral offset observed along the northern part of DST (see 5.3).

Thus, qualitatively the observed structures can produce the effects expected from plate kinematic considerations, but like in many places in the world is difficult to constrain. Therefore it cannot be ascertained that the observed structures indeed agree quantitatively with the plate kinematic predictions, especially regarding the northern part of the DST (e.g. the ca. 30–40 km shortening perpendicular to the DST across the Lebanon segment or the apparent northward decrease of the lateral offset). This difficulty led to many discussions, but it cannot be taken as an argument against the occurrence of large lateral motion along the northern part of the DST (as suggested e.g. by Dubertret 1970). Rather it reflects incomplete quantitative understanding of the te observed structures – a frequently encountered difficulty. This remains a major challenge for future research.

The northernmost (Turkish) segment of the DST requires separate treatment. Since the structures of the Turkish segment are the direct continuation of the DST farther south, and they are the only known structures that could have taken up the lateral DST offset documented just south of them, they are considered to have been in integral part of the DST during its early history. However, ca. 4 Ma ago the plate configuration and motions in that region changed when the westward extrusion of Anatolia became important and the East Anatolian FZ formed. This affected the northernmost portion of the DST, so now it is no longer a part the Sinai-Arabia plate boundary. Though the young kinematics in this area is insufficiently known, it appears that the Amanos Block – its western flank – was decoupled from the Sinai plate and tended to join the Anatolian plate (i.e. the Cyprus arc). Originally this block was probably attached, though perhaps not rigidly, to the western DST flank, given the Miocene age of the Hatay graben and the Nahr el-Kebir graben south of it. The young tectonic changes obscured the structural relations in this region during the early stages of the DST history, including its interaction with the Alpine orogenic belt. Much more data are required to resolve these issues.

In summary, the Sinai-Arabia plate kinematics provides a useful framework for analyzing the history and deformation along the DST and its flanks, though many issues need further study. However, full understanding of the structure – e.g. the fault arrangement, the partitioning of deformation between different structures, the width of the deformed zones, and changes in the structure with time – requires dynamic modeling that takes into consideration the mechanical properties of the crust (and the entire lithosphere), the regional stress field, igneous activity, and other processes in the underlying mantle.

Acknowledgements I am very grateful to X. Le Pichon and A. Sagy for their helpful reviews.

References

Adiyaman Ö, Chorowicz J (2002) Late Cenozoic tectonics and volcanism in the northwestern corner of the Arabian plate: a consequence of the strike-slip Dead Sea fault zone and the lateral escape of Anatolia. J Volcanol Geotherm Res 117:327–345

Al Tarazi E, Abu Rajab J, Gomez F, Cochran W, Jaafar R, Ferry M (2011) GPS measurements of near-field deformation along the southern Dead Sea fault system. Geochem Geophys Geosyst 12:Q12021. doi:10.1029/2011GC003736

Alchalbi A, 15 others (2010) Crustal deformation in northwestern Arabia from GPS measurements in Syria: slow slip rate along the northern Dead Sea fault. Geophys J Int 180:125–135. doi:10.1111/j.1365-246X.2009.04431.x

Aldersons F, Ben-Avraham Z, Hofstetter A, Kissling E, Al-Jazreen A (2003) Lower-crustal strength under the Dead Sea basin from local earthquake data and rheological modeling. Earth Planet Sci Lett 214:129–142. doi:10.1016/S0012-821X(03)00381-9

Al-Maleh K, Delaune M, Mouty M, Parrot JF (1992) Relations du front de la nappe ophiolitique du nord-ouest syrien avec son substratum de la part et d'autre de la faille du levant: Baer-Bassit, Kurd-Dagh. Compt Rendus Acad Sci Paris 314:1195–1202

Al-Zoubi A, ten-Brink US (2002) Lower crustal flow and the role of shear in basin subsidence: an example from the Dead Sea Basin. Earth Planet Sci Lett 199:67–79

Al-Zoubi A, Heinrich T, Qabbami I, Ten-Brink US (2007) The northern end of the Dead Sea basin: geometry from reflection seismic evidence. Tectonophysics 434:55–69

ArRajehi A, 16 others (2010) Geodetic constraints on present-day motion of the Arabian plate: implications for Red Sea and Gulf of Aden rifting. Tectonics 29:TC3011. doi:10.1029/2009TC002482

Avni Y, Garfunkel Z, Bartov Y, Ginat H (1994) Pleistocene fault system in the central and southern Negev and its implications for the tectonic and geomorphic history of the Arava rift margin. Geol Surv Isr Curr Res 9:51–58

Avni Y, Bartov Y, Garfunkel Z, Ginat H (2000) The Arava Formation – A Pliocene sequence in the Arava Valley and its western margin, southern Israel. Isr J Earth Sci 50:101–120

Avni Y, Segev A, Ginat H (2012) Oligocene regional denudation of the northern Afar dome: pre- and syn- breakup stages of the Afro-Arabian plate. Geol Soc Am Bull 124:1871–1897

Baer G, Hirsch F, Ron H (1998) The Newe Ativ graben, Mt. Hermon: its origin and implications for the regional structure. Isr J Earth Sci 46:61–78

Bandel K (1981) New stratigraphic and structural evidence for lateral dislocation in the Jordan rift connected with description of the Jurassic rock column in Jordan. N Jb Geol Paläont (Abh) 161:271–308

Bandel K, Khouri H (1981) Lithostratigraphy of the Triassic in Jordan. Facies 4:1–26

Bartov Y (1974) A structural and paleogeographic study of the central Sinai faults and domes. Dissertation, Hebrew Univ Jerusalem (in Hebrew, Eng abs), 143 p

Bartov Y (1994) The geology of the Arava. Geol Sur Isr report GSI/4/94, 16 p. (in Hebrew)

Bartov Y, Sagy A (2004) Late Pleistocene extension and strike-slip in the Dead Sea basin. Geol Mag 141:565–572

Bartov Y, Avni Y, Calvo R, Frieslander U (1998) The Zofar Fault – A major intra-rift feature in the Arava rift valley. Geol Surv Isr Curr Res 11:27–32

Basha S (1982) Stratigraphy, paleogeography and oil possibilities of the Azraq-sirhan-Turayf basin (Jordan-Saudi Arabia). Dirasat 9:85–106

Bein A, Gvirtzman G (1977) A Mesozoic fossil edge of the Arabian plate along the Levant coast and its bearing on the evolution of the Eastern Mediterranean. In: Biju-duval B, Montadert L (eds) Symposium on the history of the Mediterranean basins. Split. Edit Technip, Paris, pp 95–109

Belitzky S (1987) Tectonics of the Korazim saddle. Dissertation, Hebrew Univ Jerusalem 94 p (in Hebrew)

Ben-Avraham Z (1985) Structural framework of the Gulf of Elat (Aqaba), northern Red Sea. J Geophys Res 90:703–726

Ben-Avraham Z, Almagor G, Garfunkel Z (1979) Sediments and structure of the Gulf of Elat. Sediment Geol 23:239–267

Ben-Avraham Z, Garfunkel Z (1986) Character of transverse faults in the Elat pull-apart basin. Tectonics 5:1161–1169

Ben-Avraham Z, Hanel R, Vilinger H (1978) Heat flow through the Dead Sea rift. Mar Geol 28:253–269

Ben-Avraham Z, Ten Brink U, Bell R, Reznikov M (1996) Gravity field over the Sea of Galilee: evidence for a composite basin along a transform fault. J Geophys Res 101:533–544

Ben-Avraham Z, Garfunkel Z, Lazar M (2008) Geology and evolution of the southern Dead Sea fault with emphasis on subsurface structure. Ann Rev Earth Planet Sci 36:357–387

Ben-Avraham Z, Lazar M, Garfunkel Z, Reshef M, Ginzburg A, Rotstein Y, Frieslander U, Bartov Y, Shulman H (2012) Chapter 17: Structural styles along the Dead Sea fault. In: Roberts DG, Bally AW (eds) Regional geology and tectonics: phanerozoic passive margins, cratonic basins and global tectonic maps, vol 1C. Elsevier, Burlington, pp 617–634

Bender F (1974) Geology of Jordan. Gebrüder Borntraeger, Berlin/Stuttgart, 196 p

Bentor YK (1985) The crustal evolution of the Arabo-Nubian massif with special reference to the Sinai Peninsula. Precambian Res 28:1–74

Beydoun ZR (1977) Petroleum prospects of Lebanon: reevaluation. Am Assoc Petrol Geol Bull 61:43–64

Bosworth W (2005) The Red Sea and Gulf of Aden. J Afr Earth Sci 43:334–378

Boulton SJ, Robertson AHF, Ünlügenç U (2006) Tectonic and sedimentary evolution of the Cenozoic Hatay Graben, southern Turkey: a two phase model for graben formation. Geol Soc Lond Spec Publ 260:613–634

Bozkurt E (2001) Neotectonics of Turkey – a synthesis. Geodinamica Acta 14:3–30. doi:10.1016/S0985-3111(01)01066-X

Brew G, Barazangi M, Al-Maleh K, Sawaf T (2001a) Tectonic and geologic evolution of Syria. GeoArabia 6:573–616

Brew G, Lupa J, Barazangi M, Sawaf T, Al-Imam A, Zaza T (2001b) Structure and tectonic development of the Ghab basin and the Dead Sea fault system, Syria. J Geol Soc 158:665–674

Butler RWH, Spencer S, Griffiths HM (1998) The structural response to evolving plate kinematics during transpression: evolution of the Lebanese restraining bend of the Dead Sea transform. Geol Soc Lond Spec Publ 135:81–106

Calvo R (2002) Stratigraphy and petrology of the Hazeva formation in the Arava and the Negev. Geol Surv Isr report GSI/22/02, 264 p (in Hebrew, Engl abst)

Carton H, Singh SC, Tapponnier P, Elias A, Briais A, Sursock A, Jomaa R, King GCP, Daeron M, Jacques E, Barrier E (2009) Seismic evidence for Neogene and active shortening offshore of Lebanon (Shalimar cruise). J Geophys Res 114:B07407. doi:10.1029/2007JB005391

Cetin H, Guneyli H, Mayer L (2003) Paleoseismology of the Palu-Lake Hazar segment of the East Anatolian fault zone, Turkey. Tectonophysics 374:163–197

Chaimov TA, Barazangi M, Al-Saad D, Sawaf T, Gebran A (1990) Balanced cross sections and shortening in the Palmyride fold belt of Syria and implications for movement along the Dead Sea fault system. Tectonics 9:1369–1386

Chorowicz J, Dhont D, Ammar O, Rukieh M, Bilal A (2005) Tectonics of the Pliocene Homs basalts (Syria) and implications fro the Dead Sea fault zone activity. J Geol Soc 162:259–271

Chu D, Gordon RG (1998) Current plate motions across the Red Sea. Geophys J Int 135:313–328

Cochran JR (1983) A model for the development of the Red Sea. Am Assoc Petrol Geol Bull 67:41–49

Cochran JR (2005) Northern Red Sea: nucleation of an oceanic spreading center within a continental rift. Geochem Geophys Geosys 6(3). doi:10.1029/2004GC000826

Cole GA, Abu-Al MA, Colling EL, Halpern HI, Carrigan WJ, Savage GR, Scolaro RJ, Al-Sharidi SH (1995) Petroleum geochemistry of the Midyan and Jaizan basins of the Red Sea, Saudi Arabia. Mar Pet Geol 12:597–614

Coskun B (1994) Oil possibilities of duplex structures in the Amik-Reyhanli basin, SE Turkey. J Pet Geol 17:461–472

Crowell JC (1974) Origin of late Cenozoic basins in southern California. Soc Econ Paleont Mineralog, Spec Publ 22:190–204

Druckman Y (1974) The stratigraphy of the Triassic sequence in southern Israel. Isr Geol Surv Bull 64:1–92

Dubertret L (1932) Les formes structurales de la Syria et de la Palestine: leur origine. Compt Rendus Acad Sci Paris 195:65–67

Dubertret L (1955) Carte géologique du Liban, scale 1:200 000, with explanatory notes. Ministry of Public Works, Beirut

Dubertret L (1970) Review of structural geology of the Red Sea and surrounding areas. Phil Trans R Soc London A267:9–20

Dullo WC, Hötzl H, Jado AR (1983) New stratigraphical results from the Tertiary sequence of Midyan area, NW Saudi Arabia. Newslett Stratigr 12:83

Eaton S, Robertson AHF (1993) The Miocene Pakhna Formation, southern Cyprus, and its relationship to the Neogene tectonic evolution of the Eastern Mediterranean. Sediment Geol 86:273–296

Eckstein Y (1979) Review of the heat flow data from the Eastern Mediterranean region. Pure Appl Geophys 117:150–179

Ehrhardt A, Hubscher C, Ben-Avraham Z, Gajewski D (2005) Seismic study of pull-apart-induced sedimentation and deformation in the Northern Gulf of Aqaba (Elat). Tectonophysics 396:59–79

Elias A, Tapponnier P, Singh SC, King GCP, Briais A, Daëron M, Carton H, Sursock A, Jacques E, Jomaa R, Klinger Y (2007) Active thrusting offshore Mount Lebanon: source of the tsunamigenic A.D. 551 Beirut-Tripoli earthquake. Geology 35:755–758. doi:10.1130/G23631A.1

Ellenblum R, Marco S, Agnon A, Rockwell T, Boas A (1998) Crusader castle torn apart by earthquake at dawn, 20 May 1202. Geology 26:303–306

Eyal M, Eyal Y, Bartov Y, Steinitz G (1981) The tectonic development of the western margin of the Gulf of Elat (Aqaba) rift. Tectonophysics 80:39–66

Fleischer E (1968) The subsurface geology of the HulA Valley and the Korazijm area. Geol Surv Isr report 15 p (in Hebrew)

Förster A, Forster HJ, Massarwe R, Masri A, Tarawneh K (2007) The surface heat flow of the Arabian shield in Jordan. J Asian Earth Sci 30:271–284

Freund R (1965) A model for the development of Israel and adjacent areas since the Upper Cretaceous times. Geol Mag 102:189–205

Freund R (1970) Plate tectonics of the Red Sea and East Africa. Nature 228:453

Freund R, Garfunkel Z, Zak I, Goldberg M, Weissbrod T, Derin B (1970) The shear along the Dead Sea rift. Phil Trans R Soc London A267:107–130

Frieslander U (2000) The structure of the Dead Sea transform emphasizing the Arava. Dissertation, Hebrew Univ Jerusalem, 106 p (in Hebrew English abstract)

Frieslander U, Ben-Avraham Z (1989) Magnetic field over the Dead Sea and vicinity. Mar Pet Geol 6:148–160

Fuis GS, Mooney WD, Heaky JH, McMechan GA, Lutter WJ (1984) A seismic refraction survey of the Imperial Valley region, California. J Geophys Res 89:1165–1189

Galli P (1999) Active tectonics along the Wadi Araba-Jordan valley transform fault. J Geophys Res 104:2777–2796

Gardosh M, Reches Z, Garfunkel Z (1990) Holocene tectonic deformation along the western margins of the Dead Sea. Tectonophysics 180:123–137

Gardosh M, Kashai E, Salhov S, Shulamn H, Tennenbaum E (1997) Hydrocarbon exploration in the southern Dead Sea area. In: Niemi T, Ben-Avraham Z, Gat JR (eds) The Dead Sea. Oxford University Press, New York, pp 57–72

Gardosh MA, Garfunkel Z, Druckman Y, Buchbinder B (2010) Neotethyan rifting and its role in shaping the Levant Basin and margin. Geol Soc Lond Spec Publ 341:9–36. doi:10.1144/SP341.2
Garfunkel Z (1981) Internal structure of the Dead Sea leaky transform (rift) in relation to plate kinematics. Tectonophysics 80:81–108
Garfunkel Z (1987) Post-Precambrian sediments. In: Bentor YK, Eyal M (eds) Jebel Sabbagh sheet, explanatory note. Israel Academy of Sciences and Humanities, Jerusalem, pp 386–392
Garfunkel Z (1988) The pre-Quaternary geology of Israel. In: Yom-Tov Y, Tchernov E (eds) The zoogeography of Israel. W Junk, Dordrecht/Holland, pp 7–34
Garfunkel Z (1989) Tectonic setting of Phanerozoic magmatism in Israel. Isr J Earth Sci 38:51–74
Garfunkel Z (1998) Constraints on the origin and history of the Eastern Mediterranean basin. Tectonophysics 298:5–36
Garfunkel Z (1999) History and paleogeography during the Pan-African orogen to stable platform transition: reappraisal of the evidence from the Elat area and the northern Arabo-Nubian Shield. Isr J Earth Sci 48:135–157
Garfunkel Z (2011) The long- and short-term lateral slip and seismicity along the Dead Sea transform: an interim reevaluation. Isr J Earth Sci 58:217–235. doi:10.1560/IJES.58.3-4.217
Garfunkel Z, Bartov Y (1977) The tectonics of the Suez rift. Isr Geol Surv Bull 71:1–44
Garfunkel Z, Ben-Avraham Z (1996) The structure of the Dead Sea basin. Tectonophysics 266:155–176
Garfunkel Z, Ban-Avraham Z (2001) Basins along the Dead Sea Transform. In: Ziegler PA, Cavazza W, Robertson AHF, Crasquin-Soleau S (eds) Peri-Tethys Memoir 6: Peri Tethyan rift/wrench basins and passive margins, Mémoires du Muséum national d'histoire naturelle 186. Publications scientifiques du Muséum, Paris, pp 607–627
Garfunkel Z, Beyth M (2006) Constraints on the structural development of Afar imposed by the kinematic of the major surrounding plates. In: Yirgu G, Ebinger CJ, Maguire PKH (eds) The Afar volcanic province within the East African System, Geological Society special publication 259. Geological Society, London, pp 23–42
Garfunkel Z, Derin B (1984) Permian-Early Mesozoic tectonism and continental margin formation in Israel and its implications for the history of the eastern Mediterranean. Geol Soc Lond Spec Publ 17:187–201
Garfunkel Z, Ron H (1985) Block rotation and deformation by strike slip fault 2. The properties of a type of macroscopic discontinuous deformation. J Geophys Res 90:8589–8602
Garfunkel Z, Zak I, Freund R (1981) Active faulting in the Dead Sea rift. Tectonophysics 80:1–26
Gaulier JM, LePichon X, Lyberis N, Averdik F, Geli L, Moretti I, Deschamps A, Hafez S (1988) Seismic study of the crustal thickness, Northern Red Sea and Gulf of Suez. Tectonophysics 153:55–88
Ginat H, Avni Y (1994) The Arava conglomerate: a Pliocene valley deposit crossing the Dead Sea rift. Geol Surv Isr Curr Res 9:59–62
Ginzburg A, Ben-Avraham Z (1997) A seismic refraction study of the north basin of the of the Dead Sea, Israel. Geophys Res Lett 24:2063–2066
Ginzburg A, Reshef M, Ben-Avraham Z, Schattner U (2006) The style of transverse faulting in the Dead Sea basin from seismic reflection data: the Amatzyahu fault. Isr J Earth Sci 55:129–139
Glickson YA (1966) The lacustrine Neogene of Kefar Giladi area. Isr J Earth Sci 15:135–154
Goldberg M, Beyth M (1991) Tiran Island: an internal block at the junction of the Red Sea rift and Dead Sea transform. Tectonophysics 198:261–273
Gomez F, Meghraoui M, Darkal AN, Tabet C, Khawlie M, Khair K, Sbeinati R, Darawcheh R, Khair K, Barazangi M (2001) Coseismic displacements along the Serghaya fault: an active branch of the Dead Sea fault system in Syria and Lebanon. J Geol Soc Lond 158:405–408
Gomez F, Khawlie M, Tabet C, Darkal AN, Khair K, Barazangi M (2006) Late Cenozoic uplift along the northern Dead Sea transform in Lebanon and Syria. Earth Planet Sci Lett 241:913–931
Gomez F, Karam G, Khawlie M, McClusky S, Vernant P, Reilinger R, Jaafar R, Tabet C, Khair K, Barazangi M (2007) Global Positioning System measurements of strain accumulation and slip

transfer through the restraining bend along the Dead Sea fault system in Lebanon. Geophys J Int 168:1021–1028. doi:10.1111/j.1365-246X.2006.03328.x

Gregor CB, Mertzman S, Nairn AEM, Negendank J (1974) Paleomagnetism and the Alpine tectonics of Eurasia. V. The paleomagnetism of some Mesozoic and Cenozoic volcanic rocks from the Lebanon. Tectonophysics 21:375–396

Gregory JW (1921) The rift Valleys and geology of East Africa. Seeley, London, 479 p

Griffiths HM, Clark RA, Thorp MK, Spencer S (2000) Strain accommodation at the lateral margin of an active transpressive zone: geological and seismological evidence from the Lebanese restraining bend. J Geol Soc Lond 157:289–302

Guiraud R, Issawi B, Boswoth W (2001) Phanerozoic history of Egypt and surrounding areas. In: Ziegler PA, Cavazza W, Robertson AHF, Crasquin-Soleau S (eds) Peri-Tethys Memoir 6: Peri Tethyan rift/wrench basins and passive margins, Mémoires du Muséum national d'histoire naturelle 186. Publications scientifiques du Muséum, Paris, pp 469–509

Hall J, Aksu AE, Calon TJ, Yasar D (2005a) Varying tectonic control on basin development at an active microplate margin: Latakia Basin, Eastern Mediterranean. Mar Geol 221:15–60

Hall J, Calon TJ, Aksu AE, Meade SR (2005b) Structural evolution of the Latakia Ridge and Cyprus Basin at the front of the Cyprus Arc, Eastern Mediterranean Sea. Mar Geol 221:261–297

Hancock PL, Atiya MS (1979) Tectonic significance of mesofracture systems associated with the Lebanese segment of the Dead Sea transform faults. J Struct Geol 1:143–153

Harash A, Bar Y (1988) Faults, landslides and seismic hazards along the Jordan River gorge, northern Israel. Eng Geol 25:1–15

Hardenberg MF, Robertson AHF (2007) Sedimentology of the NW margin of the Arabian plate and the SW-NE trending Nahr El-Kebir half-graben in northern Syria during the latest Cretaceous and Cenozoic. Sediment Geol 201:231–266

Harding TP, Vierbuchen RC, Christie-Blick N (1985) Structural styles, plate tectonic settings and hydrocarbon traps of divergent (transtensional) wrench faults. Soc Econ Paleontol Mineral 37:51–77

Hatcher RD, Zeiz I, Reagan RD, Abu-Ajameh M (1981) Sinistral strike-slip motion on the Dead Sea rift: confirmation from new magnetic data. Geology 9:458–462

Heimann A, Braun D (2000) Quaternary stratigraphy of the Kinnarot Basin, Dead Sea transform, northeastern Israel. Isr J Earth Sci 49:31–44

Heimann A, Ron H (1993) Geometrical changes of plate boundaries along part of the northern Dead Sea transform: geochronological and paleomagnetic evidence. Tectonics 12:477–491. doi:10.1029/92TC01789

Heimann A, Steinitz G (1989) $^{40}Ar/^{39}Ar$ total gas ages of basalts from Notera #3 well, Hula Valley, Dead Sea Rift: stratigraphic and tectonic implications. Isr J Earth-Sci 38:173–184

Heimann A, Eyal M, Eyal Y (1990) The evolution of Barahta rhomb-shaped graben, Mount Hermon, Dead Sea Transform. Tectonophysics 141:101–110

Heimann A, Steinitz G, Mor D, Shaliv G (1996) The Cover Basalt Formation, its age and its regional and tectonic setting: implications from K-Ar and $^{40}Ar/^{39}Ar$ geochronology. Isr J Earth Sci 45:55–71

Heimann A, Zilberman E, Amit R, Frieslander U (2009) Northward migration of the southern diagonal fault of the Hula pull-apart basin, Dead Sea Transform, northern Israel. Tectonophysics 476:496–511

Hempton MR (1987) Constraints on Arabian plate motion and extensional history of the Red Sea. Tectonics 6:687–705

Henry B, Homberg C, Mroueh M, Hamdan W, Higazi F (2010) Rotations in Lebanon inferred from new palaeomagnetic data and implications for the evolution of the Dead Sea Transform system. Geol Soc Lond Spec Publ 341:269–285

Heybroek F (1942) La Geologie d'une partie du Liban sud. Leidsche Geol Medded 12:251–470

Hofstetter R, Klinger Y, Amrat AQ, Rivera L, Dorbath L (2007) Stress tensor and focal mechanisms along the Dead Sea fault and related structural elements based on seismological data. Tectonophysics 429:165–181

Hofstetter R, Gitterman Y, Pinsky V, Kraeva N, Feldman L (2008) Seismological observations of the northern Dead Sea basin earthquake on 11 February 2004 and its associated activity. Isr J Earth Sci 57:101–124. doi:10.1560/IJES.57.2.101

Homberg C, Barrier E, Mroueh M, Muller C, Hamdan W, Higazi F (2010) Tectonic evolution of the central Levant domain (Lebanon) since Mesozoic time. Geol Soc Lond Spec Publ 341:245–268

Horowitz A (1973) Development of the Hula basin. Isr J Earth Sci 22:107–139

Horowitz A (1987) Palynological evidence for the age and rate of sedimentation along the Dead Sea rift, and structural implications. Tectonophysics 141:107–115

Horowitz A, Horowitz M (1985) Subsurface late Cenozoic palynostratigraphy of the Hula Basin. Pollen Spores 27:365–390

Hubert-Ferrari A, King G, Van DerWoerd J, Villa I, Altunel E, Armijo R (2009) Long-term evolution of the North Anatolian Fault: new constraints from its Eastern termination. Geol Soc Lond Spec Publ 311:133–154

Hughes GW, Perincek D, Grainger DJ, Abu-Bshait AJ, Jared ARM (1999) Lithostratigraphy and depositional history of part of the Midyan region, northwestern Saudi Arabia. GeoArabia 4:503–542

Hurwitz S, Garfunkel Z, Ben-Gai Y, Reznikov M, Rotstein Y, Gvirtzman H (2002) The tectonic framework of a complex pull-apart basin: seismic reflection observations in the northern Kinarot-Beit-Shean Basin, Dead Sea Transform. Tectonophysics 359:289–306

Jennings CW, compiler (1973) State of California, fault and geologic map, scale 1:750 000. California Division of Mines Geology, Sacramento

Joffe S, Garfunkel Z (1987) The plate kinematics of the circum Red Sea – A reevaluation. Tectonophysics 141:5–22

Karabacak V, Altunel E, Meghraoui M, Akyüz HS (2010) Field evidences from northern Dead Sea Fault Zone (South Turkey): new findings for the initiation age and slip rates. Tectonophysics 480:172–182

Karig DE, Kozlu H (1990) Late-Palaeogene-Neogene evolution of the triple junction region near Maras, south-central turkey. J Geol Soc Lond 147:1023–1034

Kashai EL, Crocker PF (1987) Structural geometry and evolution of the Dead Sea-Jordan rift system as deduced from new subsurface data. Tectonophysics 141:33–60

Khair K (2001) Geomorphology and seismicity of the Roum fault as one of the active branches of the Dead Sea fault system in Lebanon. J Geophys Res 106:4233–4245

Kopp ML, Leonov YG (2000) Tectonics. In: Leonov YG (ed) Outline of geology of Syria, Russ Acad Sci Geol Inst Trans 526, pp 7–104, Nauka, Moscow (in Russian)

Krasheninnikov VA (2005) F: Neogeve. In: Krasheninnikov VA, Hall JK, Hirshc F, Benjamini H, Flexer A (eds) Geological framework of the Levant, V.1 Cyprus and Syria. Historic al Productioin-Hall, Jerusalem, pp 343–392

Larsen BD, Ben-Avraham Z, Shulman H (2002) Fault and salt tectonics in the southern Dead Sea basin. Tectonophysics 346:71–90

Lazar M, Ben-Avraham Z, Schattner U (2006) Formation of sequential basins along a strike-slip fault – Geophysical observations from the Dead Sea basin. Tectonophysics 421:53–69

Le Beon M, Klinger Y, Amrat AQ, Agnon A, Dorbath L, Baer G, Ruegg JC, Charade O, Mayyas O (2008) Slip rate and locking depth from GPS profiles across the southern Dead Sea Transform. J Geophys Res 113:B11403. doi:10.1029/2007JB005280

Le Pichon X, Kreemer C (2010) The Miocene to present kinematic evolution of the eastern Mediterranean and Middle East and its implication for dynamics. Ann Rev Earth Planet Sci 38:323–351

Le Pichon X, Gaulier JM (1988) The rotation of Arabia and the Levant fault system. Tectonophysics 153:271–294

Le Pichon X, Bergerat F, Roulet MJ (1988) Plate kinematics and tectonics leading to the Alpine belt formation. A new analysis. Geol Soc Am Spec Pap 218:111–131

Le Pichon X, Francheteau J, Bonin J (1973) Plate tectonics. Elsevier, Amsterdam 303 p

Mahmoud Y, Masson F, Meghraoui M, Cakir Z, Alchalbi A, Yavasoglu H, Yönlü O, Daoud M, Ergintav S, Inan S (2013) Kinematic study at the junction of the East Anatolian fault and the Dead Sea fault from GPS measurements. J Geodyn 67:30–39

Mann P, Hempton MR, Bradley DC, Burke K (1983) Development of pull-apart basins. J Geol 91:529–554
Marco S, Rockwell TK, Heimann A, Frieslander U, Agnon A (2005) Late Holocene activity of the Dead Sea Transform revealed in 3D paleoseismic trenches on the Jordan Gorge segment. Earth Planet Sci Lett 234:189–205
Marcus E, Slager J (1986) The sedimentary-magmatic sequences of Zemah-1 well (Dead Sea rift, Israel) and its emplacement in time and space. Isr J Earth Sci 34:1–10
Mart Y, Hall JK (1984) Structural trends in the northern Red Sea. J Geophys Res 89:11352–11364
Matmon A, Wdowinski S, Hall JK (2003) Morphological and structural relations in the Galilee extensional domain, northern Israel. Tectonophysics 371:223–241
McKenzie DP, Davies D, Molnar P (1970) Plate tectonics of the Red Sea and East Africa. Nature 224:125–133
Mechie J, Abu-Ayyash K, Ben-Avraham Z, El-Kelani R, Qabbani I, Weber M, DESIRE Group (2009) Crustal structure of the southern Dead Sea basin derived from DESIRE wide-angle seismic data. Geophys J Int 78:457–478. doi:10.1111/j.1365-246X.2009.04161.x
Meghraoui M, Gomez F, Sbeinati R, Van der Voerd J, Mouty M, Darkal A, Radan Y, Layyous I, Najjar HM, Darawcheh R, Hijazi F, Al-Ghazzi R, Barazangi M (2003) Evidence for 830 years of seismic quiescence from paleoseismology, archeoseismology and historical seismicity along the Dead Sea fault in Syria. Earth Planet Sci Lett 210:35–52
Meiler M, Reshef M, Shulman H (2011) Seismic depth-domain stratigraphic classification of the Golan Heights, central Dead Sea Fault. Tectopnophysics 510:354–369
Michelson H, Lipson-Benitah S (1986) The litho- and biostratigraphy of the southern Golan Heights. Isr J Earth Sci 35:221–240
Mittlefehldt D, Slager Y (1986) Petrology of the baslats and gabbros from the Zemah-1 drill hole, Jordan rift valley. Isr J Earth Sci 35:10–22
Mouty M, Delaloye M, Fontignie D, Piskin O, Wagner JJ (1992) The volcanic activity in Syria and Lebanon between Jurassic and Actual. Schwietz Mineralog Petrogt Mitt 72:91–105
Neev D, Hall JK (1979) Geophysical investigations in the Dead Sea. Sediment Geol 23:209–238
Nemer T, Meghraoui M (2006) Evidence for coseismic ruptures along the Roum Fault (Lebanon) a possible source of the AD 1837 earthquake. J Struct Geol 28:1483–1495
Nielsen C, Thybo H (2009) Lower crustal intrusions beneath the southern Baikal rift zone: evidence from full-waveform modelling of wide-angle seismic data. Tectonophysics 470:298–318
Over S, Ozden S, Unlugenc UC, Huseyin Yilmaz H (2004) A synthesis: late Cenozoic stress field distribution at northeastern corner of the Eastern Mediterranean, SE Turkey. Compt Rendus Geosci 336:93–103
Papazachos BC, Papaioannou CA (1999) Lithospheric boundaries and plate motions in the Cyprus area. Tectonophysics 308:193–204
Perinçek D, Çemen I (1990) The structural relationship between the East Anatolian and Dead Sea fault zones in southeastern Turkey. Tectonophysics 172:331–340
Petrunin A, Sobolev SV (2006) What controls thickness of sediments and lithospheric deformation at a pull-apart basin? Geology 34:389–392
Picard L (1943) Structure and evolution of Palestine, Bulletin of the Geology Department, Hebrew University. Geology Department, Hebrew University, Jerusalem, 187 p
Picard L, Baida U (1966) Geological report on the Lower Pleistocene deposits of the Ubeidiya excavations. Isr Acad Sci Hum Proc 4:1–16
Ponikarov VP (1964) Chief editor: the geologic map of Syria, Scale 1:1 000 000. Dept Geol Min Res Syria
Ponikarov VP, Kazmin VG, Kozlov VV, Krasheninnikov VA, Mikhailov IA, Razvlyaev AV, Soulidi-Kondratyev ED, Uflyand SK, Faradzhev VA (1969) Geology and mineral resources of foreign countries: Syria. Moscow, 216 p (in Russian)
Quennell AM (1959) Tectonics of the Dead Sea rift. 20th Int Geol Cong (1956), Assoc Afr Geol Surv pp 385–405
Reilinger R, McClusky S (2011) Nubia–Arabia–Eurasia plate motions and the dynamics of Mediterranean and Middle East tectonics. Geophys J Int 186:971–979

Reilinger R, 24 others (2006) GPS constraints on continental deformation in the Africa-Arabia-Eurasia continental collision zone and implications for the dynamics of plate interactions. J Geophys Res 111:B05411. doi:10.1029/2005JB004051
Renouard G (1955) Oil prospects of Lebanon. Am Assoc Petrol Geol Bull 39:2125–2169
Reznikov M, Ben-Avraham Z, Garfunkel Z, Gvirtzman H, Rotstein Y (2004) Structural and Stratigraphic framework of Lake Kinneret. Isr J Earth Sci 53:131–149
Robertson AHF (2001a) Mesozoic-Tertiary tectonic evolution of the Easternmost Mediterranean area: integration of marine and land evidence. Proc ODP Sci Results 160:723–782
Robertson AHF (2001b) Geological evolution of Cyprus, onshore and offshore evidence. In: Malpas J, Xenophontos C, Panayides A (eds) Proceedings of 3rd international conference on the geology of the eastern Mediterranean. Geological Survey Department, Nicosia, pp 11–44
Robertson AHF, Ünlügenç U, İnan N, Tasli K (2004) Missis-Andirin Complex: mélange formation related to closure and collision of the south-Tethys, in South Turkey. J Asian Earth Sci 22:413–453
Roeser HA (1975) A detailed magnetic survey of the southern Red Sea. Geol Jahrb D13:131–153
Rojay B, Heimann A, Toprak V (2001) Neotectonic and volcanic characteristics of the Karasu fault zone (Anatolia, turkey): the transition zone between the Dead Sea transform and the East Anatolian fault zone. Geodin Acta 14:197–212
Ron H (1987) Deformation along the Yammouneh, the restraining bend of the Dead Sea transform: paleomagnetic data and kinematic implications. Tectonics 9:1421–1432
Ron H, Freund R, Garfunkel Z, Nur A (1984) Block rotation by strike-slip faulting: structural and paleomagnetic evidence. J Geophys Res 89:6256–6270
Rotstein Y, Bartov Y (1989) Seismic reflection across a continental transform: an example from a convergent segment of the Dead Sea rift. J Geophys Res 94:2902–2912
Rotstein Y, Bartov Y, Frieslander U (1992) Evidence for local shifting of the main fault and changes in the structural setting, Kinnarot Basin, Dead Sea transform. Geology 20:251–254
Rukieh M, Trifonov VG, Dodonov AE, Minini H, Ammara O, Ivanova TP, Zaza T, Yusef A, Al-Shara M, Jobaili Y (2005) Neotectonic map of Syria and some aspects of Late Cenozoic evolution of the northwestern boundary zone of the Arabian plate. J Geodyn 40:235–256
Rybakov M, Fleischer L, Ten Brink U (2003) The Hula valley subsurface structure inferred from gravity data. Isr J Earth Sci 52:113–122
Ryberg T, Weber M, Garfunkel Z, Bartov Y (2007) The shallow velocity structure across the Dead Sea Transform fault, Arava Valley, from seismic data. J Geophys Res 112:B08307. doi:10.1029/2006JB004563
Sadeh M, Hamiel Y, Bock Y, Fang P, Wdowinski S (2012) Crustal deformation along the Dead Sea Transform and the Carmel Fault inferred from GPS measurements. J Geophys Res 117:B08410. doi:10.1029/2012JB009241
Saint-Marc P (1974) Etude stratigraphiqueet micropaleontologique de l'Albien, du Cenomanien et du Turonien du Liban. Notes et Memoirs de Moyen Orient 8:8–342
Şaroĝlu F, Emre O, Kuşçu I (1992) The East Anatolian fault zone in Turkey. Ann Tecton Supplement to 6:99–125
Schattner U, Weinberger R (2008) A mid-Pleistocene deformation transition in the Hula basin, northern Israel: implications for the tectonic evolution of the Dead Sea Fault. Geochem Geophys Geosys 9, 18 p. doi:10.2029/2007GC001937
Schattner U, Ben-Avraham Z, Lazar M, Huebscher C (2006) Tectonic isolation of the Levant basin offshore Galilee-Lebanon – Effects of the Dead Sea fault plate boundary on the Levant continental margin, eastern Mediterranean. J Struct Geol 28:2049–2066
Schulman N (1962) The geology of the central Jordan Valley. Dissertation, Hebrew University, Jerusalem (in Hebrew)
Schütz F, Norden B, Forster A (2012) Thermal properties of sediments in southern Israel: a comprehensive data set for heat flow and geothermal energy studies heat flow. Basin Res 24:357–376
Searle MP (1994) Structure of the intraplate Eastern Plamyride fold belt, Syria. Geol Soc Am Bull 106:1332–1350

Segev A (1984) Lithostratigraphy and paleogeography of the marine Cambrian sequence in southern Israel and southwestern Jordan. Isr J Earth Sci 33:26–33

Segev A (2009) $^{40}Ar/^{39}Ar$ and K-Ar geochronology of Berriasian-Hauterivian and Cenomanian tectonomagmatic events in northern Israel: implications for regional stratigraphy. Cretac Res 30:810–828

Senel M (ed) (2002) The Geological map of Turkey, Sheet Hatay, 1:500 000. Direct Min Res Expl (MTA), Ankara

Şengör AMC, Yilmaz Y (1981) Tethyan evolution of Turkey: a plate tectonic approach. Tectonophysics 75:181–241

Şengör AMC, Gorur N, Saroglu F (1985) Strike-slip faulting and related basin formation in zones of tectonic escape; Turkey as a case study. Soc Econ Paleontol Mineral, Spec Publ 37:227–264

Şengör AMC, Görür N, Şaroĝlu F (2005) Strike slip faulting and related basin formation in zones of tectonic escape: Turkey as a case study. Soc Econ Plaeont Minerol Spec Publ 37:227–264

Seyrek A, Demir T, Pringle MS, Yurtmen S, Westaway RWC, Beck A, Rowbotham G (2007) Kinematics of the Amanos Fault, southern Turkey, from Ar/Ar dating of offset Pleistocene basalt flows: transpression between the African and Arabian plates. Geol Soc Lond Spec Publ 290:255–284. doi:10.1144/SP290

Shaliv G (1991) Stages in the tectonic and volcanic history of the Neogene basin in the Lower Galilee and the Valleys. Geol Surv Isr Report GSI/11/91, 100 p. (in Hebrew, Eng abs)

Sharkov EV (2000) Mesozoic and Cenozoic basaltic magmatism. In: Leonov, Yu G (eds) Outline of Geology of Syria. Rus Acad Sci, Geol Inst Trans, 526. pp 177–200. Nauka, Moscow (in Russian)

Shulman H, Reshef M, Ben-Avraham Z (2004) The structure of the Golan Heights and its tectonic linkage to the Dead Sea transform and the Palmyrides folding. Isr J Earth Sci 53:225–238

Smit J, Brun JP, Cloetingh S, Ben-Avraham Z (2010) The rift-like structure and asymmetry of the Dead Sea Fault. Earth Planet Sci Lett 290:74–82

Smit J, Brun JP, Cloetingh S, Ben-Avraham Z (2011) The rift-like structure and asymmetry of the Dead Sea Fault. Earth Planet Sci Lett 290:74–82

Sneh A, Weinberger R (2003) Geology of the Metulla quadrangle, northern Israel: implications for the offset along the Dead Sea rift. Isr J Earth Sci 52:123–138

Sobolev SV, Petrunin A, Garfunkel Z, Babeyko AY (2005) Thermo-mechanical model of the Dead Sea transform. Earth Planet Sci Lett 238:78–95

Steinitz G, Bartov Y (1991) The Miocene-Pliocene history of the Dead Sea segment of the rift in light of K-Ar ages of basalts. Isr J Earth Sci 40:199–208

Steinitz G, Bartov Y, Eyal M, Eyal Y (1981) K-Ar age determination of Tertiary magmatism along the western margin of the Gulf of Elat. Geol Surv Isr Curr Res 1980:27–29

Stern RJ (1994) Arc assembly and continental collision in the neoproterozoic E African Orogen: implications for the consolidation of Godwanaland. Ann Rev Earth Planet Sci 22:319–351

Suess E (1891) Die Brüche des östlichen Africa. Denkschrifte Akademie Wissenschaft Wien 58:555–584

Tatar O, Piper JDA, Gürsoy H, Heimann A, Koçulut F (2004) Neotectonic deformation in the transition zone between the Dead Sea Transform and the East Anatolian fault zone, southern Turkey: a palaeomagnetic study of the Karasu Rift volcanism. Tectonophysics 385:17–43

Ten Brink US, Ben-Avraham Z (1989) The anatomy of a pull-apart basin: seismic reflection observations of the Dead Sea basin. Tectonics 8:333–350

Ten Brink US, Flores CH (2012) Geometry and subsidence history of the Dead Sea basin: a case for fluid-induced mid-crustal shear zone? J Geophys Res 117:B01406. doi:10.1029/2011JB008711

Ten Brink US, Ben-Avraham Z, Bell RE, Hassouneh M, Coleman DF, Andersen G, Tibor G, Coakley B (1993) Structure of the Dead Sea pull-apart basin from gravity analysis. J Geophys Res 98:21877–21894

Ten Brink US, Rybakov M, Al Zoubi AS, Hassouneh M, Frieslander U, Batayneh AT, Goldschmidt V, Daoud MN, Rotstein Y, Hall JK (1999) Anatomy of the Dead Sea transform: does it reflect continuous changes in plate motion? Geology 27:887–890. doi:10.1130/0091-7613

Ten Brink US, Al Zoubi AS, Flores CH, Rotsterin Y, Qabbani I, Harder SH, Keller RG (2006) Seismic imaging of deep low velocity zone beneath the Dead Sea basin and transform faults: implications for strain localization and crustal rigidity. Geophys Res Lett 33:L24314. doi:10.1029/2006GL027890

Trifonov VG, Trubikhin VM, Adjamian J, Jallad Z, El-Khair Y, Ayad H (1991) The Levant fault zone in northwestern Syria. Geotectonics 2:63–75

Vidal N, Alvarez-Marron J, Klaeschen D (2000) Internal configuration of the Levantine Basin from seismic reflection data (Eastern Mediterranean). Earth Planet Sci Lett 180:77–89

Walley CD (1983) The paleoecology of the Callovian-Oxfordian strata of Majdal Shams (Syria) and its implications for Levantine palaeogeography and tectonics. Palaeogeogr Palaeoclim Palaeoecol 42:323–340

Walley CD (1988) A braided strike-slip model for the northern continuation of the Dead Sea fault and its implications for Levantine tectonics. Tectonophysics 1456:63–72

Walley CD (1998) Some outstanding issues in the geology of Lebanon and their importance in the tectonic evolution of the Levantine region. Tectonophysics 298:37–62

Weber M, DESERT Working Group (2004) The crustal structure of the Dead Sea transform. Geophys J Int 156:655–681. doi:10.1111/j.-1365/246X-2004.020143.x

Weber M, DESERT working Group (2009) Anatomy of the Dead Sea transform from lithospheric to microscopic scale. Rev Geophys 47:RG2002. doi:10.1029/2008RG000264

Weinberger R, Sneh A, Harlavan Y (2003) The occurrence of Middle-Miocene volcanism at Mount Hermon, northern Israel. Isr J Earth Sci 52:179–184

Weinberger R, Gross MR, Sneh A (2009) Evolving deformation along a transform plate boundary: example from the Dead Sea Fault in northern Israel. Tectonics 28:TC5005. doi:10.1029/2008TC002316

Weinberger R, Schattner U, Medvedev B, Frieslander U, Sneh A, Harlavan Y, Gross MR (2011) Convergent strike–slip across the Dead Sea Fault in northern Israel, imaged by high-resolution seismic reflection data. Isr J Earth Sci 58:203–216. doi:10.1560/IJES.58.3-4.203

Weinstein Y (2012) Transform faults as lithospheric boundaries, an example from the Dead Sea Transform. J Geodyn 54:21–28

Westaway R (2003) Kinematics of the Middle East and Eastern Mediterranean. Turk J Earth Sci 12:5–46

Westaway R, Arger J (1996) Gölbaşi basin, southeastern Turkey: a complex discontinuity in a major strike-slip fault zone. J Geol Soc Lond 153:729–744

Westaway R, Demir T, Seyrek A, Beck A (2006) Kinematics of active left-lateral faulting in SE Turkey from offset Pleistocene river gorges: improved constraint on the rate and history of relative motion between the Turkish and Arabian plates. J Geol Soc Lond 163:149–164. doi:10.1144/0016-764905-030

Wilson JT (1965) A new class of faults and their bearing on continental drift. Nature 207:343–347

Wu JE, Ken McClay K, Whitehouse A, Dooley T (2009) 4D analogue modelling of transtensional pull-apart basins. Mar Pet Geol 6:1608–1623

Yilmaz Y (1993) New evidence and model on the evolution of the southeast Anatolian orogen. Geol Soc Am Bull 105:251–271

Zak I (1967) The geology of Mount Sedom. Thesis Hebrew University of Jrusalem, 208p (in Hebrew, English abstract)

Zanchi A, Crosta GB, Darkal AN (2002) Paleostress analyses in NW Syria: constraints on the Cenozoic evolution of the northwestern margin of the Arabian plate. Tectonophysics 357:255–278

Chapter 6
Pleistocene Strain Partitioning During Transpression Along the Dead Sea Transform, Metulla Saddle, Northern Israel

Ram Weinberger

Abstract The geometry and kinematics of deformation in the Metulla Saddle, a ~4 km-wide and ~9 km-long highly deformed structure across the Dead Sea Transform (DST) between the Hula basin and the Lebanese restraining bend (LRB) is reviewed. The variety of structures exposed provides a unique opportunity to investigate the style and sequence of deformation associated with a transform plate boundary through geological mapping, meso-structural and anisotropy of magnetic susceptibility (AMS) analyses, interpretation of high resolution seismic reflection profiles, and radiometric dating. Cretaceous and Tertiary rocks within the Metulla Saddle are faulted and folded into broad anticlines and synclines, with more intense localized shortening manifested by tight folds and thrust duplexes. There are several structural features that attest to the dominance of a N-S component of shear imposed on the plate boundary zone, including nearly vertical fault planes, horizontal striae, alternating throw from one side of the fault to the other, and an assemblage of en-echelon folds and "positive flower structures". Kinematic analysis of fault data and principal AMS axes reveals strain axes, which are compatible with simple shear deformation along ~ N-S striking left-lateral faults. Numerous structures revealing the transpressive nature of the Metulla Saddle, including N-S striking reverse faults, reactivated normal faults with reverse motion, and copious N-S trending folds of different scales. Kinematic analyses of these features as well as that of principal AMS axes support distributed E-W shortening within the deformed blocks. Numerous ~ E-W striking calcite-filled veins emplaced in the Metulla Saddle manifest a perpendicular N-S extension response. The U-Th radiometric ages of calcite precipitates obtained and K-Ar dating of folded basalts indicate that Pleistocene

R. Weinberger (✉)
Division of Geological Mapping and Tectonics, Geological Survey of Israel, 30 Malkhe Israel St., Jerusalem 95501, Israel

Department of Geological and Environmental Sciences, Ben-Gurion University of the Negev, Beer-Sheva 84105, Israel
e-mail: Rami.weinberger@gsi.gov.il

Z. Garfunkel et al. (eds.), *Dead Sea Transform Fault System: Reviews*, Modern Approaches in Solid Earth Sciences 6, DOI 10.1007/978-94-017-8872-4_6,

activity occurred interchangeably along N-S trending strike-slip faults, N-S trending contractional structures, and ~ E-W calcite-filled veins. The structural analysis provides evidence for the transition from an early (pre-Pleistocene) phase of pure (non-partitioned) strike-slip motion to a late (Pleistocene) phase of convergent strike slip. The latter phase is characterized by strain partitioning during transpression, which is demonstrated by discrete left-lateral strike-slip motion across weak N-S faults and the development of a (mini) fold-thrust belt in response to transform-normal shortening. The contractional structures that evolved in the Metulla Saddle are best interpreted as indicating increased convergence along the DST from the Hula basin northward during the Pleistocene. The increased convergence might be attributed to a more easterly position of the Pleistocene pole of rotation between the Arabian plate and the Sinai sub-plate relative to the pole position describing the total motion in the last 5 Ma.

Keywords Convergent strike slip • Strike-slip partitioning • Transpression • Anisotropy of magnetic susceptibility • Hula basin

6.1 Introduction

Strike-slip deformations commonly deviate from simple shear due to a component of strain orthogonal to the deformation zone. This leads to transpression (more aptly named "convergent strike slip") or transtension, which are deformations of simple shear and orthogonal shortening or extension, respectively (Harland 1971). Transpression or transtension occur in a wide variety of tectonic settings and scales during deformation of the Earth's lithosphere. On the largest scale, these types of deformations are an inevitable consequence of relative plate motion on a spherical surface (Le Pichon et al. 1973; Dewey et al. 1998), i.e., plate slip vectors are not precisely orthogonal or parallel to plate boundaries. Hence, plate boundary zones experience oblique plate convergence or divergence at some time during their history along some part of their length (Dewey 1975). For transforms, transpression or transtension evolves wherever the trace of the transform does not follow a small circle of rotation exactly and are mainly pronounced along restraining and releasing bends. Kinematic models based on homogeneous transpression describe the three-dimensional nature of continental deformation (e.g., Sanderson and Marchini 1984), but assume that no major discontinuities exist within the deforming zones (Fig. 6.1a). Kinematic models of strike-slip partitioning take into account the existence of major strike-slip faults in the deforming zones, showing the relation between plate motion, instantaneous strain axes and the degree of strike-slip partitioning during oblique plate convergence (Teyssier et al. 1995). In non-partitioned transpression, the strike-slip and shortening components of strain are coupled, resulting in an oblique faulting (Norris et al. 1990), en-echelon folds with axes striking ~20–40° to the main transform (Aydin and Page 1984; Sylvester 1988), strike-slip duplexes (Woodcock and Fischer 1986), and maximum shortening oriented 35–45° to the main transform (Tikoff and Teyssier 1994). In contrast, for partitioned transpression the strike-slip and shortening components are decoupled into discrete zones of deformation

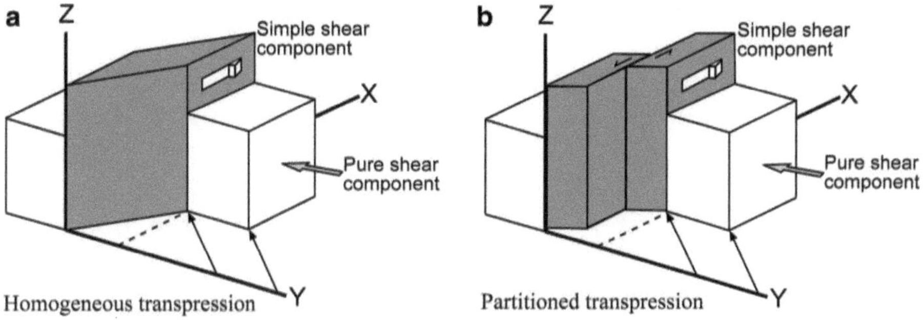

Fig. 6.1 (**a**) Homogeneous (non-partitioned) transpression in which homogenous strain is accommodated between two obliquely converging ridged blocks. Simple shear component is parallel to X-axis, orthogonal shortening is parallel to Y-axis (pure shear component), and extension is parallel to Z-axis (After Sanderson and Marchini 1984); (**b**) Partitioned transpression in which a significant simple shear component parallel to X-axis is accommodated by a discrete sinistral strike-slip fault (After Teyssier et al. 1995; Dewey et al. 1998)

(Fig. 6.1b). For complete partitioning, the strike-slip component is accommodated by the bounding, transform-parallel strike-slip faults and the shortening by transform-normal deformation (reverse faults, folds) distributed within the intervening blocks (Mount and Suppe 1987; Teyssier et al. 1995; Dewey et al. 1998).

The ~1,000 km long Dead Sea Transform (DST) accommodates strike-slip motion between the Arabian plate and the Sinai sub-plate. It connects the seafloor spreading in the Red Sea with the continental collision at the Eurasian plate (i.e., the Bitlis suture). A host of stratigraphic, structural and geochronological evidence indicates ~105 km of left-lateral offset across the DST since the Early to Middle Miocene (e.g., Quennell 1959; Freund et al. 1970; Garfunkel 1981; Joffe and Garfunkel 1987; Sneh and Weinberger 2003a). The DST is divided into three main sectors (Fig. 6.2): (1) the ~520 km long N-S to NNE-SSW trending southern section, linking the Red Sea rift to the Hula basin; (2) the ~140 km long NE-SW trending section, defined by the Yammunneh fault in Lebanon; and (3) the ~340 km long, N-S trending northern section from the Ghab basin in Syria into southern Turkey. In the southern section of the DST, the main faults are commonly shifted to the west through a series of overlapping, left-stepping, en-echelon segments. This geometry resulted in the initiation of several rhomb-shaped basins, including the Dead Sea and Hula basins, and the formation of low, sub-aerial topographic depressions. In the central section of the DST, the right bending of the NNE-striking Yammunneh fault corresponds to restraining geometry that uplifted the Lebanon and Anti-Lebanon mountains to 3,000 m above sea level (e.g., Beydoun 1977; Daeron et al. 2004; Gomez et al. 2006). The transition from the Hula basin to the Lebanese restraining bend (LRB) is complex, with contractional deformation distributed across several faults, fractures and folds. The main objective of this paper is to describe the geometry, kinematics and temporal sequence of deformation in this critical juncture through geological mapping, meso-structural and anisotropy of magnetic susceptibility (AMS) analyses, interpretation of high resolution seismic reflection profiles, and radiometric dating. Further, the variety of structures exposed

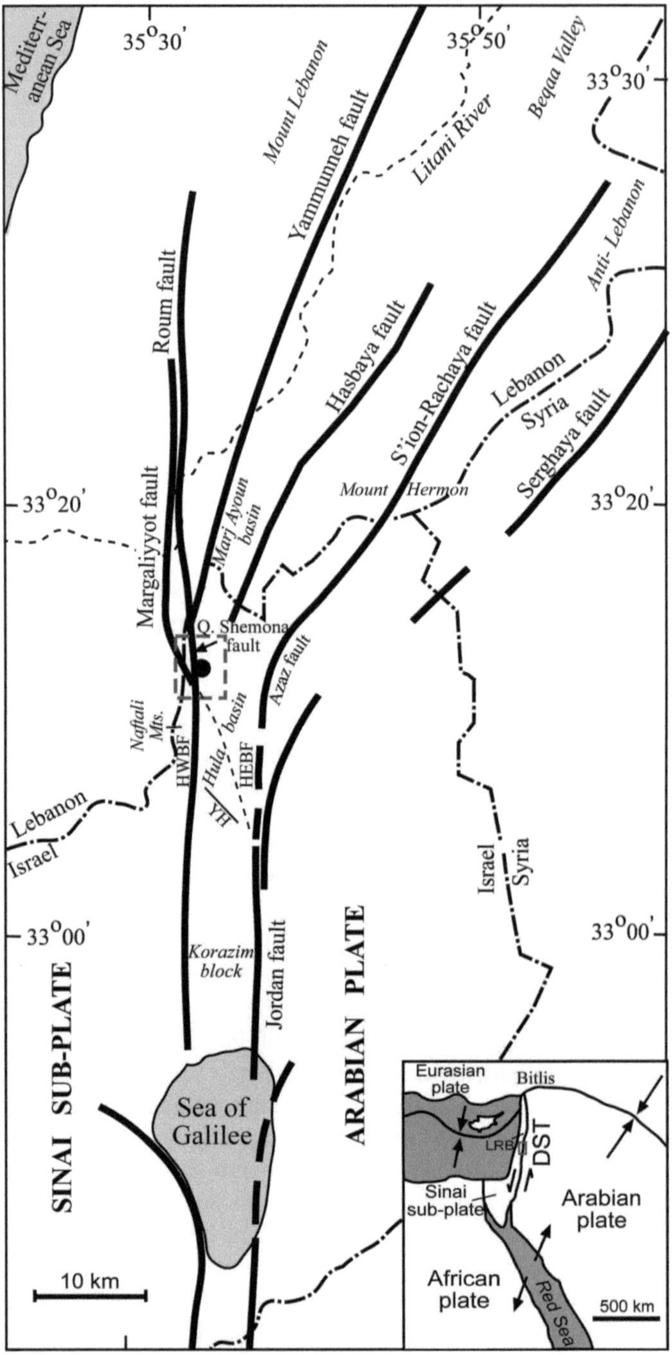

Fig. 6.2 Map of the main fault segments of the Dead Sea Transform (DST) zone in northern Israel and southern Lebanon (After Sneh and Weinberger 2003a). *Dashed rectangle* marks the Metulla Saddle study area near the town of Qiryat Shemona. Hula Western Border fault (HWBF) and Hula Eastern Border fault (HEBF) bound the Hula basin (the present Hula Valley). The concealed Hula diagonal fault crosses the basin and is marked by a *dashed line*. *Solid circle* – Qiryat Shemona town; *YH* Yesod HaMa'ala lineament. *Inset* – plate tectonic configuration resulting in left-lateral motion across the DST. *LRB* Lebanese restraining bend

both along discrete fault zones and within intervening blocks provides important geological constraints for models of transpression and strike-slip partitioning proposed for continental transforms. This paper mainly synthesizes studies that have been published by Weinberger and co-workers over the last decade (Sneh and Weinberger 2003a, b, 2006; Weinberger and Sneh 2004; Schattner and Weinberger 2008; Weinberger et al. 2009, 2010; Levi and Weinberger 2011; Nuriel et al. 2010, 2012a, b). It focuses mainly on the Pleistocene phase of deformation, in which strain partitioning during transpression prevailed along the DST in northern Israel.

6.2 Geologic Setting

The Hula basin is a graben that began as a pull-apart basin at ~4 Ma (Heimann 1990; Heimann et al. 2009). The basin is bounded by two prominent N-S striking border faults, the Hula Eastern Border Fault (HEBF) and the Hula Western Border Fault (HWBF) on its eastern and western margins, respectively (Fig. 6.1). At the southeastern end of the Hula basin the Jordan fault zone is the major active strand of the DST (Garfunkel et al. 1981; Marco et al. 1997; Sneh and Weinberger 2006). Hence, the HEBF connects the Jordan fault at the south with the Azaz fault at the north (Fig. 6.1). The HWBF extends northward to the Qiryat Shemona fault. The NW-striking Margaliyyot fault branches from the HWBF-Qiryat Shemona line at the northwestern part of the Hula basin. South of the basin the elevated block of Korazim separates the basin from the Sea of Galilee and other subsided basins along the southern sector of the DST.

Previous studies show a possible occurrence of several normal faults bounding the southern end of the Hula basin (Heimann 1990; Rybakov et al. 2003) but no clear bounding faults at the northern part of the basin (Heimann 1990; Sneh and Weinberger 2003a; Heimann et al. 2009). A NW-striking Yesod HaMa'ala lineament is expressed as a 1–3 m high scarp in the southern part of the basin. According to Heimann et al. (2009) this lineament is the present southern bounding fault of the basin, linking the Jordan fault zone in the east with the HWBF in the west. Schattner and Weinberger (2008) analyzed seismic reflection and borehole data from the Hula basin using the ~1 Ma Hazbani Basalt (Mor 1993) as a geological marker. They encountered several minor NW-striking transverse normal faults at the northeastern part of the basin, but neither the Pleistocene Hazbani Basalt nor sediments overlying these faults are displaced by them. Schattner and Weinberger (2008) demonstrated that a sub-vertical NNW-trending left-lateral strike-slip fault diagonally ruptures the basin, displacing the Hazbani Basalt. They suggested that during the mid-Pleistocene the basin entered a new geodynamic phase, in which deformation was controlled by both the N-S striking border faults and the inter-basin diagonal fault.

Further splaying occurs northwards, where the HWBF-Qiryat Shemona line branches into the Roum and Yammunneh faults in southern Lebanon (Figs. 6.2 and 6.3; Sneh and Weinberger 2003a, b). Physiographically, the LRB consists of two distinct mountain ranges: Mount Lebanon and the Anti-Lebanon ranges separated

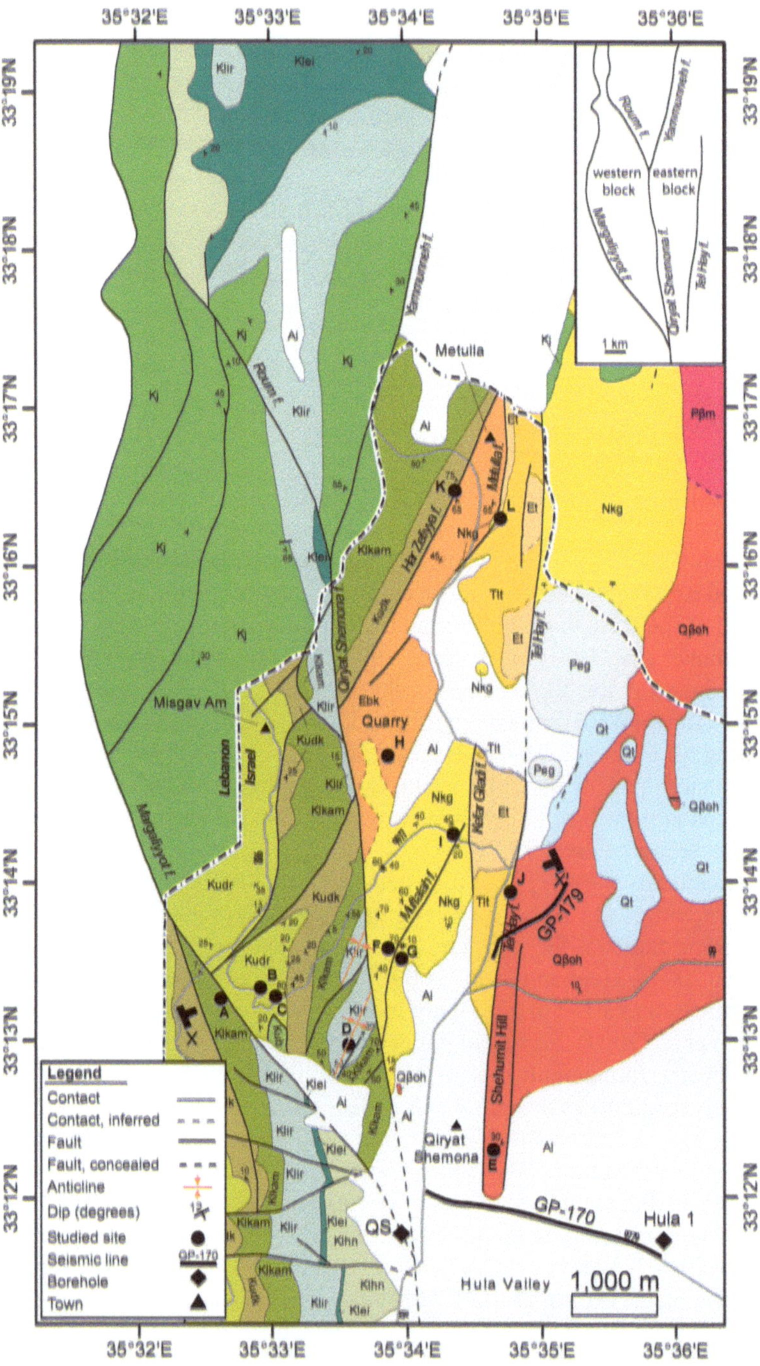

Fig. 6.3 Geological map of the study area (Sneh and Weinberger 2003b) showing rock units, faults (*solid lines*; *dashed lines* when concealed), two prominent en-echelon anticlines (*solid orange lines*); locations of sites referred to in the text (*black circles*), location of cross Section X-X' (Fig. 6.4a), locations of seismic profiles (*thick solid lines*), boreholes (*solid diamonds*), and main roads (*gray lines*). *Inset*: Division into structural blocks and the bounding faults in the Metulla

by the Beqaa Valley (Fig. 6.1). The Yemmunneh fault delineates Mount Lebanon from the east, and the Serghaya fault delineates the Anti-Lebanon range from the west. The total width of the restraining bends exceeds 80 km and it might accommodate regional shortening of 10–15 % (Hancock and Atiya 1979). More details about the transpressional style of deformation of the LRB are given by Freund (1965), Ron (1987), Butler et al. (1998), Griffiths et al. (2000), Gomez et al. (2006, 2007a, b), Homberg et al. (2010) and Henry et al. (2010). Mount Hermon is another highly deformed block situated along an oblique segment of the DST known as the S'ion-Rachaya fault at the northeastern tip of the HEBF (Freund et al. 1970; Garfunkel 1981; Ron et al. 1990). The small Marjayoun valley (Fig. 6.2) juxtaposed on the southern tip of the Yammunneh fault is considered by Freund et al. (1968) a pull-apart basin.

The Qiryat Shemona fault subdivides the area into two main blocks: a western block, between the Qiryat Shemona and Margaliyyot faults; and an eastern block between the Qiryat Shemona and Tel-Hay faults (Picard 1952; Glikson 1966; Weinberger and Sneh 2004; Fig. 6.3). These blocks, known together as the Metulla Saddle and consisting of highly deformed Cretaceous to Neogene beds, show characteristics of transpression and strike-slip partitioning and are the main focus of this paper. Figure 6.4a shows a W-E cross section through the western and eastern blocks, based on a geologic mapping (Sneh and Weinberger 2003a). The width of the blocks, with an average of ~2 km, reaches a maximum of about 3 km. A set of faults that strike approximately NE-SW divide the blocks into sub-blocks (Fig. 6.3), each displaying highly deformed beds, which are sub-vertical in places (Dubertret 1951; Picard 1952; Glikson 1966; Ron et al. 1997; Weinberger and Sneh 2004). Strata exposed in the western block belong to the carbonate rocks of the Cretaceous Kurnub and Judea groups. The rocks are folded into a broad syncline, with moderate dips along the eastern limb and vertical beds with tight parasitic folding in the core of the syncline (Fig. 6.4b). Strata exposed in the eastern block consist of the Cretaceous Judea Group as well as the Paleogene Taqiye, Timrat, and Bar Kokhba formations, and the Neogene Kefar Giladi Formation. The latter formation consists of >400 m of conglomerate and lacustrine sediments (Glikson 1966) that accumulated in a depression adjacent to the HWBF-Qiryat Shemona line during the Late Miocene – Early Pliocene (Sneh and Weinberger 2003a).

Fig. 6.3 (continued) Saddle. Rock unit abbreviations: *Klhn* Hatira and Nabi Sa'id fms. (sandstone, marl, Neocomian-Barremian), *Klei* Ein el Assad Fm. (limestone, Aptian), *Klir* Hidra and Rama fms. (limestone, marl, chalk, Aptian-Albian), *Klkam* Kamon Fm. (dolostone, Albian-Cenomanian); *Kudk* Deir Hanna Fm., Karkara Mbr. (dolostone, Cenomanian), *Kudr* Deir Hanna Fm., Rosh Haniqra Mbr. (chalk, limestone, marl, Cenomanian), *Kub* Bina Fm. (limestone, Turonian), *Kj* Judea Group (limestone and dolostone, undivided, Aptian-Turonian), *Tlt* Taqiye Fm. (marl, Paleocene), *Kums* Mount Scopus Group (chalk, undivided, Senonian-Paleocene), *Et* Timrat Fm. (chalk, limestone, Eocene), *Ebk* Bar Kokhba Fm. (limestone, Eocene), *Nkg* Kefar Gil'adi Fm. (conglomerate, limestone, clay, Neogene), *Qhu* Hula Group, *Peg* Egel Gravel (Plio-Pleistocene), *Pβm* Meshki Basalt (Pliocene), *Qβoh* Hazbani Basalt (Pleistocene), *Qt* Tufa (Pleistocene), *Al* Alluvium, terrace conglomerates and landslides (Pleistocene – Holocene)

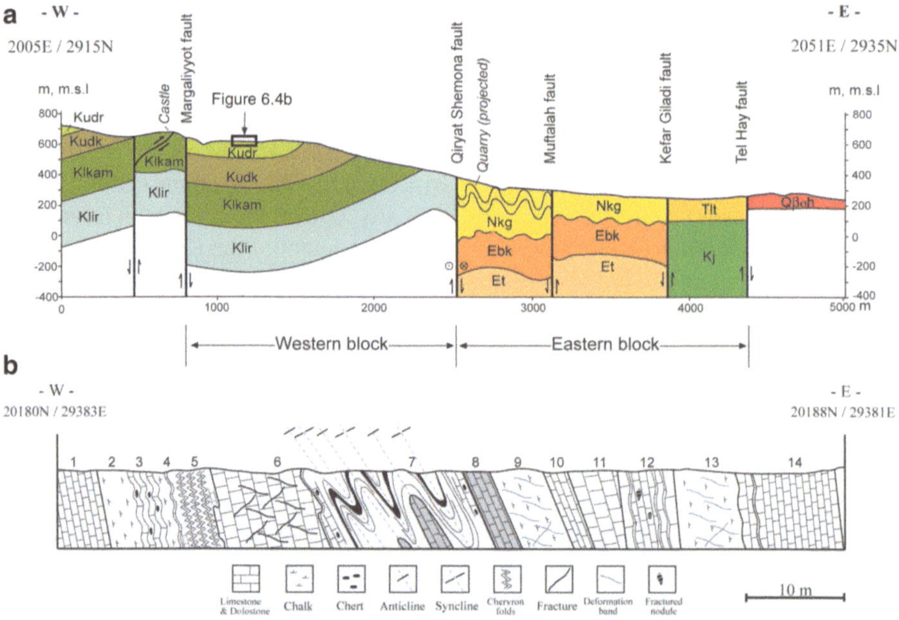

Fig. 6.4 (**a**) E-W cross section (line X-X' on Fig. 6.3) based on geologic mapping (Note the subdivision into eastern and western blocks. See rock units in Fig. 6.3). (**b**) Cross section at the core of the western block (Site B, Fig. 6.3; see location in Fig. 6.4a) showing numbered mechanical units and their associated structures. *Horizontal* and *vertical* scales are equal (After Weinberger et al. 2009)

6.3 Indications for a Distributed N-S Shear

Apart from the large, plate-scale observations and arguments in favor of left-lateral offset across the DST (Quennell 1959; Freund et al. 1970; Garfunkel 1981; Joffe and Garfunkel 1987; Le Pichon and Kreemer 2010), there are several structural features in the Metulla Saddle that attest to the N-S component of shear imposed on the plate-boundary zone in this area. These features include nearly vertical fault planes, horizontal striae, alternating throw from one side of the fault to the other, and an assemblage of en-echelon folds and positive "flower structures". In addition, kinematic analysis of fault data and AMS axes in the Metulla Saddle reveal strain axes, which are compatible with simple shear deformation along ~ N-S striking faults. Those features and analyses are described in more detail below.

6.3.1 Fault Data

Numerous subsidiary faults within the Eocene Bar Kokhba limestone are exposed in the Kefar Giladi Quarry (Site H, Fig. 6.3), forming a wide zone of brittle deformation to the east of the Qiryat Shemona master fault. Many of the faults are vertical to sub-vertical polished surfaces with associated breccia and gouge zones (Fig. 6.5a).

Fig. 6.5 Photographs of structures observed in the field area. (**a**) Vertical faults in the Eocene Bar Kokhba Formation exposed in the Kefar Giladi Quarry. The faults strike N-S with sub-horizontal undulations and slickenside lineations (Site H, Fig. 6.3). (**b**) Low-angle Metulla thrust fault with Lower Eocene Timrat Formation in the hanging wall and Neogene Kefar Giladi Formation in the footwall (at Site L, Fig. 6.3). (**c**) Oblique photo of folded and thrusted Cenomanian dolostone beds (Kamon Fm.) exposed in the north facing wall of the Chateau Neff Crusader castle (Site A, Fig. 6.3). *Solid lines* mark main thrusts. For detailed interpretation see Fig. 5 in Weinberger et al. (2009). (**d**) Polished hand sample taken from a calcite-filled vein within the Bar Kokhba Formation (Site H, Fig. 6.3), showing bands of calcite parallel to the vein wall (*top*) and brecciated host rock (*bottom*)

Lineations on the fault surface appear as either undulating fault grooves or fault striae (slickenside lineations). Occasionally, pull-aparts, slickolites, and wear-grooves indicate the sense of motion along the faults. Data were collected from 119 faults where accurate orientations could be measured, and where the fault striae and grooves allowed for the measurement of lineations (Site H, Fig. 6.3; see details in Weinberger et al. 2009). The most abundant set of vertical to sub-vertical faults strikes ~ N-S (Fig. 6.6; see also Fig. 6a in Ron et al. 1997). A less abundant set of sub-vertical faults strikes approximately E-W. There are also some shallowly dipping faults and faults that do not fall within the two main trends. The overwhelming

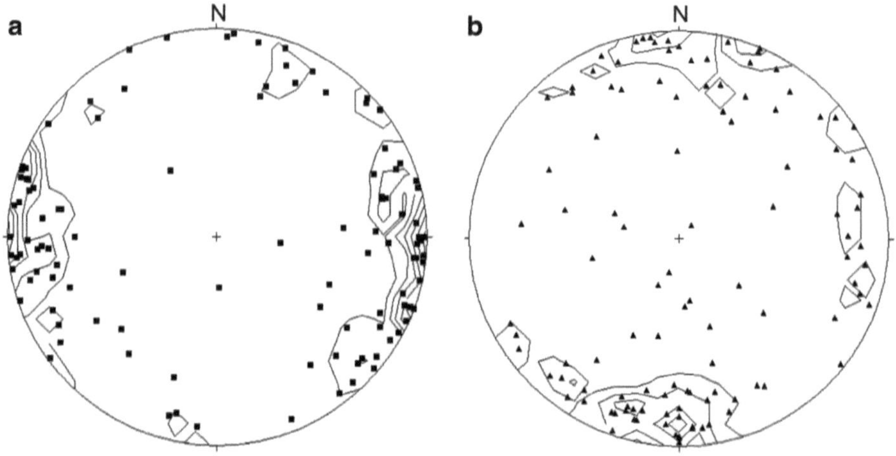

Fig. 6.6 Fault data collected from the Eocene Bar Kokhba Formation in the Kefar Giladi Quarry (Site H, Fig. 6.3). Contour intervals are 2 % per 1 % area (From Weinberger et al. 2009). (**a**) Contoured poles to fault planes (squares; n = 119) showing that most faults are *vertical* and *strike* between NNE and NNW directions. (**b**) Contoured fault striae (triangles; n = 119) (Note that most fault striae are sub-horizontal)

majority (~80 %) of fault striae are sub-horizontal (<30°), indicating a dominant strike-slip motion, mostly on the ~ N-S striking faults.

A kinematic analysis of the fault data was performed using the *FaultKin* program, which calculates P and T axes for each fault based on fault orientation, striae orientation and the sense of motion (Allmendinger et al. 2001). The mean P and T axes approximate the infinitesimal shortening ($\dot{S}_3$) and extension ($\dot{S}_1$) directions, respectively, for a population of faults. The analysis included 94 out of 119 faults that have reliable indications for the sense of motion. The results yield a definitive fault plane solution with sub-vertical theoretical fault planes striking N-S and E-W (Fig. 6.7), individual P and T axes generally confined to separate quadrants, and infinitesimal strain axes characterized by sub-horizontal NW-SE shortening (mean P-axis of 131°/06°) and sub-horizontal NE-SW extension (mean T-axis of 221°/02°). The infinitesimal strain axes resolve into a left-lateral strike-slip motion along the ~ N-S striking fault plane and a right-lateral strike-slip motion along the E-W striking fault plane. This nearly perfect strike-slip solution along vertical planes provides a first-order indication for a fault-parallel, simple shear component along the N-S striking Qiryat Shemona fault zone.

The Tel-Hay fault is a nearly vertical N-S striking fault, bounding the eastern block on the east. In the southern portion of the fault adjacent to Shehumit Hill (Fig. 6.3), the Hazbani Basalt in the western side is thrown downward by ~50 m (Weinberger et al. 2010). This throw decreases northward up to the point where it alternates, and the Hazbani Basalt in the eastern side is thrown downward (Site J, Fig. 6.3). This throw alternation is typical for strike-slip faulting. The Tel-Hay fault

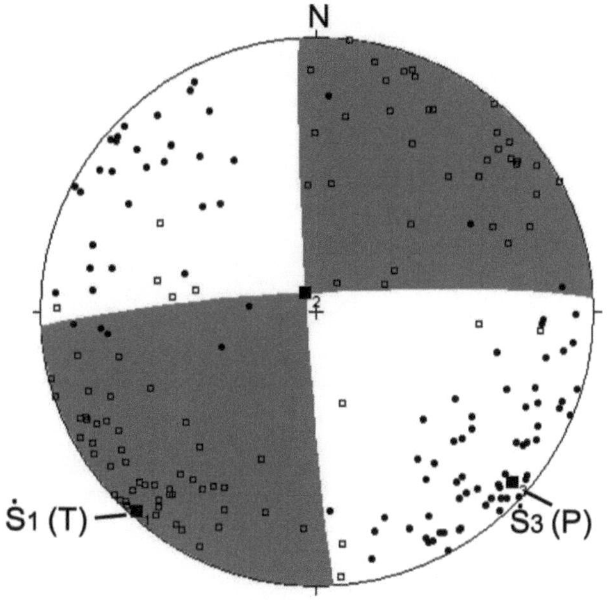

Fig. 6.7 Results of kinematic analysis of fault data collected at the Kefar Giladi Quarry (Site H, Fig. 6.3). A fault plane solution is based on kinematics of 94 faults. Infinitesimal shortening ($\dot{S}_3$) and extension ($\dot{S}_1$) directions are indicated (From Weinberger et al. 2009)

is recognized as a prominent vertical discontinuity in several high-resolution seismic reflection profiles (e.g., GP-179; Fig. 6.8). In these profiles, curved reflectors toward the fault plane attest to the strike-component of motion along this fault in addition to the observed alternating dip-slip component. Fault strands that branch from the parent fault (e.g., at ~150 ms TWT in Fig. 6.8) form faint positive flower structures, which are also typical for a strike-slip assemblage.

6.3.2 *Fold Data*

Two prominent NNE-SSW trending, left-stepping, en-echelon anticlines with wavelengths of several hundred meters are exposed at the southeastern part of the western block (Site D, Fig. 6.3). The exposed cores of the anticlines are composed of the Albian Rama Formation, whereas the flanks expose the Albian-Cenomanian Kamon Formation (Weinberger et al. 2009). The anticlines terminated against (or are cut by) the Qiryat Shemona fault, forming an angle of ~30° to the fault line. Folds associated with strike-slip faults are typically arranged in en-echelon patterns oblique to the principal direction of shears (Sylvester 1988). The arrangement of the two prominent anticlines here is compatible with a left-lateral motion along the Qiryat Shemona master fault. A population of small-scale NNE-SSW trending folds with wavelengths of centimeters to meters in the western block might also have been formed coevally due to simple shear (Weinberger et al. 2009).

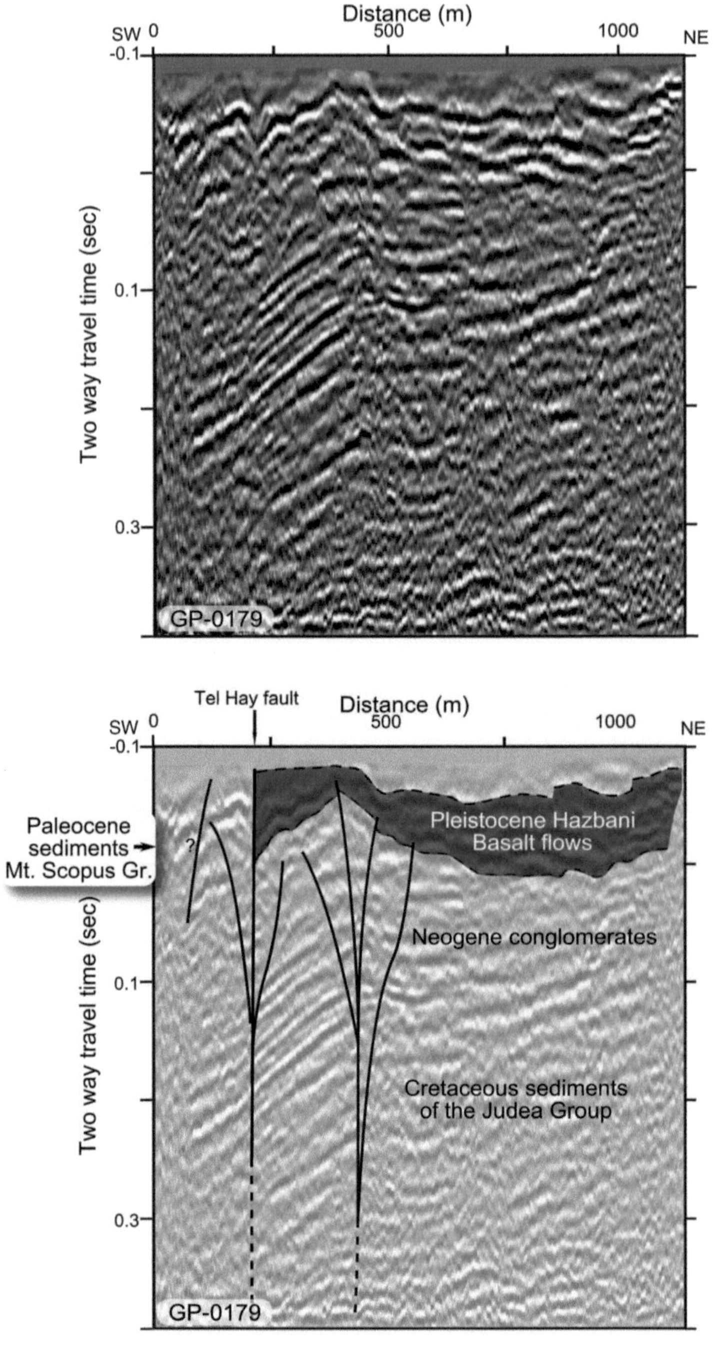

Fig. 6.8 Multi-channel seismic reflection profile GP-179 showing the Tel-Hay fault south of Site J. A *gray layer* on the lower panels marks the Hazbani Basalt. See Fig. 6.3 for location of the profile (After Weinberger et al. 2010)

6.3.3 *AMS Data*

The magnetic susceptibility of rocks results from the combined contribution of ferromagnetic, paramagnetic, and diamagnetic phases. The minerals are commonly magnetically anisotropic, depending on their preferred crystallographic orientation or grain shape. This gives rise to the anisotropy of magnetic susceptibility (AMS) of rocks, which provides a petrofabric tool by averaging out the orientation-distribution of all minerals and sub-fabrics in a rock specimen (e.g., Borradaile and Jackson 2010). AMS is describable as a magnitude ellipsoid, with maximum, intermediate and minimum principal susceptibilities (k_1, k_2, and k_3) as its axes. In many deformed settings the directions of the AMS axes show a good correlation with the directions of the principal strain axes (e.g., Borradaile and Henry 1997; Parés et al. 1999). In diamagnetic rocks, the magnetic fabrics are generally controlled by the alignment of calcite *c*-axes. The minimum principal AMS axes k_3 are parallel to the *c*-axes, indicating the principal shortening directions in the studied rocks (Borradaile and Henry 1997). Levi and Weinberger (2011) exploited the potential of using diamagnetic and weak paramagnetic AMS axes as indicators of the directions of principal strain axes in calcite-bearing rocks, isolating the AMS axes in Eocene and Neogene rocks that crop out in the eastern block. They found that AMS-based NW-SE shortening axes were preserved in three sites of weak paramagnetic rocks south of the branching point between the Har Zefiyya and Qiryat Shemona faults (Fig. 6.9). The AMS results show a good agreement between finite strain axes inferred from meso-scale NNE-trending folds and those inferred from k_3 AMS axes. The NW-SE shortening axes and are compatible with left-lateral motion along N-S striking faults (Levi and Weinberger 2011).

6.4 Indications for Imposed E-W Shortening

There are several structural features in the studied blocks that indicate E-W shortening due to transpression. The highly deformed beds in these blocks attracted the attention of previous workers (Dubertret 1951; Picard 1952; Glikson 1966) and were first related to E-W shortening by Ron et al. (1997) and later by Weinberger and Sneh (2004). These features include reverse faults, reactivated normal faults with reverse motion, and numerous N-S trending folds of different scales. In addition, kinematic analysis of fault data and AMS axes in the eastern block reveal E-W shortening as follows.

6.4.1 *Fault Data*

The ~ N-S striking Metulla fault is a thrust fault dipping 30° to the east, with the Eocene Timrat Formation overriding lacustrine beds of the Neogene Kefar Giladi Formation on the western side of the fault (Site L, Figs. 6.3 and 6.5b). This 'Andersonian' thrust

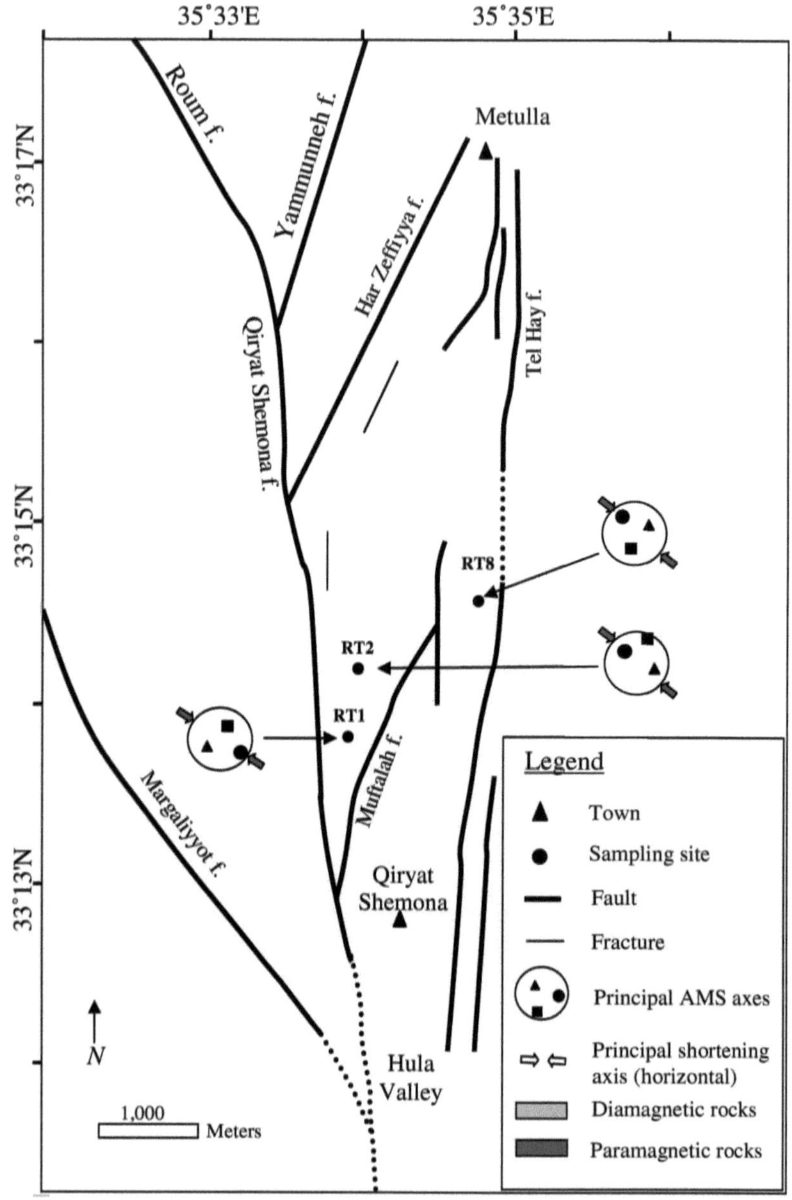

Fig. 6.9 Map of the main fault segments of the DST in the study area and sites with principal AMS axes (After Levi and Weinberger 2011). The stereograms are lower-hemisphere, equal-area projections of principal AMS axes k_1, k_2 and k_3 marked by *solid squares*, *triangles*, and *circles*, respectively. *Arrows* mark the inferred NW-SE principal shortening (*horizontal*) at the sites studied, all of which are based on paramagnetic AMS fabrics

fault indicates an E-W shortening. The NE-striking Har Zefiyya fault dips ~70° southeastward (Site K, Fig. 6.3), showing normal separation with the Eocene Bar Kokhba limestone in the hanging wall and the Cenomanian Karkara beds of the Deir Hanna Formation exposed in the footwall. However, kinematic indicators (dip-slip slickenside lineations and small-scale pull-apart basins filled with calcite) suggest a younger reverse motion along this fault. The reactivation of the Har Zefiyya fault due to E-W shortening is plausible. Another three prominent N-S striking reverse faults and several small-scale (~1 m) thrust and back-thrust faults within an Albian-Cenomanian massive dolostone unit are observed in the "Castle site" (Site A, Figs. 6.3 and 6.5c). The dip-slip sense of motion along these faults combined with results from fold analysis attest to E-W finite shortening at this site (Fig. 6.10; see below).

The abundance of faults with a left-lateral sense of motion may hide infinitesimal strain axes that were associated with faults characterized by either right-lateral or dip-slip motion. Hence, Weinberger et al. (2009) analyzed separately 18 faults with right-lateral senses of motion, as well as 14 faults with striae plunging more than 45° (dominantly dip-slip motion). The average P-axis of the faults with right-lateral motion is ESE-WNW, but contouring the P-axes reveals two maxima, one approximately E-W and the other NW-SE (Fig. 6.11a). The average P-axis derived from faults with striae plunging more than 45° is ESE-WNW and results in a thrust-fault plane solution (Fig. 6.11b).

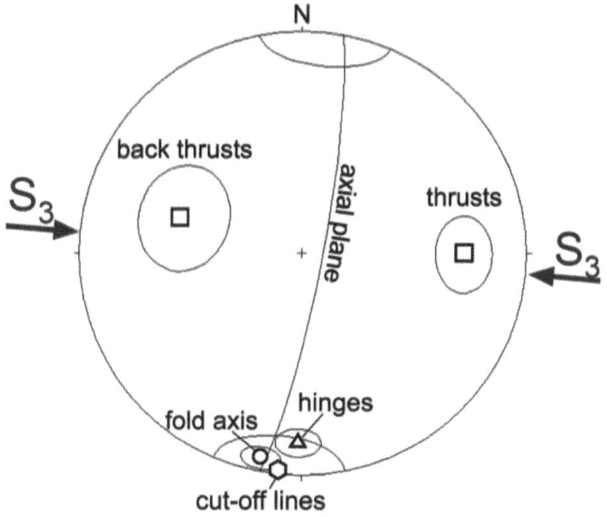

Fig. 6.10 Summary of structural data collected at Chateau Neff Crusader castle (site A, western block; Fig. 6.3). *Small circles* represent 95 % confidence intervals. Open symbols are means. *Rectangular* – thrusts (n=10) and back thrusts (n=7); *triangle* – hinges of small-scale folds (n=18); *circle* – fold axis based on π-diagram of poles to bedding measured throughout the castle exposure (n=26); *hexagon* – cutoff lines (the intersections of bedding exposed on a fault surface) (n=7). *Great circle* represents the mean axial plane of small-scale folds (n=13). *Arrows* indicate the inferred direction of finite shortening S_3 (After Weinberger et al. 2009)

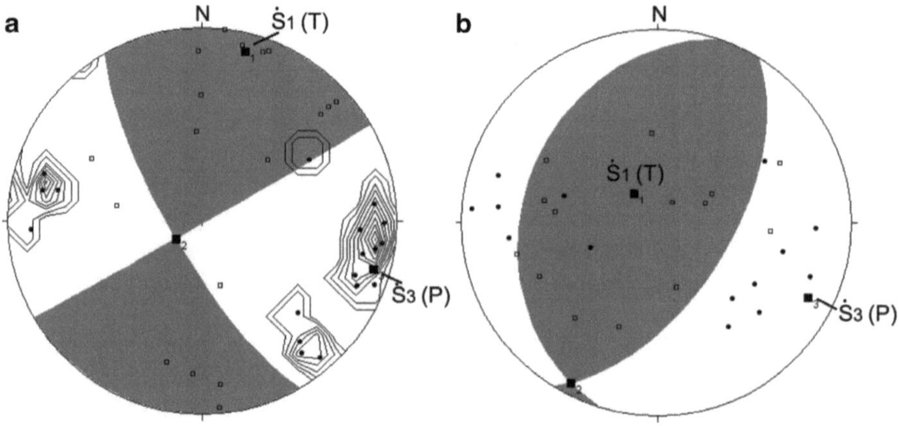

Fig. 6.11 Results of kinematic analysis of fault data collected at the Kefar Giladi Quarry (Site H, Fig. 6.3). (**a**) Contoured $\dot{S}_3$ axes and fault plane solution for 18 of the 94 faults with a component of right-lateral motion. Contour intervals are 2 % per 1 % area; (**b**) Infinitesimal shortening $\dot{S}_3$ and extension $\dot{S}_1$ axes and fault plane solution for 14 of the 94 faults with striae plunging more than 45° (From Weinberger et al. 2009)

6.4.2 Fold Data

The rocks in the western block are folded into a broad (~1.5 km wide) N-S trending syncline with moderate dips along the eastern limb and vertical beds with tight parasitic folding in the core of the syncline (Fig. 6.4b). In the "Castle site" (Site A, Fig. 6.3), the axis of the anticline is 190°/06° based on a cylindrical best-fit of poles to bedding measured along the length of the castle moat. The small-scale folds at the foot of the castle are tight, with typical wavelengths of several to tens of centimeters, and a mean hinge direction of 181°/15°. The measured fold geometries (hinges, axial planes, large-scale fold axis) at that site are internally consistent with overlapping confidence intervals, indicating a roughly sub-horizontal axis trending ~ N-S associated with ~ E-W shortening (Fig. 6.10). Ten medium to large-scale folds in the Eocene Bar Kokhba and Neogene Kefar Giladi formations, ranging in wavelength from several up to several hundreds of meters, were analyzed in the eastern block based on bedding measurements collected across each structure. The analysis suggests a bimodal distribution, with a dominant cluster centered about 177°/07° (Weinberger et al. 2009), indicating also E-W shortening.

Shehumit Hill is a north–south trending ridge about 2,200 m long and 230–400 m wide, uplifted ~85 m above the surrounding plains of the Hula basin (Site E, Fig. 6.3). The hill consists of several flows of the Hazbani Basalt dated to 1–1.5 Ma (Weinberger et al. 2010 and references therein). Previous studies interpreted the elevated ridge of Shehumit Hill as either a fissure eruption (Picard 1952), a tilted block (Schulman 1966) or a local push-up swell (Heimann and Ron 1987). High-resolution seismic reflection Profile GP-170 (see location in Fig. 6.3) crosses the southern front of Shehumit Hill (Fig. 6.12). Due to urban development, the western edge of the profile

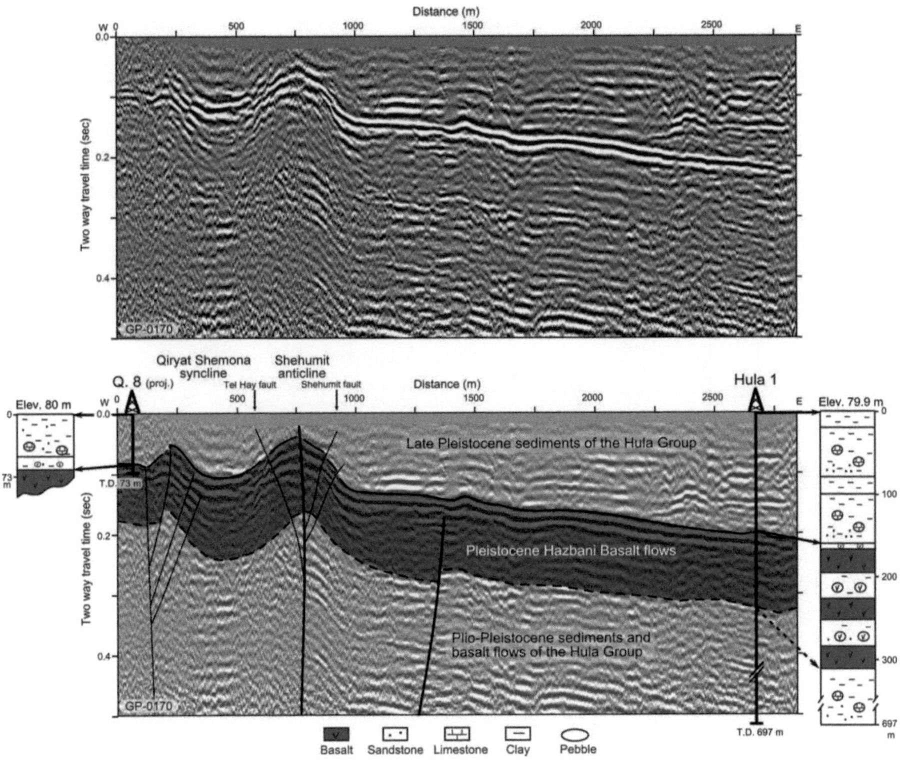

Fig. 6.12 Multi channel seismic reflection profile GP-170 crossing the buried southern continuation of the Shehumit structure. A *gray layer* on the lower panel marks the Hazbani Basalt. Contractional deformation is prominent between 0 and 300 m at ~100 ms TWT (next to Qiryat Shemona fault 100 m to the west of the profile) and between 600 and 1,000 m (Shehumit anticline). See Fig. 6.4 for location. Data of Hula 1 and Qiryat Shemona (Q8) boreholes are presented (From Weinberger et al. 2010)

does not cross the Qiryat Shemona fault but reaches to ~100 m away from it. One prominent reflector is correlated with the top of the upper flow of the Hazbani Basalt at a depth of 180 m in the Hula 1 borehole. A second prominent reflector is correlated with the lower flow of the Hazbani Basalt (Fig. 6.12). Contractional deformation in the form of a faulted anticline is prominent between 0 and 300 m at ~100 ms TWT. A positive flower structure south of Shehumit Hill between 600 and 1,000 m (at ~100 ms TWT) forms the Shehumit anticline. In between these two structures and beneath the town of Qiryat Shemona, the Hazbani Basalt is deformed into a syncline (hereafter, the Qiryat Shemona syncline). A contour map of the top surface of the Hazbani Basalt was produced by combining topographic outcrop elevations with subsurface depths based on high-resolution seismic reflection data (Fig. 6.13; Weinberger et al. 2010). The map shows that in the northwestern rim of the Hula basin basalt flows are folded into the two prominent structures, the N-S trending Shehumit anticline and a south-plunging Qiryat Shemona syncline. Both structures indicate E-W shortening together with strike-slip motion along the related faults (Fig. 6.14).

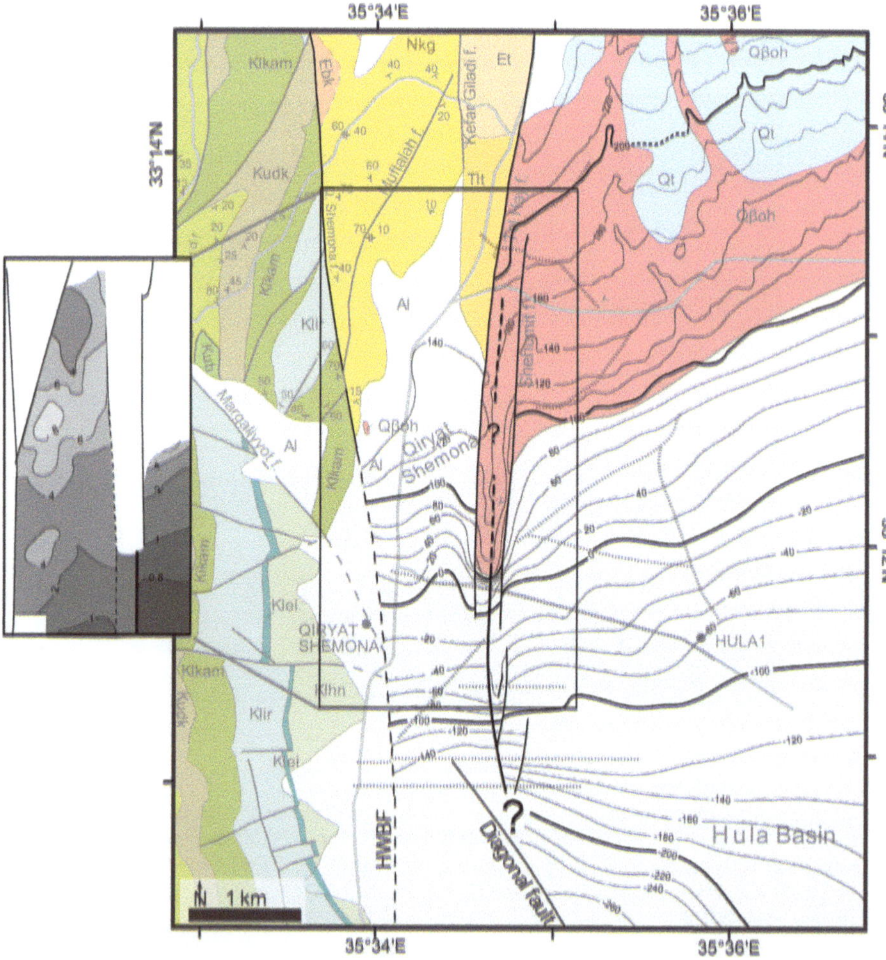

Fig. 6.13 A contour map of the (possibly eroded) top surface of the Hazbani Basalt (0 m.s.l. datum) overlying the geological map of the study area. The surface descends in a gentle slope southward from more than 240 m above m.s.l (outcrops) to 260 m below m.s.l (subsurface). Deformations of the Hazbani Basalt is concentrated in the western part of the Hula basin, where it is folded, forming a N-S trending anticline at Shehumit Hill, and a south-plunging syncline underneath the town of Qiryat Shemona. The locations of seismic reflection profiles are marked by *gray dotted lines*. *Inset*: Contour map of fundamental resonance frequencies (Zaslavsky 2010) (Note the discontinuity of the frequencies along the Shehumit fault and its southward extension into the Hula Valley (From Weinberger et al. 2010))

In light of the new imaging of the Shehumit structure in the subsurface and additional dip measurements of the Hazbani Basalt in outcrops, it is more appropriate to refer to the Shehumit structure as a fold rather than a tilted block (Schulman 1966). Based on Heimann and Ron (1987), the western flank of Shehumit Hill is not faulted, while its eastern flank is bounded by the Shehumit fault. They attributed the formation of the Shehumit structure to push-up due to left-lateral motion along the

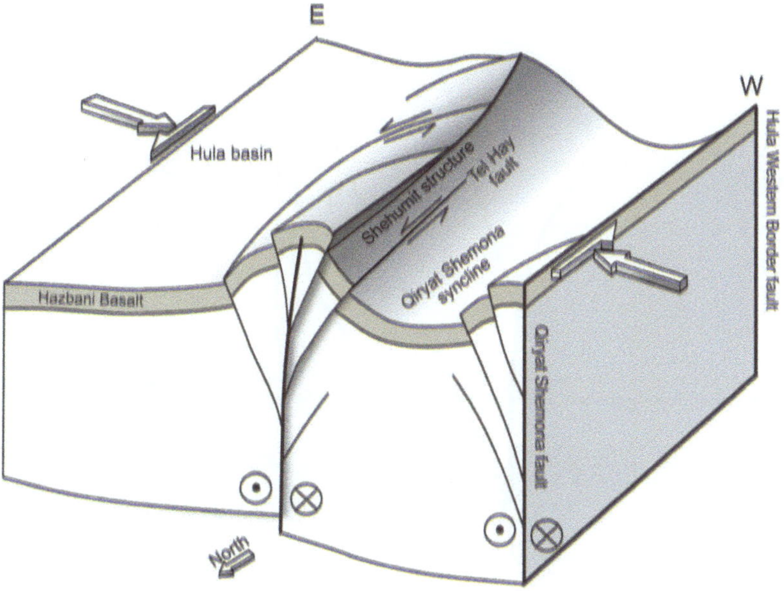

Fig. 6.14 Schematic representation of the Shehumit structure and related faults viewed from north to south. *Arrows* indicate shear and shortening strain components (After Weinberger et al. 2010)

right-bending Shehumit fault. This study shows two faults rather than one running along the foot of the hill (Tel Hay and Shehumit faults; Fig. 6.3). Based on the seismic data, these faults continue southward into the Hula basin and are interpreted as strands of a positive flower structure that extends beneath the Shehumit structure and several kilometers to the south. The possibility that a NE-trending fault located at the southern foot of the Shehumit Hill could not be rejected. Such a fault could be part of the splays within the flower structure (Fig. 6.14).

6.4.3 AMS Data

Levi and Weinberger (2011) obtained AMS-based E-W shortening axes in four sites distributed throughout the study area (Fig. 6.15). They show that the pole of the great circle along which the W-E oriented k_3 axes lie is geometrically and kinematically related to the ~N-S fold axes in eastern block and, hence, to E-W shortening.

6.5 Indications for Imposed N-S Extension

Two types of opening-mode fractures, calcite-filled veins and clastic dikes, are exposed in the Kefar Giladi Quarry (Site H, Fig. 6.3) and Har Zefiyya (Site K, Fig. 6.3) (Weinberger et al. 2009; Nuriel et al. 2010, 2012a, b). The veins are

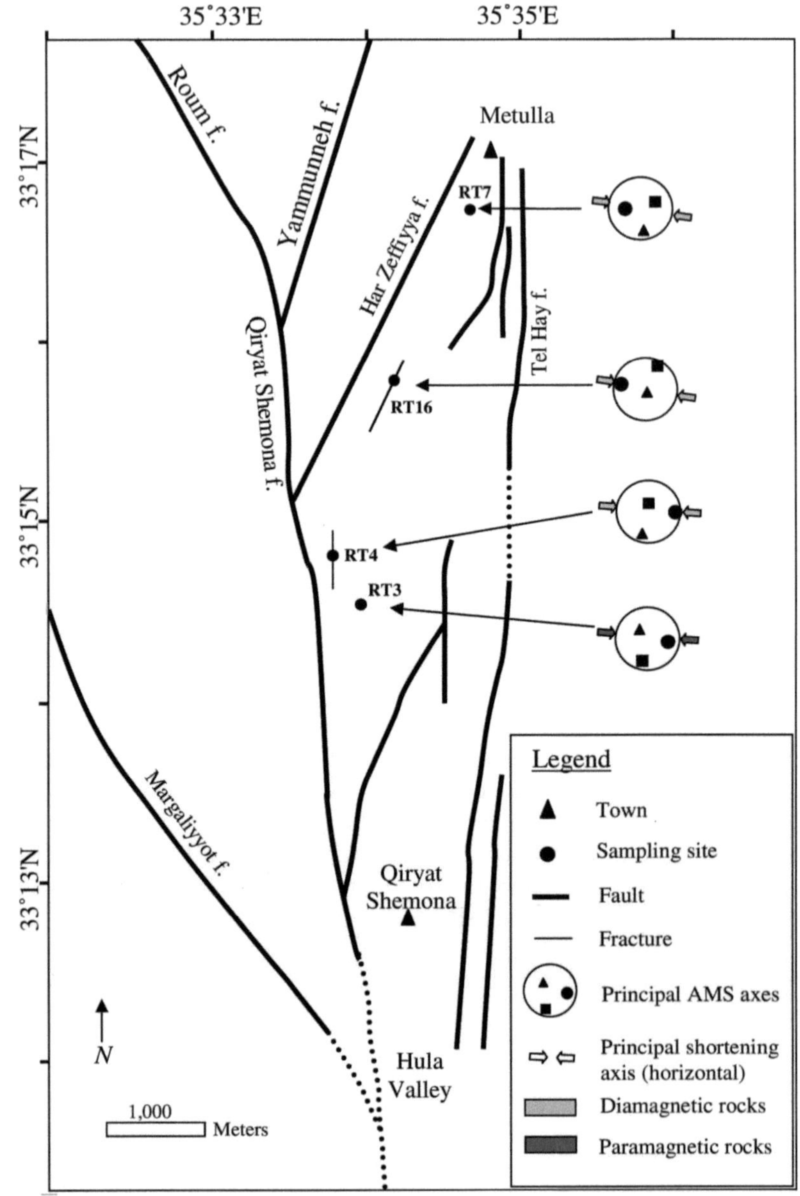

Fig. 6.15 Map of the main fault segments of the DST in the study area and sites with principal AMS axes. *Arrows* mark the inferred E-W (*horizontal*) principal shortening at the sites studied. See legend in Fig. 6.9 (From Levi and Weinberger 2011)

2–60 cm thick and up to >20 m high. The veins are characterized by vertical bands of calcite, aligned parallel to vein walls (Fig. 6.5d). Many veins consist of fibrous calcite crystals aligned perpendicular to the vein walls with a tabular geometry. There is no evidence for shear along the vein walls or internal shear displacement.

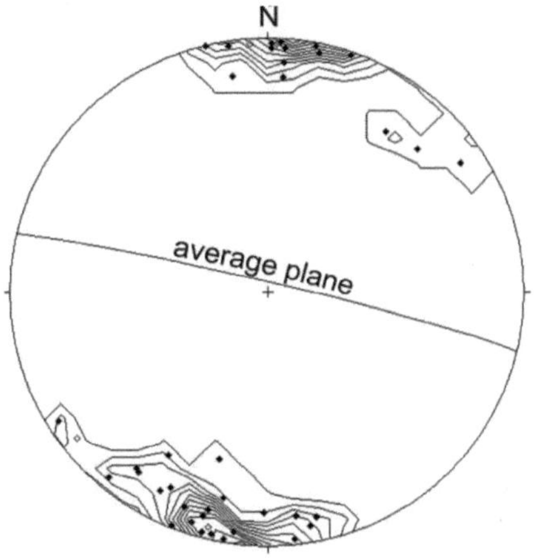

Fig. 6.16 Poles to opening-mode calcite-filled veins (n = 38) and clastic dikes (n = 5). The mean plane is 87°/014°. Contouring the data suggests that two sets exist; a dominant E-W striking set (mean of poles at 04°/188°) and a second possible set that strikes NW-SE (After Weinberger et al. 2009; Nuriel et al. 2012a)

Angular fragments of the Eocene host rock surrounded by calcite cement are occasionally found within the veins, or between the vein material and the wall rock, implying a dilational origin (Fig. 6.5d). The clastic dikes are filled with clay or soil consisting of gravel, silt and pebbles.

The mean plane of the combined population of calcite-filled veins and clastic dikes in the quarry is 87°/014°. Contouring the data suggests that two sets exist, a dominant E-W striking set and a second possible set that strikes NW-SE (Fig. 6.16). The dominant set of veins and clastic dikes thus reflect ~ N-S extension. In some cases the veins appear to be "utilizing" a pre-existing fault (i.e., plane of weakness), which may be responsible for some scatter in their orientations and the occasional NW-SE striking veins.

6.6 Discussion

6.6.1 Timing and Significance of Distributed N-S Shearing

The major structural elements in the study area are two ~ N-S trending strike-slip faults, the Tel Hay and Qiryat Shemona faults. The latter is associated with numerous subsidiary faults exposed in the Kefar Giladi Quarry, whose kinematics attests to the dominance of left-lateral strike-slip motion. This result provides an independent indication for the plate-scale kinematics of relative left-lateral motion between Arabia plate and Sinai sub-plate. Field observations, however, lack direct signs for the exact timing and duration of the strike-slip motion within the Metulla Saddle, since rocks of Cretaceous up to Pleistocene age are affected by it. It is postulated that the strike-slip motion was began in the (Early to Middle) Miocene (Bartov et al. 1980; Eyal et al. 1981), in accordance with the long-term kinematic framework of the DST (Joffe and Garfunkel 1987).

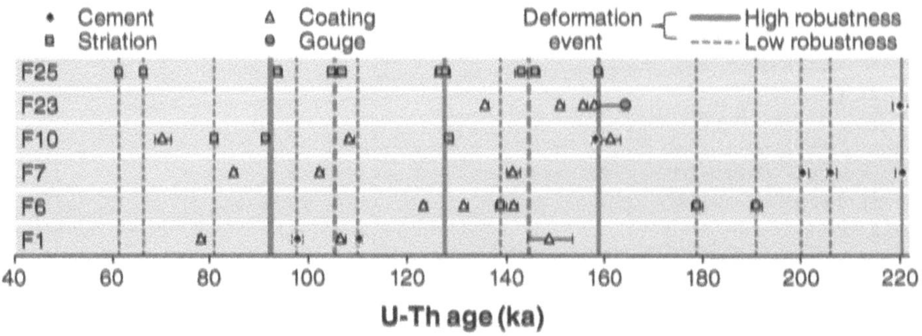

Fig. 6.17 U-Th ages of calcite precipitates within striated fault planes. Sample names are denoted at the left side. U-Th ages have ±2σ error bars (symbols may be larger than error bars). *Solid lines* indicate well-defined (high-robustness) deformation events characterized by coeval development of syntectonic precipitates (calcite cement and striations). Samples from the calcite coating layer were formed during interseismic periods. *Dashed lines* indicate moderately defined (low-robustness) deformation events based on single-sample age constraints. The timing of the deformation events is based on the age error of the dated samples, with a maximum of ±1 ka, and is represented by the *line thickness* (From Nuriel et al. 2012b)

Nuriel et al. (2012a) give more information on the late-Pleistocene strike-slip activity along the DST in the Metulla Saddle. They constrained the timing of fault activity by studying the microstructural, geochemical and geochronological of calcite precipitates associated with the Qiryat Shemona striated fault planes. They recognized four types of coexisting calcite precipitates, including calcite cement in dilation breccia, calcite in horizontal striated groove morphology, calcite gouge, and calcite coating of the fault surface. Carbon-oxygen isotopes, $^{87}Sr/^{86}Sr$ ratios, and rare earth element and yttrium (REY) patterns indicate various precipitation mechanisms associated with formation of syntectonic (calcite cement and striations), coseismic (calcite gouge), and interseismic (calcite coating) precipitates in the fault zone. Using U-Th dating of samples (e.g., calcite extracted from horizontal fault striations) from fault planes, they delineated several strike-slip events in the late Pleistocene (Fig. 6.17). Current GPS studies also demonstrate the importance of strike-slip motion along this sector of the DST (e.g., Wdowinski et al. 2004; Reilinger et al. 2006; Le Beon et al. 2008). Hence, imposed N-S shear has been distributed at the juncture between the Hula basin and the LRB over the history of the DST.

6.6.2 *Timing and Significance of Imposed E-W Shortening and N-S Extension*

Field observations and measurements attest to the dominantly contractional (transpressive) nature of the deformation in the Metulla Saddle. The folds measured at a variety of scales in the study area group into two distinct sets based on their orientations;

NW-SE trending and N-S trending folds (Weinberger et al. 2009). The latter set of folds includes large-scale folds (e.g., the castle anticline adjacent to the western block, the broad syncline of the western block and the Shehumit anticline and Qiryat Shemona syncline in the eastern block) as well as small-scale folds. These folds manifest E-W contraction associated with E-W pure shear. The younger rocks exposed in the area, the 1.5–0.8 Ma Hazbani Basalt, are affected only by N-S trending folds. Hence, folding of this set postdates ~0.8 Ma, indicating evolving E-W contraction during that interval of time.

Another consequence of E-W contraction is a perpendicular response of N-S extension during pure shear. The abundance of ~ E-W calcite-filled veins and clastic dikes oriented at high angles to the Qiryat Shemona fault demonstrates this component of strain within the Metulla Saddle. Nuriel et al. (2012b) presented structural, geochemical and U-Th geochronological data that constrain the mechanism and timing of the vein formation. U-Th ages correspond to the time of growth episode associated with major opening events. The oldest ages indicate that veins seldom formed prior to 500 ka (Nuriel et al. 2010). This implies that the E-W contraction was began not very long before that during the Pleistocene. The overwhelming majority of the veins were emplaced and developed during the period between 358 and 17 ka.

The obtained U-Th ages and radiometric dating of the folded basalts indicate that the Pleistocene activity occurred interchangeably (or coevally) along N-S trending strike-slip faults, N-S trending contractional structures, and E-W calcite-filled veins.

6.6.3 Strain Partitioning During Transpression

The two prominent NNE-SSW trending, en-echelon anticlines (Site D, Fig. 6.3) are terminated against (or cut by) the Qiryat Shemona fault, forming an angle of ~30° to the strike of the Qiryat Shemona fault. Formation of such folds is commonly attributed to premature stages of strike-slip deformation, in which simple shear is dominant (Harding and Lowell 1979; Sylvester 1988). The initial fold axes may have rotated counterclockwise during the Neogene up to ~15° due to progressive simple shear along the transform. This is indicated by the angle between the current trend of the fold axes and the expected trend based on the infinitesimal strain axes associated with pure strike-slip deformation (Fig. 6.7). These folds and some other parallel small-scale folds are attributed to an early phase (pre-Pleistocene) of non-partitioned strike-slip deformation within the Metulla Saddle (Weinberger et al. 2009).

There is a marked parallelism between the strike-slip faults, the ~ N-S trending Pleistocene folds and several N-S striking thrust faults in the Metulla Saddle. This parallelism indicates the coexistence of a strike-slip and contractional deformation in the form of a (mini) fold-thrust belt adjacent to the bounding N-S trending strike-slip faults. The coexistence of those structures is best explained by partitioned transpression along this sector of the DST. This assemblage of structures is attributed to a late phase of partitioning strike-slip deformation within the Metulla Saddle during the Pleistocene (Weinberger et al. 2009). The late Pleistocene emplacement of calcite-filled veins perpendicular to the DST zone is another manifestation of strain partitioning during transpression. This style of strain partitioning resembles the deformation adjacent to the San Andreas fault in portions of central California (e.g., Mount and Suppe 1987, 1992; Zoback et al. 1987; Tavarnelli 1998; Bawden et al. 2001; Fuis et al. 2003), and in the LRB (Griffiths et al. 2000; Gomez et al. 2007a).

The transpressional-related E-W contraction across the Metulla Saddle has been necessarily accompanied by compensatory uplift of rocks in the fault zone, as observed in other transpressional zones worldwide (Sylvester 1988; Mann 2007). This is demonstrated by localized shortening of up to 30 % in certain regions (Weinberger and Sneh 2004; Weinberger et al. 2010), leading to a topographic rise due to folding of at least ~80 m in the last ~1 Ma at Shehumit Hill. More than 400 m of the Kefar Giladi conglomerate and lacustrine sediments accumulated in a depression adjacent to the HWBF-Qiryat Shemona fault line during the Late Miocene – Early Pliocene. While the exact location of the basin's depoceneter cannot be pinpointed precisely, the accumulated sediments have certainly been displaced, uplifted and folded, forming the ~700 m high Metulla Saddle. On the basis of the assigned age of the Kefar Giladi Formation, this tectonic inversion took place post Early-Pliocene, coevally with the transpressive phase of deformation along the DST in northern Israel. Furthermore, the similar degree of tilting of the Kefar Giladi conglomerate and the younger Hazbani Basalt in the town of Qiryat Shemona (Sneh and Weinberger 2003b; Weinberger et al. 2010) indicates that this tectonic inversion took place during the Pleistocene and near the present location of the outcrops.

6.6.4 Transpression During the Convergence Across the Arabian-Sinai Plate Boundary

The observed strain partitioning during transpression cannot be regarded a product of Pleistocene north–south relative plate motion along the Metulla Saddle, because it contains N-S striking faults that show no obvious bend/jump to the right. Furthermore, the contractional deformation extends and even intensifies next to the NW-SE striking Margaliyyot fault (Fig. 6.3). These observations corroborate the possibility that the contractional deformation is a product of increased north-northwestward obliquity of plate motion relative to the north to north-northwestward strike of the studied Tel Hay and Qiryat Shemona faults (i.e., the exact strike of the Qiryat Shemona fault is 353°).

It is well demonstrated that strain partitioned during transpression evolves along obliquely convergent plate boundaries (Fossen 2010). It has been shown that even a minor component of convergence ($\alpha < 5°$, where α is the angle between the strike of the plate boundary and the plate motion velocity vector) resulting from changes in relative plate motion can lead to a high degree of strain partitioning (Teyssier et al. 1995). Hence, oblique convergence across the Arabian-Sinai plate boundary (i.e., Qiryat Shemona segment of the DST) seems to be a viable kinematic explanation for the partitioned transpression observed in the Metulla Saddle during the Pleistocene. The degree of strain partitioning is nearly complete (see Fig. 3 in

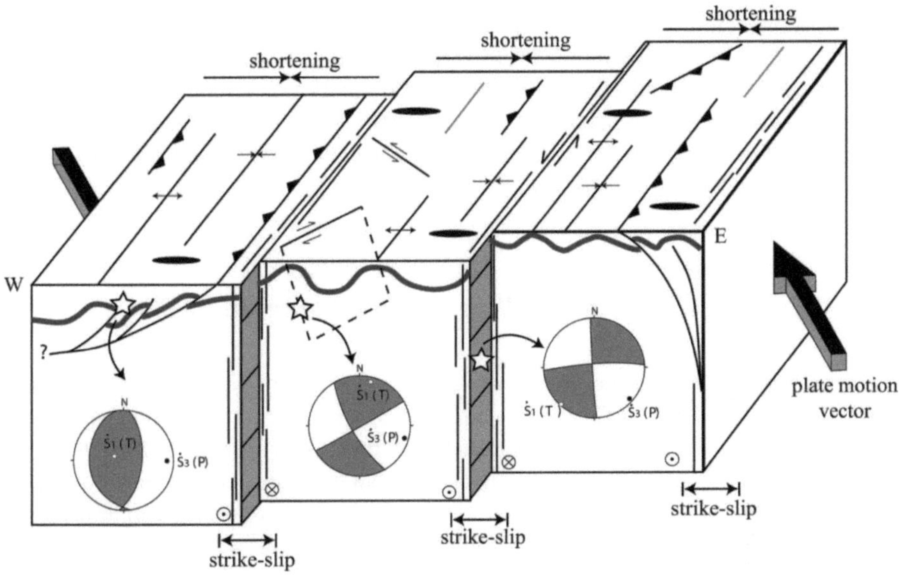

Fig. 6.18 Proposed model for the Pleistocene (current) phase of strain partitioning during transpression across the DST in the Metulla Saddle. The strike-slip component of plate motion is localized to discrete N-S fault zones whereas the intervening blocks are subjected to E-W shortening, resulting in the development of a (mini) fold-thrust belt along the transform margins. Elongated solid ellipses represent calcite-filled veins associated with N-S extension during strain partitioning. Potential seismicity (stars and the associated focal plane solutions) arises from slip along the weak N-S strike-slip faults, reactivation of faults within the blocks, or actively growing blind thrust faults. Fault segments (*vertical* and *horizontal lines* drawn next to the edges of the blocks) form brittle fault zones adjacent to the N-S bounding faults. *Gray lines* represent locked faults (After Weinberger et al. 2009)

Teyssier et al. 1995) considering that $\alpha > 10°$ and the 70–90° angle between the plate boundary and the directions of infinitesimal shortening (as manifested by fault kinematics in Fig. 6.7) and finite shortening (as manifested by young folds, thrust faults and veins in Figs. 6.10 and 6.16). Accordingly, the convergent strike-slip motion produces a major component of N-S shearing accommodated by ~ N-S striking faults, a significant component of E-W shortening perpendicular to the DST accommodated by N-S trending folds, and a minor N-S extension as manifested by numerous E-W striking veins (Fig. 6.18).

What makes this case study of the Metulla Saddle unique, and potentially valuable for understanding transform-margin deformation globally, is the evidence for the progression through time from almost pure strike-slip (non-partitioned) to partitioned transpressive deformation. Weinberger et al. (2009) envision an early phase of almost pure strike slip distributed across the DST. At that point the plate boundary consisted of a limited number of poorly connected fault segments and the upper crust was strong. With time, slip accrued along the plate boundary and expedited increasing convergence. The former led to the development of weak, through-going and well-connected fault zones in the upper crust, whereas the latter led to transpression along these zones within the Metulla Saddle and northward. As a consequence, the late (and current) phase of deformation is partitioned into strike-slip motion along the weak N-S striking major vertical fault zones and the development of a (mini) fold-thrust belt due to E-W shortening adjacent to the transform (Fig. 6.18).

6.6.5 *Evolving Convergence Due to Possible Changes in Relative Plate Motion Along the DST*

Previous studies divided the history of plate motion along the DST into two main stages (Garfunkel 1981; Joffe and Garfunkel 1987). In the first stage, pre 5 Ma, pure left-lateral strike-slip motion prevailed. In the second stage, post 5 Ma, the left-lateral motion was accompanied by a small component of transverse extension at the southern sector of the DST (and consequently, a small component of transverse contraction at the central sector of the DST). The date 5 Ma serves as a convenient time marker for dividing the history of the Arabian-Sinai plate motion into two intervals in which the "average" motions are distinct. It is based on the initiation time of seafloor spreading in the northern part of the Red Sea during that time (Garfunkel 1981; Joffe and Garfunkel 1987). The transition between these two stages was most likely a consequence of continuous rather than abrupt changes in plate motion. This is based on Euler's theorem that continuous changes of relative plate motion in a multi-plate system are a foreseeable consequence of plate motion on a spherical surface (Le Pichon et al. 1973, p. 34). In practice, researchers had to assume that for some finite time interval the motion between two plates could be described by a single pole of rotation. The Arabian-Sinai plate motion described by relocation (migration eastward) of the Euler pole of rotation is based mainly on studies of fault geometry along the southern sector of the DST. From the Hula basin northward along the central sector of the DST, the pole inferred by Garfunkel (1981) predicts increased convergence across the LRB in the last 5 Ma. Notably, the increasing convergence is also expected to intensify the contractional deformation evolved along the Yammunneh restraining bend (Gomez et al. 2007b).

Several observations support a more easterly position of the Pleistocene pole relative to the pole position describing the total motion in the last 5 Ma (Weinberger et al. 2009; Garfunkel 2010). Late Pliocene – early Pleistocene normal faulting and joints in the southern Negev indicates increased extension along the Arava Valley in the southern sector of the DST (Avni et al. 2000; Levi 2003). Concurrently, contractional structures in the Metulla Saddle presented in this paper are best interpreted as indicating increased convergence during the Pleistocene from the Hula basin northward. These mutually-related features (divergence-type features at the south and convergence-type features at the north) indicate an eastward migration of the Euler pole position. This migration continued the older (Miocene to Pliocene) trend and suggests that this situation was attained already in the early to mid-Pleistocene. As noted by Garfunkel (2010), some GPS results (e.g., Reilinger et al. 2006), even if approximate, raise the possibility that the present Euler pole is located distinctly east of the poles describing the earlier motions across the DST (see also Butler et al. 1998). If so, then the eastward migration described of the Euler pole before 5 Ma ago continued in more recent times. The consequence of the pole migration is associated with plate convergence across the Qiryat Shemona segment of the DST, providing a viable kinematic explanation for the Pleistocene transpression and strain partitioning presented in this study.

Evidence for a Pleistocene tectonic transition might be recognized in a series of locations along the DST (see summary in Schattner and Weinberger 2008). The most striking example is the tectonic transition in the Hula basin, which formed as a pull-apart basin ~4 Ma, but entered a new geodynamic phase during the mid-Pleistocene. At that time a left-lateral, through-going, diagonal strike-slip fault developed within the basin (Schattner and Weinberger 2008). Gomez et al. (2007b) showed that the LRB went through a two-stage tectonic evolution, the later (current) of which is characterized by strain partitioning during transpression. The transition between these phases was attributed to changes in plate motion, with a northward relative motion in the early phase (pre-Pleistocene), and increasing convergence of the plates during the current phase. At the north end of the DST in the Karasu Valley, Seyrek et al. (2007) showed crustal thickening and folding due to transpression, which they attributed to changes in relative plate motion during the Pliocene.

6.7 Summary

This paper reviews the style and sequence of deformation associated with an actively convergent section of the DST across the Metulla Saddle. A major structural element in the study area is the Qiryat Shemona strike-slip faults. It is associated with numerous subsidiary faults, whose kinematics attests to the dominance of left-lateral motion along this sector of the DST. This result provides an independent indication for the plate-scale kinematics of relative left-lateral motion between the Arabian plate and the Sinai sub-plate in northern Israel. Using U-Th dating of samples from fault planes, several deformation events were delineated, documenting a prolonged history of late-Pleistocene strike-slip activity in the Metulla Saddle. The strike-slip activity was accompanied by intense folding and thrusting. A set of N-S trending folds including large-scale folds as well as small-scale folds manifest significant E-W contraction. Because the younger rocks exposed in the area, the 1.5–0.8 Ma Hazbani Basalt, are affected only by N-S trending folds, the folding of this set postdates ~0.8 Ma. The U-Th ages obtained suggest that the Pleistocene activity occurred interchangeably along N-S trending strike-slip faults, N-S trending contractional structures, and ~ E-W calcite-filled veins. The coexistence of a strike-slip and contractional deformation in the form of a (mini) fold-thrust belt adjacent to the bounding N-S trending strike-slip faults is best explained by partitioned transpression along this sector of the DST during the Pleistocene. The partitioned transpression cannot be regarded a product of Pleistocene northward relative plate motion, because the Metulla Saddle contains ~ N-S striking faults that show no obvious bend/jump to the right. The present observations corroborate the possibility that the contractional deformation is the result of increased north-northwestward obliquity of plate motion relative to the ~ N-S strike of the studied Qiryat Shemona and Tel Hay faults. The oblique convergence across the Arabian-Sinai plate boundary seems to be a viable kinematic explanation for the partitioned transpression observed in the Metulla Saddle during the Pleistocene. The degree of strain partitioning is nearly complete. Accordingly, the convergent strike-slip motion produced a major component of N-S shearing accommodated by N-S bounding left-lateral faults, a significant component of E-W shortening perpendicular to the DST accommodated by N-S trending folds and reverse faults, and a minor N-S extension as manifested by numerous E-W striking veins. Several observations support a more easterly position of the Pleistocene–Recent pole of rotation of the Arabian-Sinai relative to the pole position describing the total motion in the last 5 Ma. The consequence of the pole migration is plate convergence across the Qiryat Shemona segment of the DST, providing a feasible kinematic explanation for the Pleistocene transpression and strain partitioning presented in this study.

Acknowledgments This study was supported by grant no. 2004232 from the United States-Israel Bi-national Science Foundation (BSF) and grants from the Israeli Ministry of Energy and Water Resources. I am grateful to Zvi Garfunkel for encouraging me to write this review and for illuminating the way in the study of the Dead Sea Transform. I am indebted to Amihai Sneh, Michael Gross, Uri Frieslander, Uri Schattner, Beny Medvedev, Perach Nuriel, Gideon Rosenbaum, Tsafrir Levi, Yehudit Harlavan, and Mira Bar-Matthews for helpful discussions, fruitful work, and an excellent partnership during the course of this study. Constructive reviews of Ian Alsop and an anonymous reviewer significantly help to improve the paper. The structural data were plotted by using Rick Allmendinger's *Stereonet* and *FaultKin* programs.

References

Allmendinger RW, Marrett RA, Cladouhos TT (2001) FaultKinWin: a program for analyzing fault slip data for Windows™ computers. (http://www.geo.cornell.edu/geology/faculty/RWA/programs/)

Avni Y, Bartov Y, Garfunkel Z, Ginat H (2000) The Arava Formation – A Pliocene sequence in the Arava Valley and its western margin, southern Israel. Isr J Earth Sci 50:101–120

Aydin A, Page BM (1984) Diverse Pliocene-Quaternary tectonics in a transform environment, San Francisco Bay region, California. Geol Soc Am Bull 95:91–105

Bartov Y, Steinitz G, Eyal M (1980) Sinistral movement along the Gulf of Aqaba – Its age and relation to the opening of the Red Sea. Nature 285:220–222

Bawden GW, Thatcher W, Stein RS, Wicks C, Hudnut K, Peltzer G (2001) Tectonic contraction across Los Angeles after removal of groundwater pumping effects. Nature 412:812–815

Beydoun ZR (1977) Petroleum prospects of Lebanon. Am Assoc Pet Geol Bull 61:43–64

Borradaile GJ, Henry B (1997) Tectonic applications of magnetic susceptibility and its anisotropy. Earth Sci Rev 42:49–93

Borradaile GJ, Jackson M (2010) Structural geology, petrofabrics and magnetic fabrics. J Struct Geol 32:1519–1551

Butler RWH, Spencer HM, Griffiths HM (1998) The structural response to evolving plate kinematics during transpression: evolution of the Lebanese restraining bend of the Dead Sea Transform. In: Holdsworth RE et al (eds) Continental transpressional and transtensional tectonics, Geological Society of London special publication 135. Geological Society, London, pp 81–106

Daeron M, Benedetti L, Tapponnier P, Sursock A, Finkel RC (2004) Constraints on the post 25-ka slip rate of the Yammouneh fault (Lebanon) using in situ cosmogenic 36Cl dating of offset limestone-clast fans. Earth Planet Sci Lett 227:105–119

Dewey JF (1975) Finite plate evolution: some implications for the evolution of rock masses at plate margins. Am J Sci 275A:260–284

Dewey JF, Holdsworth RE, Strachan RA (1998) Tranpression and transtension zones. In: Holdsworth RE et al (eds) Continental transpressional and transtensional tectonics, Geological Society of London special Publication 135. Geological Society, London, pp 1–14

Dubertret L (1951) Carte geologique au 1:50,000 feuille de Mardjayoun. Rep. Libanaise, Min. des Travaux Pub., Beyrouth. Notice explicative (1952), p 64

Eyal M, Eyal Y, Bartov Y, Steinitz G (1981) Tectonic development of the western margin of the Gulf of Elat (Aqaba) rift. Tectonophysics 80:39–66

Fossen H (2010) Structural geology. Cambridge University Press, New York, 463p

Freund R (1965) A model of the structural development of Israel and adjacent areas since Upper Cretaceous times. Geol Mag 102:189–205

Freund R, Zak I, Garfunkel Z (1968) Age and rate of sinistral movement along the Dead Sea rift. Nature 220:253–255

Freund R, Garfunkel Z, Zak I, Goldberg M, Weissbrod T, Derin B (1970) The shear along the Dead Sea rift. Phil Trans R Soc London A-267:107–130

Fuis GS et al (2003) Fault systems of the 1971 San Fernando and 1994 Northridge earthquakes, southern California: relocated aftershocks and seismic images from LARSE II. Geology 31:171–174

Garfunkel Z (1981) Internal structure of the Dead Sea leaky transform (rift) in relation to plate kinematics. Tectonophysics 80:81–108

Garfunkel Z (2010) The long- and short-term lateral slip and seismicity along the Dead Sea Transform: an interim evaluation. Isr J Earth Sci 58:217–235

Garfunkel Z, Zak I, Freund R (1981) Active faulting in the Dead Sea rift. Tectonophysics 80:1–26

Glikson YA (1966) The lacustrine Neogene in the Kefar Gil'adi area, northern Jordan Valley. Isr J Earth Sci 15:85–100

Gomez F, Khawlie M, Tabet C, Darkal AN, Khair K, Barazangi M (2006) Late Cenozoic uplift along the northern Dead Sea Transform in Lebanon and Syria. Earth Planet Sci Lett 241:913–931

Gomez F, Karam G, Khawlie M, McClusky S, Vernant P, Reilinger R, Jaafar R, Tabet C, Khair K, Barazangi M (2007a) Global Positioning System measurements of strain accumulation and slip transfer through the restraining bend along the Dead Sea fault system in Lebanon. Geophys J Int 168:1021–1028

Gomez F, Nemer T, Tabet C, Khawlie M, Meghraoui M, Barazangi M (2007b) Strain partitioning of active transpression within the Lebanese restraining bend of the Dead Sea fault (Lebanon and SW Syria). In: Cunningham WD, Mann P (eds) Tectonics of strike-slip restraining and releasing bends, Geological Society of London special publication 209. Geological Society, London, pp 285–303

Griffiths HM, Clark RA, Thorp KM, Spencer S (2000) Strain accommodation at the lateral margin of an active transpressive zone: geological and seismological evidence from the Lebanese restraining bend. J Geol Soc 157:289–302

Hancock PL, Atiya MS (1979) Tectonic significance of mesofracture systems associated with the Lebanese segment of the Dead Sea Transform fault. J Struct Geol 1:143–153

Harding TP, Lowell JD (1979) Structural styles, their plate-tectonic habitats, and hydrocarbon traps in petroleum provinces. AAPG Bull 63:1016–1058
Harland WB (1971) Tectonic transpression in Caledonian Spitzbergen. Geol Mag 108:27–42
Heimann A (1990) The development of the Dead Sea rift and its margins in northern Israel during the Pliocene and the Pleistocene. Geol Surv Rep GSI/28/90, p 83, Isr Geol Surv Jerusalem (in Hebrew; English summary)
Heimann A, Ron H (1987) Young faults in the Hula pull apart basin, central Dead Sea Transform. Tectonophysics 141:117–141
Heimann A, Zilberman E, Amit R, Frieslander U (2009) Northward migration of the southern diagonal fault of the Hula pull-apart basin, Dead Sea Transform, northern Israel. Tectonophysics 476:496–511
Henry B, Homberg C, Mroueh M, Hamdan W, Higazi F (2010) Rotations in Lebanon inferred from new palaeomagnetic data and implications for the evolution of the Dead Sea Transform system. In: Homberg C, Bachmann M (eds) Evolution of the Levant margin and western Arabia platform since the Mesozoic, Geological Society of London special publication 341. Geological Society, London, pp 269–285
Homberg C, Barrier E, Mroueh M, Muller C, Hamdan W, Higazi F (2010) Tectonic evolution of the central Levant domain (Lebanon) since Mesozoic time. In: Homberg C, Bachmann M (eds) Evolution of the Levant margin and western Arabia platform since the Mesozoic, Geological Society of London special publication 341. Geological Society, London, pp 245–268
Joffe S, Garfunkel Z (1987) Plate kinematics of the circum Red Sea – A re-evaluation. Tectonophysics 141:5–22
Le Beon M, Klinger Y, Amrat AQ, Agnon A, Dorbath L, Baer G, Ruegg JC, Charade O, Mayyas O (2008) Slip rate and locking depth from GPS profiles across the southern Dead Sea Transform. J Geophys Res 113:B11403. doi:10.1029/2007JB005280
Le Pichon X, Kreemer C (2010) The Miocene-to-present kinematic evolution of the Eastern Mediterranean and Middle East and its implications for dynamics. Ann Rev Earth Planet Sci 38:323–351
Le Pichon X, Francheteau J, Bonnin J (1973) Plate tectonics. Elsevier, Amsterdam, p 300
Levi T (2003) Joint set as a tool for analyses the tectonic deformation in the central Arava western rift margin. M.Sc. thesis, Ben Gurion University, Beer Sheba, p 109 (in Hebrew, English abstract)
Levi T, Weinberger R (2011) Magnetic fabrics of diamagnetic rocks and the strain field associated with the Dead Sea fault, northern Israel. J Struct Geol 33:566–578
Mann P (2007) Global catalogue, classification and tectonic origins of restraining- and releasing bends on active and ancient strike-slip fault systems. In: Cunningham WD, Mann P (eds) Tectonics of strike-slip restraining and releasing bends, Geological Society of London special publication 209. Geological Society, London, pp 13–142
Marco S, Agnon A, Ellenblum R, Eidelman A, Basson U, Boas A (1997) 817-year-old walls offset sinistrally 2.1 m by the Dead Sea Transform, Israel. J Geod 24:11–20
Mor D (1993) A time-table for the Levant Volcanic Province, according to K-Ar dating in the Golan Heights. J Afr Earth Sci 16:223–224
Mount VS, Suppe J (1987) State of stress near the San Andreas fault: implications for wrench tectonics. Geology 15:1143–1146
Mount VS, Suppe J (1992) Present-day stress orientations adjacent to active strike-slip faults: California and Sumatra. J Geophys Res 97:11995–12013
Norris R, Koons PO, Cooper AF (1990) The obliquely convergent plate boundary in the South Island of New Zealand. J Struct Geol 12:715–725
Nuriel P, Rosenbaum G, Uysal TI, Zhao J, Golding SD, Weinberger R, Karabacak V, Avni Y (2010) Formation of fault-related calcite precipitates and their implications for dating fault activity in the East Anatolian and Dead Sea fault zones. In: Fagereng A, Toy VG, Rowland JV (eds) Geology of the earthquake source: a volume in honour of Rick Sibson, Geological Society of London spec publication 359. Geological Society, London, pp 229–248. doi:10.1144/SP359.13

Nuriel P, Weinberger R, Rosenbaum G, Golding SD, Zhao J, Uysal TI, Bar-Matthews M, Gross MR (2012a) Timing and mechanism of late-Pleistocene calcite vein formation across the Dead Sea fault Zone northern Israel. J Struct Geol 36:43

Nuriel P, Rosenbaum G, Zhao J, Golding SD, Feng Y, Villemant B, Weinberger R (2012b) U-Th dating of striated fault planes. Geology 40:647–650

Parés JM, van der Pluijm BA, Turell JD (1999) Evolution of magnetic fabrics during incipient deformation of mudrocks (Pyrenees, northern Spain). Tectonophysics 307:1–14

Picard L (1952) The geology of Kefar-Giladi. Bull Israel Explor Soc "Lif" memorial vol B, pp 73–77 (in Hebrew)

Quennell AM (1959) Tectonics of the Dead Sea rift. XX Int. Geol Congr Assoc Serv Geol Afr 385–405

Reilinger R et al (2006) GPS constraints on continental deformation in the Africa-Arabia-Eurasia continental collision zone and implications for the dynamics of plate interactions. J Geophys Res 111:B05411. doi:10.1029/2005JB004051

Ron H (1987) Deformation along the Yammuneh, the restraining bend of the Dead Sea Transform: Paleomagnetic data and kinematic implications. Tectonics 6:653–666

Ron H, Nur A, Eyal Y (1990) Multiple strike-slip fault sets; a case study from the Dead Sea Transform. Tectonics 9:1421–1431

Ron H, Shamir G, Eyal Y (1997) Deformation of Margaliot block between Roum and Margaliot faults. Isr Geol Soc Ann Meet Kefar Gil'adi field trip guidebook, pp 33–45 (in Hebrew)

Rybakov M, Fleischer L, ten Brink US (2003) The Hula Valley subsurface structure inferred from gravity data. Isr J Earth Sci 52:113–122

Sanderson DJ, Marchini WRD (1984) Transpression. J Struct Geol 6:449–458

Schattner U, Weinberger R (2008) A mid-Pleistocene deformation transition in the Hula basin, northern Israel: implications for the tectonic evolution of the Dead Sea fault. Geochem Geophys Geosyst 9:Q07009. doi:10.1029/2007GC001937

Schulman N (1966) The Qiryat Shemona (northern Jordan Valley) basalt ridge: a tilted fault block. Isr J Earth Sci 15:161–164

Seyrek A, Demir T, Pringle MS, Yurtmen S, Westaway RWC, Beck A, Rowbotham G (2007) Kinematics of the Amanos Fault, southern Turkey, from Ar/Ar dating of offset Pleistocene basalt flows: transpression between the African and Arabian plates. In: Cunningham WD, Mann P (eds) Tectonics of strike-slip restraining and releasing bends, Geological Society of London special publication 209. Geological Society, London, pp 255–284

Sneh A, Weinberger R (2003a) Geology of the Metulla quadrangle, northern Israel: implications for the offset along the Dead Sea rift. Isr J Earth Sci 52:123–138

Sneh A, Weinberger R (2003b) Geological map of Israel, sheet 2-II Metulla, scale 1:50,000. Isr Geol Surv, Jerusalem

Sneh A, Weinberger R (2006) Geological map of Israel, sheet 2-IV Rosh Pinna, scale 1:50,000. Isr Geol Surv, Jerusalem

Sylvester AG (1988) Strike-slip faults. Geol Soc Am Bull 100:1666–1703

Tavarnelli E (1998) Tectonic evolution of the Northern Salinian Block, California, USA: Paleogene to Recent shortening in a transform fault-bounded continental fragment. In: Holdsworth RE et al (eds) Continental transpressional and transtensional tectonics, Geological Society of London special publication 135. Geological Society, London, pp 107–118

Teyssier C, Tikoff B, Markley M (1995) Oblique plate motion and continental tectonics. Geology 23:447–450

Tikoff B, Teyssier CB (1994) Strain modeling of displacement-field partitioning in transpressional orogens. J Struct Geol 16:1575–1588

Wdowinski S, Bock Y, Baer G, Prawirodirdjo L, Bechor N, Naaman S, Knafo R, Forrai Y, Melzer Y (2004) GPS measurements of current crustal movements along the Dead Sea fault. J Geophys Res 109:B05403. doi:10.1029/2003JB002640

Weinberger R, Sneh A (2004) The geology of Qiryat Shemona region. Isr Geol Soc Annu Meet HaGoshrim field trips guidebook, pp 55–78 (in Hebrew)

Weinberger R, Gross MR, Sneh A (2009) Evolving deformation along a plate boundary transform: example from the Dead Sea fault in northern Israel. Tectonics 28:TC5005. doi:10.1029/2008TC002316
Weinberger R, Schattner U, Medvedev B, Frieslander U, Sneh A, Harlavan Y, Gross RM (2010) Convergent strike-slip across the Dead Sea fault in northern Israel imaged by high-resolution seismic reflection data. Isr J Earth Sci 58:203–216
Woodcock NH, Fischer M (1986) Strike-slip duplexes. J Struct Geol 8:725–735
Zaslavsky Y (2010) Ambient noise measurements for seismic response and exploration of sedimentary layers: case of Kiryat Shemona. Geophys Inst Rep 500/522/10 Lod, p 46
Zoback MD, Zoback ML, Mount VS, Suppe J, Eaton J, Healy JE, Oppenheimer D, Reasenberg P, Jones L, Raleigh CB, Wong IG, Scotti O, Wentworth C (1987) New evidence on the state of stress of the San Andreas fault system. Science 238:1105–1111

Chapter 7
Review of On-Fault Palaeoseismic Studies Along the Dead Sea Fault

Shmuel Marco and Yann Klinger

Abstract The aim of this short note is to provide a summary over on-fault palaeoseismic works on the behavior of the Dead Sea fault (DSF). Key achievements of these studies include: the determination of slip rate of the DSF across different space and time resolutions, which converges at around 4–5 mm/a, confirmation of the sinistral relative motion between the Arabia Plate and the Sinai Subplate and of thrust motion and normal faulting associated with restraining bends and pull apart grabens respectively, and the verification of historical accounts on several strong earthquakes that were associated with surface ruptures. The comparison between the state-of-the-art slip-rate determinations along the DSF and the total motion accommodated by the known historical and instrumental earthquakes shows that current seismicity rates cannot account for the full slip rate. As previously proposed, the short-term rate of seismicity is not necessarily representative of the long-term seismic activity along the DSF. Assuming the historical records of the last two millennia are complete for strong earthquakes, we note long periods of quiescence in the sections of the northern Yammouneh, the Jordan Valley, and the southern Araba.

Keywords Dead Sea Transform fault • Tectonics • Active faulting • Earthquakes • Palaeoseismology • Middle East

---

S. Marco (✉)
Department of Geophysical, Atmospheric, and Planetary Sciences, Tel Aviv University, Ramat Aviv, Tel Aviv 69978, Israel
e-mail: shmulikm@tau.ac.il

Y. Klinger
Institut de Physique du Globe de Paris, Sorbonne Paris Cité, Univ. Paris Diderot, UMR 7154 CNRS, Paris, France
e-mail: klinger@ipgp.fr

Z. Garfunkel et al. (eds.), *Dead Sea Transform Fault System: Reviews*, Modern Approaches in Solid Earth Sciences 6, DOI 10.1007/978-94-017-8872-4_7,

7.1 Introduction

As early as 1869 Lartet suggested that Arabia and Africa have drifted apart to open up the Red Sea. Dubertret (1932) followed this idea suggesting a 160 km sinistral shear along the Dead Sea Fault associated with a 6° rotation between Arabia and Africa. Wellings (cited in Willis 1938) noted that this hypothesis corresponds to the offset of the marine Cambrian and Jurassic beds across the rift south of the Dead Sea. Willis (1938) rejected this hypothesis. During the following years, it was largely ignored until systematic research by Quennell (1956) provided evidence for 107 km sinistral slip. Later, with the advent of the plate tectonics, Freund and collaborators (Freund 1965; Freund et al. 1970, 1968) and Wilson (1965) recognized the Dead Sea rift as a transform fault zone that transfers the opening at the Red Sea to the collision zone at the Taurus-Zagros mountain belt.

The evidence for left-lateral shear along the Dead Sea Fault (DSF) since the early-middle Miocene is based on observations from four independent sources: regional plate tectonics, local geology, seismology, and geodesy. The plate tectonics shows that the opening of the Red Sea, where the Arabian plate is breaking away from Africa, is transferred to the collision with Eurasia via sinistral shear along the DSF (Freund 1965; Garfunkel 1981; Joffe and Garfunkel 1987; Quennell 1956). Sinistral motion explains the systematic offset of numerous pre-Miocene geologic features by a total of ~105 km, south of the Lebanon Restraining Bend (LRB) (Bartov et al. 1980; Freund 1965; Quennell 1956). Garfunkel et al. (1981) mapped the active fault traces of the DSF and associated these scarps with historical earthquakes that are reported from the region. Later research revealed prehistoric seismicity as well (Table 7.1 and Fig. 7.1). Focal mechanisms of moderate-to-large earthquakes show sinistral motion along the DSF (e.g. (Baer et al. 1999; Klinger et al. 1999; Salamon et al. 1996), Hofstetter et al. 2007). Geodetic measurements for the very short-term deformation are consistent with overall geologic observations of a sinistral slip-rate of 4–5 mm/year (Le Beon et al. 2008; McClusky et al. 2003; Reilinger et al. 2006; Wdowinski et al. 2004). In a review of the slip and seismicity of the DSF, Garfunkel (2011) concludes that the slip rate is slowing from an average rate of 6–7 mm/year over the last 5 Ma to 4–5.5 mm/year in the Pleistocene together with a slight eastward shift of the Euler pole of rotation between Sinai and Arabia.

Several authors noted explicitly that the detailed shape of the DSF had changed through time e.g. (Garfunkel 1981; Heimann and Ron 1987, 1993; Rotstein et al. 1992; ten Brink et al. 1999; ten-Brink and Ben-Avraham 1989). For the section south of the LRB, the widest zone of distributed faulting is about 50 km wide. It is found in the Galilee, where the early-stage (Miocene) faults were associated with formation of basins (Freund et al. 1970; Shaliv 1991) and with rotation of rigid blocks about sub-vertical axes (Ron et al. 1984), although the linkage to the transform movement is not well established. In this region, subsequent post-Miocene deformation took place mostly in the form of normal faulting on E-W trending faults with some strike-slip motion currently localized in a very narrow zone.

Table 7.1 List of on-fault palaeoseismic investigations at the DSF system arranged by time of publication

# in Fig. 7.1	References	Segment	Observations	Fault type	Achievement Earthquakes/ Slip Rate (SR)/Last Event (LE)/Recurrence (Re)
1	Reches and Hoexter (1981)	S. Jordan Valley	Trenches	N	31 BCE, 747 CE
2	Marco and Agnon (1995)	Dead Sea	Outcrops	N	Re
3	Amit et al (1996, 1999, 2002)	Southern Arava	Trenches	N	Re
4	Marco et al.; Ellenblum et al. (1998)	Jordan Gorge	Trenches, archaeology	LL	1202, 1759
5	Galli (1999)	Arava-Jordan-Hula Valley	Outcrops, archaeology	LL	
6	Enzel et al. (2000)	Dead Sea	Outcrops	N	Re
7	Klinger et al. (2000a)	Northern Arava	Archaeology, outcrops	LL	1212
8	Klinger et al. (2000b)	Northern Arava	Outcrops	LL	SR
9	Niemi et al. (2001)	Northern Arava	Outcrops	LL	SR
10	Zilberman et al. (2000)	Hula Valley	Trenches	N	
11	Gomez et al. (2001, 2003)	Serghaya Fault	Trenches, Outcrops	LL	1705 or 1759/
12	Meghraoui et al. (2003)	Misyaf, Yammouneh	Trenches, archaeology	LL	
13	Marco et al. (2003)	Sea of Galilee	Archaeology	N	749
14	Daëron et al. (2004, 2005, 2007)	South Yammouneh	Trenches	LL	1202
15	Zilberman et al. (2005)	South Arava	Trenches	N	3/1068/
16	Marco et al. (2005)	Jordan Gorge	Trenches	LL	1202, 1759/LE/SR
17	Chorowicz et al. (2005)	Yammouneh	Outcrops	LL	SR
18	Akyuz et al. (2006)	Northern Yammouneh	Trenches, Outcrops	LL	859, 1408, 1872
19	Nemer and Meghraoui (2006)	Roum Fault	Trenches, Outcrops	LL	1837, SR
20	Haynes et al. (2006)	Northern Arava	Archaeology	LL	634 or 659/660, 873, 1068, and 1546
21	Elias et al. (2007)	Lebanon thrust	Outcrops	R	551 CE/
22	Thomas et al. (2007)	Aqaba	Archaeology	LL	
23	Ferry et al. (2007)	Jordan Valley	Outcrops	LL	SR
24	Le Beon et al. (2008)	Arava	Outcrops	LL	

(continued)

Table 7.1 (continued)

# in Fig. 7.1	References	Segment	Observations	Fault type	Achievement Earthquakes/ Slip Rate (SR)/Last Event (LE)/Recurrence (Re)
25	Nemer et al. (2008)	Rachaya and Serghaya faults	Outcrops, trenches	LL	1759
26	Makovsky et al. (2008)	Elat Fault	Submarine geophysics	LL	SR
27	Altunel et al. (2009)	S. Turkey	Trenches, archaeology	LL	SR
28	Le Beon et al. (2010)	Arava	Outcrops	LL	SR
29	Karabacak et al. (2010)	Northern Yammouneh	Outcrops	LL	SR
30	Ferry et al. (2011)	Jordan Valley	Trenches, archaeology	LL	SR, Re, LE
31	Le Beon et al. (2012)	Arava	Outcrop	LL	SR

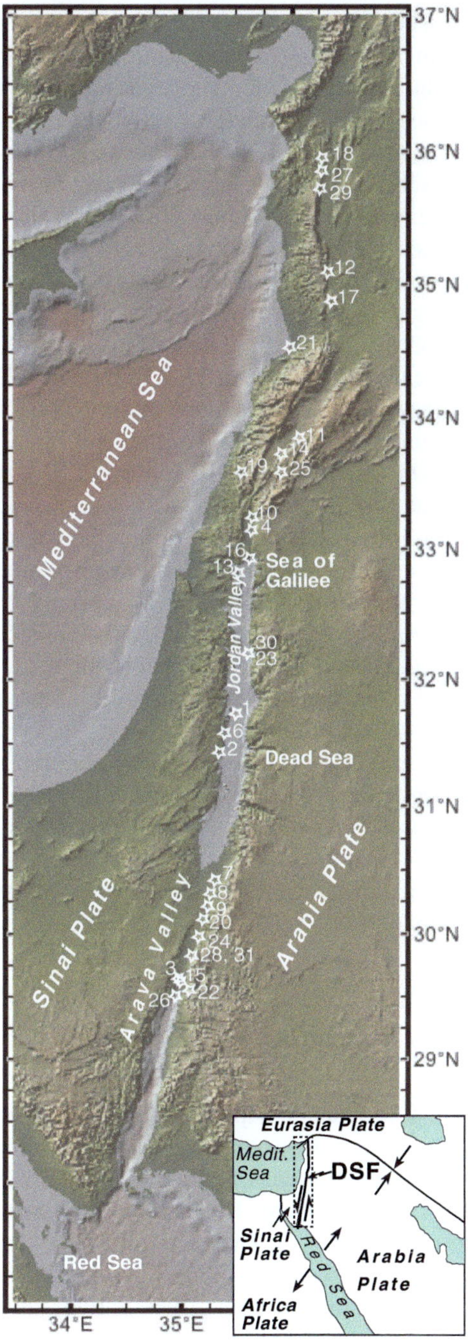

Fig. 7.1 Location of palaeoseismic studies along the Dead Sea Fault. Numbers refer to entries in Table 7.1

The deformation further south is characterized by a 20–30-km-wide zone with primarily strike-slip and some normal slip on faults trending sub-parallel to the main transform fault. The location of the active fault strands along the Dead Sea Transform fault zone (DSF) changed through time. In the western margins of Dead Sea basin, the early activity began a few kilometers west of the present shores and moved toward the center of the basin in four stages. Similar basinward migration of faulting is apparent in the Hula Valley north of the Sea of Galilee as well as in the Negev and the Sinai Peninsula. In the Arava Valley, seismic surveys reveal a series of buried inactive basins whereas the current active strand is located along the eastern margin. In general, the faulting along the DSF became localized by the end of the Miocene. The subsidence of fault-controlled basins in the early stage, stopped at the end of the Miocene. During the Plio-Pleistocene, new faults were formed in the Negev west of the main transform, possibly manifesting another cycle that has begun with the widening of the fault zone (Marco 2007).

This paper reviews the contribution of recent studies on the most recent activity of the DSF (mostly its southern section, south of the LRB) and its subsidiary faults. These studies contributed to the recognition of the active strands in the fault zone, constrained the slip rates and recurrence intervals of the various fault segments, and determined the time of the last slip event after which the segment had remained locked. The offset geological and archaeological bodies confirm the plate tectonics paradigm and provide an independent examination of the structural complexities such as restraining bends, pull-apart basins, and distributed slip among fault branches. The fault was characterized as a "leaky transform" because of these pull-aparts (Garfunkel 1981). The temporal overlap of geological and historical-archaeological information provides important crosschecks of the sources of data. On-fault research requires the identification of faults. This is relatively easy where the fault zone is narrow and exposures are good. It is less complete where the fault zone is wide and consists of many branches, or where the faulting activity shifts location and young sediments or basalt flows cover the faults.

7.1.1 Detailed Mapping of the Fault Zone

The first detailed maps of the DSF zone that emphasized the offset of Pleistocene to sub-recent units were published by Garfunkel et al. (1981). Various structures such as pull apart grabens, restraining bends, branching faults, and folds, which comprise the fault zone were also described and analyzed at the same time (Eyal et al. 1981; Garfunkel 1981). The recognition of small displacements of young geological features such as alluvial fans or lake deposits opened the road for neotectonics and palaeoseismic research. These studies showed that the fault consists of numerous segments separated by discontinuities or sharp changes in their strike. A GIS-based map of the faults suspected as being "active" (Bartov et al. 2002) was generated by defining active faults as those that either cross or bound Pliocene and younger stratigraphic units. It is not clear yet whether the strong earthquake ruptures are arrested at the mapped segment ends. This question is important because the length of the ruptures correlates with the earthquake size (e.g., Wells and Coppersmith 1994). The detailed maps reveal that the structure of the DSF is variable (described from south to north): The southernmost part is the Gulf of Aqaba-Elat, where a series of left-stepping, en-echelon fault arrangement forms three elongate deep pull-apart basins separated by shallow thresholds (Ben-Avraham and Garfunkel 1979). The largest earthquake (M7.2) ever recorded instrumentally occurred in the gulf on November 22, 1995 (Klinger et al. 1999; Pinar and Turkelli 1997; Shamir et al. 2003). The next section toward the north is the Araba Valley, characterized by a narrow fault zone that offsets a series of Pleistocene-Holocene alluvial fans (Garfunkel et al. 1981). Several palaeoseismic studies quantify the displacements (Klinger et al. 2000a; Le Beon et al. 2010, 2012; Niemi et al. 2001). Seismic reflections reveal that the subsurface structure of the Araba Valley includes buried Miocene grabens (Frieslander 2000). The next section is the pull-apart basin of the Dead Sea, where two sub-parallel faults bound the deepest place on the Earth continents. The Dead Sea pull-apart structure is bounded by the Amaziahu Fault on the south, where a sharp scarp in Late Pleistocene lake deposits reaches a height of up to 30 m. The northern boundary does not have a surface expression. The western fault continues due north along the Jordan Valley, where it is recognized as a narrow fault zone in Pleistocene sediments (Ferry and Meghraoui 2008; Ferry et al. 2007; Garfunkel et al. 1981). Two faults branch off the main fault zone. One at the northeastern end of the Dead Sea a NE-striking fault scarp manifests normal slip component, and the second, further north, a NW-striking fault branches off the Bet She'an Valley, also has a normal slip component with Mount Gilboa at the footwall. The northern end of the Jordan Valley section is where the fault zone consists of two parallel faults, which form the graben structure of Kinarot and the Sea of Galilee further north. The western boundary fault of the southern Sea of Galilee bends westward, making the northern part of the basin wider. This branch crosses the town of Tiberias (Hazan et al. 2004; Marco et al. 2003). The eastern boundary fault bends northeastward. In contrast to the dual fault at the south there is a single fault to the north of the Sea of Galilee, known as the Jordan Gorge Fault, which offsets manmade structures (Ellenblum et al. 1998) and Holocene stream channels (Marco et al. 2005; Wechsler et al. 2011). The Hula pull-apart basin to the north of the Jordan Gorge section is also where the fault system splays into several branches, namely (east to west) the Rachaia Fault, Serghaia Fault, Yammouneh Fault, and Roum Fault (Fig. 2). The Yammouneh, which takes up most of the plate motions, continues northward to the triple junction in southern Turkey, where it joins the East Anatolian Fault on the northeast and the Eastern Mediterranean collision zone on the west.

The on-fault studies resolve the debate revolving around the identification of the active branch. The suggestions that the main active fault is the Carmel Fault (Girdler 1990) or the Roum Fault (Butler et al. 1997, 1999) are not supported by the observations. Although they do take a small portion of the plate motions, the Yammouneh exhibits the major activity as well as the most prominent topographic feature, with some local contribution of the Lebanon Bend-related Rachaya-Serghaia fault branches (Daëron et al. 2005, 2007; Fleury et al. 1999; Gomez et al. 2003). Palaeoseismic records from the DSF in Syria and in southern Turkey also reveal major earthquake ruptures along the northern extension of the DSF (Akyuz et al. 2006; Altunel et al. 2009; Karabacak et al. 2010; Meghraoui et al. 2003). The analysis of three GPS campaigns between 1996 and 2008 shows an oblique motion along the Carmel Fault with about 0.7 mm/year left-lateral and about 0.6 mm/year N-S extension (Sadeh et al. 2012).

7.1.2 Test Plate Tectonic Paradigm

The debates on the nature of movement along the DSF (e.g., Vroman 1973) have been replaced by widely-agreed consensus on its left-lateral sense. Euler's theorem provides the means for resolving the relative motion on any one of three plate boundaries in a triple junction given the motions of the other two. The DSF connects the northern Red Sea spreading center to the collision zone located in South Eastern Turkey. If the fault were a pure transform, i.e., along a perfect small circle, the calculated DSF motion should be the same when solved for either junction, and this should conform to results from the DSF itself. In practice, the determination of accurate slip rates at both the southern end of the DSF, where it connects with the Red Sea and the Gulf of Suez, and at the northern end, where it connects with the East Anatolian fault and the subduction zone of the Cyprus arc, are poorly constrained and do not allow for such theoretical proof. Qualitative and quantitative confirmation for sinistral movement along the DSF, however, are validated by numerous observation of sinistral displacements of geological and archaeological bodies, as well as normal displacements of strata at the margins of pull-apart basins and all along the DSF itself. At present, the slip rates determined by the palaeoseismic studies alone (Table 7.3) are too variable to provide a definite quantitative confirmation of the plate tectonic models. Many of the on-fault palaeoseismic studies document the vertical components of slip. The few locations with suitable markers for measuring strike slip invariably confirm the sinistral nature of the motion. These data include offset alluvial fans (Garfunkel et al. 1981; Klinger et al. 2000a; Le Beon et al. 2010, 2012; Niemi et al. 2001), stream channels (Ferry et al. 2007; Marco et al. 2005; Wechsler et al. 2011), and archaeological structures (Table 7.1).

7.1.3 Earthquake Activity – History, Prehistory

Instrumental seismology along the DSF is limited to only one strong earthquake, the M_W7.2, 1995 Nuweiba, and one moderate, M_L6.1, 1927 Jericho earthquake. In order to recover earthquake history we must combine historical records, archaeological observations, and geological evidence for earthquake fault ruptures. All three kinds of data can either relate to off-fault or on-fault phenomena. The off-fault effects are reviewed by A. Agnon in Chapter 8. The on-fault research essentially relies on identification of surface, or near-surface, disruptions of geomorphic features, soil layers and man-made structures, either naturally outcropping or exposed in man-made trenches.

The first palaoeseismic trench was opened near Jericho across the fault trace identified by Garfunkel et al. (1981). Offset strata were dated using indicative archaeological artifacts (Reches and Hoexter 1981) and interpreted as the ruptures of the historically-recorded earthquakes of 31 BCE and 747 CE. Another set of pioneering palaeoseismic research trenches explored slip on normal faults and the development desert soils as a means to decipher tectonics at the western border of the fault zone in the southern Arava (Amit et al. 1995; Gerson et al. 1993). Several studies were aided with geophysical detection of faults, e.g., by GPR (Basson et al. 2002), high-resolution seismic reflection (Marco et al. 2005; Agnon et al. 2006), and magnetic field (Altunel et al. 2009). These, and later on-fault palaeoseismic observations, are summarized in Table 7.1.

7.2 Achievements of DST On-fault Studies

7.2.1 Test Reliability of Historical Records

Abundant accounts on historical earthquakes have been catalogued (e.g., Ambraseys 2009; Ambraseys and Finkel 1995; Ambraseys et al. 1994; Amiran et al. 1994; Guidoboni and Comastri 2005; Guidoboni et al. 1994; Russell 1985; Sbeinati et al. 2005). The catalogues include descriptions of damage to property, natural phenomena, human reactions, and fatalities. Reference to fault ruptures are usually missing. It is assumed that this lack of reference is not because there were no surface ruptures but rather because the contemporary reporters of these accounts were not aware of the association between earthquakes and faulting. Hence, pairing an historical earthquake to a specific fault often remains difficult. A major uncertainty is related to dating because the commonly used methods in palaeoseismic research, namely radiocarbon and luminescence, have large error margins and these need to be correlated with the often-uncertain dates of reported earthquakes. Several authors have pointed out the pitfalls of potential circular reasoning (e.g., Ambraseys 2005; Marco 2008; Rucker and Niemi 2010). Commonly, the geologists who find evidences for past earthquakes look for records of historical earthquakes listed in catalogues whose dates fall within the geological dating ranges. This practice works fine for the section south of the Hula Valley, where the fault zone is relatively simple. It is less certain for the fault zone farther north, where several branches, in addition to the Yammouneh fault that continues northward to Turkey, may be associated with strong earthquakes.

Table 7.2 Historical DSF earthquakes confirmed by surface ruptures observed in palaeoseismic investigations

Locations Fig. 7.1	Date	Segment	References
1	31 BCE	Jordan Valley	Reches and Hoexter (1981)
12	115 CE	Misyaf, Yammouneh	Meghraoui et al. (2003)
21	551	Lebanon thrust	Elias et al. (2007)
13, 1	749	Jordan Valley	Marco et al. (2003) and Reches and Hoexter (1981)
18	859	Northern Yammouneh	Akyuz et al. (2006)
15	3/1068	South Arava	Zilberman et al. (2005)
12	1170	Missyaf, Yammouneh	Meghraoui et al. (2003)
14, 4	1202	South Yammouneh	Daëron et al. (2005), Ellenblum et al. (1998), and Marco et al. (1997, 2005)
8, 9	1212	Northern Arava	Klinger et al. (2000b) and Niemi et al. (2001)
18	1408	Northern Yammouneh	Akyuz et al. (2006)
8	1458	Northern Arava	Klinger et al. (2000b)
11	1705	Serghaya Fault	Gomez et al. (2001)
4, 16	10/1759	Jordan Gorge	Ellenblum et al. (1998), Marco et al. (1997), and Marco et al. (2005)
14, 11, 25	11/1759	Serghaya Fault	Daëron et al. (2005), Gomez et al. (2001), and Nemer et al. (2008)
19	1837	Roum Fault	Nemer and Meghraoui (2006)
18	1872	Northern Yammouneh	Akyuz et al. (2006)

An example of successful matching between geological and archeological data is found at the site of Ateret, where structures were built on top of the active fault. There, accurate measurements of the slip associated with the earthquakes of 1202 and 1759 CE were possible (Ellenblum et al. 1998; Marco et al. 1997). Palaeoseismic trenches dug some 10 km southward confirmed these archaeoseismic results (Marco et al. 2005; Wechsler et al. 2011), while trenches along the Yammouneh fault (Daëron et al. 2005, 2007) were crucial to determine the northward extension of the ground ruptures associated with these two events. The rupture sizes and locations are remarkably in accord with independent analysis of historical reports on these earthquakes (Ambraseys and Barazangi 1989; Ambraseys and Melville 1988; Sieberg 1932). Table 7.2 lists all the historical records that were confirmed by direct observations. The observations include faulted strata and archaeological structures. The evidence for surface ruptures indicates that the earthquakes magnitudes were greater than M6, and that the historical records are largely reliable in such magnitude range.

Assuming the historical earthquake catalogues of earthquakes that were not confirmed yet by geological studies are also reasonably reliable, we illustrate their locations along the DSF system (Fig. 7.2). We interpret the locations to be close to the maximum reported damage area, although ideally it would be better if more data were available and more robust objective methods could be used (e.g., Bakun and Wentworth 1998; Sirovich et al. 2002; Zohar and Marco 2012).

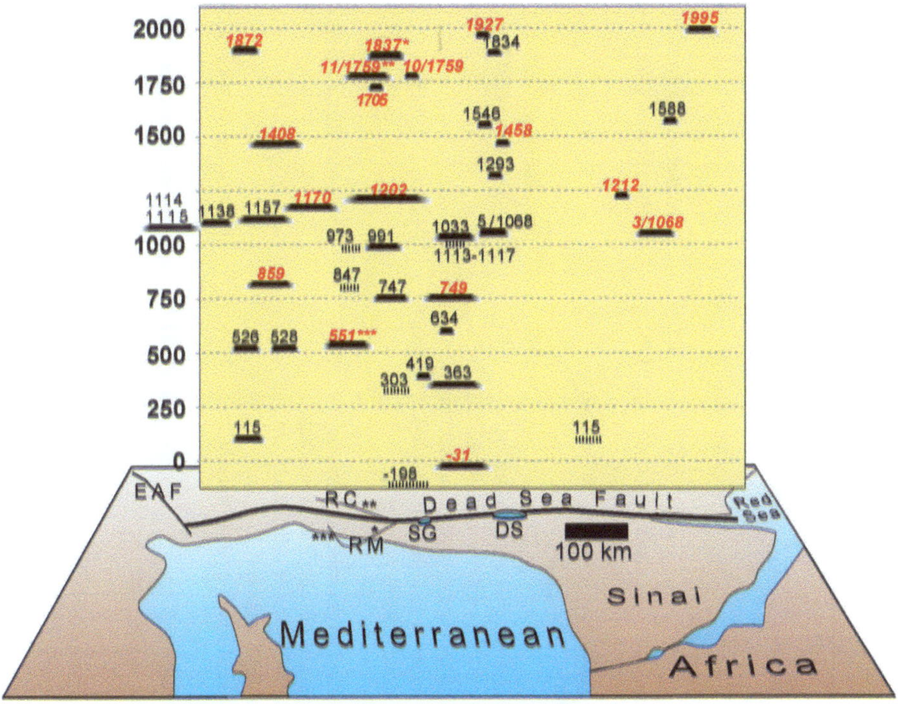

Fig. 7.2 Historical earthquakes distribution in space and time. Each bar corresponds to an approximate earthquake location along the DSF at the map. *Red dates* are earthquakes whose ruptures were found in palaeoseismic studies (see Table 7.2). Abbreviations: *EAF* East Anatolia Fault, *RC* Rachaya Fault, *RM* Roum Fault, *SG* Sea of Galilee, *DS* Dead Sea (Partly based on Garfunkel et al. 1981)

7.2.2 Slip Rate – Miocene to Holocene

Dated geological slip markers of two kinds are used for constraining slip rate (Table 7.3). One kind is pre-Miocene, which determines the total offset and thus the long-term slip rate. The other kind uses features that formed while the DSF was active, which determine short-term rates of individual sections of the fault. Palaeoseismic studies aim in particular at the Pleistocene-Recent activity. The short-term slip rate may be used for estimating the slip deficit by assuming that the total slip is represented either by the long-term slip rate, and/or the geodetically measured velocity away from the plate boundary. Along the southern section of the DSF, a minimum long-term rate is determined by dividing the 107-km total slip that is determined from the offset of pre-DSF geologic features visible both in the Negev and in Jordan, by the time of earliest faulting, about 20–25 Ma ago. The youngest rock unit that is offset by the full 107 km are 20-Ma dikes exposed in Sinai and Arabia (Bartov et al. 1980), providing a minimum long slip rate of about 5.35 mm/year. The motion post-dates the dikes but the precise initiation time is unclear yet.

Table 7.3 Slip rate estimates of the DSF

Period	Span, Ma	Rate mm a^{-1}	Max	Min	Data	Published	Reference
	?				Geological	1947	Dubertret (1947)
Miocene	?	5			Geological	1956	Quennell (1956)
Late Pleistocene-Recent	0.1	10			Geological	1968	Freund et al. (1968)
7–10 Ma	10	5±1	4	6	Geological	1970	Freund et al. (1970)
Last 1,000 years	0.001	0.8–1.7			Historical seismicity, magnitudes estimates based on extent of damage, seismic moment based on M-Mo empirical relation in California	1981	Garfunkel et al. (1981)
Plio-Pleistocene	5	7–10			Geological	1981	Garfunkel et al. (1981)
Last 4,500 years	0.0045	2.2			Estimates of slip associated with historical seismicity. Based on the assumption of GR relation extrapolated to high M earthquakes	1981	Ben-Menahem (1981)
20 Ma	20	5.1±0.3	4.8	5.5	Total offset, initiation based on pre-DSF dikes. Minimum rate. Dating – composite isochron	1981	Eyal et al. (1981)
1733 years section in the Late Pleistocene	0.0017	6.4±0.4	6	6.8	Recurrence based on deformed uppermost Lisan beds, rate of seismicity, assuming GR $\log N = 5.24 – 0.68\,M$ calculated from the 1733-year-long section	1986	El-Isa and Mustafa (1986)
Plio-Pleistocene	5	20			Geological, based on offset travertines in northern Hula Valley	1986	Steinitz and Bartov (1986)
Plio-Pleistocene	5	6 (0.283°/Ma)			Plate kinematics	1987	Joffe and Garfunkel (1987)
Holocene	0.01	>0.7			Geological, secondary deformation	1990	Gardosh et al. (1990)
Plio-Pleistocene	5	5.4–6.1			Geological	1990	Heimann (1990)
Plio-Pleistocene	5	3–7.5			Drainage systems, Arava Fault	1998	Ginat et al. (1998)

Late Pleistocene-Holocene	0.1	≥10			Offset fans	1999	Galli (1999)
Pleistocene	2	2–6, prefer 4			Alluvial fans, N. Arava	2000	Klinger et al. (2000a)
Pleistocene	2	4.7 ± 1.3			Alluvial fans, Arava	2001	Niemi et al. (2001)
Holocene	0.01	1–2			Serghaya Fault	2001	Gomez et al. (2001)
Last 2,000 years	0.002	6.9 ± 0.1			Displaced aqueduct; Paleo and Archaeoseismology, Missyaf (DSF in Syria). Assuming straight original shape	2003	Meghraoui et al. (2003)
1996–1999	0	2.6 ± 1			Geodesy, GPS	2003	Pe'eri et al. (2002)
Survey-Mode GPS	0	5.6–7.5			Increase from south to north	2003	McClusky et al. (2003)
Holocene	0.01	1.4 ± 0.2			Serghaya Fault	2003	Gomez et al. (2003)
1996–2003	0	3.3 ± 0.4			Geodesy, GPS results mostly on W side of fault. Based on model assuming fit to arctangent and locking depth	2004	Wdowinski et al. (2004)
25 ka	0.025	3.8–6.4	3.8	6.4	Geological, Lebanon	2004	Daëron et al. (2004)
Last 5,000 years	0.005	≥3	3	–	Single stream channel, Jordan Gorge. Northern margin offset 15 m, southern margin 9 m. C14 age of bulk organic matter. Minimum for that point, fault zone may be wider than trench	2005	Marco et al. (2005)
Last 6 Ma	6	c. 3.3			Offset c. 6 Ma NW margin of the Shin Volcano by c. 20 km	2005	Chorowicz et al. (2005)
1996–2003	0	4.4 ± 0.3			GPS	2005	Mahmoud et al. (2005)
Last 47.5 k years	0.0475	4.7–5.1 mm/year			Offset channels incised into the Lisan, Jordan Valley. Main channel predates highest stand	2007	Ferry et al. (2007) Comment/Reply: Ferry and Meghraoui (2008) and Klein (2008)
4 years	0		4	5	GPS	2007	Gomez et al. (2007)

(continued)

Table 7.3 (continued)

Period	Span, Ma	Rate mm a^{-1}	Max	Min	Data	Published	Reference
1999–2005	0	4.9±1.4			GPS	2008	Le Beon et al. (2008)
Holocene	0.01	2.7±1.5			Offset reef	2008	Ma kovsky et al. (2008)
Last 2,000	0.002	>2.8±0.2			Offset Hellenistic walls, Ateret,	–	Agnon et al. (2010)
7 kyears	0.007	6			Offset archaeology	2009	Altunel et al. (2009)
8 years	0	1.8–3.3			GPS	2010	Alchalbi et al. (2010)
25 ka	0.025	5			Offset channels, archaeology	2011	Ferry et al. (2011)
141 ka	0.141	5.4±2.7			Offset channels	2011	Le Beon et al. (2010)
		4.5±0.9					
		8.1±2.9					
Last 300 ka	300 ka	5–7			Offset channels	2012	Le Beon et al. (2012)

The consistency of instantaneous GPS rates (Le Beon et al. 2008), Quaternary rates (Klinger et al. 2000a; Niemi et al. 2001; Le Beon et al. 2010, 2012) and several-millions-years-long slip rates (Bartov et al. 1980) stands out. However it contrasts with the large variability of rates determined by paleoseismological studies and archaeological markers. For example, offset of a Roman period aqueduct is interpreted to show 7 mm/a (Meghraoui et al. 2003) whereas offset walls from the late twelfth century are offset only 2.1 m (Ellenblum et al. 1998). This may be explained either by distributed deformation unaccounted for by on-fault palaeoseismic investigations or, more probably, by too short time-windows of observation along some fault segments spans.

The discrepancy between short-term slip rates calculated by adding earthquake ruptures, and long-term rates is referred to as slip deficit. The long-term rate of 105 km slip in 20–25 Ma is 4–5 mm/year, in agreement with GPS results. Garfunkel et al. (1981) assigning sinistral slip to most of the historical earthquakes along the DSF, estimated that the sum of the seismic slip accounts for about two thirds of the long-term slip. Salamon et al. (1996) focusing only on the twentieth century seismicity found that only 7 % of the long-term slip was accommodated by earthquakes, leaving 83 % of the slip needed to match the long term slip-rate unexplained. Hence, such calculations show with little ambiguity that one needs to be able to document earthquake time series significantly longer than one earthquake cycle to be able to know something sensible about slip-rates from paleoseismology (Wechsler et al. 2011). This is particularly evident where the seismic activity might not be homogeneous but rather clustered, as it has been suggested for the DSF (Marco et al. 1996). And yet, the results of the palaeoseismic studies on the main strand of the DSF, which span enough time, are also in agreement with the geodetic and with the long-term rates (Ferry et al. 2011; Klinger et al. 2000a; Le Beon et al. 2010; Niemi et al. 2001).

7.2.3 Structural Details, Pull-Aparts, Thrusts

Structurally complex sections along the DSF include overlapping segments, commonly associated with pull-apart basins, right jogs where push-up swells occur, and splay faults of various strike directions. These complexities give rise to a variety of types of faults.

Thrust faults are mostly common along the Lebanese Restraining Bend (Elias et al. 2007) and the nearby Palmyride folds range, in Syria (Abou Romieh et al. 2012; Alchalbi et al. 2010; Chaimov and Barazangi 1990). Also the slip rates on the different thrust faults outcropping in these two areas is still subject to active discussion, the thrust associated to the LRB seems to be currently the most active with several large historical earthquakes associated with these structures.

Normal faults are usually associated with the occurrence of pull-apart basins (Garfunkel 1981), although several normal faults are also visible in the region of Galilee. Morphological analyses and palaeoseismic trench studies at the margins of the southern Arava confirm the normal nature of the fault on the west (Amit et al. 2002, 1999; Zilberman et al. 2005) and on the east (Thomas et al. 2007). Holocene activity of normal faults has been documented at the eastern boundary fault of the

Hula pull-apart valley (Zilberman et al. 2000). In Tiberias city, on the western shore of the Sea of Galilee, a normal fault offset early eighth century CE buildings, whereas late eighth century buildings located on top of the fault are not affected, bracketing an earthquake during the eighth century (Marco et al. 2003). Late Pleistocene normal fault zones were documented as active during the deposition of the Lisan Formation, at the western margins of the Dead Sea Basin (Bartov and Sagy 2004; Marco and Agnon 1995, 2005). On the western bounding fault of the Gulf of Aqaba Shaked et al. (2004, 2012) reconstruct vertical movements by dating buried coral reefs and submerged archaeological site. These palaeoseismic studies confirm the "leaky" nature of the DSF (Garfunkel 1981) and the paradigm of pull-apart structures (rhomb grabens) along strike-slip faults.

The structural role and history of activity of apparently "incongruent" faults that strike NE but show normal displacement instead of reverse motion that is expected to conform with NW-SE shortening, are not explored yet. These include the Sheikh Ali Fault at the NE end of the Sea of Galilee, a fault at the NE corner of the Dead Sea, and the NE corner of the Gulf of Aqaba. The NW-SE extension there is incompatible with the maximum horizontal compression that is inferred on the basis of analysis of meso-structures in the region (Eyal and Reches 1983).

7.3 Discussion

On-fault studies have confirmed the location of the main active strands of the DSF. In many cases, those studies have also brought new information about past earthquakes, documented, or not, in historical records. Building on these observations, we try to interpret the space-time distribution of earthquakes over the last two millennia (Fig. 7.2) and to suggest that a short-term pattern, periods without significant seismicity alternating with periods of intense activity, might exist.

Several earthquakes appear to define a seismic sequence rupturing from north to south. The most conspicuous series starts with the 1114/1115 CE earthquakes on the East Anatolian Fault and continues with the earthquakes of 1157, 1170, and 1202 CE. If these sequences indeed happen as outlined above, we do not see a similar or repeat pattern in the past. If this is correct, the famous North Anatolian Fault sequence of the 20th century, when a sequence of strong earthquake ruptures began on the east and propagated westward (Barka 1996; Stein et al. 1997) may not be a recurring sequence either.

Striking quiescence periods, of the order of several hundreds of years, appear to indicate imminent ruptures of several section of the fault. The segment between the Sea of Galilee and the Dead Sea ruptured in 31 BCE, 363 CE, 749 AD, and 1033 CE, followed by a millennium-long quiescence in which only moderate earthquakes occurred. The northern segment in Syria has been quiet for over 8 centuries (Meghraoui et al. 2003). Finally, the southern segment of the DSF along the Wadi Araba, has apparently not ruptured significantly for at least six hundred years. One might wonder if the Mw7.3 Nuweiba earthquake in 1995 in the Gulf of Aqaba marks the beginning of a new seismic sequence that could rupture a longer section of the DSF in the near future.

The estimated Late Pleistocene – Holocene slip rates seem to converge between 4 and 5 mm/year. The long-term rates that are based on offset Miocene and Pliocene geological bodies are in agreement with plate motion rates as determined geodetically. We therefore regard the variation in palaeoseismically-determined rates as indications of insufficient temporal coverage for some segments. The conversion of modern seismicity to slip also does not amount neither to the long-term slip nor to the geodetic slip rate (Garfunkel 2011), most probably because it represents a short time.

7.3.1 Future Targets

More data are needed to examine whether ruptures stop at mapped segment boundaries. We can at least confirm that several large earthquakes, such as the 1202 rupture, went through segment boundaries. Smaller earthquakes such as the October 30, 1759 were probably confined to a single segment but there are no data from the adjacent segments that could definitely rule out ruptures there as slip partitioning occurs where parallel or sub-parallel fault segments occur, e.g., the bounding faults of pull-apart grabens and where there are branches that split off the main fault, e.g., the Carmel Fault, Serghaya Fault, and Roum Fault.

The rupture history of transverse faults that connect overlapping faults at the boundaries of pull-apart basins is also unknown. To date none of the palaeoseismic studies addressed faults such as the Amatzyahu Fault that forms the southern boundary of the Dead Sea. These faults may either act independently or as rupture terminations of longer strike-slip faults. Comparisons of palaeoseismic records can reveal which is correct.

The structural role of the extension features at the NE corners of the Gulf of Aqaba, the Dead Sea, and the Sea of Galilee is unclear yet. Exploring the NE-SW-striking normal faults that appear in all these locations and express "incongruent" ~ NW-SE extension can shed light on these structures.

Acknowledgments Partial funding was provided by the Israel Science Foundation grant 1736/11 to SM.

References

Abou Romieh M, Westaway R, Daoud M, Bridgland DR (2012) First indications of high slip rates on active reverse faults NW of Damascus, Syria, from observations of deformed Quaternary sediments: implications for the partitioning of crustal deformation in the Middle Eastern region. Tectonophysics 538–540:86–104

Agnon A, Migowski C, Marco S (2006) Intraclast breccia layers in laminated sequences: recorders of paleo-earthquakes. In: Enzel Y, Agnon A, Stein M (eds) New Frontiers in Dead Sea paleoenvironmental research. Geological Society of America Special Publication, Boulder, pp 195–214

Agnon A, Marco S, Sagy A, Ellenblum R (2010) Discrepancy between GPS (5 yrs) and archaeoseismic (3 kyr) slip rate across the Ateret site (Dead Sea fault): secular variations versus distributed slip. In: Proceedings AGU Fall meeting abstracts, San Francisco, CA, vol 1. American Geophysical Union, p 2206

Akyuz HS, Altunel E, Karabacak V, Yalciner CC (2006) Historical earthquake activity of the northern part of the Dead Sea Fault Zone, southern Turkey. Tectonophysics 426:281–293

Alchalbi A, Daoud M, Gomez F, McClusky S, Reilinger R, Romeyeh MA, Alsouod A, Yassminh R, Ballani B, Darawcheh R, Sbeinati R, Radwan Y, Masri RA, Bayerly M, Ghazzi RA, Barazangi M (2010) Crustal deformation in northwestern Arabia from GPS measurements in Syria: slow slip rate along the northern Dead Sea Fault. Geophys J Int 180:125–135

Altunel E, Meghraoui M, Karabacak V, Akyuz SH, Ferry M, Yalciner CC, Munschy M (2009) Archaeological sites (Tell and Road) offset by the Dead Sea Fault in the Amik Basin, Southern Turkey. Geophys J Int 179:1313–1329

Ambraseys NN (2005) The seismic activity in Syria and Palestine during the Middle of the 8th century; an amalgamation of historical earthquakes. J Seismol 9:115–125

Ambraseys NN (2009) Earthquakes in the Mediterranean and Middle East: a multidisciplinary study of seismicity up to 1900. Cambridge University Press, Cambridge, UK

Ambraseys NN, Barazangi M (1989) The 1759 earthquake in the Bekaa valley: implications for earthquake hazard assessment in the eastern Mediterranean region. J Geophys Res 94:4007–4013

Ambraseys NN, Finkel CF (1995) The seismicity of Turkey and adjacent areas, a historical review, 1500–1800. Eren, Istanbul

Ambraseys NN, Melville CP (1988) An analysis of the eastern Mediterranean earthquake of 20 May 1202. In: Lee WKH, Meyers H, Shimazaki K (eds) Historical seismograms and earthquakes of the world. Academic, San Diego, pp 181–200

Ambraseys NN, Melville CP, Adams RD (1994) The seismicity of Egypt, Arabia, and the Red Sea: a historical review. Cambridge University Press, Cambridge, UK

Amiran DHK, Arieh E, Turcotte T (1994) Earthquakes in Israel and adjacent areas: Macroseismic observations since 100 B.C.E. Isr Explor J 44:260–305

Amit R, Harrison JBJ, Enzel Y (1995) Use of soils and colluvial deposits in analyzing tectonic events – The Southern Arava Rift, Israel. Geomorphology 12:91–107

Amit R, Harrison JBJ, Enzel Y, Porat N (1996) Soils as a tool for estimating ages of Quaternary fault scarps in a hyperarid environment – The southern Arava Valley, the Dead Sea rift, Israel. Catena 28:21–45

Amit R, Zilberman E, Porat N, Enzel Y (1999) Relief inversion in the Avrona Playa as evidence of large-magnitude historical earthquakes, southern Arava Valley, Dead Sea rift. Quat Res 52:76–91

Amit R, Zilberman E, Enzel Y, Porat N (2002) Paleoseismic evidence for time dependency of seismic response on a fault system in the southern Arava valley, Dead Sea rift. Isr Geol Soc Am Bull 114:192–206

Baer G, Sandwell D, Williams S, Bock Y, Shamir G (1999) Coseismic deformation associated with the November 1995, Mw=7.1 Nuweiba earthquake, Gulf of Elat (Aqaba), detected by synthetic aperture radar interferometry. J Geophys Res 104:25221–225232

Bakun WH, Wentworth CM (1998) Estimating earthquake location and magnitude from seismic intensity data. Bull Seismol Soc Am 87:1502–1521

Barka A (1996) Slip distribution along the North Anatolian Fault associated with the large earthquakes of the period 1939 to 1967. Bull Seismol Soc Am 86:1238–1254

Bartov Y, Sagy A (2004) Late Pleistocene extension and strike-slip in the Dead Sea Basin. Geol Mag 141:565–572

Bartov Y, Steinitz G, Eyal M, Eyal Y (1980) Sinistral movement along the Gulf of Aqaba – Its age and relation to the opening of the Red Sea. Nature 285:220–221

Bartov Y, Sneh A, Fleischer L, Arad V, Rosensaft M (2002) Map of suspect active faults in Israel. The Geological Survey of Israel, Jerusalem

Basson U, Ben-Avraham Z, Garfunkel Z, Lyakhovsky V (2002) Development of recent faulting in the southern Dead Sea rift according to GPR imaging. EGS Stephan Mueller Spec Publ Ser 2:1–23

Ben-Avraham Z, Garfunkel Z (1979) Continental breakup by a leaky transform: the Gulf of Elat (Aqaba). Science 206:214–216

Ben-Menahem A (1981) Variation of slip and creep along the Levant Rift over the past 4500 years. Tectonophysics 80:183–197

Butler RWH, Spencer S, Griffiths HM (1997) Transcurrent fault activity on the Dead Sea Transform in Lebanon and its implications for plate tectonics and seismic hazard. J Geol Soc 154:757–760

Butler RWH, Spencer S, Griffiths HM (1999) Discussion on transcurrent fault activity on the Dead Sea Transform in Lebanon and its implications for plate tectonics and seismic hazard – Reply. J Geol Soc 156:1246–1248

Chaimov TA, Barazangi M (1990) Crustal shortening in the Palmyride fold belt, Syria, and implications for movement along the Dead Sea Fault system. Tectonics 9:1369–1386

Chorowicz J, Dhont D, Ammar O, Rukieh M, Bilal A (2005) Tectonics of the Pliocene Homs basalts (Syria) and implications for the Dead Sea Fault Zone activity. J Geol Soc 162:259–271

Daëron M, Benedetti L, Tapponnier P, Sursock A, Finkel RC (2004) Constraints on the post 25-ka slip rate of the Yammouneh fault (Lebanon) using in situ cosmogenic 36Cl dating of offset limestone-clast fans. Earth Planet Sci Lett 227:105–119

Daëron M, Klinger Y, Tapponnier P, Elias A, Jacques E, Sursock A (2005) Sources of the large AD 1202 and 1759 Near East earthquakes. Geology 33:529–532

Daëron M, Klinger Y, Tapponnier P, Elias A, Jacques E, Sursock A (2007) 12,000-year-long record of 10 to 13 paleoearthquakes on the Yammouneh Fault, Levant Fault system, Lebanon. Bull Seismol Soc Am 97:749–771

Dubertret L (1932) Les formes structurales de la Syrie et de la Palestine. Académie des Sciences Comptes Rendus 195:65–67

Dubertret L (1947) Problemes de la geologic du Levant. Bull Soc Geol Fr Ser 5:3–31

Elias A, Tapponnier P, Singh SC, King GCP, Briais A, Daëron M, Carton H, Sursock A, Jacques E, Jomaa R, Klinger Y (2007) Active thrusting offshore Mount Lebanon: source of the tsunamigenic A.D. 551 Beirut-Tripoli earthquake. Geology 35:755–758

El-Isa ZH, Mustafa H (1986) Earthquake deformations in the Lisan deposits and seismotectonic implications. Geophys J R Astron Soc 86:413–424

Ellenblum R, Marco S, Agnon A, Rockwell T, Boas A (1998) Crusader castle torn apart by earthquake at dawn, 20 May 1202. Geology 26:303–306

Enzel Y, Kadan G, Eyal Y (2000) Holocene earthquakes inferred from a fan-delta sequence in the Dead Sea graben. Quat Res 53:34–48

Eyal Y, Reches Z (1983) Tectonic analysis of the Dead Sea rift region since the Late-Cretaceous based on mesostructures. Tectonics 2:39–66

Eyal M, Eyal Y, Bartov Y, Steinitz G (1981) The tectonic development of the western margin of the Gulf of Elat (Aqaba) rift. Tectonophysics 80:39–66

Ferry M, Meghraoui M (2008) Reply to the comment of Dr M. Klein on: "A 48-kyr-long slip rate history for the Jordan Valley segment of the Dead Sea Fault". Earth Planet Sci Lett 268:241–242

Ferry M, Meghraoui M, Abou Karaki N, Al-Taj M, Amoush H, Al-Dhaisat S, Barjous M (2007) A 48-kyr-long slip rate history for the Jordan Valley segment of the Dead Sea Fault. Earth Planet Sci Lett 260:394–406

Ferry M, Meghraoui M, Abou Karaki N, Al-Taj M, Khalil L (2011) Episodic behavior of the Jordan Valley section of the Dead Sea fault inferred from a 14-ka-long integrated catalog of large earthquakes. Bull Seismol Soc Am 101:39–67

Fleury J, Chorowicz J, Somma J (1999) Discussion on transcurrent fault activity on the Dead Sea Transform in Lebanon and its implications for plate tectonics and seismic hazard. J Geol Soc 156:1243–1248

Freund R (1965) A model of the structural development of Israel and adjacent areas since Upper Cretaceous times. Geol Mag 102:189–205

Freund R, Zak I, Garfunkel Z (1968) Age and rate of the sinistral movement along the Dead Sea Rift. Nature 220:253–255

Freund R, Garfunkel Z, Zak I, Goldberg M, Weissbrod T, Derin B (1970) The shear along the Dead Sea rift. Philos Trans R Soc Lond A 267:107–130

Frieslander U (2000) The structure of the Dead Sea Transform emphasizing the Arava using new geophysical data. The Hebrew University of Jerusalem, p 101

Galli P (1999) Active tectonics along the Wadi Araba-Jordan Valley transform fault. J Geophys Res 104:2777–2796
Gardosh M, Reches Z, Garfunkel Z (1990) Holocene tectonic deformation along the western margins of the Dead Sea. Tectonophysics 180:123–137
Garfunkel Z (1981) Internal structure of the Dead Sea leaky transform (rift) in relation to plate kinematics. Tectonophysics 80:81–108
Garfunkel Z (2011) The long- and short-term lateral slip and seismicity along the Dead Sea Transform: an interim evaluation. Isr J Earth Sci 58:217–235
Garfunkel Z, Zak I, Freund R (1981) Active faulting in the Dead Sea rift. Tectonophysics 80:1–26
Gerson R, Grossman S, Amit R, Greenbaum N (1993) Indicators of faulting events and periods of quiescence in desert alluvial fans. Earth Surf Proc Land 18:181–202
Ginat H, Enzel Y, Avni Y (1998) Translocation of Plio-Pleistocene drainage system along the Dead Sea Transform, south Israel. Tectonophysics 284:151–160
Girdler RW (1990) The Dead-Sea Transform-Fault system. Tectonophysics 180:1–13
Gomez F, Meghraoui M, Darkal AN, Sbeinati R, Darawcheh R, Tabet C, Khawlie M, Charabe M, Khair K, Barazangi M (2001) Coseismic displacements along the Serghaya fault: an active branch of the Dead Sea Fault system in Syria and Lebanon. J Geol Soc 158:405–408
Gomez F, Meghraoui M, Darkal AN, Hijazi F, Mouty M, Suleiman Y, Sbeinati R, Darawcheh R, Al-Ghazzi R, Barazangi M (2003) Holocene faulting and earthquake recurrence along the Serghaya branch of the Dead Sea fault system in Syria and Lebanon. Geophys J Int 153:658–674
Gomez F, Karam G, Khawlie M, McClusky S, Vernant P, Reilinger R, Jaafar R, Tabet C, Khair K, Barazangi M (2007) Global Positioning System measurements of strain accumulation and slip transfer through the restraining bend along the Dead Sea fault system in Lebanon. Geophys J Int 168:1021–1028
Guidoboni E, Comastri A (2005) Catalogue of earthquakes and tsunamis in the Mediterranean area from the 11th to the 15th century. Istituto Nazionale di Geofisica, Bologna
Guidoboni E, Comastri A, Traina G (1994) Catalogue of ancient earthquakes in the Mediterranean area up to the 10th century. Istituto Nazionale di Geofisica, Bologna
Haynes JM, Niemi TM, Atallah M (2006) Evidence for ground-rupturing earthquakes on the Northern Wadi Araba fault at the archaeological site of Qasr Tilah, Dead Sea Transform fault system, Jordan. J Seismol 10:415–430
Hazan N, Stein M, Marco S (2004) Lake Kinneret levels and active faulting in the Tiberias area. Isr J Earth Sci 53:199–205
Heimann A (1990) The development of the Dead Sea rift and its margins in the northern Israel during the Pliocene and the Pleistocene. Golan Research Institute and Geological Survey of Israel, Jerusalem
Heimann A, Ron H (1987) Young faults in the Hula pull-apart basin, central Dead Sea Transform. Tectonophysics 141:117–124
Heimann A, Ron H (1993) Geometric changes of plate boundaries along part of the northern Dead Sea Transform: geochronologic and paleomagnetic evidence. Tectonics 12:477–491
Hofstetter R, Klinger Y, Amrat AQ, Rivera L, Dorbath L (2007) Stress tensor and focal mechanisms along the Dead Sea fault and related structural elements based on seismological data. Tectonophysics 429(3–4):165–181
Joffe S, Garfunkel Z (1987) Plate kinematics of the circum Red Sea- a re-evaluation. Tectonophysics 141:5–22
Karabacak V, Altunel E, Meghraoui M, Akyüz H (2010) Field evidences from northern Dead Sea Fault Zone (South Turkey): new findings for the initiation age and slip rate. Tectonophysics 480:172–182
Klein M (2008) A comment on: "A 48-kyr-long slip rate history for the Jordan Valley segment of the Dead Sea Fault" EPSL 260 (2007) 394–406. Earth Planet Sci Lett 268:239–240
Klinger Y, Rivera L, Haessler H, Maurin JC (1999) Active faulting in the Gulf of Aqaba: new knowledge from the Mw7.3 earthquake of 22 November 1995. Bull Seismol Soc Am 89:1025–1036
Klinger Y, Avouac JP, Abou-Karaki N, Dorbath L, Bourles D, Reyss JL (2000a) Slip rate on the Dead Sea Transform fault in northern Araba Valley (Jordan). Geophys J Int 142:755–768

Klinger Y, Avouac JP, Dorbath L, Abou-Karaki N, Tisnerat N (2000b) Seismic behaviour of the Dead Sea Fault along Araba Valley. Jordan Geophys J Int 142:769–782

Le Beon M, Klinger Y, Amrat AQ, Agnon A, Dorbath L, Baer G, Ruegg JC, CharadeO, Mayyas O (2008) Slip rate and locking depth from GPS profiles across the southern Dead Sea Transform. J Geophys Res Solid Earth 113:B11403, pp 19

Le Beon M, Klinger Y, Al-Qaryouti M, Meriaux AS, Finkel RC, Elias A, Mayyas O, Ryerson FJ, Tapponnier P (2010) Early Holocene and Late Pleistocene slip rates of the southern Dead Sea Fault determined from Be-10 cosmogenic dating of offset alluvial deposits. J Geophys Res Solid Earth 115:B11414, pp 24

Le Beon M, Klinger Y, Meriaux AS, Al-Qaryouti M, Finkel RC, Mayyas O, Tapponnier P (2012) Quaternary morphotectonic mapping of the Wadi Araba and implications for the tectonic activity of the southern Dead Sea fault. Tectonics 31:TC5003, pp 25

Mahmoud S, Reilinger R, McClusky S, Vernant P, Tealeb A (2005) GPS evidence for northward motion of the Sinai Block: implications for E. Mediterranean tectonics. Earth Planet Sci Lett 238:217–224

Makovsky Y, Wunch A, Ariely R, Shaked Y, Rivlin A, Shemesh A, Ben Avraham Z, Agnon A (2008) Quaternary transform kinematics constrained by sequence stratigraphy and submerged coastline features: the Gulf of Aqaba. Earth Planet Sci Lett 271:109–122

Marco S (2007) Temporal variation in the geometry of a strike-slip fault zone: examples from the Dead Sea Transform. Tectonophysics 445:186–199

Marco S (2008) Recognition of earthquake-related damage in archaeological sites: examples from the Dead Sea fault zone. Tectonophysics 453:148–156

Marco S, Agnon A (1995) Prehistoric earthquake deformations near Masada, Dead Sea graben. Geology 23:695–698

Marco S, Agnon A (2005) High-resolution stratigraphy reveals repeated earthquake faulting in the Masada Fault Zone, Dead Sea Transform. Tectonophysics 408:101–112

Marco S, Agnon A, Ellenblum R, Eidelman A, Basson U, Boas A (1997) 817-year-old walls offset sinistrally 2.1 m by the Dead Sea Transform, Israel. J Geodyn 24:11–20

Marco S, Hartal M, Hazan N, Lev L, Stein M (2003) Archaeology, history, and geology of the A.D. 749 earthquake, Dead Sea Transform. Geology 31:665–668. doi:10.1130/G19516.19511

Marco S, Rockwell TK, Heimann A, Frieslander U, Agnon A (2005) Late Holocene slip of the Dead Sea Transform revealed in 3D palaeoseismic trenches on the Jordan Gorge segment. Earth Planet Sci Lett 234:189–205

Marco S, Stein M, Agnon A, Ron H (1996) Long-term earthquake clustering: a 50,000-year paleoseismic record in the Dead Sea Graben. J Geophys Res-Solid Earth 101(B3):6179–6191

McClusky S, Reilinger R, Mahmoud S, Sari DB, Tealeb A (2003) GPS constraints on Africa (Nubia) and Arabia plate motions. Geophys J Int 155:126

Meghraoui M, Gomez F, Sbeinati R, derWoerd JV, Mouty M, Darkal AN, Radwan Y, Layyous I, Najjar HA, Darawcheh R, Hijazi F, Al-Ghazzi R, Barazangi M (2003) Evidence for 830 years of seismic quiescence from palaeoseismology, archaeoseismology and historical seismicity along the Dead Sea fault in Syria. Earth Planet Sci Lett 210:35–52

Nemer T, Meghraoui M (2006) Evidence of coseismic ruptures along the Roum fault (Lebanon): a possible source for the AD 1837 earthquake. J Struct Geol 28:1483–1495

Nemer T, Meghraoui M, Khair K (2008) The Rachaya-Serghaya fault system (Lebanon): evidence of coseismic ruptures, and the AD 1759 earthquake sequence. J Geophys Res 113:B05312

Niemi TM, Zhang H, Atallah M, Harrison BJ (2001) Late Pleistocene and Holocene slip rate of the Northern Wadi Araba fault, Dead Sea Transform, Jordan. J Seismol 5:449–474

Pe'eri S, Wdowinski S, Shtibelman A, Bechor N (2002) Current plate motion across the Dead Sea Fault from three years of continuous GPS monitoring. Geophys Res Lett 29:42–45. doi:10.1029/2001GL013879

Pinar A, Turkelli N (1997) Source inversion of the 1993 and 1995 Gulf of Aqaba earthquakes. Tectonophysics 283:279–288

Quennell AM (1956) Tectonics of the Dead Sea rift. In: Asociacion de Servicios Geologicos Africanos. Congreso Geologico Internacional, 20th sesion, Mexico, pp 385–405

Reches Z, Hoexter DF (1981) Holocene seismic and tectonic activity in the Dead Sea area. Tectonophysics 80:235–254

Reilinger R, McClusky S, Vernant P, Lawrence S, Ergintav S, Cakmak R, Ozener H, Kadirov F, Guliev I, Stepanyan R, Nadariya M, Hahubia G, Mahmoud S, Sakr K, ArRajehi A, Paradissis D, Al-Aydrus A, Prilepin M, Guseva T, Evren E, Dmitrotsa A, Filikov SV, Gomez F, Al-Ghazzi R, Karam G (2006) GPS constraints on continental deformation in the Africa-Arabia-Eurasia continental collision zone and implications for the dynamics of plate interactions. J Geophys Res Solid Earth 111:B05411, pp 26

Ron H, Freund R, Garfunkel Z, Nur A (1984) Block-rotation by strike-slip faulting: structural and paleomagnetic evidence. J Geophys Res 89:6256–6270

Rotstein Y, Bartov Y, Frieslander U (1992) Evidence for local shifting of the main fault and changes in the structural setting, Kinarot basin, Dead Sea Transform. Geology 20:251–254

Rucker JD, Niemi TM (2010) Historical earthquake catalogues and archaeological data: achieving synthesis without circular reasoning. In: Sintubin M, Stewart IS, Niemi TM, Altunel E (eds) Ancient earthquakes, The Geological Society of America special paper. The Geological Society of America, Boulder, pp 97–106

Russell KE (1985) The earthquake chronology of Palestine and northwest Arabia from the 2nd through the mid-8th century A.D. Bull Am Sch Orient Res 260:37–60

Sadeh M, Hamiel Y, Ziv A, Bock Y, Fang P, Wdowinski S (2012) Crustal deformation along the Dead Sea Transform and the Carmel Fault inferred from 12 years of GPS measurements. J Geophys Res Solid Earth 117:B08410

Salamon A, Hofstetter A, Garfunkel Z, Ron H (1996) Seismicity of the eastern Mediterranean region: perspective from the Sinai subplate. Tectonophysics 263:293–305

Sbeinati MR, Darawcheh R, Mouty M (2005) The historical earthquakes of Syria: an analysis of large and moderate earthquakes from 1365 B.C. to 1900 A.D. Ann Geophys 48:347–435

Shaked Y, Lazar B, Marco S, Stein M, Tchernov D, Agnon A (2004) Detailed evolution of fringing reefs: space and time constraints from the Gulf of Aqaba. Coral Reefs 24:165–172. DOI:10.1007/s00338-004-0454-2

Shaked Y, Lazar B, Marco S, Stein M, Agnon A (2012) Late holocene events that shaped the shoreline at the northern Gulf of Aqaba recorded by a buried fossil reef. Israel J Earth Sci 58(3–4):355–368

Shaliv G (1991) Stages in the tectonic and volcanic history of the Neogene basin in the Lower Galilee and the valleys. Geological Survey of Israel – Report (in Hebrew, English abstract) 11/91, pp 1–94

Shamir G, Baer G, Hofstetter A (2003) Three-dimensional elastic earthquake modelling based on integrated seismological and InSAR data: the Mw = 7.2 Nuweiba earthquake, gulf of Elat/Aqaba 1995 November. Geophys J Int 154:731–744

Sieberg A (1932) Erdbebengeographie, Handbuch der Geophysik, Band IV. Borntraeger, Berlin, pp 527–1005

Sirovich L, Pettenati F, Cavallini F, Bobbio M (2002) Natural-neighbor isoseismals. Bull Seismol Soc Am 92:1933–1940

Stein RS, Barka A, Dietrich JH (1997) Progressive failure on the North Anatolian fault since 1939 by earthquake stress triggering. Geophys J Int 128:594–604

Steinitz G, Bartov Y (1986) The 1985 time table for the tectonic events along the Dead Sea Transform. Terra Cognita 6:160

ten Brink US, Rybakov M, Al-Zoubi AS, Hassouneh M, Frieslander U, Batayneh AT, Goldschmidt V, Daoud MN, Rotstein Y, Hall JK (1999) Anatomy of the Dead Sea Transform: does it reflect continuous changes in plate motion? Geology 27:887–890

ten-Brink US, Ben-Avraham Z (1989) The anatomy of a pull-apart basin: reflection observations of the Dead Sea Basin. Tectonics 8:333–350

Thomas R, Niemi TM, Parker ST (2007) Structural damage from earthquakes in the second-ninth centuries at the archaeological site of Aila in Aqaba, Jordan. Bull Am Sch Orient Res 346:59–77

Vroman AJ (1973) Is a compromise between the theories of tension and of shear for the origin of the Jordan-Dead Sea trench possible? Isr J Earth Sci 22:141–156

Wdowinski S, Bock Y, Baer G, Prawirodirdjo L, Bechor N, Naaman S, Knafo R, Forrai Y, Melzer Y (2004) GPS Measurements of current crustal movements along the Dead Sea Fault. J Geophys Res 109:1–16
Wechsler N, Rockwell TK, Klinger Y, Agnon A, Marco S (2011) Testing earthquake recurrence models with 3D trenching along the Dead-Sea Transform, 2nd INQUA-IGCP-567 International Workshop on Active Tectonics, Earthquake Geology, Archaeology and Engineering, Corinth, Greece
Wells DL, Coppersmith KJ (1994) New empirical relationships among magnitudes, rupture length, rupture width, rupture area, and surface displacement. Bull Seismol Soc Am 84:974–1002
Willis B (1938) Wellings' observations of Dead Sea structure (with discussion). Geol Soc Am Bull 49:659–668
Wilson JT (1965) A new class of faults and their bearing on continental drift. Nature 207:343–347
Zilberman E, Amit R, Heimann A, Porat N (2000) Changes in Holocene paleoseismic activity in the Hula pull-apart basin, Dead Sea rift, northern Israel. Tectonophysics 321:237–252
Zilberman E, Amit R, Porat N, Enzel Y, Avner U (2005) Surface ruptures induced by the devastating 1068 AD earthquake in the southern Arava valley, Dead Sea rift, Israel. Tectonophysics 408:79–99
Zohar M, Marco S (2012) Re-estimating the epicenter of the 1927 Jericho earthquake using spatial distribution of intensity data. J Appl Geophys 82:19–29

Chapter 8
Pre-Instrumental Earthquakes Along the Dead Sea Rift

Amotz Agnon

Abstract The Dead Sea rift offers a wealth of information about pre-instrumental earthquakes. The types of potential archives include historic seismicity, archaeological sites, disturbed beds in lake deposits, rockfalls within caves as well as on free slopes, and displaced marine terraces. The rich historical archive is useful as a key for deciphering the geological archives.

Of the geological archives developed for the Dead Sea rift, lake sections stand out due to the long periods covered with high resolution. Lake deposits contain long and potentially continuous archives of the environment, and of earthquakes in particular. The Holocene drop in Dead Sea level, accentuated with a fast anthropogenic drop, have triggered incision and outcrop formation, permitting access and direct investigation of archives. The ongoing analysis of cores from lake drill-holes will augment the continuity of the archive.

The historical information spans periods that exceed the seismic cycle of individual fault segments. One of the provoking results of the comparisons of historical versus geological archives of earthquake activity is the significant difference in the apparent length of the earthquake cycle, where prehistorical data indicates long quiescence periods. This suggests that even the long historical record of the Levant does not encompass the full earthquake cycle along the entire Dead Sea fault. This result underscores the significance of paleoseismic research for the understanding of earthquake-fault mechanics and for hazard assessment.

Keywords Paleo-earthquakes • Earthquake clustering • Historic earthquakes • Dead Sea earthquakes

---

A. Agnon (✉)
The Fredy and Nadine Herrmann Institute of Earth Sciences, The Hebrew University of Jerusalem, Edmond J. Safra campus, Givat Ram, Jerusalem 91904, Israel
e-mail: amotz@huji.ac.il

Z. Garfunkel et al. (eds.), *Dead Sea Transform Fault System: Reviews*, Modern Approaches in Solid Earth Sciences 6, DOI 10.1007/978-94-017-8872-4_8,

8.1 Introduction

The study of pre-instrumental earthquakes has made considerable progress in the world during the last 40 years since the introduction of geological observations to a subject previously dominated by historic research. Paleoseismology has opened a new dimension in the study of recurrence and magnitudes of past earthquakes (Sieh 1978; Swan et al. 1980; Yeats et al. 1997; McCalpin 2009). Paleo-earthquake research is essential for the study of any fault system for which the loading-unloading cycle is longer than the period covered by history.

The historical archive for the Dead Sea rift (DSR) is extensive over the last three millennia and, at times, includes complementary sources from coexisting cultures (Guidoboni et al. 1994; Ambraseys et al. 1994; Guidoboni and Comastri 2005; Karcz 2004; Sbeinati et al. 2005; Ambraseys 2005a, b, 2009).

The value of pre-instrumental seismicity has been underscored by several authors, and several articles have reviewed preinstrumental earthquakes in the Levant (see Garfunkel 2011, for a recent review). Ambraseys (1971) has detected a pattern of interaction in catalogues of historic earthquakes along the two conjugate faults that bound the Anatolian block (NAF and EAF in Fig. 8.1). Karcz et al. (1977) have used archaeological evidence to test historical catalogues, and found a different distribution of damage, where the former exhibit higher concentration along the rift. Garfunkel et al. (1981) have used historic earthquakes to identify a spatio-temporal

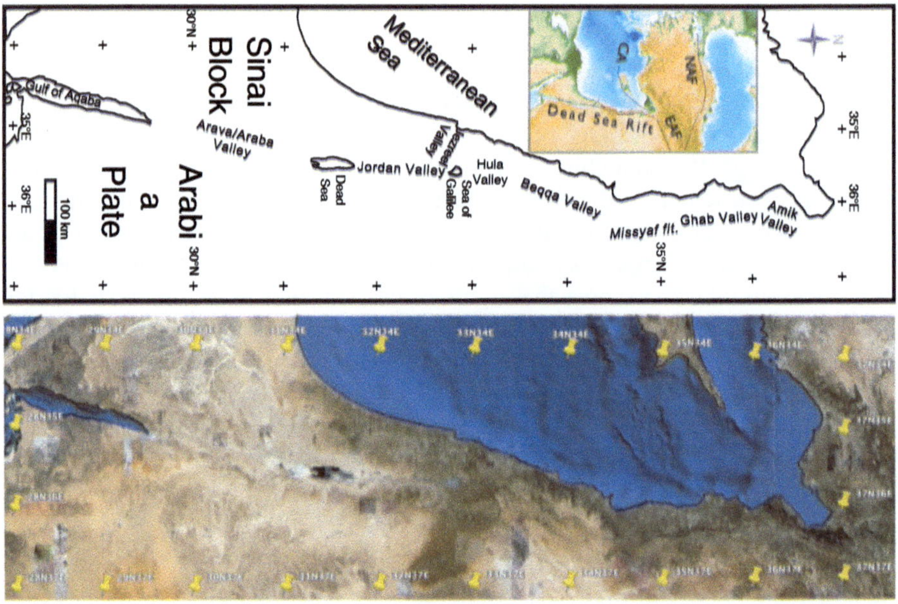

Fig. 8.1 Basins along the Dead Sea rift. *Inset* shows the adjacent plate boundaries: *EAF* East Anatolian fault, *CA* Cypriot arc, *NAF* North Anatolian fault (Generated by http://woodshole.er.usgs.gov/mapit/). Lower panel - from © 2012 GoogleEarth

pattern of seismic gaps. Stiros (2001) has used historical accounts, archaeological excavations, and geological observations for identifying a cluster of activity that spanned three centuries along four plate boundaries in the Eastern Mediterranean, including the Dead Sea rift. Agnon et al. (2006) have reviewed recent research focused on lake sediments for studying historic and pre-historic earthquakes. They reaffirmed temporal earthquake clustering with a long cycle of activity, of order 10 kyear, longer than the historic period.

The present review attempts to be more comprehensive, including historical, archaeological, and geological information, and spanning historical and prehistorical times. Comparison between alternative sources is given where applicable. We start with a systematic review of the sources that include historical catalogues, excavated archaeological ruins (on- and off-fault), and geological evidence of several types: lake seismites, coastal terraces with related sediments, and rock falls (in caves and on free slopes). A discussion of the extent and the reliability of each type follows its description. A following section presents the temporal correlations between the various sources. In the spirit of the present review, an attempt is made to cover all possible correlations. Subsequently we consider the issue of completeness of the archives, critical to the assessment of recurrence patterns, which comprises the discussion section.

8.2 Sources of Information

This review focuses on earthquakes recorded by off-fault effects, including collapsed and cracked buildings (from historical accounts and archaeological excavations), disturbed sediments, and displaced coastal features. Direct evidence of surface ruptures is reviewed in a different chapter (Marco and Klinger 2014), but on occasion such information is used here for supporting the present interpretations (Ellenblum et al. 1998; Gluck 2001; Haynes et al. 2006; Akyuz et al. 2006; Ferry et al. 2011; Sbeinati et al. 2010).

8.2.1 Historic Earthquakes

During the time of the development of the paleoseismic research in the Dead Sea rift, namely the last two decades, several comprehensive catalogues have been published under a modern standard. The catalogues of Ambraseys et al. (1994) and Guidoboni et al. (1994) mark the transition to such a standard. The extent of literature on historic earthquakes of the Levant is formidable, and often catalogues are conflicting and confusing, as pointed out by Karcz (2004). Since that review, focused on Jewish sources between the second century BCE and the eighth century CE, three additional catalogues have been published (Guidoboni and Comastri 2005; Sbeinati et al. 2005; Ambraseys 2009). The present review is written from a geological perspective by a non-historian so only recent extensive regional catalogues are used systematically. When physical evidence for earthquakes support the catalogues of Ben Menahem (1991) and Amiran et al. (1994), they are used to augment the ones mentioned above. Sbeinati et al.'s (2005) catalogue is focused on the northern Levant, whereas Ambraseys et al. (1994) focus on the Red Sea; the others include areas around the Mediterranean. Ambraseys (2006a, b) filtered the larger events for which he estimated surface wave magnitudes (Ms) from macroseismic data. Salamon (2010) and Kagan et al. (2011) compiled lists of larger and generally consensual events reported in these catalogues to have shaken the Levant, and in particular the DSR.

Figure 8.2 displays sites central to the historical discussion of earthquakes (a) and interpreted locations of historic events (b–d). Some of the events require special attention due to possible bias in the historic documents and interaction with other disciplines (e.g. Karcz 2004; Ambraseys 2005a, b). Archaeological evidence for demise has been associated with historic earthquakes even when the dating of damaged structures was poorly constrained. Thus questionable dating could be presented as infallible. This approach can potentially introduce spurious interpretations of archaeological as well as historical data.

Kagan et al. (2011) have recently compiled a list of historic events that could potentially affect the Dead Sea basin, and tested for correlations between each event and dated lake seismites. They have used an attenuation relation that describes the decay of macroseismic intensity with distance from the source in the Eastern Mediterranean and Middle East (Ambraseys and Jackson 1998):

$$M_S = -1.54 + 0.65(I_i) + 0.0029(R_i) + 2.14\log(R_i) + 0.32p, \tag{8.1}$$

where M_S is estimated surface-waves magnitude and I_i is the Medvedev-Sponheuer-Karnik intensity. $R_i = \sqrt{r_i^2 + r_0^2}$ with r_i being the mean isoseismal radius of intensity I_i, and r_0=9.7 km. Equation (8.1) was based on 488 isoseismal contours that were fit to about 9,000 intensity points originating from 123 shallow instrumental earthquakes. Ambraseys (2006a, b; 2009) studied different seismogenic zones, the DSR being one of them. He adjusted the coefficients of Eq. (8.1) to macroseismic data of 59 instrumental DSR events, and obtained:

$$M_S = -0.138 + 0.554(I_i) + 0.0033(r_i) + 1.54\log(r_i) + 0.31p, \tag{8.2}$$

where r_i (in km) in the near-field is the distance from a point with the i th intensity to the source or its nearest rupture. Ambraseys (2006a, b, 2009) used Eq. (8.2) to assess source location and magnitude of 80 pre-instrumental events. This allowed him to calculate frequency-size distribution and rate of moment release.

Hough and Avni (2011) used extensive data by Avni (1999) from the M6.3 1927 Jericho earthquake to calibrate the attenuation relation for the region using the schemes of Bakun and Wentworth (1997). They have augmented the calibration by "did you feel it" reports for the M5.0 2005 Lebanon earthquake (Atkinson and Wald 2007). Hough and Avni (2011) extend the attenuation relation derived by Malkawi

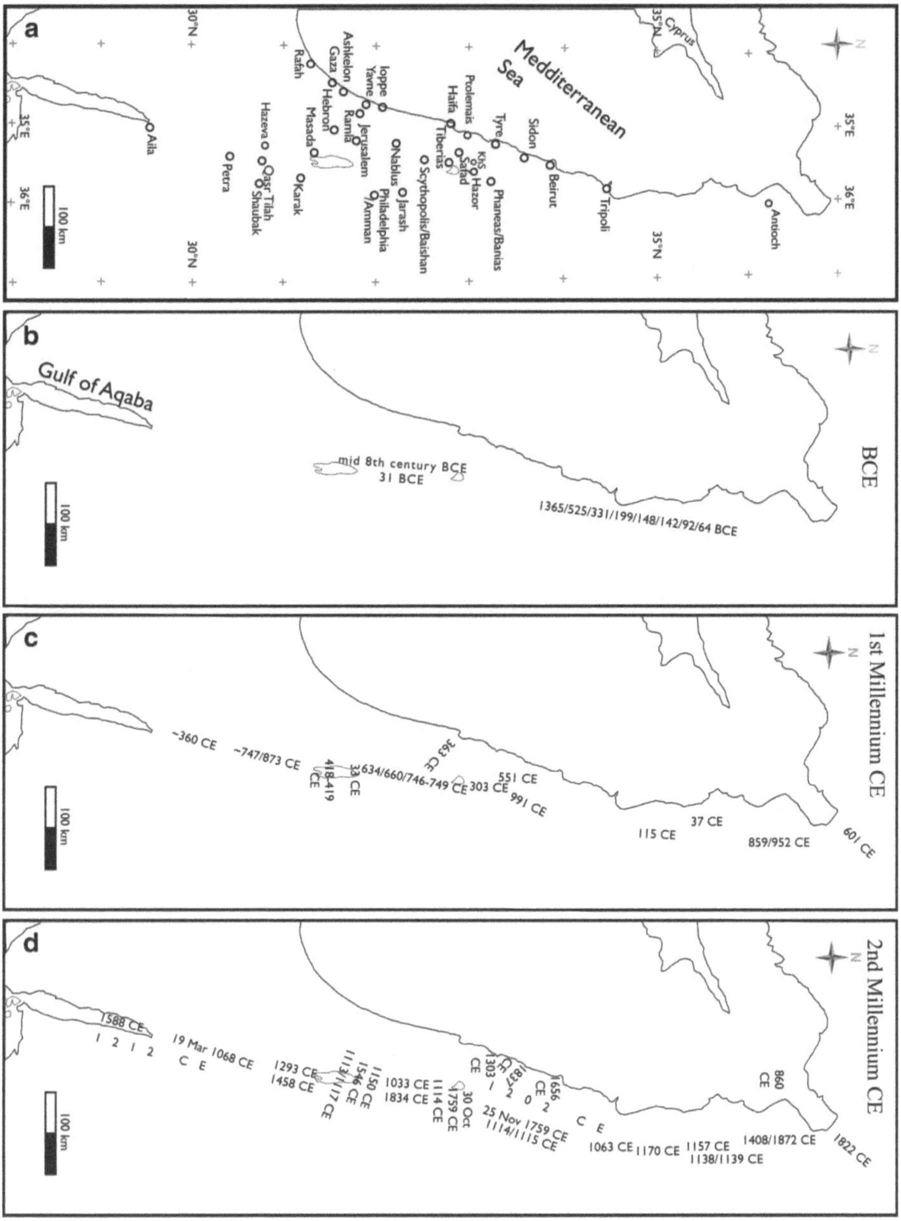

Fig. 8.2 Sites of documented historic earthquakes (**a**) and inferred source zones for BCE (**b**), 1st Millennium CE (**c**) and second Millennium CE (**d**)

and Fahmi (1996) (see review by Al-Qaryouti 2008) to include the M7.3 1995 Gulf of Aqaba event. Rearranged in the form of Eqs. (8.1 and 8.2), a preliminary calibrated attenuation relation for DSR is given by:

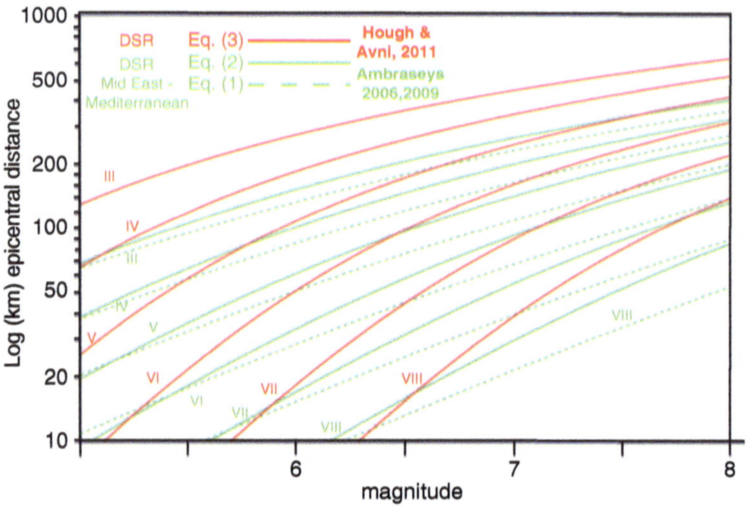

Fig. 8.3 Attenuation relations used in this and previous studies that correlate historical macroseismic data with physical evidence for earthquakes. The *curves* represent Eqs. (8.1, 8.2 and 8.3), where for Eq. (8.1) r_0 was set to null in accordance with Ambraseys (2006a, b, 2009). Setting r_0 to 9.7 km makes a minor difference

$$M = 0.388 + 0.588(I_i) + 0.00282(r_i) + 0.98\log(r_i). \tag{8.3}$$

Hough and Avni (2011) use Eq. (8.3) to assess the magnitudes of two historic earthquakes, 1170 and 1202 CE, as 6.6 and 7.6, respectively (compared with Ambraseys (2006a, b) who estimates 7.3 and 7.2 respectively).

Figure 8.3 displays Eqs. (8.1, 8.2 and 8.3) where iso-intensity lines are plotted on (log r, M) field. At small epicentral distances (<20 km) or at moderate magnitudes (<6.5) the three equations predict similar intensities (to within the scatter of the data). At large distances, Eq. (8.3) predicts intensities higher by one (low magnitude) to two (high magnitude) units. A general observation is that the curves are progressively more convex (downward) from (8.1, 8.2 and 8.3). We will return to these attenuation relations for correlation of historical seismicity and physical evidence.

8.2.2 Archaeological Evidence of Earthquakes

The DSR and its surroundings are dotted with numerous archaeological excavations that have generated reports of earthquake damage tied to particular strata and hence often associated with historic events (Fig. 8.4) (Karcz et al. 1977). As is generally the case for seismology and paleoseismology (Yeats et al. 1997; McCalpin 2009), in archaeoseismic studies the distinction between on-fault and off-fault phenomena is

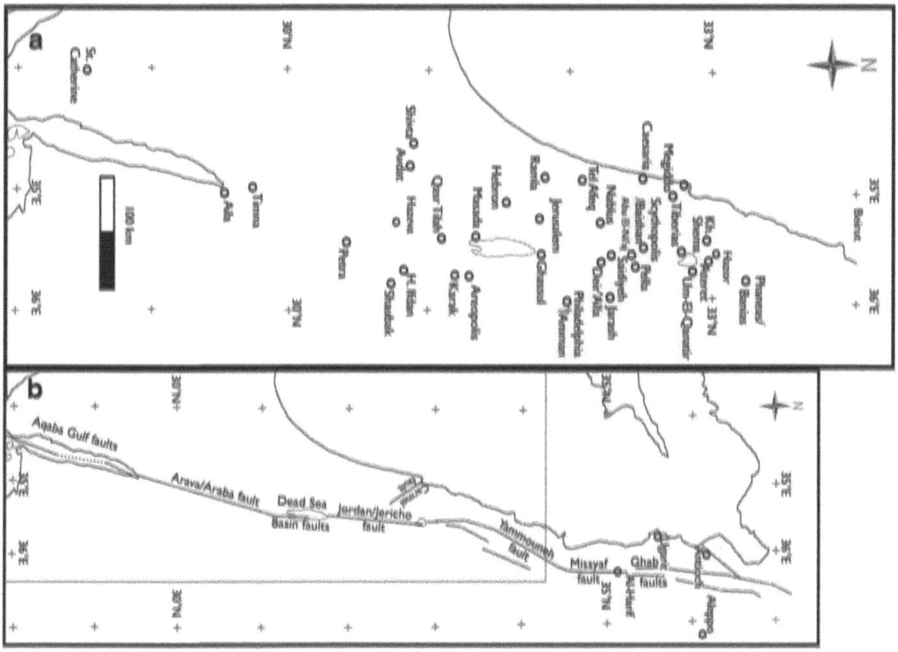

Fig. 8.4 Archaeological sites where earthquake evidence has been reported (**a**) The main fault segments forming the Dead Sea Rift are shown (**b**)

central. The DSR is unique in that it offers clear cases of on-fault archaeoseismic sites. Marco (2008) has recently reviewed archaeoseismic studies along the Dead Sea rift. In this section we present cases that are pertinent to the present discussion.

The study of earthquakes via the archaeology of ruins should benefit from close collaboration between geologists and archaeologists (and historians if the period in question is documented in chronicles). Coauthoring of papers with archaeologists increases the prospects for professional reading of the archaeology (Karcz et al. 1977; Marco et al. 1997, 2006; Ellenblum et al. 1998; Marco et al. 2003; Shaked et al. 2004; Thomas et al. 2007; Wechsler et al. 2009; Sbeinati et al. 2010; Ferry et al. 2011). The first example described below is a unique case where excavations have been systematically driven and steered in collaboration between a historian and earthquake geologists, where the on-site archaeologists made their operational decisions in order to address seismological issues.

8.2.2.1 Tel Ateret

The most striking example of an on-fault archaeoseismic study is the excavations of Tel-Ateret (Fig. 8.5). Marco et al. (1997) have documented the Crusader castle of Vadum Iacob (Tel Ateret) over which an Ottoman mosque had been constructed, respectively offset sinistrally 2.1 and 0.5 m (Fig. 8.5). Ellenblum et al. (1998) have

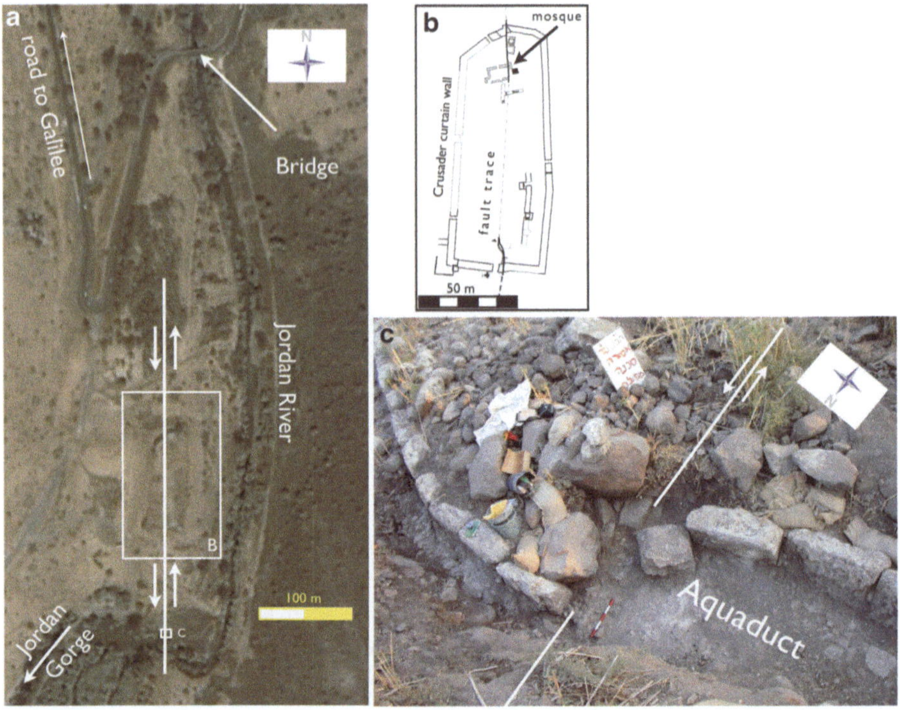

Fig. 8.5 Tel Ateret. (**a**) A satellite photo (GoogleEarth) of the Ateret – Benot Ya'aqov bridge. Vadum Iacob castle straddles the fault trace, that runs through an aqueduct system south of the castle. (**b**) The outline of Vadum Iacob Crusader castle (curtain wall) and the Ottoman mosque. (**c**) Aqueduct photo taken down and to the south-south-west. The *red*-and-*white scale* is 0.5 m aligned north to south

assigned individual earthquakes to the two offset events: 20 May 1202 CE (1.6 m) and 30 October 1759 (0.5 m). These earthquakes have been independently assigned macroseismic magnitudes of 7.2 (Ambraseys 2006b) and 6.5 (Ambraseys and Barazangi 1989). The macroseismic magnitudes are compatible with displacement-magnitude systematics (Wells and Coppersmith 1994).

The advantage of archaeology is its precision in determining the amounts slip. When combined with history, archaeology can resolve the timing of slip events as no other method can. The Crusader's curtain wall was meticulously laid, allowing a precision approaching 1 cm. The Ottoman wall allowed ~0.1 m precision. Very similar and synchronous offsets, albeit with lower precisions in amount and date, have been recorded by paleoseismic trenching of fluvial channels some 10 km south of Tel Ateret (Marco et al. 2005; Marco and Klinger 2014).

Meghraoui et al. (2003) and Sbeinati et al. (2010) reconstruct an offset aqueduct at Al-Harif, where they assign an offset of 13 m and an age of ~2 ka (Fig. 8.4b). The challenge with deciphering offset from aqueducts is often the lack of control on initial, pre-deformation structure. In addition, archaeological dating of damage to an aqueduct is not as precise as dating of living quarters, where one may find a wealth of well dated artifacts, and floors that demonstrate a stratigraphic context, constraining the age of the vertical features. Not only the destruction of aqueducts is difficult to date, their construction date is typically vague for similar reasons. We have excavated an aqueduct at the site of Tel Ateret and, ironically, it is dated based on its left-lateral offset of 2 m, similar to the offset of the nearby Crusader's castle (Fig. 8.5c).

The dating of the earthquakes associated with the offset of Vadum Iacob are exceptionally well constrained by history, documented both by Crusaders and Muslims. The castle was erected during 11 months and, before completion, conquered and demolished by Saladin in late August, 1179 CE (Ellenblum 2007). The floor at the time of construction and conquest stands out due to its conspicuous lime color and richness with metallic weapons and construction tools. In the subsequent 22.5 years, thin (several-cm scale) soil accumulated on the rubble to be ruptured during the earthquake of 20 May 1202 CE. A meter (or more) of soil has accumulated during the last eight centuries. A fraction of this soil layer (30 % or more) was ruptured during the 30 Oct 1759 M ~ 6.5 event.

8.2.2.2 Qasr Tilah

An additional case where the DSR or its branches offset archaeological structures have been documented, where the archaeological context jointly with radiocarbon dates allow particular earthquakes to be suggested. A water reservoir and an associated aqueduct in Qasr Tilah, south of the Dead Sea (Fig. 8.4), are offset by about 2 m (Klinger et al. 2000; Haynes et al. 2006). The dates of damage from the last four rupture events was correlated by Haynes et al. (2006) to the historic earthquakes of 634 or 659/660, 873, 1068, and an Ottoman Period event. The assignment to a seventh century CE earthquake was based on dates of construction and repair during the occupation of the site, that has seemed to be abandoned during later centuries. Haynes et al. (2006) have assigned a historic earthquake from 873 CE to the second rupture at a paleoseismic trench. The epicenter of 873 CE earthquake was placed by Ambraseys et al. (1994) well in the Arabian Plate, yet Haynes et al. (2006) suggest that due to sparse population the catalogue is biased. Haynes et al. (2006) disregard the possibility that one of the mid eighth century earthquakes was candidate. Bikai (2002) attributes collapse of the Blue Chapel to the mid-eighth century AD earthquake. Eklund (2008) infers widespread destruction by a mid-eighth century earthquake, and this may suggest rupture during this time south of the Dead Sea.

The third rupture in the trench, being the penultimate event, crosses a layer of seventh–tenth century CE dated by Haynes et al. (2006), who suggest the earthquake of 18 March 1068. Haynes et al. (2006) select the earthquake of 1546 as the best candidate for the most recent event to rupture Qasr Tilah, since the layer cut during this rupture is dated to the Ottoman period (1515–1918 CE). The earthquake of 1546 was played down by Ambraseys and Karcz (1992) as one that affected only Jerusalem, hence being exaggerated. An alternative is the earthquake of 1834 with reported damage from Karak to Caesarea (Fig. 8.2) and asphalt emissions in the

Dead Sea (Amiran et al. 1994; Ambraseys 2009). Garfunkel et al. (1981) interpreted this event as a rupture in the Dead Sea and northern Arava, based on the then available catalogues. Ben-Menahem (1991) placed the 1834 epicenter at the southern dead sea, noting toppling of structures east of the Dead Sea. More data is needed for ruling between the historic earthquakes of 1546 and 1834 as the ultimate event in Qasr Tilah.

8.2.2.3 Fallen and Cracked Off-fault Structures

Severe ground shaking having caused structural destruction can be recorded and subsequently unearthed in archaeological sites. Marco (2008) has surveyed some cases of fallen and cracked masonry structures to exemplify useful diagnostics for earthquake related damage. Additional cases, some mentioned by Russell (1980) include Petra, Sephourias, Bet Shearim, and Scythopolis for the 363 earthquake, and several others. Distinct features that testify for earthquake damage include lack of evidence for alternative causes of damage, and special features such as human skeletons in positions indicating attempts for self-protection (Nur 2008).

Archaeology often extends the seismic record beyond the historic period. An ongoing collaboration between archaeologists and geologists enabled the interpretation of destruction layers and masonry damage in Megiddo (Fig. 8.4), perpetually reconstructed for twenty-six centuries (e.g. Porat et al. 2012). This site offers unique data on the Carmel fault branch of the Dead Sea rift. Marco et al. (2006) have documented 18 instances of damage, of which 16 may have resulted from seismic shaking. They denote ten of the cases as "probably catastrophic", of which seven indicate shaking (four of them horizontal). In a single case, a liquified sand bed is taken as clear evidence of catastrophic shaking. Altogether, the damage seems to had happened in six events, where at least two of them had been caused by earthquakes: shortly before 5 ka and between 2.8 and 2.9 ka. Another likely event preceded the latter by about a century. Evidence for additional events lacks confidence in source of damage or dating, or both. The likely source of the earthquake damage is rupture on the Carmel fault that runs apparently underneath the site (Fig. 8.4), yet the Jordan fault, less than 40 km away, is a viable candidate.

Russell (1980) and Hammond (1980) have based their arguments for a large 363 earthquake, damaging the entire southern sector of the continental DSR, on the collapse of the Temple of the Winged Lions in Petra and destruction at the Main Theatre (Fig. 8.6a). This dramatic finding matched well with the concept of a drastic decline of Petra during the fourth–fifth century CE up to a complete abandonment following destruction by a mid sixth century earthquake, namely 551 CE. The excavations of 1991–1993, that have casted doubt on damage by the 551 CE event. These excavations have demonstrated that the site continued to function to a final destruction during the very late eighth century or the ninth century CE (Eklund 2008). The impressive collapse of a colonnade from the Great Temple, likely due to an earthquake (Fig. 8.6b) is yet to be dated.

Fig. 8.6 colonnades (**a**) Temple of Winged Lions; (**b**) Great Temple; (**c**) Hippos SE church; (**d**) Hippos NW church

Some archaeological sites show damage in more than a single layer. Tsafrir and Foerster (1997) documented impressive damage from a 363 CE earthquake to Scythopolis. Earthquake damage from 363 CE has been cited by many authors for many sites (e.g. Russell 1980). Scythopolis was hit again early-to-mid seventh century CE before seismogenic demolition (under the Arab name Baishan) during the mid-eighth century CE. Tsafrir and Foerster (1997) used numismatic evidence to date the latter event as 749 CE. Karcz (2004) pointed out that this find does not rule out an earlier event (747 CE) while Ambraseys (2009) argued that the case is not closed without ruling out later events (757 or 768–775 CE). All these discussions underscore the precision available in historical documents buttressed by archaeological evidence which allow accuracy not attainable in purely geochronological studies. However, geological evidence can help when the resolution is sufficient, as discussed below in the "Correlation" section.

The attribution of damage in ruins to earthquakes in general and to particular historical temblors is often challenged. Magness (1997) has argued against the assignments of a pair of earthquakes, 306 CE and 419 CE, to damage in Khirbet Shema' (Fig. 8.2a) (Meyers et al. 1976). The argument is based on a lack of positive evidence of earthquake destruction at 306 CE; Meyers et al. (1976) merely present precise numismatic evidence for prominent occupation between 306 and 341 CE, interpreting it as a building phase following a putative earthquake. Magness (1997) have accepted the evidence for earthquake destruction of the top layer, but assigns it to the mid-eighth century CE, long after abandonment of the site. This case demonstrates the sensitivity of earthquake chronologies derived from archaeology to the assumptions at the basis of the interpretation.

Fig. 8.7 Qal'at Subayba keystone

Several authors have commented on the circular reasoning and feed back between archaeological, historical, and geological interpretations, giving rise to false identifications (Karcz and Kafri 1978; Karcz 2004; Ambraseys 2005b, 2009; Rucker and Niemi 2010). The case of Qal'at Subayba is a useful example for the pitfalls (Figs. 8.3 and 8.7). Nur (2008) have attributed the damage in Qal'at Al-Subayba ("Qal'at Namrud") to the 1202 CE earthquake, whilst the damaged building was only erected later around 1230 and buttressed only towards 1260 (Ellenblum 1989). This precise dating is based on massive inscriptions in the site (Amitai 1989). The damage, erroneously attributed by Nur (2008) to the 1202 CE earthquake, is likely dated to the earthquake of 30 October 1759 CE (Hartal 2001).

A comprehensive and critical review of historic earthquake catalogues versus archaeological and geological evidence from DSR, a long overdue foundation for earthquake research and hazard assessment, is beyond the scope of this paper. Yet the Correlation section below discusses some promising research opportunities.

8.2.3 *Lake Seismites*

Seilacher (1969) pioneered the systematic identification of beds deformed at the water-sediment interface during earthquakes. He has coined the genetic term "seismite" for sedimentary rocks displaying structures that can be interpreted to result from earthquake shaking. Several types of textures in lacustrine facies have been since recognized:

1. Liquefied and fluidized beds: structures indicating folding and intrusion of coarse grain layers (Sims 1973, 1975).
2. Intraclast breccias (mix layers): beds comprising of clasts derived from the local pre-shaking sediment (Marco et al. 1996; Agnon et al. 2006).
3. Homogenites: Homogeneous massive intervals in the otherwise layered to laminated section; the composition within the homogenite is identical to a homog-

enous mixture of the compositions of the different layers or laminae (Chapron et al. 1999, following Cita et al. 1984).

4. Turbidites and silt layers: graded bedding, typically a thin bed of coarser material within fines (Siegenthaler et al. 1987, and Doig 1990, following Heezen and Ewing 1952).

Sims (1973) has underscored the potential in the study of earthquakes in active continental zones from liquefaction structures in lacustrine beds. He subsequently analyzed late Quaternary sections and inferred recurrence intervals (Sims 1975). Hempton and Dewey (1983) described a sequence of five seismites in the East Anatolia fault (EAF, Fig. 8.1), and interpreted one to reflect higher shaking intensity. With the absence of age data they used spacing for conjecturing a non regular recurrence pattern. Siegenthaler et al. (1987) have attributed a homogenite to slumping during seiche triggered by a historic earthquake in Lake Lucerne, Switzerland. They identified such homogenites in cores and seismic reflection profiles and opened the way to reconstruction of recurrence intervals and patterns. Doig (1990, 1991) had observed discoloration in small lakes at the epicentral zone of an instrumentally recorded earthquake in Canada. He inferred that graded silty layers penetrated in short cores represent earthquake induced slumping in the catchment (Doig 1998). Correlation with historic events based on constant sedimentation rate and extrapolation to prehistorical times revealed variable recurrence rates.

Marine paleoseismology has developed in parallel to lacustrine paleoseismology (Kastens 1984; Adams 1990; Roep and Everts 1991; Cita et al. 1996; Moretti et al. 1999; McHugh et al. 2006; Monecke et al. 2006; Moretti and Sabato 2007), where the distinction between seismites and other structures reflecting high mechanical energy has remained a challenge (Cita et al. 1984). In a recent issue devoted to soft sediment deformation triggered by earthquakes (Owen et al. 2011, and references therein), Gibert et al. (2011) have cast doubts on the inference of earthquake histories from one of the most common sedimentary structures used to infer earthquake shaking, namely load structures: multiple superimposed liquified layers may result from a single shaking event. Therefore, seismites that evidently formed at the water sediment interface are more useful for determining the timing of individual events and recurrence patterns.

While lacustrine paleoseismology has flourished elsewhere (e.g. Davenport and Ringrose 1987; Jones and Omoto 2000; Waldmann et al. 2011), certain properties of the sediments in the Dead Sea and its ancestral lakes (Lake Lisan in particular) have enabled unique contributions to the discipline.

8.2.3.1 Prehistoric Dead Sea Seismites

Laminated lacustrine sediments, common around the Dead-Sea Basin (Fig. 8.8), provide a recorder of high energy events at the lake bottom (Manspeizer 1985). The last global glacial cycle and the ensuing rise of lake level have deposited laminar, perhaps varved, sedimentary sections dominated by alternation of detrital and pure aragonite laminae (Begin et al. 1974; Katz et al. 1977; Barkan et al. 2001). Aragonite,

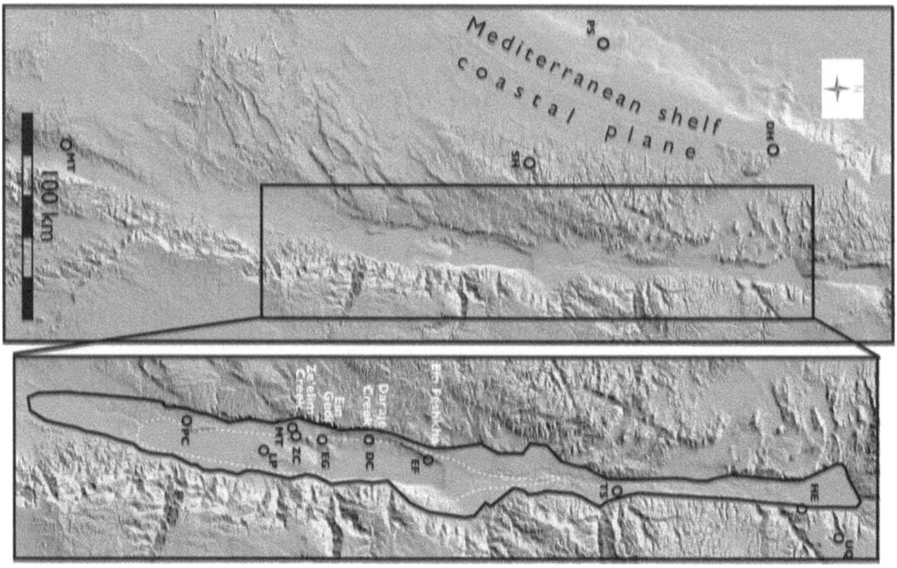

Fig. 8.8 Extent of late Pleistocene lakes and locations of paleoseismic study sites in the area. *White curve* Lisan Formation, *White broken curve* Ze'elim Formation, *PC* Pratzim Creek, *LP* Lisan Peninsula, *MT* Masada Terrace, *ZC* Ze'elim Creek, *DC* Daraje Creek, *EG* Ein Gedi, *EF* Ein Feshkha, *HE* Ha'Onn Escarpment, *SH* Soreq and Hartuv Caves, *DH* Denia Cave at Haifa, *PS* Palmahim slump. The *shaded* relief map is by Hall (1996)

a polymorph of calcium carbonate, has the advantage of preserving the primary chemistry and isotopic composition (Katz et al. 1977), hence allowing radiometric dating (Haase-Schramm et al. 2004; Stein 2011). A typical facies in the Lisan Formation comprises alternations of dark detritus versus pure white aragonite, deposited during periods of a stratified lake (Katz et al. 1977). These can be explained as seasonal laminae, or varves, where the detritus was transported to the lake during flash floods and subsequent evaporation led to aragonite precipitation (Katz et al. 1977; Barkan et al. 2001).

Laminated sections are found up to levels of ~200 m below sea level (b.s.l., where during the twentieth century A.D. the level declined below 400 m b.s.l. (Bookman et al. (2004))). The late Pleistocene Lisan Formation line the valley floor on a stretch 275 km long; the Holocene Ze'elim Formation stretches about 170 km (Fig. 8.8). The Lisan Formation (Shaw 1947) has long been recognized for unique study opportunities due to lithological uniformity, high resolution stratigraphy, and well exposed beach deposits (Bowman 1971; Bartov et al. 2002). The unit has been studied extensively in recent years (e.g. Haase-Schramm et al. 2004; Enzel et al. 2006; Ron et al. 2007; Belmaker et al. 2007; Torfstein et al. 2008; Prasad et al. 2009). The distortion and destruction of lamination in the otherwise laminated sediments of Lisan Formation has been used as an indicator for paleo-earthquakes by several authors (Pettijohn and Potter 1964; Seilacher 1984; El-Isa and Mustafa 1986; Marco and Agnon 1995; Agnon et al. 2006; Heifetz et al. 2005; Katz et al.

2009; Wetzler et al. 2010; Alsop and Marco 2011). The Ze'elim Formation (Yechieli 1993) shows similar facies, with beach ridges limited to levels lower than 370 m b.s.l., and laminar aragonite lower than 400 m (Bookman et al. (2004)). The unit exhibits seismites similar to those described in Lisan Formation (Ken-Tor et al. 2001; Migowski et al. 2004; Agnon et al. 2006; Kagan et al. 2011). The upper part of the section corresponds to historic times, enabling correlation of seismites with historic earthquakes (Ken-Tor et al. 2001; Migowski et al. 2004; Agnon et al. 2006; Kagan et al. 2011).

El-Isa and Mustafa (1986) have recognized the scientific potential of intraformational structures indicating post depositional seismic shaking (seismites). They listed three categories of processes generating structures that can be used to study paleo-earthquakes. The first category includes local faulting, cracking, and slumping. The faults are normal or horizontal, of various size and age. The second category comprises effects of liquefaction: sand boils in coarser sediments and destruction of lamination in fine grain sediments. The latter is associated with loss of cohesion and "ultimate mixing of the particles". The third category of processes was the focus of El-Isa and Mustafa's (1986) work, namely folding over décollement surfaces. Marco and Agnon (1995, 2005) used detailed stratigraphy to establish a link between faulting events and phenomena of the likes of El-Isa and Mustafa's (1986), with emphasis on their first and second categories.

The Dead Sea near-shore facies exhibits the more common types of seismites. Enzel et al. (2000) have documented 11 liquified layers in fan delta Holocene sediments of Daraje Creek (DC, Fig. 8.8). These include ball-and-pillow structures, clastic dike intrusions, convolute lamination, and micro-faults.

8.2.3.2 Seismites in Beds from the Historic Period

Several authors approached late Holocene beds exposed by recent retreat of the Dead Sea lake, attempting to correlate seismites with historic earthquakes. These correlations are discussed in a separate section below. Enzel et al. (2000) identify the ultimate rupture exposed in Daraje Creek with the M=6.2 1927 event. Ken-Tor et al. (2001) have identified eight seismites in the fan deposits of Ze'elim Creek (ZC, Fig. 8.8). They have extracted 24 radiocarbon dates from the 7 m section that featured two unconformities representing significant hiatuses. They dated these to eighth–tenth and late thirteenth–fifteenth centuries CE, paving the way for high accuracy lake level curves (Bookman et al. (2004)). They were able to correlate all eight seismites from the section studied to historic earthquakes, although some ambiguities could not be resolved with the data. In particular, a seismite deposited around 400 CE could be interpreted as 363 or alternatively ad 419 CE (see their Figs. 8.4 and 8.3b respectively). We return to the correlation between individual seismites and particular historical events in a later section.

To circumvent the hiatuses, and to access the lacustrine laminated facies appropriate for high resolution studies, Migowski et al. (2004) have collected continuous cores from the Dead Sea shores. Three coring sites along the western shore were designed to provide lateral coverage (Fig. 8.8): Ein Feshkha (EF), Ein Gedi (EG), and Ze'elim Creek, which allowed outcrop control with Ken-Tor et al. (2001) data. They studied a core from Ein Gedi in considerable detail. The lacustrine facies, with alternations of detritus and chemical precipitates (aragonite with, in places, halite), builds 3 m of the section of Ein Gedi core, corresponding to a historic period.

Migowski et al. (2004) had counted laminae from a 2.2 m continuously laminated section under a microscope for construction of a chronological model. The age model was constructed under the following assumptions: (i) each cycle of detritus/chemical laminae represents a year of deposition; (ii) the top of IBLs correspond to years of historic earthquakes; each calibrated radiocarbon age represents a *terminus post quem* for the deposition. They resolved breccia layers as thin as 2 mm; the thickest seismite they report is 9 cm and the average 1.4 cm. They counted 22 seismites within the otherwise laminated interval and correlated them to earthquakes between the 140 BCE and 1293 CE events.

Given the uneven distributions of earthquakes in history and of IBLs in the core, a unique age model anchored by seismite-earthquake correlation to the absolute time scale seemed plausible. The correlation was constrained by four radiocarbon dates that yielded ages 100–250 years older than the historic dates of the earthquakes. This was explained by the long time of deposition for organic matter through the dense brine formed the hypolimnion. Such a dense hypolimnion, separated from the surface water, is expected for the counted interval with its facies of interlaminated aragonite and detritus (Barkan et al. 2001).

Only two deformed layers, correlative to dates of ~90 CE and ~175 CE, could not be matched with historic events. A striking example of the high resolution of this study was given by Agnon et al. (2006) who show that the 1202 CE earthquake can be resolved from the 1212 CE earthquake.

As a test for the correlation, Migowski et al. (2004) and Agnon et al. (2006) have evaluated the local intensity of historic earthquakes in the Ein Gedi site based on the historical epicenter and magnitude, with the aid of attenuation relations (Fig. 8.3). All correlated earthquakes had high expected intensities (right-lower side of the distance-magnitude diagram), whereas earthquakes that are missing from the record were expected to have low intensity (left upper side). Some medium intensity earthquakes were identified, and some were missing for the Ein Gedi core. This test corroborates the correlation of deformed layers with earthquakes and at the same time lends support to the assumption of annual laminae cycles.

Four prominent historic events did not show in the interval where laminae were counted; these events could have been masked by successive events (1063 followed by 1068 CE), or had been duplicated in the historical catalogues due to spurious reading of multiple calendars (1032 versus 1033 CE, Ambraseys et al. 1994). Agnon et al. (2006) suggested that the earthquake of 1202 CE left a faint mark, barely resolved in the successive IBL assigned by Migowski et al. (2004) to the 1212 CE event.

Agnon et al. (2006) have revised the analysis of Ken-Tor et al. (2001) using a uniform deposition rate between unconformities. With such a model they found a unique match between each of the eight seismites and a historic event. They have accounted for the masking of event horizons by subsequent earthquakes and analyzed the resolving power of the section (see below).

Kagan et al. (2011) have studied an extended section in the more lacustrine facies of Ze'elim Creek and added seismites from sections that fill in the hiatuses in the original section. Ongoing retreat of the lake has enhanced the exposure, affording a continuous section of the last three millennia. They also collected data from an additional section at Ein Feshkha in the northernmost Dead Sea (Fig. 8.8) and analyzed the outcrop data jointly with Migowski et al.'s (2004) data from Ein Gedi core.

Altogether, Kagan et al. (2011) study over a 100 seismites, half of which in the northern (truncated) Ein Feshkha section. The relative abundance of seismites in the northern section has been attributed to preferential recording of the northern extent of the rift, perhaps by waves guided through the Jordan/Jericho fault. The comparison of the three sites allows to single out events that have affected the entire basin, generating wide spread deformation dubbed "intra-basin seismites" (IBS).

Levi et al. (2006) have studied the magneto-fabrics of clastic dikes that intrude the Lisan Formation (Marco et al. 2002). They conclude that some of the dikes, in particular those showing evidence of horizontal propagation, were injected laterally during earthquake shaking. Porat et al. (2007), based on optically stimulated luminescence (OSL) dating of dike material, infer ages of between 15 and 7 ka for injection. With the caution that the mechanism for resetting the OSL signal is not yet known, they suggest that such ages can be of use in paleoseismology.

8.2.3.3 Convolute Lamination (Intraformational Folds)

Convolute lamination – a manifestation of intraformational folding in the laminated Lisan Formation – provides a spectacular manifestation for the mechanical energy that reach the normally quiet lake bottom (Fig. 8.9). Such structures have long attracted the eye of geologists who typically associate the deformation with earthquake shaking (Pettijohn and Potter 1964). El-Isa and Mustafa (1986) were the first to systematically approach the recurrence of intra-formational folds in a partial section of the Lisan Formation ~17 m thick, east of the Lisan Peninsula (Fig. 8.8). They have documented folded beds with amplitudes between 1 and 15 cm. They have assumed a relation between the amplitude of the fold and ground acceleration, and estimated respective magnitudes. The frequency-size relation derived should be regarded with care for three reasons: (1) without knowledge of the epicentral distance the magnitude estimates are minimal; (2) disregard for breccia layers may trim the strongest earthquakes from the archive; (3) the frequency of recurrence is a lower bound, as multiple events may be erroneously amalgamated (Alsop and Marco 2011). An additional source of uncertainty arises from the possibility that folding occurs at finite depth in the sediment (Gibert et al. 2011).

Heifetz et al. (2005) have noted the geometrical similarity between the convolute beds and deformation structures that develop between fluid layers. They differentiated between symmetric folds and asymmetric billow-like structures, and conjectured that the billows are precursors to a state of total turbulence, represented by breccia layers. This suggested a mechanism for the formation of folds: Kelvin Helmholtz instability (KHI) caused by vorticity transfer into a pair of layers of contrasting density (and possibly viscosity) sliding horizontally at different speeds in response

Fig. 8.9 Intraformational folds in Lisan Formation, Peratzim Creek

to a seismic shock. Such layering is bound to form in compacting muds due to fluid expulsion, hindered settling, and formation of sedimentation fronts (Thacker and Lavelle 1977).

Heifetz et al. (2005) have linearized the KHI problem and inferred a dependence of inception of billows on driving wave and sediment properties. The higher peak horizontal acceleration, the shorter the time required for fold growth; the longer the period of the driving wave, the longer the time available for fold growth; the thinner the bed undergoing folding, the smaller the acceleration required for onset of folding instability. The density difference has two contradicting effects, ultimately resulting in an inverse proportion with the threshold for billowing. Assuming small (<10 %) density contrasts (merely due to progressive dewatering) and estimating the viscosity from the seismic wave attenuation, the threshold for any folding of a layer 0.1–1 m thick under a driving frequency of 1Hz is 0.2–0.7 g (where $g \sim 10$ m/s^2). Such conditions are possible near the epicenter of a M6.2 earthquake in the Dead Sea (Oth et al. 2007). Under bottom gradients the critical acceleration is smaller.

Wetzler et al. (2010) have corroborated the KHI approach by noting a scaling in the power spectrum of the folds which is identical to that observed in other KHI systems. They have furthered the link between horizontal peak ground acceleration and the onset of folding. The KHI approach is a promising avenue for further research, and laboratory experiments on mud will be very useful for testing the application.

Two aspects of mud, a two-phase fluid, are very different from experimentally tested single phase fluids that can develop turbulence in layers via KHI. The first aspect is that mud can develop localized states that do not propagate. Lioubashevski et al. (1999) have shown that thin mud suspensions subject to vertical shaking develop localized sub-harmonic states (oscillons) known previously only in dry granular media. A more essential aspect is the expulsion of the suspending fluid (brine in the case of Dead Sea – Lisan sediments) from the grains which fossilizes the folding. This is the process that enables us to access the geometry of these dynamic states long after stabilization.

8.2.3.4 IBLs ("Mixed Layers")

Seilacher's (1969) type example for a seismite comprised sequences of three sub-units from bottom: block faulted, rubble, and liquefied. The contacts between sub-units are gradual and so is the bottom of the sequence, in contrast to the top contact that is sharp. These features are replicated in the Dead Sea seismites identified by Marco and Agnon (1995, 2005). Owing to the seasonal lamination, the Dead Sea lacustrine seismites stand out in outcrop and drill holes, allowing calibration with historic earthquakes (Ken-Tor et al. 2001; Migowski et al. 2004; Kagan et al. 2011) and offering a unique insight into the long term seismic behavior of an active plate boundary (Marco et al. 1996; Agnon et al. 2006).

Agnon et al. (2006) renamed the "mix layers" of Marco and Agnon (1995) to "intraclast breccia layers" (IBLs), a descriptive rather than a genetic term. In the typical laminated facies of the lacustrine Dead Sea Quaternary sections, these breccias form conspicuous massive intervals, where in places fragments of laminae float in a fine grain matrix (Fig. 8.10). The top contact is sharp where a single laminae can be traced overlying the breccia layer for the entire extent of outcrops (tens of meters and possibly more). By contrast, the bottom contact is gradual, in places involved with intraformational folding (Fig. 8.10). The research focus on IBLs in the late Quaternary Dead Sea initiated due to their relations with intraformational faults (Fig. 8.11). Marco and Agnon (1995) have noticed the stratigraphic relations indicating that IBLs form simultaneously with small scale faulting and formation of micro topography (scale 0.1–1 m) on the lake bed. Marco and Agnon (2005) have documented these temporal relations in detail, and Fig. 8.11 represents visually one of their examples. Striking correlation between columnar sections on faulted blocks is limited to intervals in the hanging wall (h). Three intervals are not correlated and signify events of fast subsidence of the hanging wall followed by sedimentation that fills the tectonic micro-bathymetry. Of six IBLs in a 9 m section (7 m on the footwall), three are followed directly by the three differential subsidence events. Marco and Agnon (1995, 2005) have interpreted these relations as evidence for a causative relation between faulting and IBL. Such cases of clear thickening of IBL in the hanging wall, and slumping down the fault plane, are common near Masada and in a number of additional sites. These indicate that the resuspended material settled right after local faulting, so local faulting might have driven local resuspension.

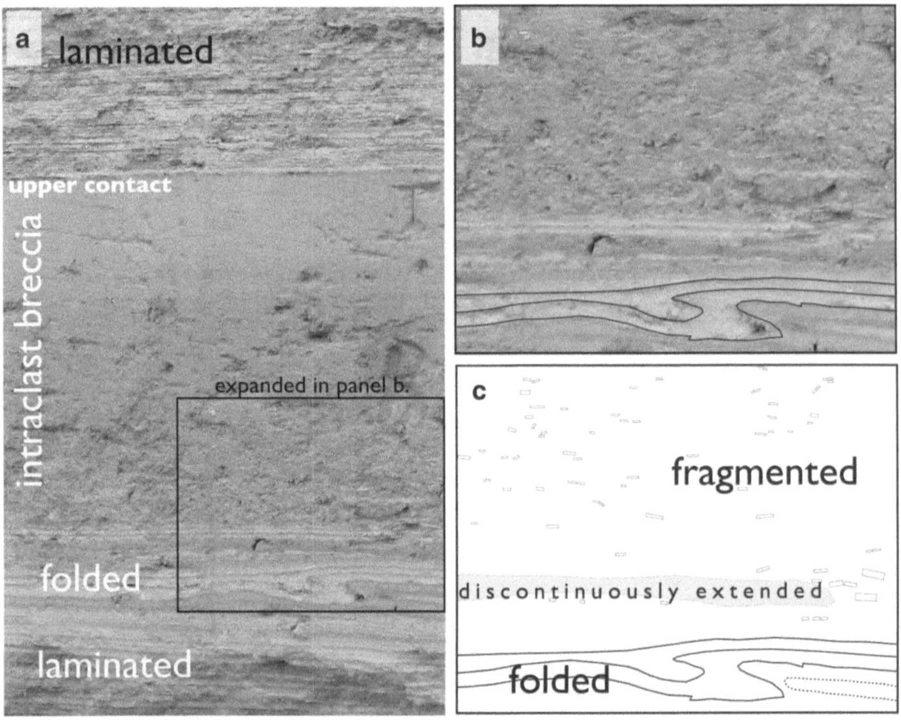

Fig. 8.10 A photograph of a sequence containing an intraclast breccia layer (**a**). (**b**) shows detail of lower contact with upward transition from folded to fragmented structures as traced in (**c**). Masada Terrace

Alternatively, faulting on a different but not too distant fault caused re-suspension and was followed by an aftershock that was related to the local fault that cause the bathymetry. In the more common setting as featured in Fig. 8.11, IBLs do not exhibit significant difference between adjacent fault blocks, so the sediment of the IBL had resuspended and resettled due to somewhat remote faulting.

The mechanism of formation of the IBLs is not yet clear, but a tentative explanation was offered by Heifetz et al. (2005) together with the mechanism for formation of convolute lamination discussed above. According to this suggestion, the transition from folding to brecciation corresponds to the transition from billows to turbulence. Such a transition is the extreme case of Kelvin-Helmholtz instability, where the drive (seismic waves) overcomes the damping due to gravity and viscosity (see Section "Intraformational folds" above).

We note the fundamental difference between KHI and Rayleigh-Taylor instability that more commonly cause seismites (e.g. Hempton and Dewey 1983): the latter depends on a gravitational drive. In Rayleigh-Taylor instability an inverse density gradient (denser on top) is rectified where the role of shaking is to overcome static friction and initiate the flow. In the case of the intraclast breccias of Lisan Formation the density gradient is normal due to compaction, and it acts to attenuate the instability. So seismic energy is required during the development of the instability, cascading

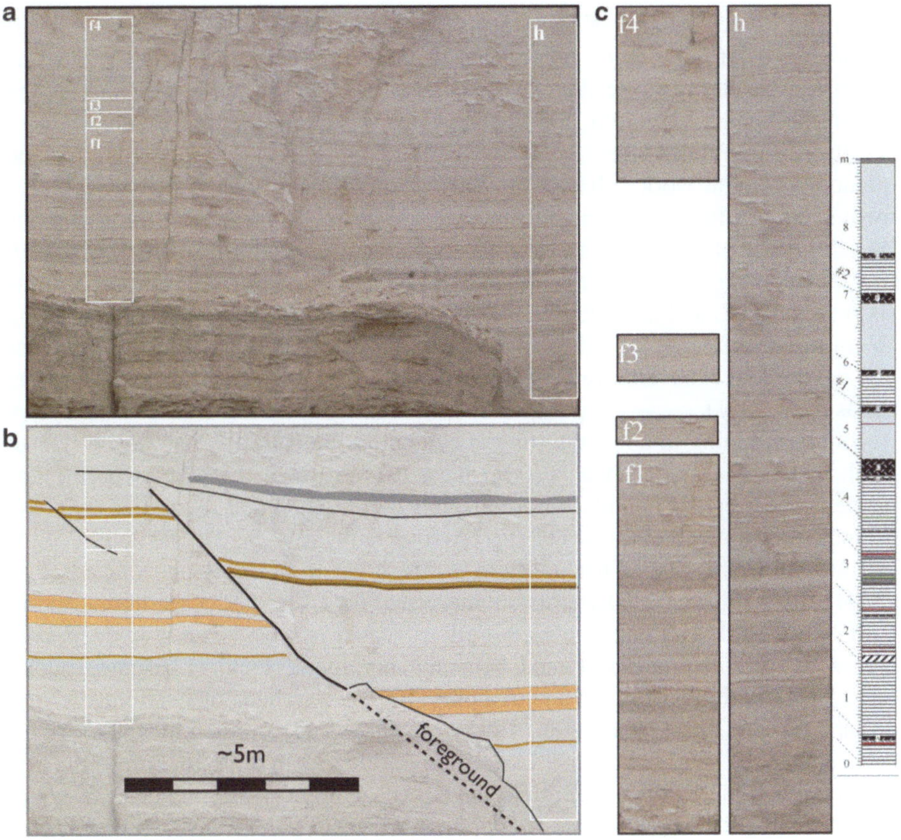

Fig. 8.11 An intraformational fault, Lisan Formation, Masada terrace. (**a**) A photograph with demarcations of selected sections expanded in panel c. (**b**) A line tracing of panel a. (**c**) Selected sections taken respectively on the foot wall (*1* and *2*) and the hanging wall (*3*). Sections are laid out to emphasize the remarkable visual correlations between the faulted blocks at the bottom and top, where the middle section of 3 is missing between 1 and 2. For lithological and attempted temporal correlation, see Marco and Agnon (2005)

into smaller scales down to the laminar scale required for complete brecciation (and to the grain scale for forming homogenites). Some gravitational energy may become available during instability because sliding on a slope is involved. The relative role of gravitational and seismic energy depends on the initial slope. The initial slope can be constrained from measurements of lateral variations in thickness of deformed units.

8.2.4 Coastal Marine Environment

DSR is one of a few transform boundaries tearing continental crust and allowing seawater to inundate some of the basins. The connection with the sea provides a datum – the global sea level, and as the case is for the DSR, coral and vermetid reef tables allow precise dating of these levels. In addition to their use as reference levels for precise determination of vertical throw, paleo-shorelines can provide piercing points for measuring strike slip. As the DSR is leaky in the south and restrained in its central (Lebanon) section, contrasting styles of slip partitioning are displayed.

Slip partitioning in a restraining setting is manifested in the coastal marine and offshore Lebanon work (Elias et al. 2007). Sonar images of the bottom revealed fresh scarps in the soft sediment associated with a >100 km long trace of a the Beirut thrust that is partitioned from the Yammouneh transform segment in the rift (Fig. 8.4). Abrupt emergence of a vermetid reef by ~0.8 m has been dated to the sixth century CE and attributed to the 551 CE earthquake (Morhange et al. 2006). Geomorphic similarity of three successively higher benches suggests a recurrence of a similar M ~ 7.5 event every ~1.6 kyear, since the Holocene sea level maximum (Elias et al. 2007).

Slip partitioning in a leaky setting is evident in the Gulf of Eilat-Aqaba (Deves et al. 2011). Makovsky et al. (2008) have demonstrated that slip is partitioned between the western boundary normal fault (Eilat fault) and a parallel fault that accommodates much of the strike-slip (Avrona fault) (Fig. 8.12). The details of the seismotectonic history of the northern gulf are yet to be resolved. Shaked et al. (2004, 2011) inferred catastrophic submergence for a reef flat in the northernmost Gulf of Aqaba (Fig. 8.12). The seemingly truncated reef table is 4–5 m deep, covered with a few decimeters of sand. Most notably, loose blocks of basement rocks, up to 50 cm in dimension, are laid in a plan similar to archaeological nomadic sites on land (Avner 1998; Avner et al. 1994) (***NS*** in Fig. 8.12). Such loose blocks get carried away by seasonal southern storms at bottom depth of up to 2 m, so the persistent arrangement in a plan reminiscent of archaeological sites indicates fast, perhaps catastrophic submergence. A support for catastrophic events associated with submergence was unearthed onshore: a buried reef with corals in living position, preserving fine skeletal details without any bioturbation or reworking, topped by an unsorted conglomerate (***ET*** near Elat fault in Fig. 8.12). The buried reef preserved evidence for an abrupt sedimentation event ~4.7 ka and a terminal event associated with subsidence and massive sedimentation at ~2.3 ka. A ~4 m vertical displacement is suggested by the submerged reef table (Shaked et al. 2004). At least 2 m and likely the entire 4 m were catastrophically thrown, corresponding to M7¨ ±0.5 (Wells and Coppersmith 1994). For comparison, trenches in the fan of Sh'horet Creek (Fig. 8.12, ***ST***) reveal 0.1–1.3 vertical displacement events during the same period (Amit et al. 2002). The difference may signify the transition from the Arava/Araba valley to the gulf, yet it may also be related to the difference of seismic response between fan deposits and coral beach.

Additional dating of coral terraces, together with dating turbidites in the deep facies of the Gulf of Aqaba, might complement the history of earthquakes in the southernmost segment of the Dead Sea rift. Efforts to date such deposits are on the way. A control for earthquake dates is provided by Thomas et al. (2007) who report seven archaeological layers damaged by earthquake faulting (some subsequently repaired) since the second century CE. They conclude that historical catalogues are incomplete in this region.

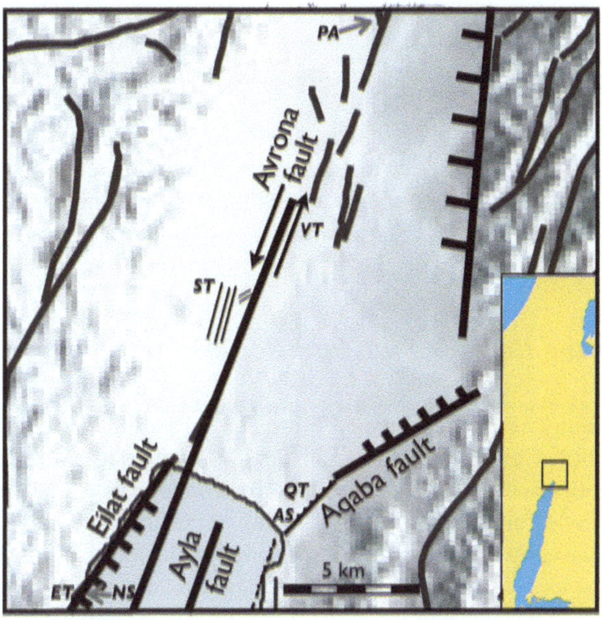

Fig. 8.12 The tectonic setting of the southern Arava/Araba Valley and northern Gulf of Aqaba. The background topography and the curving faults are from Kesten et al. (2008.) Eilat Fault (Modified after Garfunkel et al. 1981). Shaked et al. (2004) have trenched Holocene marine sediments (***ET***) that together with a submerged nomadic site (***NS***) constrain vertical coseismic slip of 4 m ~2.3 ka. The plate boundary northward was mapped by Garfunkel et al. (1981) and strike slip was established based on observation of a small pull-apart basin (PA) when this concept was quite young (Zak and Freund 1966). Holocene sediments crossing this segment, renamed Avrona fault after a playa carrying this name, were trenched by Amit et al. (2002) (***VT***). They report trenching of Pleistocene sediments cut by nearby marginal normal faults (***ST***). Zilberman et al. (2005) surveyed an irrigation system from early Islamic times that was deformed co-seismically during the 1068 CE March event. Makovsky et al. (2008) have identified this fault in the gulf and established left lateral motion of ~30 m on a submerged reef terrace at ~65 m depth. Using this beach line as a piercing point and eustatic curves with estimated age of ~11 ka they inferred a strike slip rate of 2.7±0.5 mm/year. Slater and Niemi (2003) trenched Holocene sediments across Aqaba fault (***QT***). Thomas et al. (2007) have analyzed the adjacent archaeological Ayla sites (***AS***) and identified seven damaged layers, attributing them to earthquakes. Some of the layers show faulting, and some walls record minor strike slip. An earthquake following 360 CE was identified with the 19 March 363 CE event. Tibor et al. (2010) have collected multibeam bathymetric data and verified Makovsky et al. (2008) estimate of offset submerged reef terrace. They have also identified a conspicuous lineament splitting the northern gulf (Ayla fault)

In addition to the direct effect of the Dead Sea rift on the tongue of the Red Sea in the Gulf of Aqaba, and the effect of the Beirut thrust that branches from the rift, earthquakes from the rift affect the entire continental slope of the Mediterranean (Fig. 8.1). Slumps triggered by DSR earthquake show on the bathymetry of the Mediterranean shelf and slope (PS, Fig. 8.8) and seem to generate local sea waves on the Mediterranean coast (Almagor and Garfunkel 1979; Ambraseys and Melville 1988; Ambraseys and Synolakis 2010; Salamon et al. 2007, 2011). Elias et al.

(2007) argue that, for the Lebanese coast, tsunamis are generated by submarine thrusting. High energy deposits in the port of Caesarea (Fig. 8.3), together with destruction of the port, have also been associated with a remote DSR earthquake – the 115 CE event (Fig. 8.2c) (Reinhardt et al. 2006). A sea wave in Yavne was reported for this earthquake (Fig. 8.2a) (Ambraseys et al. 1994). A strong earthquake at Antioch was reported for the same year (Guidoboni et al. 1994). The likely mechanism for sea waves generated by an earthquake rupturing the northern DSR is submarine slumping.

8.2.5 Damage Recorded in Hard Rock

Earthquake shaking often shifts suspended rock masses so they can fall and activate a geological clock. Calcite precipitated in caves, if containing sufficient radioactive Uranium, provides such a clock. Free slope rock falls may become clocks if they contain minerals useful for cosmogenic and luminescence dating, or if they bury such minerals. Indirect clocks can form by modification of the local drainage and accumulating of sediments behind fallen rocks or slides. Researchers of the Dead Sea rift have used all these instances in the quest of dating paleo-earthquakes and assessing their magnitudes.

8.2.5.1 Cave Deposits (Speleothems)

Cave deposits are potentially a useful recorder of ground shaking. The cave protects the deposits from much of the variations in the external environment, and speleothems often show fine lamination, enabling high resolution sampling for dating and for reconstructions of the physico-chemical conditions. In cases where the deposits are datable with sufficient resolution, and deposition is continuous, damage due to earthquakes can be dated. The optimal climate zone for this study should be sufficiently humid to prevent desiccation and sufficiently warm to prevent freezing. The Judean Hills, abutting the Dead Sea rift, proves ideal for the development of this method: deposition in caves is continuous through the dating range of the U-Th method (Bar-Matthews et al. 2000). This has motivated Kagan et al. (2005) to study two adjacent caves in the Judean Hills (Fig. 8.8, SH), where they have documented preferred orientation of severed stalagmites. They have argued that north–south and east–west dominant orientations of fallen stalagmites is consistent with westward propagating surface wave fronts, where the respective polarization of Love and Rayleigh waves would accord with the orientations of fallen stalagmites. Kagan et al. (2005) seem to be the first to present an extended archive of earthquakes based on cave deposits, spanning the last 185 kyear. They have identified 38 datable speleothems seemingly damaged by earthquakes ("speleoseismites"), most of which showing micro-stratigraphic evidence of renewed calcite precipitation following damage. Such evidence together with dating laminae predating and postdating the damage, bracket the age of damage. Some samples could not be bracketed from both ends, but belonged to clusters of ages; these clusters where taken as a limit for the age of an event. Because Kagan et al. (2005) dated the last damaging event to around 5–6 ka, they inferred that these pre-historic earthquakes were stronger than any historic event.

Braun (2009) has studied Denya Cave in Haifa (Fig. 8.8), a city straddling the Carmel fault (Fig. 8.4). This active branch of the Dead Sea rift bounds the Jezre'el Valley from south (Fig. 8.1). The fault runs closest to heavily populated areas, yet the maximum magnitude expected from the fault and the recurrence rates are poorly constrained. Braun (2009) has devised an isochron method for mitigating sensitivity to the anomalous initial isotope ratios in samples from this cave.

Kagan (2011) and Braun et al. (2011) combined results from the two speleoseismic sites to assess coupling between the Carmel branch and the main Dead Sea rift during the Holocene. They find that the last event to have been well recorded in the Judean Hills, around 5 ka, is the ultimate at Denya Cave, Haifa. The penultimate event hit the two sites respectively at 8–9 ka and 10–11 ka. The authors then use these examples together with data from the Tell Megiddo (Marco et al. 2006) and from the rift valley to lay out possible scenarios for coupling between segments and branches of faults.

8.2.5.2 Free Slope Failure

As in other active regions (McCalpin 2009), landslides and rockfalls in the Dead Sea rift have been used to constrain dates and intensities of seismic shaking. Recently, Katz et al. (2011) have studied landslides triggered by large earthquakes along Ha'Onn Escarpment (Fig. 8.8), east of Lake Kinneret (Sea of Galilee). Measuring optically simulated luminescence (OSL) of buried quartz sand they obtain the following dates for land sliding events: before 65 ka, around 65 ka, 6 ka, and 5 ka. In addition, paleoseismic trenching yielded the following OSL dates for earthquake ruptures: 45, 40, 35, 10, 5 ka, and a younger (<5 ka) event. The 5 ka slide was likely triggered by the earthquake rupture observed in the trench. This work followed Yagoda-Biran et al. (2010) who calculated peak ground accelerations of 0.15–0.5 g required for mobilizing three slides around Lake Kinneret.

Katz and Crouvi (2007) have considered the effect of foundations in ancient settlements: they found that the town Safad/Zefat (Fig. 8.2a) that has been inhabited for more than two millennia, suffers from reduced stability of foundations. Buildings founded in archaeological debris are more susceptible to landslides induced by earthquakes and to amplification of ground shaking. The authors caution that sites like Safad, having been demolished by numerous historic earthquakes, might have biased the historical catalogues towards high magnitudes.

Wechsler et al. (2009) have described a landslide that displaced archaeological remains including an aqueduct in Umm-El-Qanatir, east of Ha'Onn Escarpment (Fig. 8.3). They attribute the damage to the 749 CE earthquake based on the in-site finds limited to sixth to mid-eighth century CE. All these studies manifest the role of earthquakes in triggering land sliding around Lake Kinneret.

A novel approach to earthquake driven gravitational collapse on slopes is given by Matmon et al. (2005). They use cosmogenic dating together with OSL for dating collapse of several meter large sandstone boulders in Timna (Fig. 8.8). They identify three distinct events around 3–4.5 ka, 15 ± 1 ka, and 31 ± 5 ka respectively. Notably the oldest event recorded is associated with a ~5 m boulder displaced 20 m horizontally from its source cliff, with merely 2 m vertical drop. The boulder is found due north of its north-facing scar, suggesting a significant role of horizontal acceleration parallel to the slip vector. Matmon et al. (2005) point out that the cycle of boulder shedding off the cliff may be limited by either of the steps: undercutting and ground shaking. The apparent cycle of 12–15 ka may be determined by the rate of undercutting after a large earthquake. On the other hand, a similar period (yet with a different phase) seems to emerge from Kagan et al. (2005) cave deposits (discussed in the following section). Further research is required for ruling between these proposed mechanisms for determining the period of boulder shedding events.

8.3 Correlation of Historical, Archaeological, and Geological Evidence

The foregoing sections demonstrate that the DSR stands out in the richness of historical, archaeological, and geological evidence of earthquakes. The availability of these three independent and interdisciplinary sources of paleo-earthquake data in the DSR enable cross tests between individual data sets. As pointed out by several authors, the mutual independency of these sources is not always warranted, and circular reasoning should carefully be avoided (Karcz 2004; Ambraseys 2006a; Thomas et al. 2007; Rucker and Niemi 2010).

8.3.1 Prior to the First Millennium BCE

Kagan et al. (2005) have discussed the correlation of speleoseismites and lake seismites. The possibility to match lake seismites to each of the fewer speleoseismites (during times of deposition of the appropriate facies in the lake) has suggested that the cave acts as a filter recording only very strong events. This is in line with the absence of speleoseismites in the historical period. An event dated to earlier than 5.1–6.3 ka can be correlated with a lake seismite dated ~6 ka by Migowski et al. (2004). The correlation between Judean Hills speleoseismites and Lisan seismites was noted by Kagan et al. (2005). Out of four speleoseismic event in the interval 70–15 ka, one was dated to a time of hiatus in the Lisan formation record (event xv).

Table 8.1 Correlation of selected prehistoric earthquakes inferred from archaeologic sites with geochronologically dated seismites

Range [ka]	Archaeological years BCE	Ruin	Geochr-onol. [ka]	Geological site	Refs. notes
5.9–6	Fortieth century	Ghassul	5.9–6.3	Ein Gedi core	a, b, c/1
5.9–6	Fortieth century	Ghassul	>5.3–6.1	Judean Hills caves	a, b, d
5.0–5.1	End 4th mill.	Megiddo	4.2–5.6	Judean Hills caves	d, e
5.0–5.1	End 4th mill.	Megiddo	4.2–5.6	Denya Cave, Mt. Carmel	e, f
5.0–5.1	End 4th mill.	Megiddo	4.7–5.3	Sea of Galilee – Ha-'On	e, g
4.6–4.7	Twenty-seventh century	Ai	4.4–5.0	Ein Gedi core	b, c/1
4.6–4.7	Twenty-seventh century	Ai	3.7–5.5	Tamar Creek	b, c, h/2
4.2–4.3	Twenty-seventh century	H. Ifdan	4.2–4.0	Ein Gedi core	b, c, i/1
4.2–4.3	Twenty-seventh century	H. Ifdan	3.7–5.5	Tamar Creek	i, h/2
3.3–3.4	~ 1365	Ugarit	3.3–3.5	Ein Gedi core	c, j/1
3.3–3.4	~ 1350	Pella	3.3–3.5	Ein Gedi core	c, k/1
3.1–3.2	~ 1150	Pella	3.0–3.2	Ein Feshkha	c, k/1
3.1–3.2	Twelfth century	Deir'Alla	3.0–3.2	Ein Feshkha	l, 1
3.0–3.1	~1050	Timna	3.0–3.2	Ein Feshkha	m, n
3.0–3.1	~1050	Timna	2.0–4.2	Mt. Timna	n, o/3
2.75–2.76	~760–750	various	2.75–2.85	Ein Gedi, Ein Feshkha	c, m, p/1
2.75–2.76	~760–750	Deir'Alla	2.75–2.85	Ein Gedi, Ein Feshkha	c, m, l, p/1
2.65–2.75	~700	Various	2.65–2.75	Ein Gedi, Ein Feshkha	c, m/1
2.65–2.75	~700	Deir'Alla	2.65–2.75	Ein Gedi, Ein Feshkha	c, m, l/1

References: (a) Hennessey (1969). (b) Karcz et al. (1977); Karcz and Kafri (1978). (c) Migowski et al. (2004). (d) Kagan et al. (2005). (e) Marco et al. (2006). (f) Braun (2009); Braun et al. (2011). (g) Katz et al. (2011). (h) Gluck (2001). (i) Levi et al. (2002). (j) Schaeffer (1948); Hanfmann (1951). (k) Bourke et al. (1999); Bourke (2004). (l) Ferry et al. (2011), based on Franken (1992). (m) Kagan et al. (2011). (n) Ben-Menahem (1991). (o) Matmon et al. (2005). (p) Ambraseys (2009); Austin et al. (2000)

Comments: (1) Range calculated from sedimentation rates in Ein Gedi core constrained by radiocarbon dates. (2) Large uncertainty in geochronologic date based on OSL. (3) Large uncertainty in geochronologic date based on cosmogenic nuclides

For the other three speleoseismic events, the uncertainty in dating disallows correlation to individual Lisan seismites, yet they seem to correlate to periods of high recurrence rate (see below in the Recurrence section).

There is little data for correlating prehistorical archaeology with geological observations of earthquake damage (Table 8.1). Migowski et al. (2004) correlated the ~6 ka event with archaeological destruction at the Tells of Ghassul (Fig. 8.3). This is one of a series of destructions from the Bronze Age attributed to earthquakes

(Hennessey 1969; Karcz et al. 1977). The speleoseismic date has also been correlated with a 5.3–5.4 ka breccia layer from Ein Gedi core (Migowski et al. 2004; Kagan et al. 2005).

Braun et al. (2011) correlate the latest speleoseismite in the Carmel cave with damage recorded at Megiddo (Marco et al. 2006). This date – ~5 ka – is common to the Judean Caves (Kagan et al. 2005), the ruins of Ai (Karcz et al. 1977), surface rupture documented south-west of the Dead Sea (Gluck 2001), and lake seismite from Ein Gedi core (Migowski et al. 2004) (Table 8.1).

Migowski et al. (2004) correlate a breccia layer from Ein Gedi core dated 4.7 ka with archaeoseismic damage of Ai (Karcz et al. 1977; Karcz and Kafri 1978). This archaeoseismic event also correlates with ground rupture documented south-west of the Dead Sea (Gluck 2001). The uncertainty of dating this rupture, based on OSL, allows correlation also with earthquake evidence 4.3 ~ 4.2 from Khirbet Hamrat Ifdan in Wadi Araba (Levi et al. 2002).Ferry et al. (2011) cite archaeological reports from ongoing excavations and conclude that an earthquake hit the central Jordan Valley around 4.3 BCE.

Ben-Menahem (1991) assigns the Biblical destruction of Jericho to an earthquake, and notwithstanding the delay of seven centuries in writing the text, he calculates a date of 1560 BC for this event. Ambraseys (2009) notes that the biblical story mentions no earthquake. He rejects the archaeological evidence for the biblical Jericho event for which he adopts a date of ~1400 BCE. A seismite from Ein Gedi core is dated to ~3.3–3.4 ka (Migowski et al. 2004) so, in the absence of positive evidence for contemporary earthquake damage in Jericho, one might conjecture that seismic damage elsewhere has inspired the biblical stories from that era.

Migowski et al. (2004) correlate the seismite dated ~3.3–3.4 ka with an earthquake that perhaps destroyed the fortifications of Ugarit (Fig. 8.3) and was arguably referred to by a contemporary tablet in El Amarna in Egypt and by a letter from Abimilki of Tyre (Schaeffer 1948; Hanfmann 1951; Ben-Menahem 1991).

Nur and Cline (2000) suggest that much of the damage to Aegean and East Mediterranean sites around 1200 BCE is due to a 50-year sequence of earthquakes ("storm"). Incidentally, this time window is one of quiescence in the Ein Gedi core (Migowski et al. 2004).

8.3.2 Tenth to Second Centuries BCE

For the last three millennia it is possible to estimate the extent of rupture of historic earthquakes with the aid of archaeological and geological data. The completeness of this information is improving for each consecutive millennium, and for the first half of this period it is clearly insufficient. Figure 8.13 displays interpreted locations of ruptures based chiefly on historical data, with complementary information from archaeology (gray bars) and Dead Sea lake seismites (stars) as they correlate with the historic data. We mark earthquakes matched to Dead Sea lake seismites in Ze'elim Creek, Ein Feshkha (Kagan et al. 2011), Ein Gedi (Migowski et al. 2004), and Daraje Creek (Enzel et al. 2000). The format follows a preliminary attempt by

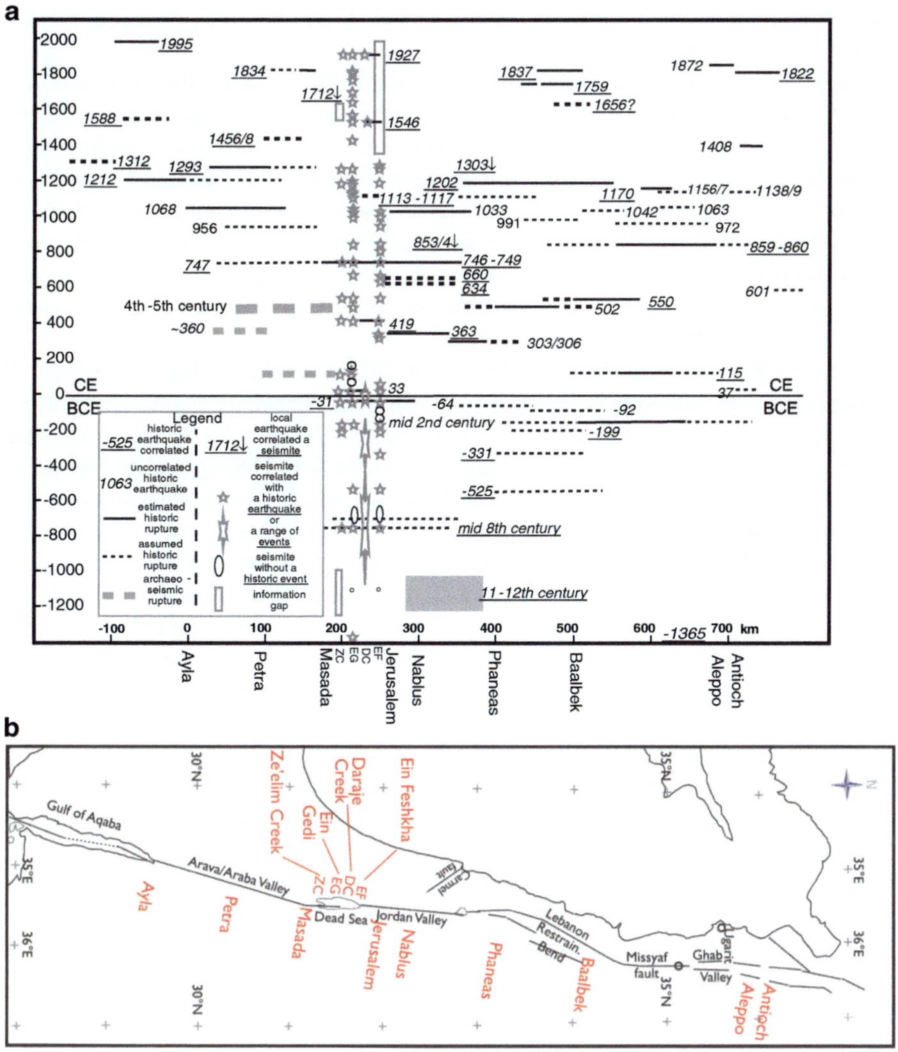

Fig. 8.13 (**a**) Estimated spatial extent of ruptures from the historic periods along the DSR. The extent of rupture (*horizontal lines*) is based on documents and physical evidence. (**b**) Map showing principal archeoseism and paleoseism sites indicated in (a) projected on the main segments of DSR

Garfunkel et al. (1981), which did not have geological data to compare. This update is based on recent catalogues, and includes historical information from older periods, starting with the Biblical earthquake of 759/760 BCE (Guidoboni et al. 1994; Ambraseys 2009). The present version also extends 200 km south-south-west to include the Gulf of Aqaba, where the latest strong earthquake ruptured the DSR: the Earthquake of Nueiba on 22 Nov 1995 amounted to M7.2 or more (Klinger et al. 1999; Shamir et al. 2003; Baer et al. 2008). The interpretation is naturally biased by the preconceptions of the present author, and it should be taken as a preliminary rather than a definite summary.

As is apparent from the panels of Fig. 8.3 and from Fig. 8.13, the information is sparser for older times and southern locations, in accord with population trends. However, the ability to match a great majority of Dead Sea seismites with historic earthquakes may suggest that the completeness of the historic and seismite archives are not so dissimilar. In this sense, the uneven distribution of ruptures in space and time exhibited in Fig. 8.13 may reflect genuine spatio-temporal clustering of earthquakes in historical times. As Garfunkel et al. (1981) argue, the relative quiescence in Wadi Araba/Arava during the instrumental period supports a different seismic behavior along different segments of the rift.

We start the history of the first Millennium BCE with the earthquake of the mid eighth century as this event is reflected in numerous archaeological sites (Ambraseys 2009). However, note that the Iron Age stratigraphy is being hotly debated, including the role of the earthquake in the cultural transitions (e.g. Herzog and Singer-Avitz 2004; Fantalkin and Finkelstein 2006). Whether all that damage is the result of a single seismogenic rupture remains to be determined. The dates referred to in the Bible enable researchers to name 759 (Ben-Menahem 1979; Ambraseys 2009) as the year of the event. This earthquake is correlated in all three Dead Sea paleoseismic sites (stars in Fig. 8.13). Kagan et al. (2011, Appendix C) point out that each of the three seismite sites exhibits two eighth century events. This is in line with the large areal extent of damaged archaeological sites that led researchers to infer a very high magnitude single event. Austin et al. (2000) assign M8.2 to this earthquake. Ambraseys (2005a) notes that such a high magnitude earthquake would completely demolish Jerusalem, where historical or archaeological evidence for destruction is not established.

The assumption of a regular repeat time of a century for Dead Sea earthquakes is central to the Austin et al. (2000) argument. And indeed, a superficial inspection of the catalogues for the last millennium can yield an approximately uniform rate of repeat times for destructive earthquakes in the Holy Land, namely once a century. However, a closer look shows pairs of events that often span a wide section along the transform. Salamon (2010) has reviewed sequences of historic earthquakes and identified 13 cases where successive events did not necessarily form as aftershocks. We note two pairs of earthquakes separated by a decade or less: 1202 CE ruptured Lebanon, severely shaking Syria and northern Israel; 1212 hit the Gulf of Aqaba. These events with magnitudes estimated 7.0–7.5 could erroneously be taken as a single M > 8 event if we did not have the temporal resolution. The pair of 1834 CE (M ~ 6) and 1837 CE (M7.0, Ambraseys 1998) at the Jordan and Dead Sea Valleys and Lebanon, respectively, would be taken as a single M ~ 7.5 event.

In addition to the unjustified assumption of a century long seismic cycle, criticism of Austin et al. (2000) comes from archeological grounds. Fantalkin and Finkelstein (2006) argue that some of the damage purported to originate from seismic shaking could have resulted from static loads.

The successive seismite found in Ein Gedi core (Migowski et al. 2004) and at Ein Feshkha (Kagan et al. 2011) is correlated to 525 BCE. This earthquake is listed by Ben-Menahem (1991) and Sbeinati et al. (2005), with damage to the southern

Lebanese towns Sur and Sidon, and a local sea wave. Morhange et al. (2006) document variable uplift of paleo-shore-lines from this period, and suggest multiple seismic events. The 551 CE event, with perhaps larger macroseismic effects, (Elias et al. 2007) is recorded as well at both these seismite sites, yet in addition it is correlated to a seismite at the southern (further form the epicenter) Ze'elim Creek site. The fact that 550 BCE is not correlated in Ze'elim Creek may be due to a smaller source, or due to less favorable sedimentary conditions (e.g. lack of lamination) in this particular interval. Additional studies in a lake-ward section, being exposed due to the ongoing level drop, may resolve this issue.

The three overlying clear seismites in Ein Feshkha (Kagan et al. 2011) were not correlated, largely due to sparse historical evidence from that period (deformed layer is considered "questionable"). The correlation of an overlying seismite to 331 BCE in Syria is based on a nineteenth century chronicle ("many victims and destruction") interpreted by Sbeinati et al. (2005) as intensity $I = VI$. Enzel et al. (2000) date a seismite from Daraje Creek to 2.0–2.4 ka, correlative with the 331 BCE Ein Feshkha seismite, or with one of the adjacent uncorrelated seismites.

Two additional seismites at Ein Feshkha are uncorrelated between 331 and 199–198 BCE, yet the error in dating could allow respective correlation with the adjacent seismites. The latter date corresponds to a damaging earthquake in the southern Lebanese coast, with more than half of Sidon destroyed by a landslide, for which Guidoboni et al. (1994) and Sbeinati et al. (2005) respectively assign intensity *X* and *VIII*. This event was not recorded in Ein Gedi core and Ze'elim Creek.

Of the 30 deformed layers in Ein Feshkha section overlying the one correlated to 199–198 BCE, spanning second century BCE to fourteenth century CE (the sequence is truncated), three are labeled "questionable" seismites. All 30 deformed layers could be matched with historic earthquakes to within he dating error (Kagan et al. 2011). While each individual deformed layer can be matched to one or several earthquakes, a unique match of the entire set is not possible: the best fit Bayesian model of radiocarbon ages anchored to historic earthquakes (Kagan et al. 2010) leaves 2 out 30 seismites unmatched at respective model dates (95 % confidence) of 101–42 BCE and 365–595 CE (the latter considered "questionable").

Except for the last century in this sixteenth centuries long interval, the Ein Gedi core is particularly detailed as it corresponds to the chronologic model based on laminae counting anchored to 22 seismites (Migowski et al. 2004). Two of these are not matched to historic earthquakes, at respective model dates of ~90 and ~175 CE. Note that the two pairs of unmatched events in from each section considered are not correlated mutually, suggesting four local shocks that were not recorded historically.

Several earthquakes from the counted laminae interval are worth noting. These are large historical events that are expected to affect the Dead Sea basin.

The information about a mid-second century BCE events is scant and comes from secondary but relatively reliable sources. Several dates between 148 and 130 BCE are mentioned in the sources, and the confusion is yet to be resolved (Ambraseys 2009; Guidoboni et al. 1994; Karcz 2004; Sbeinati et al. 2005). Significant destruction of the city of Antioch was reported (Guidoboni et al. 1994; Ambraseys et al. 1994; Guidoboni and Comastri 2005; Sbeinati et al. 2005; Ambraseys 2009). Historical accounts cannot distinguish between a Dead Sea rift source and adjacent plate boundaries. A tsunami hit the Lebanese coast near Sidon. A collapse of a village "above Sidon" (Karcz 2004, and historical sources therein) suggest ground shaking and a regional earthquake. For a single event to be responsible to the damage from Antioch to Ptolemais, some 400 km (Fig. 8.2), it should be M ~ 8. Yet a pair of events, each around M7.5, is consistent with the geological data (Kagan et al. 2011) and with the historical evidence for two separate earthquakes. Ze'elim Creek and Ein Gedi core feature each a seismite from the mid-second century BCE (Migowski et al. 2004; Agnon et al. 2006; Kagan et al. 2011). Ein Feshkha section features two clear seismites from that time, separated by a 4 cm undisturbed sequence. The calculated mean rate of deposition – 0.27 ± 0.03 cm/year, yields a minimum of 13–17 years between events. Including an additional 1 cm of sequence embodied in the upper seismite would give 17–20 years. This is similar to the upper range of difference between various interpretations of historic dates.

8.3.3 First Century BCE to Seventh Century CE

A regional earthquake is reported for 92 BCE with tsunami in Lebanon and flooding in Egypt, but its source location is not clear. This event is correlated to seismites in Ein Feshkha and in Ein Gedi core, while Ze'elim Creek sections show undisturbed laminar facies. We will use this information for a comparison with more recent events for which we have more information.

The 64 BCE earthquake was studied in detail by Karcz (2004). He concluded that "it is unlikely that local intensity was much in excess of human perception". Sbeinati et al. (2005) have reevaluated this event (65 BCE) and classify the completeness of data as "acceptable". An intensity VI is assigned to Jerusalem, where the source is likely in western Syria (*VIII* > *I* > *VII* in Antioch). While Ken-Tor et al. (2001) identify this event with a seismite in Ze'elim Creek, Agnon et al. (2006) preferred a mid-second century source for that seismite. They based their correlation on a deposition model with a uniform sedimentation rate except in hiatuses. Accepting this model, the 64 BCE event would be reworked and "masked" by the subsequent 31 BCE earthquake. Migowski et al. (2004) were the first to acknowledge that the subsequent 31 BCE event would mask 64 BCE. Recently however, Kagan et al. (2011) have found a seismite at this period in the Ein Feshkha section, in agreement with Sbeinati et al. (2005) intensity assessment. If we are to accept Karcz' (2004) assessment that the Syrian earthquake was only felt around the Dead Sea, we miss a local event during that period that left a mark on the Dead Sea sediments, and perhaps was erroneously amalgamated with the large 64/65 BCE earthquake that hit Antioch.

The 31 BCE and the subsequent 33 CE earthquakes are well known due to their contemporaneity with dramatic political and military events. Kagan et al. (2011) classify these earthquakes as two of the nine forming "intra-basin seismites", identified in the three sections considered in detail. In addition, a seismite dated ~2 ka in Daraje creek (Enzel et al. 2000) could be correlated to one of these earthquakes.

The 362/363 CE earthquake deserves special attention, as since Russell's (1980) work it is considered one of the largest events on DSR (e.g. Ambraseys 2009). Kagan et al. (2011) suggest that two events of M ~ 6.5, one from 363 north of the Dead Sea and one from a close date south of the Dead Sea, had been erroneously amalgamated to a single M > 7 event. The sites north of the Dead Sea are skewed significantly to the west, suggesting a Carmel Fault source. The Seismite C in Ze'elim Creek section of Ken-Tor et al. (2001) was constrained by radiocarbon dates to correlate with either the 363 or 419 CE earthquake. A uniform deposition rate model of Agnon et al. (2006), anchored to historic earthquakes, ruled in favor of the latter date. Likewise, an Ein Gedi core seismite from the laminae counting interval was correlated with 419 CE, where 363 CE was not correlated (Migowski et al. 2004). Kagan et al. (2010, 2011) section in Ze'elim Creek was fit better to models with a seismite from 419 CE, where again 363 CE was not correlated. By contrast, their models for Ein Feshkha section show correlation of seismites to both earthquakes. They suggest that the Ein Feshkha records a moderate earthquake that hit the north, whereas the southern earthquake did not generate seismites in the study sites. A single large earthquake that demolish the entire region from the Galilee to Petra and Aila would correspond to M7.4 (Ambraseys 2006b, 2009; Russell 1980; Thomas et al. 2007). This would result in broad scale damage on the east of the rift, in contrast to the observations (Russell 1980). Kagan et al. (2011) interpret the historical and the physical observation as a pair of consecutive smaller events, one in northern Israel (perhaps on the Carmel fault) and one in the Arava/Araba Valley. This issue requires additional research.

The earthquake of 419 CE, generating "intra basin seismites" according to Kagan et al. (2011), is likely a local moderate event. The limited reports of damage do not allow a precise location of source, but it may be similar to 1927 Jericho M6.2 earthquake. By contrast, the 551 CE earthquake that is also considered by Kagan et al. (2011) an "intrabasin seismite" is a remote and strong event: it ruptured a putative thrust under Lebanon with an estimated M ~ 7.5 (Elias et al. 2007).

Rucker and Niemi (2010) speculate that a strong earthquake hit east of the Dead Sea 597–598 CE, based on an inscription in Areopolis. Since historic documentation from Petra is missing after 597 CE, such an event might have escaped the chronicles. Ambraseys (2009) also lists an earthquake earlier than 597 in Areopolis. Hence the correlation of all sixth century seismites with the Lebanon event is still not certain.

The Jordan Valley has been struck by three earthquakes during the seventh century CE: one at 634 and two at 660 (Amiran et al. 1994; Guidoboni et al. 1994). Tsafrir and Foerster (1997) dated earthquake damage in Scythopolis/Baishan to between the end of the sixth century and the second half of the seventh. Additional archaeoseismic evidence was reported nearby in Ein Hanatziv (Karcz et al. 1977). Kagan et al. (2011) found two seismites in Ein Feshkha, while a laminated sequence was deposited in Ze'elim Terrace. Migowski et al. (2004) correlated the single seismite from Ein Gedi core with the latter year. This is consistent with northerly local sources, and a stronger shaking in the Dead Sea from the 660 event that is said to had been strong at Jericho (Amiran et al. 1994).

The high resolution of late Holocene sequences exposed during the recent retreat of the Dead Sea lake is used by Neumann et al. (2009) for assessing possible geobotanic response to large earthquakes. They have investigated the Ein Feshkha section, exposed on the north west corner of the lake (Fig. 8.7), and found inconclusive evidence for possible seismic effects on olive pollen.

8.3.4 Eighth Century CE

Karcz (2004) and Ambraseys (2005a) studied the historical sources of this period and suggest that damage reports for a 749 CE earthquake were accentuated by confusion with an earlier event (~747 CE); amalgamation of these two (and perhaps more) events gave an overestimation of the size of the 749 CE event. The mid-eighth century CE events have been reportedly recognized in a number of additional sites (Marco et al. 2003). The assignment of the spectacular parallel pair of fallen colonnades of a Byzantine church in Hippos (Fig. 8.6c) to the 749 CE event (e.g. Nur 2008) is not based on archaeological evidence. A fallen colonnade from another Byzantine church on the site can be tied to the mid-eighth century events, based on an inscription from the late sixth century CE on one of the columns (Fig. 8.5d) (Marco 2008; Segal 2007). In any case, such spectacular manifestations of ground shaking may be misleading in that they strongly depend on the masonry, and the information they give on ground motion turns out to be limited (Hinzen 2011). The above mentioned earthquakes during the seventh century CE, while less famous, can be associated with the Hippos damage.

Kagan et al. (2011) report two seismites from the mid eighth century. This may seem consistent with a pair of earthquakes at Jerusalem at 746 and 757 CE, reportedly damaging Al Aksa mosque (Ambraseys 2009). However, their correlation hinges on the assignment of the underlying seismite to 873 CE discussed below.

The rupture in Qasr Tilah trenched and dated seventh to tenth century CE by Haynes et al. (2006) could be associated with a mid eighth century event. Support to this hypothesis is given by Bikai (2002) and Eklund (2008), who infer terminal destruction of Petra due to an earthquake at the mid eighth century.

8.3.5 Ninth to Eleventh Centuries CE

This period is marked by the transition of the center of the Muslim empire from Damascus to Baghdad, and a consequent deterioration of the quantity and quality of chronicles. This is evident in the confusion regarding some of the earthquake reports from the period, and in particular for the years 873, 956, and 1068 CE. Despite becoming a periphery, the Levant is mentioned in numerous earthquake reports (Ambraseys 2009). During this period the Ein Feshkha paleoseismic sites show high rate of activity, and it is difficult to identify individual events.

853/4 CE: An earthquake with dramatic mass movements and tragic consequences had been recorded in Tiberias and vicinity (Sbeinati et al. 2005; Ambraseys 2009). This event is likely recorded in one of the numerous seismites from this period, yet no particular correlation has been suggested (Table 4 of Kagan et al. 2011).

873 CE: The historical evidence originated from well within the Arabian Plate (Ambraseys et al. 1994; Ambraseys 2009). Haynes et al.'s (2006) conjecture that this earthquake originated from a surface rupture at Qasr Tilah constrained to between the seventh and tenth centuries. In order to accept this conjecture, we need to accept that "due to the sparse population of the Hejaz and Wadi Araba, the historical record of the earthquake maybe distorted or biased, thus the catalogue may be incorrect". As suggested above, the Qasr Tilah rupture might be associated with one of the mid-eight century events. Kagan et al. (2011) tentatively identify the 873 earthquake in the northern (Ein Feshkha) and central (Ein Gedi) sections of the Dead Sea, but not in the southern (Ze'elim Creek) (Fig. 8.13). Barring directivity effects, this could indicate a lower sensitivity of the southern site. More likely, and considering the historical indications for a remote source, the 873 event is not recorded in the Dead Sea.

Tenth century: Only a few seismites are apparent in the sections correlated to the tenth century: two in Ein Feshkha (Kagan et al. 2011) and one in Ein Gedi (Migowski et al. 2004). Likewise, only a few earthquakes appear in historical catalogues covering that period in the Levant (Sbeinati et al. 2005; Ambraseys 2009). As pointed out by Ambraseys (2009), the earthquakes of 952 (on the junction of the DSR with EAF) and 956 (Eastern Mediterranean with damage in Egypt) had been confused by a contemporary chronicler. The single seismite identified in Ein Feshkha at the mid-tenth century (Kagan et al. 2011) reflects one of these large events or a smaller one not recorded historically.

1033 CE: This event ruptured the surface in the Jordan Valley and had been designated $7.0 \leq Ms < 7.8$ (Ambraseys and Jackson 1998). More recent work excluded that earthquake from the set of $Ms \geq 7.0$ events on the DSR (Ambraseys 2009). Ferry et al. (2011) associate the ultimate rupture trenched in the Jordan Valley with this event, reasserting that this is one of the largest historic earthquakes recorded on this segment. Migowski et al. (2004) identify a 7.4 thick breccia layer with the 1033 CE earthquake, for which Kagan et al. (2011) correlate a 1.5 cm seismite. The absence of a seismite report from Ze'elim Creek may be due to sandy facies and difficult access.

1063 CE: Significant damage in Tripoli (VII-VIII, Sbeinati et al. 2005) led Elias et al. (2007) to infer rupture on a local fault that they mapped traversing the coast and bounding the Lebanon Thrust from north. A rupture of the entire mapped length, 65 km, would generate $M \sim 7$ earthquake, which is a likely upper bound for this earthquake. Such a magnitude would be consistent with Eq. 8.3 and the systematics of Wells and Coppersmith (1994). This earthquake is correlated to a seismite in Ein Feshkha. Migowski et al. (2004) considered this earthquake to be masked by the subsequent 1068 CE earthquake. However, using the resolving power criterion of Agnon et al. (2006), it seems that this earthquake might be resolved in case that it generated a seismite: the 1068 seismite is only 4 mm thick; the resolution for this thickness varies between 2 and 5 years during the time between the eighth and thirteenth centuries. If the absence of a seismite in the laminae counting interval in Ein Gedi core can be corroborated in nearby sections, it can be used for calibrating the attenuation relation as discussed below. Yet this potential can be realized only if it can be shown that no small, local, and forgotten event is responsible for the Ein Feshkha seismite.

1068 CE: The latter earthquake was also identified with archaeological damage in the southern Arava (Zilberman et al. 2005). These authors detected vertical displacement of up to a meter across an early islamic irrigation system. Using accepted magnitude – rupture length – slip systematics (Wells and Coppersmith 1994), Zilberman et al. (2005) suggest an M>6.6. Accepting that the rupture extended to Qasr Tilah, about 150 km to the north, this earthquake could have reached M7.2–7.8 m, consistent with several m average lateral slip. The event is recorded both in Ein Gedi core (Migowski et al. 2004) and Ein Feshkha, where the Ze'elim Creek sections are in sandy facies with difficult access (Kagan et al. 2011).

8.3.6 Twelfth to Fifteenth Centuries CE

Historically, this period from the Crusades to the Ottoman conquest is very well represented by chroniclers from both the Frankish and the Muslim cultures. Historical records of seven large earthquakes found in catalogues are correlated to Ein Feshkha seismites (Kagan et al. 2011): 1114/1117, 1150, 1170, 1202, 1212, 1293, 1312. Of these, four are correlated in the laminated section of Ein Gedi core (Migowski et al. 2004; Agnon et al. 2006): 1114/1117, 1202, 1212, 1293. Only two thirteenth century events are correlated in Ze'elim Terrace (Kagan et al. 2011, following Ken-Tor et al. 2001). The southward diminishing abundance of seismites may reflect the observation that the historic earthquakes from the twelfth and fourteenth centuries ruptured north of the Jordan Valley (Ambraseys 2006a, b, 2009; Sbeinati et al. 2005). The 1202 CE event is a special case: it was very large (M ~ 7.5) and closer to the Dead Sea than the earthquakes from adjacent centuries. Also, 1202 is so close in time to 1212 that the two seismites are not always resolved from each other (Agnon et al. 2006). So a generalization may be made that large earthquakes from north of the Jordan Valley may be sufficiently intense at the north of the Dead Sea to generate a seismite, but the intensity decays significantly to the south of the lake. We use this observation in calibrating the attenuation relation below.

8.3.7 Sixteenth to Nineteenth Centuries CE

This period is not represented in the Ein Feshkha section due to local truncation, and future work on nearby outcrops can fill this gap. Ein Gedi core shows eight seismites correlated to the historic earthquakes of 1456/1458, 1546, 1588, 1868, 1712, 1759,

1822, 1834/1837. Ze'elim Creek shows a 10 cm thick seismite for the first and a 25 cm thick one for the last of these events (Kagan et al. 2011). The facies in the sixteenth-seventeenth centuries of this site does not allow rejection of sedimentary disturbance. Yet the eighteenth to early nineteenth centuries left a laminar and undisturbed deposit, suggesting that the 1712, 1759, and 1822 earthquakes were not intense there. This is in agreement with the northerly epicenters of these events (Sbeinati et al. 2005; Ambraseys 2009).

A comment on 1546 CE: As mentioned above, Enzel et al. (2000) have identified the topmost liquified layer in Daraje Creek with the instrumental 11 Jul 1927 M6.2 Jericho earthquake. Ambraseys and Karcz (1992) note that the 1546 earthquake was similar in damage distribution to the 1927 event, so the penultimate rupture in Daraje Creek (dated mid-2nd Millennium CE) could arguably associated with this rupture. Kagan et al. (2011) assign both earthquakes to the Kalia fault that traverses the basin some 20 km north of the Daraje Creek rupture. Yet it is not inconceivable that the ruptures recorded in the fan delta are of large aftershocks: a M6.2 event typically triggers aftershocks up to M5.2. Wells and Coppersmith (1994) give this as the lower limit for breaking the surface by normal faults, with rupture lengths of several km. The age of the penultimate event determined by Enzel et al. (2000) is 400–500 years, in agreement with the 1546 CE date of the historic event similar in damage distribution to the 1927 Jericho earthquake (Ambraseys and Karcz 1992). The penultimate Daraje seismite can again be interpreted as an aftershock of a M ~ 6.2 event on Kalia fault. Alternatively, a shore-parallel fault segment in the north-west of the lake was the source for either or both these moderate events.

8.4 Completeness

All the records used by seismologists are incomplete in one or more aspects. Historical, archaeoseismc, and paleoseismic archives have the advantage of going back in time, as compared with the more systematic instrumental archives. However, the non-instrumental archives suffer from inherent problems of completeness. The more complete the archive, the more one can use it as an independent data set to correlate with others.

Historical archives are limited by the availability of cultures writing history and by preservation of these writings. During several periods the Levant was populated by coexisting (often competing) literate cultures, leaving complementary archives that can be mutually cross-checked. This was the case at Medieval times, when Muslim and Crusader historians described the same earthquakes (e.g. Ellenblum et al. 1998; Guidoboni and Comastri 2005) or during the Roman times, for which Roman as well as Jewish sources are available (e.g. Karcz 2004). When the center of the Islamic empire moved from Damascus to Baghdad, during the mid eighth century CE, the reports of earthquakes became vague. This process culminated with the 1068 CE earthquakes. Al-Bana, a contemporary chronicler who lived in Baghdad, reported two events for the same year – 18 March and 29 May. The two reports are very similar- except that the latter states a complete devastation for Ramla and only Ramla. Makdisi (1956) suggested that Al-Bana got two versions of the same event from two independent sources, where the date records the time of arrival of the report or of its recording in writing. The former report was accurate (an earthquake in the south corroborated by additional sources) and the latter was spurious (devastation of Ramla). Perhaps it confused the more likely devastation of Ramla in the 1,033 event which was confirmed by other sources (Ambraseys 2009). This case exemplifies the effect of political geography, demography, and in particular distribution of literate population. A well known general observation is that the southern parts of the rift, that bisects large deserts, is clearly less represented in the historical archive. Further advance in the use of historical data in seismological research requires a systematic assessment of the completeness of the archives.

8.4.1 Archaeological Archives

Archaeological archives are possibly incomplete due to discontinuous occupation and also due to poor conservation or lack of excavations. For example, Tel Ateret records the 1202 CE earthquake during the Ayyubid times when it was used as a shrine, shortly after the site was constructed by Crusaders and conquered by Saladin (Ellenblum et al. 1998; Ellenblum 2007). Archaeology sets a limit also on the time of the ultimate earthquake: it had ruptured during the Ottoman times, after the introduction of tobacco. This is due to the discovery of smoking pipes on a floor in context with a wall displaced ~0.5 m (Marco et al. 1997; Ellenblum et al. 1998). During that time the site was not occupied, although it was in marginal use as a shrine. Luckily, historical considerations including reported damage could distinguish between 1759 and 1837 CE, and between the two shocks of the former year (Ambraseys and Barazangi 1989; Ambraseys 1998; Ellenblum et al. 1998).

The distribution of archaeological sites during the period of consideration is a key to the potential of retrieving a complete archive. Sparse monumental building during antiquity is reflected by localized evidence for destruction. The Chalcolithic Tells of Ghassul (Table 8.1) are a good example. As has been said for the historical archive, the southern parts of the rift that bisects large deserts is clearly less represented in the archaeological archive.

8.4.2 Lake Archives

Lake archives are limited by hiatuses, not always evident in outcrop (e.g. Machlus et al. 2000), and by facies. When layering is obliterated during diagenesis by turbation (e.g. shallow facies susceptible to wave action) there is no mark of earthquakes. In the case of the Dead Sea, seismites from evaporites such as halite and gypsum

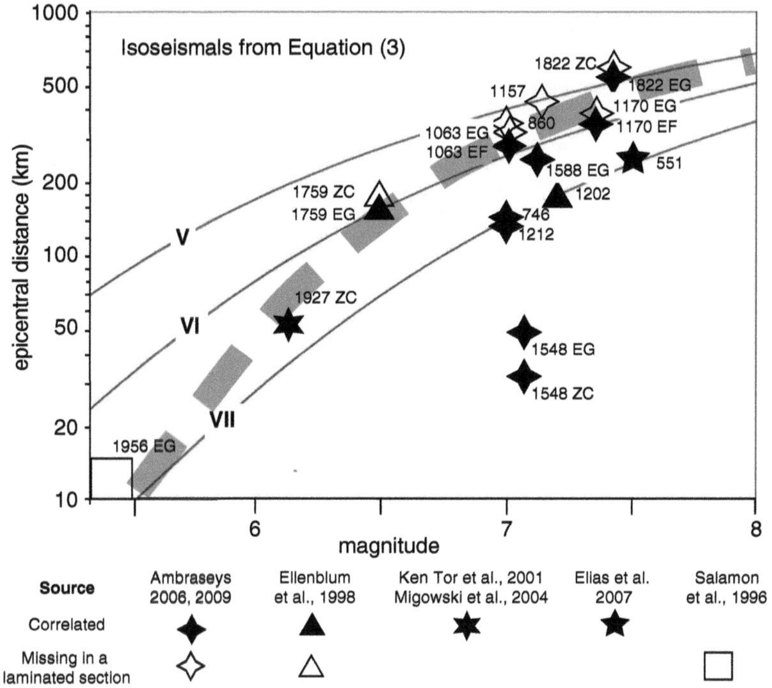

Fig. 8.14 Distances of selected historic ruptures (logarithmic) from Dead Sea seismite sites versus macroseismic magnitudes. The *empty symbols* denote sites without a correlated seismite for a given earthquakes (years CE)

have not been studied. The advantage of lake records is that superposition is dictated by gravity, and layers (down to laminae scale) are continuous and sub horizontal on outcrop scale and larger. Therefore disturbances due to high mechanical energy at lake bottom tend to stand out in outcrop and attract attention.

A method to test the completeness of the lake events during historic periods is presented in Fig. 8.14. The format is familiar from Fig. 8.3, but this time we only plot isoseismals V-VI-VII using Eq. 8.3 (Hough and Avni 2011). We plot individual earthquakes using a solid symbol for those correlated to a Dead Sea seismite, and an empty one for uncorrelated ones. For this preliminary edition we focus on events with better constrained magnitudes and locations, and in particularly on events that are correlated for one of the sites and absent in another. This helps to constrain a hypothetical critical isoseismal for the generation of seismites in the Dead Sea (gray broken curve). The critical intensity required for seismite formation seems to diminish with magnitude and distance: for a close epicenter (<40 km) and small magnitude (≤6) the critical local intensity is close to VII; for a distant epicenter (>300 km) and high magnitude (≥7.2) the local intensity is V-VI. The critical curve in Fig. 8.14 gives a preliminary criterion for the completeness of the lake record.

8.4.3 Cave Archives

Cave archives require special caution with regard to completeness. Limestone precipitation is not regular and the ensuing stratigraphy may be complex. Recently, Kagan (2011) and Braun et al. (2011) have reported a paleo-earthquake in the Judean Hills caves (SH and in Fig. 8.8) at ~8.6 ka. They go on to compare the archive from this site with the Denya Cave site (DH in Fig. 8.8), considered a proxy for activity on the Carmel fault, and with lacustrine and landslide archives. This comparison allows Kagan (2011) and Braun et al. (2011) to discuss coupling of the transform with its Carmel fault branch. For the Pleistocene however, such a discussion would critically depend on the assumption that both records are complete. A complete speleoseismic record will be achieved only at the limit where additional age determinations give no new dates of paleo-earthquakes. Another issue with the present state of the art is that five out of 18 speleoseismites reported by Kagan et al. (2005) had no dated material preceding damage, so they were biased by the use of the *terminus ante quem* as damage date.

8.4.4 Rockfall Archives

Rockfall archives also require statistically significant sampling, such that new sampling adds no new dates. With the present costs of cosmogenic dating, and with the low availability of new techniques such as *in situ* C^{14} dates, this requirement is not realistic, so rockfall archives can hardly be considered complete.

8.5 Discussion: Recurrence Patterns

A chief goal of paleo-earthquake research is resolving recurrence patterns and, in particular, recurrence intervals characteristic of the seismogenic zones in question. Goes (1996) has used historical and paleoseismological records from the American continent to evaluate *aperiodicity* (α) of large earthquakes, using a definition of α as

$$\alpha = \frac{\sigma(\tau)}{\langle \tau \rangle} \tag{8.4}$$

In Eq. (8.1) τ represents the distribution of recurrence intervals, the <brackets> denote average, and σ is the standard deviation. A perfectly periodic behavior is represented by vanishing α whereas clustered behavior yields α larger than unity. Goes (1996) also recognized the limitation imposed by the brevity of archives, one that is alleviated by the long time recorded for the DSR.

Table 8.2 Distributions of recurrence intervals at different locations and times along the Dead Sea rift

Site/segment	Window [ka]	Recurrence interval [kyear]			σ(τ) SDEV	aperiodicity α ⟨	n	Reference
		Mean	Max	Min				
Judean Hills	185–0	9	24.5	3	6.7	0.73		K05, B11
Peratzim Creek	72–25	1.6	8.5	0.25	2.9	1.8	29	M96, A06
Ein Gedi Core	4–0	0.1			0.088	0.9		M04, A06
Ein Gedi Core	2–0	0.05			0.07	0.75		M04, A06
Ein Feshkha	2.5–0	0.05						
Ze'elim Terrace								K11
Dead Sea								
Yammounneh	~12–6	1–1.2			Unknown		5–6	D07
Yammounneh	6–0	1–1.2			Unknown		5	N08
Jordan Valley	25–0	0.79	1.5	0.28	0.52	0.66	5	F11

A06 – Agnon et al. (2006), B11 – Braun et al. (2011), D07 – Daeron et al. (2007)
F11 – Ferry et al. (2011), K05 – Kagan et al. (2005), K11 – Kagan et al. (2011)
M04 – Migowski et al (2004), M96 – Marco et al. (1996), N08 – Nemer et al. (2008)
α "intra basin seismites" (IBSs), simultaneously recorded at Ein Feshkha, Ein Gedi core, and Ze'elim Terrace

The foregoing section clarifies that a discussion of recurrence based on the records we have may suffer from their incompleteness. Yet such a discussion can help point out gaps in the information needed for advancing our understanding of the nature of seismogenic zones. Table 8.2 summarizes data on recurrence from several studies that provide sufficient information for characterizing the distribution of intervals.

Marco et al. (1996) have identified 29 seismites in the dated section of Peratzim Creek and inferred a mean recurrence interval of 1.6 kyear between ~72 and ~25 ka. Agnon et al. (2006) have revised the estimate based on more extensive ages of Haase-Schramm et al. (2004). This age scheme recognized a 5 kyear hiatus in the Lisan Formation around 45 ka (Fig. 8.15). The variation, between 250 and 8,500 years, is a clear manifestation of temporal clustering of earthquakes. More accurate and extensive dating of Lisan Formation will modify the figures (Stein 2011) but clustering of seismites will likely prevail (Kagan 2011).

Figure 8.15 shows the cumulative number of seismites versus interpolated age using Haase-Schramm et al.'s (2004) age model. Anomalous lithology and hiatuses are marked following Stein (2011) who has tabulated the ages versus depth and interpolated for seismite ages. Figure 8.16 shows smoothed recurrence intervals on a logarithmic scale. Despite the periods of poor dating constraints (including hiatus 43–48 ka), Figures 8.15 and 8.16 show a pattern of long quiescence periods between quasi-periodic clusters. During each cluster of seismicity the recurrence interval is quite uniform, varying among clusters between 200 and 1,400 years. Quiescence periods may linger 3–10,000 years.

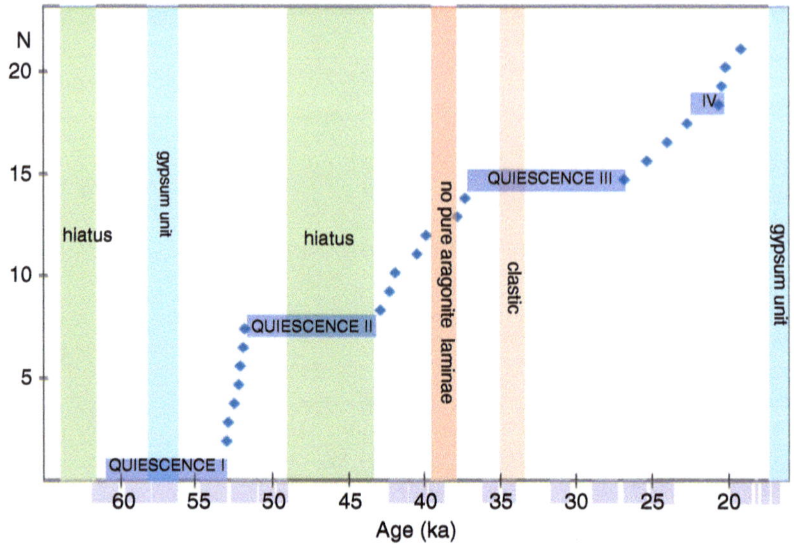

Fig. 8.15 Clustering of seismites in Lisan Formation. The *horizontal axis* shows model ages (Stein 2011) according to stratigraphic heights in PZ1 (Marco et al. 1996) of dated samples (Haase-Schramm et al. 2004). Anchors for the age model are marked by *magenta rectangles* beneath the axis. The *blue diamonds* denote cumulative number of seismites. Periods of anomalous deposition (gypsum, clastics, their aragonite-poor inter-lamination) or lack of are shown by colored shading. Four periods of apparent quiescence emerge separated by quasi-periodic clusters of activity (See Fig. 8.16)

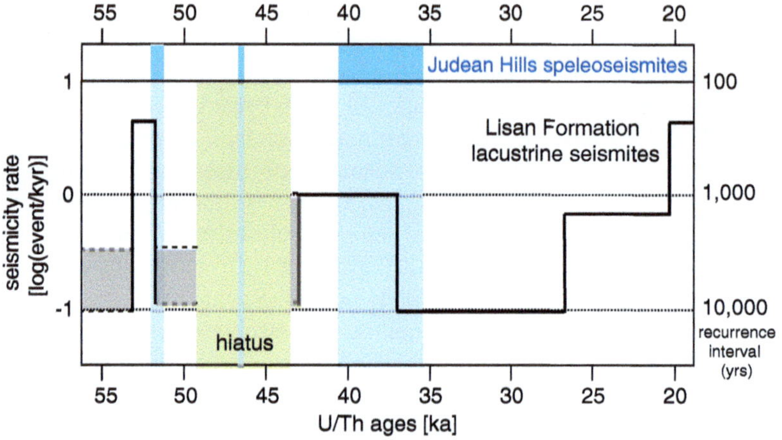

Fig. 8.16 Rates of seismicity recorded in the Lisan Formation and Judean Hills caves. The ordinate gives the logarithmic rate of Lisan seismites (event per millennium; see *right hand axis* for recurrence intervals in years). The *gray rectangles* represent the uncertainty: *upper bound* is calculated based on the intervals represented by continuous deposition of datable sediments; *lower bounds* assume no large earthquakes during non-datable intervals (data as in Fig. 8.15). *Cyan rectangles* represent date ranges of Judean Hills speleoseismites following Kagan et al. (2005). Other colored rectangles as in Fig. 8.15

The high rate of seismicity at ~52 ka (an event in ~200 years) is anomalous for the entire period represented by sediments within Lisan Formation, but a similar or higher rate (an event per ~100 years) during the last millennium is indicated in Ze'elim Creek (Migowski et al. 2004). Agnon et al. (2006) argue that the resolution of seismites in the ~52 ka cluster of seismites in Lisan Formation at Peratzim Creek would not allow recording of a rate higher than an event in 200 year. Kagan et al. (2011) study seismites from the last two millennia in three sites currently onshore the Dead Sea northern basin. They distinguish events recorded simultaneously at the northernmost, the center, and the southernmost basin ("intra basin seismites" of Kagan et al. 2011). These events, seemingly with higher local intensity around the Dead Sea, show an average recurrence interval of ~200 years, about twice that of any individual section.

Marco et al. (1996) have used the *aperiodicity* (Eq. 8.4) to illustrate and quantify earthquake clustering in the distribution of intervals: if the aperiodicity is smaller than unity the distribution is periodic, whereas exceeding unity indicates clustering. The dated section of Lisan Formation has featured aperiodicity around 1.75. Agnon et al. (2006) found that the correction of the age model to include Haase-Schramm et al. (2004) dates decreases the aperiodicity only slightly to about 1.6.

Independent indication for long-term clustering has been reported for the Arava/Araba Valley (Fig. 8.1). Amit et al. (2002) dug paleoseismic trenches and dated clastic sediments with optically simulated luminescence (OSL) up to 90 ka. The dates constrained the ages of dip-slip events on marginal faults, showing more frequent activity of smaller events during the Holocene, and migration of activity across a shear zone.

The Holocene Dead Sea archive shows a quasi-periodic behavior. Aperiodicity for Ein Gedi core varied from 0.75 for the historic period to 0.9 for the entire upper 5 m (3.8 ka to present). The changes of recurrence during the late Quaternary may indicate transitions, or mode switchings, from clustered distributions via quasi-periodic distribution to a pronounced periodicity during the late Holocene. Such mode-switching was anticipated in simulations using two different physical models (Ben-Zion et al. 1999). Lyakhovsky et al. (1997) matched the behavior of the Lisan archive reported by Marco et al. (1996) with a model that explicitly computes distributed damage in the plate. In that model, mode-switching emerges when the rate of healing of fractures in the seismogenic zone balances the rate of mechanical loading (determined by the long-term slip rate). A simpler model that displays clustered versus periodic behavior accounts for distributed damage by stress variations off-fault (Kenner and Simons 2005).

Kagan et al (2005) found a mean recurrence interval of 10–14 kyear with aperiodicity of 0.6 for the cave seismites (interpreted to represent $M \geq 7.8$) during the last 185 kyear. These are likely of the largest events along the adjacent segment of DSR. However, correlation with lake seismites (discussed in the next section) does not show a consistent relationship with their clusters (Stein 2011), and in view of the completeness problem explained above, we may be looking at a partial set of such events.

A key variable in any long term seismological analysis is the ratio between the durations of the seismic cycle and the available archive. Each segment of the plate boundary may show a cycle given simply by the interval between the largest events on that segment. However, segments may interact in a complex manner, particularly where the plate boundary is complex (Braun et al. 2011). Several cycles of individual segments may combine to show a cycle for the system. Such cyclic behavior may explain the long-term earthquake clustering apparent in Lake lisan's record (Marco et al. 1996; Lyakhovsky et al. 1997).

Migowski et al. (2004) noted another manifestation of complex cycles: alternating periods (several centuries each) of high versus low rates of seismicity recorded in the Dead Sea basin, correlating them to similar periods in the East Anatolian fault (EAF, Fig. 8.1), and out of phase with the North Anatolian fault (NAF) (Ambraseys 1971). Agnon et al. (2006) found that the increase in activity at NAF and decrease at EAF at around 500 CE is heralded by a decrease at DSR, with more complicated relations around 1000 CE. These correlations suffer from two aspects: (1) the Anatolian data is somewhat out of date; (2) the Dead Sea record is local, biased due to minor events with epicenter close to the Dead Sea. In addition, a comparison of the rate of seismic moment release is likely to be more meaningful.

The recent compilation of the Dead Sea archive of historic seismites (Kagan et al. 2011) shows a cyclic behavior similar to the Ein Gedi core archive of Migowski et al. (2004) and Agnon et al. (2006) (Fig. 8.17a): the mean recurrence rate more than doubles during the tenth–twelfth centuries CE relative to the rest of the first millennium and the rest of the second millennium. The aperiodicity α calculated for that time (Eq. 8.4) changes only slightly around 0.6. Ambraseys (2006a, b, 2009) estimated magnitudes of 139 events during the period 100–2000 CE from macro-seismic data and computed an average slip rate. He found an abrupt increase in slip rate after 1150 CE, from values that fluctuated between 0.22 and 0.26 cm/year since the 700 CE to the range 0.35–0.4 cm/year (Fig. 8.17b). Note that each of the three periods delineated in Fig. 8.17a is punctuated by a short interval of accelerated macro-seismic slip rate (Fig. 8.17b), associated with very large earthquakes. During the first period, this interval spans 530–560 CE, where the 551 CE earthquake (perhaps with associated activity) is recorded in the Dead Sea. The second period is associated with the 1156–1157 and 1170 CE earthquakes by Ambraseys' data (2006a, b, 2009). The third period is associated with the 1759 CE double event (Ambraseys and Barazangi 1989; Ellenblum et al. 1998). If this pattern is verifiable, it might have far reaching implications for the seismic behavior of the Dead Sea rift and other plate boundaries. However, a recent revised estimate of the 1170 and 1202 magnitudes (Hough and Avni 2011) might increase the acceleration of slip during this period (Table 8.3); more significantly, the revision would delay the timing of acceleration to the end of the middle period (thirteenth century). Figure 8.17 underscores the need for more paleoseismic data and additional historic macroseismic analyses.

Migowski et al. (2004) noted alternating activity cycles between south and north within DSR (see their Fig. 8.7b). Ambraseys (2006b, 2009) elaborated on this pattern (Ambraseys 2006b's Fig. 8.7). He inferred that earthquakes with $Ms \geq 7.2$

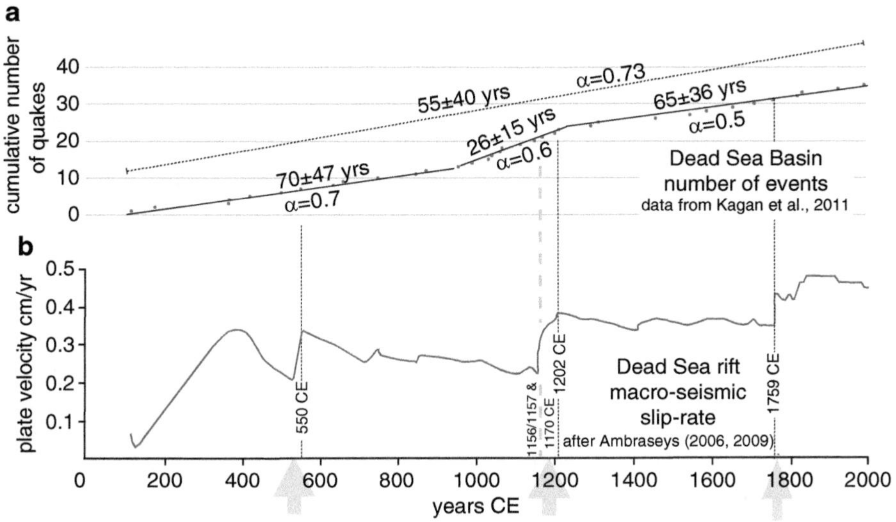

Fig. 8.17 Temporal variation of activity along the Dead Sea rift during the last nineteenth centuries. (**a**) cumulative number of events recorded by Dead Sea seismites (Fig. 8.13a; Kagan et al. 2011). The *slopes* of the graph suggest three periods with alternating recurrence intervals: 100–950 CE (70±47 years), 950–1250 (26±15 years), 1250–2000 (65±36 years). The aperiodicities of the three periods are, respectively, 0.68, 0.57, and 0.55. For comparison, the mean recurrence interval for the entire period is 55±40 years, with aperiodicity of 0.73. (**b**) Temporal variation of macro-seismic slip rate along the Dead Sea rift following the estimate by Ambraseys (2006a, b, 2009). Three periods of accelerated seismicity are discernible (*vertical gray arrows* beneath time axis), associated with the seismic episodes of the sixth-, twelfth-thirteenth-, and eighteenth-centuries CE (*vertical broken lines* denote historic earthquakes)

Table 8.3 Magnitude estimates of key historic earthquakes along the Dead Sea rift

Year CE	Ambraseys (2006b, 2009)	Other estimates	Reference
363	M 7.4	M 6.5–7	Present study[a]
551	M 7.3	M 7.4–7.6	Elias et al. (2007)
1170	M 7.3	M 7.5	Hough and Avni (2011)
1202	M 7.2	M 7.8	Hough and Avni (2011)

[a] The magnitude estimate is based on respective rupture of the Carmel fault along 60 km and the Arava/Araba Valley segment along 100 km. This choice seems appropriate for generating the damage pattern recorded by Russell (1980) and extended to account for the damage in Aila (Thomas et al. 2007)

propagate along a regular oscillatory trend. In view of the paleoseismic evidence from the Dead Sea, the assignment of M7.4 to the first event in the series (19 May 363 CE) is likely exaggerated – a result of amalgamation of the two events, one in the northern and the other in southern Holy Land. The revision of magnitudes of the 1170 and 1202 CE events (Hough and Avni 2011) further obliterates the oscillatory pattern inferred by Ambraseys (2006b, 2009).

Daeron et al. (2007) found a recurrence interval of 1.0–1.26 millennia for M ~ 7.5 ruptures on the Yammounneh fault, determined reliably for a limited period (~12–6.4 ka). Nemer et al. (2008) obtained similar results for the last six millennia. The dating uncertainty in both studies does not allow us to assess the aperiodicity.

Ferry et al. (2011) combined trenching, archaeology, and historical catalogues to study the earthquake history of the Jordan Valley (Fig. 8.1). They determined a mean recurrence interval of 790 years for the last four millennia, where intervals are smaller than age uncertainties. The standard deviation for this period is 520 years and the nominal aperiodicity at 0.66. This low figure is likely the result of the small number of intervals – (only five) and the limited period for which the dating precision is sufficient.

Ambraseys (2006b, 2009) compute frequency-size distributions for DSR. Notably, the macroseismic data covering the last two millennia are distributed significantly differently from the instrumental data. The macroseismic data taken together with the instrumental data seem to indicate a characteristic earthquake behavior with up to three times higher frequency for events of $M_s \geq 7$. Only if the assessment of M_s based on macroseismic analysis (Ambraseys 2006b, 2009) is systematically overestimated by the maximum reported uncertainty of 0.3, the distribution could follow a Gutenberg-Richter's relation. Yet if the macroseismic analysis has systematically underestimated the actual magnitudes, the Gutenberg-Richter relation underestimates the frequency of $M_s \sim 7$ by a factor around 30.

Hough and Avni (2011) construct a composite frequency-size distribution based on a modified version of the data of Ambraseys (2006b, 2009). The distribution, within the uncertainty of data, conform separate Gutenberg-Richter relations ("b-value" of unity) for either historical or instrumental data. For the M7, the historic earthquakes are four times more abundant than what would be predicted by extrapolation of the instrumental; for M7.6 the factor is 8. Ambraseys (2006b, 2009) frequency size relation for $4.5 \leq M_s \leq 5.5$ fits a Gutenberg-Richter relation with a similar "b value". The historical data then exceeds the extrapolation of instrumental data by a similar order, with a factor of about 5 between M_s6.7 and M_s7.3 .

To summarize the discussion of recurrence, the DSR shows variable recurrence in time and space (Table 8.2). The different types of archives with different sensitivities and uncertainties, and the bias of Dead Sea seismites toward local earthquakes, prevent definite statements to be made as yet. However, two broad conclusions can be drawn:

1. The historic period is represented well in the late Holocene lacustrine sections, with seemingly periodic behavior, with aperiodicity (standard deviation over mean recurrence interval) $\alpha \sim 0.6 \pm 0.1$. The common recurrence time is 65–70 years, yet between the tenth and the twelfth centuries CE the period is around 26 years.
2. While recurrence intervals for large earthquakes may be stable for periods up to several millennia, they are not stable over the 10-millennia duration scale. Yet, the longest archive of the strongest events, namely the speleoseismic archive, exhibits quasi-periodic behavior.

The latter conclusion may be a consequence of the incompleteness of this record, which is being complemented by further dating of damaged cave deposits. Alternatively, the largest earthquakes (say M>7.5) may be controlled by the broad-scale segmentation of the rift, which may be considered constant for the 10–100 millennial time scale. Lyakhovsky et al. (1997) and by Kenner and Simons (2005) have proposed mechanical models for transitions between two patterns of frequency-size statistics and for the anoalous incidence of larger earthquakes. The large scale segmentation of the rift with segment lengths of ~100 km (Fig. 8.4b) may be reflected in the high abundance of larger earthquakes.

Acknowledgments The author is grateful to G.C.P. King and S. Wesnousky for thoughtful reviews, and to his numerous collaborators for inspiring discussions over two decades of research. Special thanks to E.J. Kagan for a meticulous editorial work. This research was supported by Israel Science Foundation (grant no. 1181/12), The German-Israeli Bi-national Science Foundation (GIF), the Helmholtz Association (Virtual Institute DESERVE), and by the Earth Science Administration in the Ministry of Energy and Water.

References

Adams J (1990) Paleoseismology of the Cascadia subduction zone – Evidence from turbidites off the Oregon-Washington margin. Tectonics 9:569–683. doi:10.1029/TC009i004p00569

Agnon A, Migowski C, Marco S (2006) Intraclast breccia layers in laminated sequences: recorders of paleo-earthquakes. Geol Soc Am Spec Pap 401:195–214

Akyuz HS, Altunel E, Karabacak V, Yalciner CC (2006) Historical earthquake activity of the northern part of the Dead Sea Fault Zone, southern Turkey. Tectonophysics 426:281–293. doi:10.1016/j.tecto.2006.08.005

Almagor G, Garfunkel Z (1979) Submarine slumping in continental margin and northern Sinai. Am Assoc Pet Geol Bull 63:324–340

Alsop GI, Marco S (2011) Soft-sediment deformation within seismogenic slumps of the Dead Sea Basin. J Struct Geol 33:433–457. doi:10.1016/j.jsg.2011.02.003

Al-Qaryouti MY (2008) Attenuation relations of peak ground acceleration and velocity in the Southern Dead Sea Transform region. Arab J Geosci 1:111–117. doi:10.1007/s12517-008-0010-4

Ambraseys NN (1971) Value of historical records of earthquakes. Nature 232:375–379

Ambraseys NN (1998) The earthquake of 1 January 1837 in southern Lebanon and northern Israel. Ann Geophys 15:923–935

Ambraseys NN (2005a) The seismic activity in Syria and Palestine during the middle of the 8th century; an amalgamation of historical earthquakes. J Seismol 9:115–125

Ambraseys N (2005b) Historical earthquakes in Jerusalem – A methodological discussion. J Seismol 9:329–340

Ambraseys N (2006a) Earthquakes and archaeology. J Archaeol Sci 33:1008–1016. doi:10.1016/j.jas.2005.11.006

Ambraseys NN (2006b) Comparison of frequency of occurrence of earthquakes with slip rates from long-term seismicity data: the cases of Gulf of Corinth, Sea of Marmara and Dead Sea Fault Zone. Geophys J Int 165:516–526. doi:10.1111/j.1365-246X.2006.02858.x

Ambraseys NN (2009) Earthquakes in the Mediterranean and Middle East: a multidisciplinary study of seismicity up to 1900. Cambridge University Press, New York, 947 pp

Ambraseys NN, Barazangi M (1989) The 1759 earthquake in the Bekaa valley. Implications for earthquake hazard assessment in the eastern Mediterranean region. J Geophys Res 94:4007–4013

Ambraseys NN, Jackson JA (1998) Faulting associated with historical and recent earthquakes in the Eastern Mediterranean region. Geophys J Int 133:390–406

Ambraseys N, Karcz I (1992) The earthquake of 1546 in the Holy Land. Terra Nova 4:253–262. doi:10.1111/j.1365-3121.1992.tb00480.x

Ambraseys NN, Melville CP (1988) An analysis of the eastern Mediterranean earthquake of 20 May 1202. In: Lee WKH et al (eds) History of seismography and earthquakes of the world. Academic, San Diego, pp 181–200

Ambraseys N, Synolakis C (2010) Tsunami catalogs for the Eastern Mediterranean revisited. J Earthq Engin 14:309–330. doi:10.1080/13632460903277593

Ambraseys NN, Melville CP, Adams RD (1994) The seismicity of Egypt, Arabia and the Red Sea: a historical review. Cambridge University Press, New York, 181 pp. doi:10.1017/CBO9780511524912

Amiran D, Arieh E, Turcotte T (1994) Earthquakes in Israel and adjacent areas: macroseismic observations since 100 BCE. Isr Expl J 44:260–305. (Correction, Isr Explor J 45: 201, 1995)

Amit R, Zilberman E, Enzel Y, Porat N (2002) Paleoseismic evidence for time dependency of seismic response on a fault system in the southern Arava Valley, Dead Sea rift, Israel. Geol Soc Am Bull 114:192–206

Amitai R (1989) Notes on the Ayyūbid Inscriptions at al-Ṣubayba (Qalʿat Nimrūd). Dumbart Oaks Pap 43:113–119

Atkinson GM, Wald DJ (2007) "Did you feel it"? Intensity data: a surprisingly good measure of earthquake ground motion. Seismol Res Lett 78:362–368. doi:10.1785/gssrl.78.3.362

Austin SA, Franz GW, Frost EG (2000) Amos's earthquake: an extraordinary Middle East seismic event of 750 BC. Int Geol Rev 42:657–671

Avner U (1998) Settlement, agriculture and paleoclimate in Uvda Valley, southern Negev desert, 6th-3rd millenia BC. In: Issar AS, Brown N (eds) Water, environment and society in times of climatic change. Kluwer Academic Publishers, Dordrecht, pp 147–202

Avner U, Carmi I, Segal D (1994) Neoloithic to Bronze Age settlmentof the Negev and Sinai in light of radio- carbon dating. A view from the South- ern Negev. In: Bar-Yosef O, Kra RS (eds) Late Quaternary chronology and paleoclimates of the eastern Mediterranean. Radiocarbon, Tucson, pp 265–300

Avni R (1999) The 1927 Jericho earthquake: comprehensive Macro- seismic analysis based on contemporary sources. Ben-Gurion University of the Negev, Beersheba, 203 pp

Baer G, Funning GJ, Shamir G, Wright TJ (2008) The 1995 November 22, M(w) 7.2 Gulf of Elat earthquakecyclerevisited.GeophysJInt175:1040–1054.doi:10.1111/j.1365-246X.2008.03901.x

Bakun WH, Wentworth CM (1997) Estimating earthquake location and magnitude from seismic intensity data. Bull Seismol Soc Am 87:1502–1521

Bar-Matthews M, Ayalon A, Kaufman A (2000) Timing and hydrological conditions of sapropel events in the Eastern Mediterranean, as evident from speleothems, Soreq Cave, Israel. Chem Geol 169:145–156

Barkan E, Luz B, Lazar B (2001) Dynamics of the carbon dioxide system in the Dead Sea. Geochem Cosmochem Acta 65:355–368

Bartov Y, Stein M, Enzel Y, Agnon A, Reches Z (2002) Lake levels and sequence stratigraphy of Lake Lisan, the late Pleistocene precursor of the Dead Sea. Quat Res 57:9–21. doi:10.1006/qres.2001.2284

Begin ZB, Ehrlich A, Nathan Y (1974) Lake Lisan, the Pleistocene precursor of the dead sea. Geol Surv Isr Bull 63:30pp

Belmaker R, Stein M, Yechieli Y, Lazar B (2007) Controls on the radiocarbon reservoir ages in the modern Dead Sea drainage system and in the last glacial Lake Lisan. Radiocarbon 49:969–982

Ben-Menahem A (1979) Earthquake catalogue for the Middle East (92 B.C. to 1980 A.D.). Boll Geofis Teor Appl 21:245–313

Ben-Menahem A (1991) Four thousand years of seismicity along the Dead Sea rift. J Geophys Res 96:20195–20216

Ben-Zion Y, Damen K, Lyakhovsky V, Ertas D, Agnon A (1999) Spontaneous mode switching of earthquakes, Earth Planet. Sci Lett 172:11–21

Bikai PM (2002) The churches of Byzantine Petra, in Petra: a Royal City unearthed. Near East Archaeol 65:271–276

Bookman R, Enzel Y, Agnon A, Stein M (2004) Late Holocene lake levels of the Dead Sea. Geol Soc Am Bull 116:555–571

Bourke SJ (2004) Cult and archaeology at Pella in Jordan: excavating the Bronze and Iron Age temple precinct (1994–2001). J Proc R Soc N S W 137:1–31

Bourke S, Sparks R, Mairs L (1999) Bronze age occupation on Tell Husn (Pella): report on the University of Sydney's 1994/95 field seasons. Mediterr Archaeol 12:51–66

Bowman D (1971) Geomorphology of the shore terraces of the late Pleistocene Lisan Lake (Israel). Palaeogeogr palaeoclim Palaeoecol 9:183–209. doi:10.1016/0031-0182(71)90031-9

Braun Y (2009) Dating paleo-seismic activity on the Carmel fault using damaged cave deposits from Denya Cave, Mt. Carmel. M.Sc. thesis, The Hebrew University of Jerusalem, Jerusalem, 87 pp

Braun Y, Kagan E, Bar-Matthews M, Ayalon A, Agnon A (2011) Dating speleoseismites near the Dead Sea Transform and the Carmel fault: clues to coupling of a plate boundary and its branch. Isr J Earth Sci 58:257–273. doi:10.1560/IJES.58.3-4.257. Spec. vol., eds. Agnon A, Amit R, Michetti A, Hough S, The Dead Sea rift as a natural laboratory for neotectonics and paleoseismology

Chapron E, Beck C, Pourchet M, Deconinck JF (1999) 1822 earthquake-triggered homogenite in Lake Le Bourget (NW Alps). Terra Nova 11:86–92. doi:10.1046/j.1365-3121.1999.00230.x

Cita MB, Beghi C, Camerlenghi A, Kastens KA, McCoy FW, Nosetto A, Parisi E, Scolari F, Tomadin L (1984) Turbidites and megaturbidites from Herodotus Abyssal-Plain (Eastern Mediterranean) unrelated to seismic events. Mar Geol 55:79–101. doi:10.1016/0025-3227(84)90134-8

Cita MB, Camerlenghi A, Rimoldi B (1996) Deep-sea tsunami deposits in the eastern Mediterranean: new evidence and depositional models. Sediment Geol 104(1–4):155–173. doi:10.1016/0037-0738(95)00126-3

Davenport CA, Ringrose PS (1987) Deformation of Scottish Quaternary sediment sequence by strong earthquake motions. In: Jones ME, Preston RM (eds) Deformation of sediments and sedimentary rocks, Geological Society special publication, 29. Geological Society by Blackwell Scientific, Oxford/London, pp 299–314

Daeron M, Klinger Y, Tapponnier P, Elias A, Jacques E, Sursock A (2007) 12,000-year-long record of 10 to 13 Paleoearthquakes on the Yammoûneh Fault, Levant Fault System, Lebanon. Bull Seismol Soc Am 97:749–771. doi:10.1785/0120060106

Deves M, King GCP, Klinger Y, Agnon A (2011) Localised and distributed deformation in the lithosphere: modelling the Dead Sea region in 3 dimensions. Earth Planet Sci Lett 308:172–184. doi:10.1016/j.epsl.2011.05.044

Doig R (1990) 2300 yr history of seismicity from silting events in Lake Tadoussac, Charlevoix, Quebec. Geology 18:820–823

Doig R (1991) Effects of strong seismic shaking in lake sediments, and earthquake recurrence interval, Témiscaming, Quebec. Can J Earth Sci 28:1349–1352

Doig R (1998) 3000-year paleoseismological record from the region of the 1988 Saguenay, Quebec, earthquake. Bull Seismol Soc Am 88:1198–1203

Eklund S (2008) Stone weathering in the monastic building complex on Mountain of St Aaron in Petra, Jordan. M.A. thesis, University of Helsinki Institute for Cultural Studies Department of Archaeology, 113 p

Elias A, Tapponnier P, Singh SC, King GCP, Briais A, Daeron M, Carton H, Sursock A, Jaques E, Jomaa R, Klinger Y (2007) Active thrusting offshore Mount Lebanon: source of the tsunamigenic A.D. 551 Beirut-Tripoli earthquake. Geology 35:755–758. doi:10.1130/G23631A.1

El-Isa ZH, Mustafa H (1986) Earthquake deformations in the Lisan deposits and seismotectonic implications. Geophys J R Astron Soc 86:413–424
Ellenblum R (1989) Who built Qalʿat al-Ṣubayba? Who built Qalʿat al-Ṣubayba? Dumbart Oaks Pap 43:103–112
Ellenblum R (2007) Crusader castles and modern histories. Cambridge University Press, Cambridge, 362 pages
Ellenblum R, Marco S, Agnon A, Rockwell T, Boas A (1998) A Crusader castle torn apart by the 1202 earthquake. Geology 26:303–306
Enzel Y, Kadan G, Eyal Y (2000) Holocene earthquakes inferred from a fan-delta sequence in the Dead Sea graben. Quat Res 53:34–48
Enzel Y, Agnon A, Stein M (eds) (2006) New frontiers in Dead Sea paleoenvironmental research, Geological Society of America special paper 401. Geological Society of America, Boulder
Fantalkin A, Finkelstein I (2006) The Sheshonq I campaign and the 8th-century BCE earthquake-more on the archaeology and history of the South in the Iron I-IIa. J Inst Archaeol Tel Aviv Univ 33:18–42
Ferry M, Meghraoui M, Abou Karaki NA, Al Taj M, Khalil L (2011) Episodic behavior of the Jordan Valley section of the Dead Sea fault inferred from a 14-ka-long integrated catalog of large earthquakes. Bull Seismol Soc Am 101:39–67. doi:10.1785/0120100097
Franken HJ (1992) Excavations at Tell Deir ‘Alla. The late bronze age sanctuary. Peeters, Louvain
Garfunkel Z (2011) The long- and short-term lateral slip and seismicity along the Dead Sea Transform: an interim evaluation. Isr J Earth Sci, Spec Vol, Agnon A, Amit R, Michetti A, Hough S (eds) The Dead Sea rift as a natural laboratory for neotectonics and paleoseismology (in press)
Garfunkel Z, Zak I, Freund R (1981) Active faulting in the Dead Sea rift. Tectonophysics 80:1–26
Gibert L, Alfaro P, Garcia-Tortosa FJ, Scott G (2011) Superposed deformed beds produced by single earthquakes (Tecopa Basin, California): insights into paleoseismology. Sediment Geol 235:148–159. doi:10.1016/j.sedgeo.2010.08.003
Gluck D (2001) The landscape evolution of the south western Dead Sea Basin and the paleoseismic record of the south western marginal fault of the Dead Sea Basin and of the Carmel fault during the Late Pleistocene and Holocene. M.Sc. thesis, Hebrew University, Jerusalem, Israel [in Hebrew with English abstract]
Goes SDB (1996) Irregular recurrence of large earthquakes: an analysis of historic and paleoseismic catalogs. J Geophys Res 101:5739–5749. doi:10.1029/95JB03044
Guidoboni E, Comastri A (2005) Catalogue of earthquakes and tsunamis in the Mediterranean Area from the 11th to the 15th Century. INGV-SGA, Bologna
Guidoboni E, Comastri A, Traina G (1994) Catalogue of ancient earthquakes in the Mediterranean area up to the 10th century. INGV-SGA, Rome
Haase-Schramm A, Goldstein SL, Stein M (2004) U–Th dating of Lake Lisan aragonite (Late Pleistocene Dead Sea) and implications for glacial East Mediterranean climate change. Geochim Cosmochim Acta 68:985–1005
Hall JK (1996) Digital topography and bathymetry of the area of the Dead Sea depression. Tectonophysics 266:177–185. doi:10.1016/S0040-1951(96)00189-8
Hammond PC (1980) New evidence for the 4th-century A. D. destruction of Petra. Bull Am Sch Orient Res 238:65–67
Hanfmann GMA (1951) The Bronze Age in the Near East: a review article [Part I]. Am J Archaeol 55:355–365
Hartal M (2001) The al-Subayba (Nimrod) fortress: towers 11 and 9. With contributions by Amitai R, Boas A. (Israel Antiquities Authority reports no. 11.) pp. 130, 197 line drawings and half-tones, 22 plans in text. Israel Antiquities Authority, Jerusalem
Haynes JM, Niemi TM, Atallah M (2006) Evidence for ground-rupturing earthquakes on the Northern Wadi Araba fault at the archaeological site of Qasr Tilah, Dead Sea Transform fault system, Jordan. J Seismol 10:415–430. doi:10.1007/s10950-006-9028-9
Heezen BC, Ewing M (1952) Turbidity currents and submarine slumps, and the 1929 Grand Banks [Newfoundland] earthquake. Am J Sci 250:849–873. doi:10.2475/ajs.250.12.849

Heifetz E, Agnon A, Marco S (2005) Soft sediment deformation by Kelvin-Helmholtz instability: a case from Dead Sea earthquakes. Earth Planet Sci Lett 236:497–504
Hempton MR, Dewey JF (1983) Earthquake-induced deformational structures in young lacustrine sediments, East Anatolian Fault, southern Turkey. Tectonophysics 98:T7–T14. doi:10.1016/0040-1951(83)90294-9
Hennessey IB (1969) Preliminary report on a first season of excavation sat Telleilat Ghassul. Levant 1:1–24
Herzog Z, Singer-Avitz L (2004) Redefining the centre: the emergence of State in Judah. J Inst Archaeol Tel Aviv Univ 31:209–244
Hinzen KG (2011) Sensitivity of earthquake toppled columns to small changes in ground motion and geometry. Isr J Earth Sci 58:309–326
Hough SE, Avni R (2011) The 1170 and 1202 CE Dead Sea Rift earthquakes and long-term magnitude distribution of the Dead Sea Fault zone. Isr J Earth Sci 58:295–308. doi:10.1560/IJES.58.3-4.295. Spec. Vol., eds. Agnon A, Amit R, Michetti A, Hough S, The Dead Sea rift as a natural laboratory for neotectonics and paleoseismology
Jones AP, Omoto K (2000) Towards establishing criteria for identifying trigger mechanisms for soft-sediment deformation: a case study of Late Pleistocene lacustrine sands and clays, Onikobe and Nakayamadaira Basins, northeastern Japan. Sedimentology 47:1211–1226
Kagan E (2011) Multi-archive paleoseismology along the southern Dead Sea rift: independent recording by lake and cave sediments. PhD thesis, Hebrew University, Jerusalem
Kagan E, Bar-Matthews M, Ayalon A, Agnon A (2005) Dating large and infrequent earthquakes using damaged cave deposits. Geology 33:251–264
Kagan EJ, Stein M, Agnon A, Bronk Ramsey C (2010) Paleoearthquakes as anchor points in Bayesian radiocarbon deposition models: a case study from the Dead Sea. Radiocarbon 54(3):1018–1026
Kagan E, Stein M, Agnon A, Neumann F (2011) Intrabasin paleoearthquake and quiescence correlation of the late Holocene Dead Sea. J Geophys Res 116:B04311. doi:10.1029/2010JB007452
Kastens KA (1984) Earthquakes as a triggering mechanism for debris flows and turbidites on the Calabrian ridge. Mar Geol 55:13–33. doi:10.1016/0025-3227(84)90130-0
Karcz Y (2004) Implications of some early Jewish sources for estimates of earthquake hazard in the Holy Land. Ann Geophys 47:759–792
Karcz Y, Kafri U (1978) Evaluation of supposed archaeoseismic damage in Israel. J Archaeol Sci 5:237–253
Karcz I, Kafri U, Meshel Z (1977) Archaeological evidence for subrecent seismic activity along the Dead Sea–Jordan rift. Nature 269:234–235
Katz O, Crouvi O (2007) The geotechnical effects of long human habitation (2000 < years): earthquake induced landslide hazard in the city of Zefat, northern Israel. Eng Geol 95:57–78
Katz A, Kolodny Y, Nissenbaum A (1977) The geochemical evolution of the Pleistocene Lake Lisan–Dead Sea system. Geochim Cosmochim Acta 41:1609–1626. doi:10.1016/0016-7037(77)90172-7
Katz A, Agnon A, Marco S (2009) Earthquake-induced barium anomalies in the Lisan Formation, Dead Sea rift valley. Isr Earth Planet Sci Lett 286:219–229. doi:10.1016/j.epsl.2009.06.031
Katz O, Amit R, Yagoda-Biran G, Hatzor YH, Porat N, Medvedev B (2011) Quaternary earthquakes and landslides in the Sea of Galilee area, the Dead Sea Transform: paleoseismic analysis and implication to the current hazard. Isr J Earth Sci 58:275–294. doi:10.1560/IJES.58.3-4.275. Spec. vol, eds. Agnon A, Amit R, Michetti A, Hough S, The Dead Sea rift as a natural laboratory for neotectonics and paleoseismology
Kenner SJ, Simons M (2005) Temporal clustering of major earthquakes along individual faults due to post-seismic reloading. Geophys J Int 160:179–194. doi:10.1111/j.1365-246X.2005.02460.x
Ken-Tor R, Agnon A, Enzel Y, Marco S, Negendank JFW, Stein M (2001) High resolution geological record of historic earthquakes in the Dead Sea Basin. J Geophys Res 106:2221–2234
Kesten D, Weber M, Haberland C, Janssen C, Agnon A, Bartov Y, Rabba I, The DESERT Group (2008) Combining satellite and seismic images to analyse the shallow structure of the Dead Sea Transform near the DESERT transect. Int J Earth Sci (Geol Rundsch). doi:10.1007/s00531-006-0168-5

Klinger Y, Rivera L, Haessler H, Maurin JC (1999) Active faulting in the Gulf of Aqaba: new knowledge from the M-W 7.3 earthquake of 22 November 1995. Bull Seismol Soc Am 89:1025–1036

Klinger Y, Avouac JP, Dorbath L, Abou Karaki N, Tisnerat N (2000) Seismic behaviour of the Dead Sea fault along Araba valley, Jordan. Geophys J Int 142:769–782. doi:10.1046/j.1365-246x.2000.00166.x

Levi TE, Adams RB, Hauptmann A, Prange M, Schmitt-Schtrecker S, Najjar M (2002) Early Bronze Age metallurgy: a newly discovered copper manufactory in southern Jordan. Antiquity 76:425

Levi T, Weinberger R, Aifa T, Eyal Y, Marco S (2006) Earthquake-induced clastic dikes detected by anisotropy of magnetic susceptibility. Geology 3:69–72. doi:10.1130/G22001.1

Lyakhovsky V, Ben-Zion Y, Agnon A (1997) Distributed damage, faulting, and friction. J Geophys Res 102:27635–27649

Lioubashevski O, Hamiel Y, Agnon A, Reches Z, Fineberg J (1999) Oscillons and solitary waves in a vertically vibrated colloidal suspension. Phys Rev Lett 83:3190–3193

Magness J (1997) Synagogue typology and earthquake chronology at Khirbet Shema', Israel. J Field Archaeol 24:211–220

Makdisi G (1956) Autograph Diary of an Eleventh-Century Historian of Baghdad. Bull Sch Orient Afr Stud 1:9–31

Machlus M, Enzel Y, Goldstein SL, Marco S, Stein M (2000) Reconstruction of low-stands of Lake Lisan between 55 and 35 kyr. Quat Int 73–74:137–144. doi:10.1016/S1040-6182(00)00070-7

Makovsky Y, Wunch A, Ariely R, Shaked Y, Rivlin A, Shemesh A, Ben-Avraham Z, Agnon A (2008) Quaternary transform kinematics constrained by sequence stratigraphy and submerged coastline features: the Gulf of Aqaba. Earth Planet Sci Lett 271:109–122

Malkawi AIH, Fahmi KJ (1996) Locally derived earthquake ground motion attenuation relations for Jordan and conterminous areas. Q J Eng Geol Hydrogeol 29:309–319. doi:10.1144/GSL.QJEGH.1996.029.P4.05

Manspeizer W (1985) The Dead Sea rift: impact of climate and tectonism on Pleistocene and Holocene sediments. In: Biddle KT, Christie-Black N (eds) Strike-slip deformation, basin formation and sedimentation, Society for Economic Paleontology and Mineralogy special publication 37. The Society, Tusla, pp 143–158

Marco S (2008) Recognition of earthquake-related damage in archaeological sites: examples from the Dead Sea fault zone. Tectonophysics 453:148–156. doi:10.1016/j.tecto.2007.04.011

Marco S, Agnon A (1995) Prehistoric earthquake deformations near Massada, the Dead Sea graben. Geology 23:695–698

Marco S, Agnon A (2005) Repeated earthquake faulting revealed by high-resolution stratigraphy. Tectonophysics 408:101–112

Marco S, Agnon A, Ellenblum R, Eidelman A, Basson U, Boaz A (1997) 817-year-old walls offset sinistrally 2.1 m by the Dead Sea Transform, Israel. J Geodyn 24:11–20

Marco S, Agnon A, Finkelstein I, Ussishkin D (2006) Ch. 31: Megiddo earthquakes. In: Megiddo IV. E. and C. Yass Publications in Archaeology, Tel-Aviv, pp 569–575

Marco S, Klinger Y (2014) Review of on-fault Palaeo-seismic studies along the Dead Sea Fault. In: Garfunkel Z et al (eds) Dead Sea transform fault system: reviews, Modern approaches in solid earth sciences 6. Springer, Dordrecht

Marco S, Hartal M, Hazan N, Lev L, Stein M (2003) Archaeology, history, and geology of the A.D. 749 earthquake, Dead Sea transform. Geology 31:665–668. doi:10.1130/G19516.1

Marco S, Rockwell T, Agnon A, Heimann A, Frieslander U (2005) Late Holocene slip of the Dead Sea Transform revealed in 3D palaeoseismic trenches on the Jordan Gorge Fault. Earth Planet Sci Lett 234:189–205

Marco S, Stein M, Agnon A, Ron H (1996) Long term earthquake clustering: a 50,000 year paleoseismic record in the Dead Sea Graben. J Geophys Res 101:6179–6192

Marco S, Weinberger R, Agnon A (2002) Radial clastic dykes formed by a salt diapir in the Dead Sea Rift, Israel. Terra Nova 14:288–294

Matmon A, Shaked Y, Agnon A, Porat N, Enzel Y, Finkel R, Lifton N, Boarretto E (2005) Lessons to exposure age dating from constraining the time of earthquake induced rockfalls along the margins of the Dead Sea fault system, southern Israel. Earth Planet Sci Lett 240:803–817
McCalpin JP (ed) (2009) Paleoseismology, 2nd edn. Academic, San Diego
McHugh CMG, Seeber L, Cormier M-H, Dutton J, Cagatay N, Polonia A, Ryan WBF, Gorur N (2006) Submarine earthquake geology along the North Anatolia Fault in the Marmara Sea. Earth Planet Sci Lett 248:661–684. doi:10.1016/j.epsl.2006.05.038
Meghraoui M, Gomez F, Sbeinati R, Van der Woerd J, Mouty M, Darkal AN, Radwan Y, Layyous I, Al Najjar H, Darawcheh R, Hijazi F, Al-Ghazzi R, Barazangi M (2003) Evidence for 830 years of seismic quiescence from palaeoseismology, archaeoseismology and historical seismicity along the Dead Sea fault in Syria. Earth Planet Sci Lett 210:35–52. doi:10.1016/S0012-821X(03)00144-4
Meyers EM, Kraabel AT, Strange JF (1976) Ancient synagogue excavations at Khirbet Shema', Upper Galilee, Israel, 1970–1972, Ann Amer Sch Orient Res XLII. Duke University Press, Durham
Migowski C, Agnon A, Bookman R, Negendank JFW, Stein M (2004) Recurrence pattern of Holocene earthquakes along the Dead Sea transform revealed by varve-counting and radiocarbon dating of lacustrine sediments. Earth Planet Sci Lett 222:301–314
Monecke K, Anselmetti FS, Becker A, Schnellmann M, Sturm M, Giardini D (2006) Earthquake-induced deformation structures in lake deposits: a late Pleistocene to holocene paleoseismic record for central Switzerland. Ecol Geol Helv 99:343–362. doi:10.1007/s00015-006-1193-x
Moretti M, Alfaro P, Caselles O et al (1999) Modelling seismites with a digital shaking table. Tectonophysics 304:369–383
Moretti M, Sabato L (2007) Recognition of trigger mechanisms for soft-sediment deformation in the Pleistocene lacustrine deposits of the Sant'Arcangelo Basin (Southern Italy): seismic shock vs. overloading. Sediment Geol 196:31–45. doi:10.1016/j.sedgeo.2006.05.012
Morhange C, Pirazzoli PA, Marriner N, Montagioni LF, Nanmour T (2006) Late Holocene relative sea-level changes in Lebanon, East- ern Mediterranean. Mar Geol 230:99–114. doi:10.1016/j.margeo.2006.04.003
Nemer T, Gomez F, Al Haddad S, Tabet C (2008) Coseismic growth of sedimentary basins along the Yammouneh strike-slip fault (Lebanon). Geophys J Int 175:1023–1039
Neumann FHS, Kagan EJS, Stein MPI, Agnon API (2009) Assessment of the effect of earthquake activity on regional vegetation – High-resolution pollen study of the Ein Feshka section, Holocene Dead Sea. Rev Palaeobot Palynol 155:42–51
Nur A (2008) Apocalypse: earthquakes, archaeology, and the wrath of God. Princeton University Press, Princeton. ISBN 9780691016023
Nur A, Cline EH (2000) Poseidon's horses: plate tectonics and earthquake storms in the LateBronze Age Aegean and Eastern Mediterranean. J Archaeol Sci 27:43–63
Oth A, Wenzel F, Wust-Bloch H, Gotschammer E, Ben-Avraham Z (2007) Parameterization of a composite attenuation relation for the Dead Sea area based on 3-D modeling of elastic wave propagation. Pure Appl Geophys 164:23–37. doi:10.1007/s00024-006-0147-6
Owen G, Moretti M, Alfaro P (2011) Recognising triggers for soft-sediment deformation: current understanding and future directions. Sediment Geol 235:133–140
Pettijohn FJ, Potter PE (1964) Atlas and glossary of primary sedimentary structures. Springer-Verlag, Berlin, Science – 370pp
Porat N, Levi T, Weinberger R (2007) Possible resetting of quartz OSL signals during earthquakes-evidence from late Pleistocene injection dikes, Dead Sea basin. Isr Quat Geochronol 2:272–277. doi:10.1016/j.quageo.2006.05.021
Porat N, Duller GAT, Roberts HM, Piasetzky E, Finkelstein I (2012) OSL dating in multi-strata Tel: Megiddo (Israel) as a case study. Quat Geochronol 10:359–366. doi:10.1016/j.quageo.2011.11.011

Prasad S, Negendank JFW, Stein M (2009) Varve counting reveals high resolution radiocarbon reservoir age variations in palaeolake Lisan. J Quat Sci 24:690–696

Reinhardt EG, Goodman BN, Boyce JI, Lopez G, van Hengstum P, Rink WJ, Mart Y, Raban A (2006) The tsunami of 13 December A.D. 115 and the destruction of Herod the Great's harbor at Caesarea Maritima, Israel. Geology 34:1061–1064. doi:10.1130/G22780A.1

Roep TB, Everts AJ (1991) Pillow-beds – A new type of seismites – An example from an Oligocene turbidatefancomplex,Alicante,Spain.Sedimentology39:711–724.doi:10.1111/j.1365-3091.1992.tb02148.x

Ron H, Nowaczyk NR, Frank U, Schwab MJ, Naumann R, Striewski B, Agnon A (2007) Greigite detected as dominating remanence carrier in Late Pleistocene sediments, Lisan formation, from Lake Kinneret (Sea of Galilee), Israel. Geophys J Int 170:117–131. doi:10.1111/j.1365-246X.2007.03425.x

Rucker JD, Niemi TM (2010) Historical earthquake catalogues and archaeological data: achieving synthesis without circular reasoning. In: Sintubin M, Stewart IS, Niemi TM, Altunel E (eds) Ancient earthquakes, Geological Society of America special paper 471. Geological Society of America, Boulder, pp 97–106

Russell K (1980) The earthquake of May 19, A.D. 363. Bull Am Sch Orient Res 238:47–63

Salamon A (2010) Patterns of seismic sequences in the Levant – Interpretation of historical seismicity. J Seismol 14:339–367. doi:10.1007/s10950-009-9168-9

Salamon A, Rockwell T, Guidoboni E, Comastri E (2011) A critical evaluation of tsunami records reported for the Levant Coast from the second millennium BCE to the present. Isr J Earth Sci 58:327–354. doi:10.1560/IJES.58.2-3.327

Salamon A, Rockwell T, Ward SN, Guidoboni E, Comastri A (2007) Tsunami hazard evaluation of the eastern Mediterranean: historical analysis and selected modeling. Bull Seismol Soc Am 97:705–724

Sbeinati MR, Darawcheh R, Mouty M (2005) The historical earthquakes of Syria: an analysis of large and moderate earthquakes from 1365 B.C. to 1900 A.D. Ann Geophys 48:347–435

Sbeinati MR, Meghraoui M, Suleyman G, Gomez F, Grootes P, Nadeau MJ, Al Najjar H, Al Ghazzi R (2010) Timing of earthquake ruptures at the Al Harif Roman aqueduct (Dead Sea fault, Syria) from archaeoseismology and paleoseismology. In: Sintubin M, Stewart IS, Niemi TM, Altunel E (eds) Ancient earthquakes, Geological Society of America special paper 471. Geological Society of America, Boulder, pp 244–267

Schaeffer CFA (1948) Stratigraphic Compareé et Chronologie de l'Asie Occidental. Oxford University Press, New York, 653 pp

Seilacher A (1969) Fault-graded beds interpreted as seismites. Sedimentology 13:155–159

Seilacher A (1984) Sedimentary structures tentatively attributed to seismic events. Mar Geol 55:1–12

Segal A (2007) The Churches of Sussita, Interim Report at the End of Seven Excavation Seasons (2000–2006). http://hippos.haifa.ac.il/index.php/8-general/36-churches-hippos

Shaked Y, Agnon A, Lazar B, Marco S, Avner U, Stein M (2004) Large earthquakes kill coral reefs at the north-west Gulf of Aqaba. Terra Nova 16:133–138

Shaked Y, Lazar B, Marco S, Stein M, Agnon A (2011) Late Holocene events that shaped the shoreline at the northern Gulf of Aqaba as recorded by a buried reef. Isr J Earth Sci 58. Spec. vol, eds. Agnon A, Amit R, Michetti A, Hough S, The Dead Sea rift as a natural laboratory for neotectonics and paleoseismology (in press)

Shamir G, Baer G, Hofstetter A (2003) Three-dimensional elastic earthquake modelling based on integrated seismological and InSAR data: the M-w=7.2 Nuweiba earthquake, gulf of Elat/Aqaba 1995 November. Geophys J Int 154:731–744. doi:10.1046/j.1365-246X.2003.01978.x

Shaw SW (1947) Southern Palestine: geological map on a scale of 1:250,000 with explanatory notes. Govt. Printer, Jerusalem, 42 pp

Siegenthaler C, Finger W, Kelts K, Wang S (1987) Earthquake and seiche deposits in Lake Lucerne, Switzerland. Eclogae Geol Helv 80:241–260

Sieh K (1978) Prehistoric large earthquakes produced by slip on the San Andreas fault at Pallett Creek, California. J Geophys Res 83:3907–3939. doi:10.1029/JB083iB08p03907

Sims JD (1973) Earthquake-induced structures in sediments of sediments of Van Norman Lake, San Fernando, California. Science 182:161–163

Sims JD (1975) Determining earthquake recurrence intervals from deformational structures in young lacustrine sediments. Tectonophysics 29:153–159

Slater L, Niemi TM (2003) Detection of active faults along the Dead Sea Transform using ground penetrating radar and implications for seismic hazards within the city of Aqaba, Jordan. Tectonophysics 368:33–50

Stein M (2011) Paleo-earthquakes chronometry in the late Quaternary Dead Sea basin. Isr J Earth Sci 58:237–255. doi:10.1560/IJES.58.3-4.237. Spec. vol, eds. Agnon A, Amit R, Michetti A, Hough S, The Dead Sea rift as a natural laboratory for neotectonics and paleoseismology

Stiros SC (2001) The AD 365 Crete earthquake and possible seismic clustering during the fourth to sixth centuries AD in the Eastern Mediterranean: a review of historical and archaeological data. J Struct Geol 23:545–562. doi:10.1016/S0191-8141(00)00118-8

Swan FH, Schwartz DP, Cluff LS (1980) Recurrence of moderate to large magnitude earthquakes produced by surface faulting on the Wasatch fault zone, Utah. Bull Seismol Soc Am 70:1431–1462

Thacker WC, Lavelle JW (1977) Two-phase flow analysis of hindered settling. Phys Fluids 20:1577–1579

Thomas R, Parker ST, Niemi TM (2007) Structural damage from earthquakes in the second-ninth centuries at the archaeological site of Aila in Aqaba, Jordan. Bull Am Sch Orient Res 346:59–77

Tibor G, Niemi TM, Ben-Avraham Z, Al-Zoubi A, Sade RA, Hall JK, Hartman G, Akawi E, Abueladas A, Al-Ruzouq R (2010) Active tectonic morphology and submarine deformation of the northern Gulf of Eilat/Aqaba from analyses of multibeam data. Geomarine Lett 30:561–573. doi:10.1007/s00367-010-0194-y

Torfstein A, Gavrieli I, Katz A et al (2008) Gypsum as a monitor of the paleo-limnologicalhydrological conditions in Lake Lisan and the Dead Sea. Geochimica Cosmochimica Acta 72:2491–2509

Tsafrir Y, Foerster G (1997) Urbanism at Scythopolis-Bet Shean in the fourth to seventh centuries. Dumbart Oaks Pap 51:85–146

Waldmann N, Anselmetti FS, Ariztegui D, Austin JA, Pirouz M, Moy CM, Dunbar R (2011) Holocene mass-wasting events in Lago Fagnano, Tierra del Fuego (54 degrees S): implications for paleoseismicity of the Magallanes-Fagnano transform fault. Basin Res 23:171–190. doi:10.1111/j.1365-2117.2010.00489.x

Wechsler N, Katz O, Dray Y, Gonen I, Marco S (2009) Estimating location and size of historical earthquake by combining archaeology and geology in Um-El-Qanatir, Dead Sea Transform. Nat Hazards 50:27–40. doi:10.1007/s11069-008-9315-6s

Wells DL, Coppersmith KJ (1994) New empirical relationships among magnitude, rupture length, rupture width, rupture area, and surface displacement. Bull Seismol Soc Am 84:974–1002

Wetzler N, Marco S, Heifetz E (2010) Quantitative analysis of seismogenic shear-induced turbulence in lake sediments. Geology 38:303–306. doi:10.1130/G30685.1

Yagoda-Biran G, Hatzor HY, Amit R, Katz O (2010) Constraining regional paleo peak ground acceleration from back analysis of prehistoric landslides: example from Sea of Galilee, Dead Sea transform. Tectonophysics 490:81–89

Yeats RS, Sieh K, Allen CR (1997) The geology of earthquakes. Oxford University Press, Oxford, 576 p

Yechieli Y (1993) The effects of water level changes in closed lakes (Dead Sea) on the surrounding groundwater and country rocks. PhD thesis, Weizmann Institute, Rehovot

Zak I, Freund R (1966) Recent strike-slip movements along the Dead Sea rift. Isr J Earth Sci 15:33–37

Zilberman E, Amit R, Porat N, Enzel Y, Avner U (2005) Surface ruptures induced by the devastating 1068 AD earthquake in the southern Arava valley, Dead Sea Rift, Israel. Tectonophysics 408:79–99

Chapter 9
Instrumental Data on the Seismic Activity Along the Dead Sea Transform

Abraham Hofstetter, Catherine Dorbath, and Louis Dorbath

Abstract We analyzed the catalog of instrumental recordings of seismic activity from 1900 to 2010 along the Dead Sea Transform. The seismicity pattern reveals significant activity confined to 5 main sections of the transform. In all the sections of the transform there is a significant amount of seismic activity at depths of 9–10 km (lower part of the upper crust). The seismic activity extends to large depths of 20 km and more, where about 30 % of the seismic activity occurs in the lower crust, especially in the Dead Sea basin and the Arava Valley. The deep seismicity is correlative with previous low heat flow measurements along the transform, and thus suggesting a relatively cold crust. We analyzed more than 4,300 S-wave spectra of earthquakes in the magnitude range is $0.8 \leq M_d \leq 6.2$, with M_0 values ranging from $3.1x10^{11}$ N·m to $5.4x10^{18}$ N·m, and Brune stress drop estimates, $\Delta\sigma$, between 0.1 MPa and 15 MPa. The total seismic moment release in the years 1900–2010 in the Dead Sea Transform due to all the earthquakes, including the earthquake in 1927, is only a fraction of the expected seismic moment release.

A. Hofstetter (✉)
Geophysical Institute of Israel, Lod 71100, Israel

Ecole et Observatoire des Sciences de la Terre,
University of Strasbourg, Strasbourg 67000, France
e-mail: ramih@gii.co.il

C. Dorbath
Ecole et Observatoire des Sciences de la Terre, University of Strasbourg,
Esplanade, Strasbourg 67000, France

IRD, UMR 154, 14 Avenue Edouard Belin, Toulouse 31400, France
e-mail: c.dorbath@unistra.fr

L. Dorbath
Ecole et Observatoire des Sciences de la Terre, University of Strasbourg,
Strasbourg 67000, France
e-mail: louis.dorbath@eost.u-strasbg.fr

Z. Garfunkel et al. (eds.), *Dead Sea Transform Fault System: Reviews*,
Modern Approaches in Solid Earth Sciences 6, DOI 10.1007/978-94-017-8872-4_9,

Keywords Dead Sea Transform • Earthquakes • Catalog • Instrumental data • Seismic networks

9.1 Introduction

The Dead Sea Transform (DST), which accommodates the left-lateral motion between the Sinai Subplate and the Arabian plate, is about 1,200 km long, connecting the Taurus-Zagros compressional front, in the north, to the extensional zone of the Red Sea, in the south. Over the past few million years tectonics has shaped the Dead Sea Transfrom (e.g., Freund et al. 1970; Ben-Menahem et al. 1976; Ben-Avraham et al. 2008). The Dead Sea Transform comprises a few major basins being connected by several large faults, where the Dead Sea basin is the largest one.

Based on historical accounts and instrumental records, some describing harmful or devastating earthquakes, it is clear that the Dead Sea Transform poses a major seismic threat to the population on both its sides. Figure 9.1 illustrates the locations of all instrumentally recorded earthquakes occurring between 1900 and 2010 along the Dead Sea Transform, or in the commencement of its branching-out faults or nearby faultings, a total of about 3,800 earthquakes. The seismic activity is somewhat spread in the transform with no clear lineation and thus we include seismic activity up to about 10 km on both sides of the transform. We also inserted improved locations of epicenters in the Dead Sea basin (after Hofstetter et al. 2012), based on the tomoDD algorithm of Zhang and Thurber (2003). One notable earthquake is the July 11, 1927 earthquake, M 6.2, which occurred in the northern part of the Dead Sea basin resulting in 285 deaths, 940 wounded, and extensive damage in many towns and villages on both sides of the Dead Sea Transform (Ben-Menahem et al. 1976; Shapira et al. 1992; Avni 1998). Throughout the twentieth century, several other widely felt earthquakes occurred in 1903, 1928, 1956, 1970, 1979 and 2004, with magnitudes M ~5.0–5.5, causing no or minor damage (e.g. Arieh et al. 1982; Amiran et al. 1994). However, five out of seven relatively moderate-strong events occurred in the Dead Sea basin and two occurred just north of the basin.

In the past decades a series of intensive seismological studies improved our understanding of the seismotectonic activity of the Dead Sea Transform, i.e., Ben-Menaham et al. (1976) and Salamon et al. (1996, 2003) studied the seismicity along the Sinai sub-plate. Arie et al. (1982), Shapira and Feldman (1987), Van Eck and Hofstetter (1989, 1990), Shapira (1997) and Hofstetter et al. (2007) studied the seismic activity in the Dead Sea basin and along the Dead Sea Transform exhibiting complex focal mechanisms. Rotstein et al. (1991) and Shamir et al. (2006) studied the structure of the northern Dead Sea basin and the Jericho valley based on a series of earthquakes including the mainshock in 2004. Salamon (2005), Al-Tarazi et al. (2006) and Hofstetter et al. (2008) described the effects of the mainshock in 2004 and their tectonic implications. Below we focus our attention on instrumental data obtained from the seismic activity along the Dead Sea Transform, which is listed in the catalog from 1900 to 2010 (http://www.gii.co.il/home.html), where the latitude range is from 29.5°N (northern tip of the Gulf of Aqaba) to 33.5°N (southern

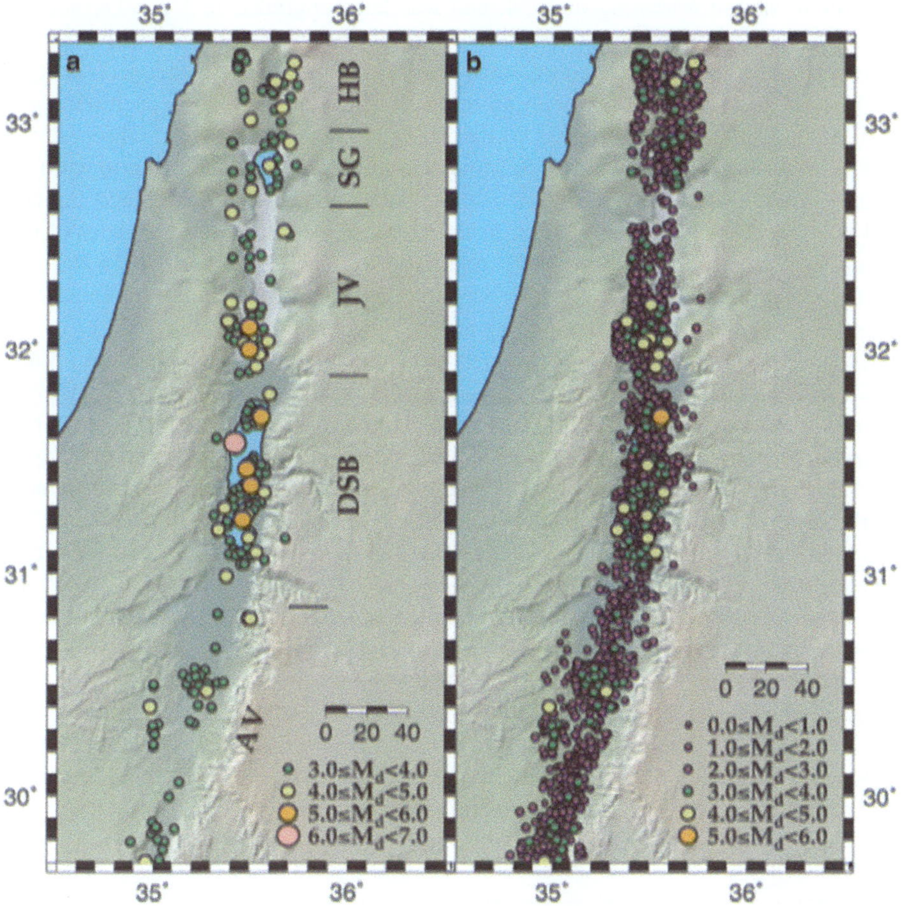

Fig. 9.1 Earthquakes that occurred along the Dead Sea Transform in the time periods: (**a**) 1900–2010, $M_d \geq 3$; (**b**) 1983–2010, all magnitudes. The Dead Sea Transform is divided to five main sections: *AV* Arava Valley, *DSB* Dead Sea basin, *JV* Jordan valley, *SG* Sea of Galilee, *HB* Hula Basin

Lebanon), as was recorded by the short period and broad band stations of Jordan and Israel, and using the coda magnitude (Shapira 1988).

9.2 Data and Acquisition Systems

The seismic monitoring in the Middle East started in the end of the nineteenth century with the installation of station Helwan (HLW) in Cairo, Egypt. From 1898 until 1912 Helwan was the only seismological station that operated in the region, along with some other more remote operating stations like Athens and Istanbul. Another important station that started operating in 1912 was Ksara, which is located in central-east part of Lebanon. Station JER of the WWSSN, located in the Hebrew University of

Jerusalem, started operating in 1954, and was later upgraded in 1963. During the middle of the twentieth century temporary stations operated for a short period of time in various part of Israel enriching somewhat the seismological catalog. Two relatively large seismic networks, with up to 35 short period stations in each network, have been installed in 1983, one in Jordan (JSO, Jordan Seismological Observatory) and the other in Israel (ISN, Israel Seismic Network), reaching stabilization within a year. With time those networks have been upgraded and several broad band stations and many accelerometers have been installed. Since that time the seismicity of the Dead Sea Transform is continuously monitored by the ISN and JSO (Fig. 9.2). In 2000 the catalogs of both networks were merged to create a common Dead Sea Transform catalog, and it is an ongoing process since then.

In addition to the above mentioned permanent networks, a few temporary networks, each time for about 1–2 years, were operating in the Dead Sea Transform or near its outskirts significantly enriching the database. The temporary network added important information, however, still the main backbone of the catalog is based on the permanent networks. Those campaigns include: (1) a portable network comprising of 6 stations that operated in the Dead Sea basin in 1983–1984 and later north of Eilat in 1984–1985; (2) a 12-station-network around the Dead Sea basin in 1986–1988; (3) a large network including up to 70 stations in Jordan and Israel south of the Dead Sea Lake, known as DESERT2000 (i.e., Weber et al. 2004), and crossing the Dead Sea Transform in the Arava Valley in a WNW-ESE line operating from April 2000 to May 2001; (4) a more condensed network in the Dead Sea basin and its immediate outskirts comprising up to 70 stations and operating from Oct. 2006 to May 2008, known as DESIRE (i.e., Weber et al. 2009a, b; Braeuer et al. 2012a, b). The seismicity presented here is based on the compilation of data, collected by all the above-mentioned networks.

9.3 Distribution of Seismicity in Space and Time

Figure 9.3a, b presents the distribution of earthquakes along the Dead Sea Transform as a function of depth and time and also the accumulated number of earthquakes as a function of depth. We avoid all depth determinations prior to 1981, which were usually obtained using a small number of stations or seismic phases. Epicentral location errors are about ±2–3 km for most of the events and about ±3–5 km for the depth. Location errors are significantly smaller in the case of events that were reported by DESIRE, or once we apply local earthquake tomography (Hofstetter et al. 2012; Braeuer et al. 2012a, b) for the events in the Dead Sea basin and its adjacent regions. The number of earthquakes in this time period is relatively small (see Fig. 9.3c) so it does not cause any major bias of the observations. We mark the limits of the five main sections of the DST: Arava (AV), Dead Sea basin (DSB), Jordan Valley (JV) north of the Dead Sea Basin and south of the Sea of Galilee Basin (SG) and Hula Basin (HB). The horizontal dashed lines (Fig. 9.3b) mark the inner schematic division of the crust into the upper and lower parts and the uppermost part of the mantle, essentially an average of several studies that were conducted along or across the Dead Sea Transform (Ginzburg et al. 1979a, b; El-Isa et al. 1987; ten

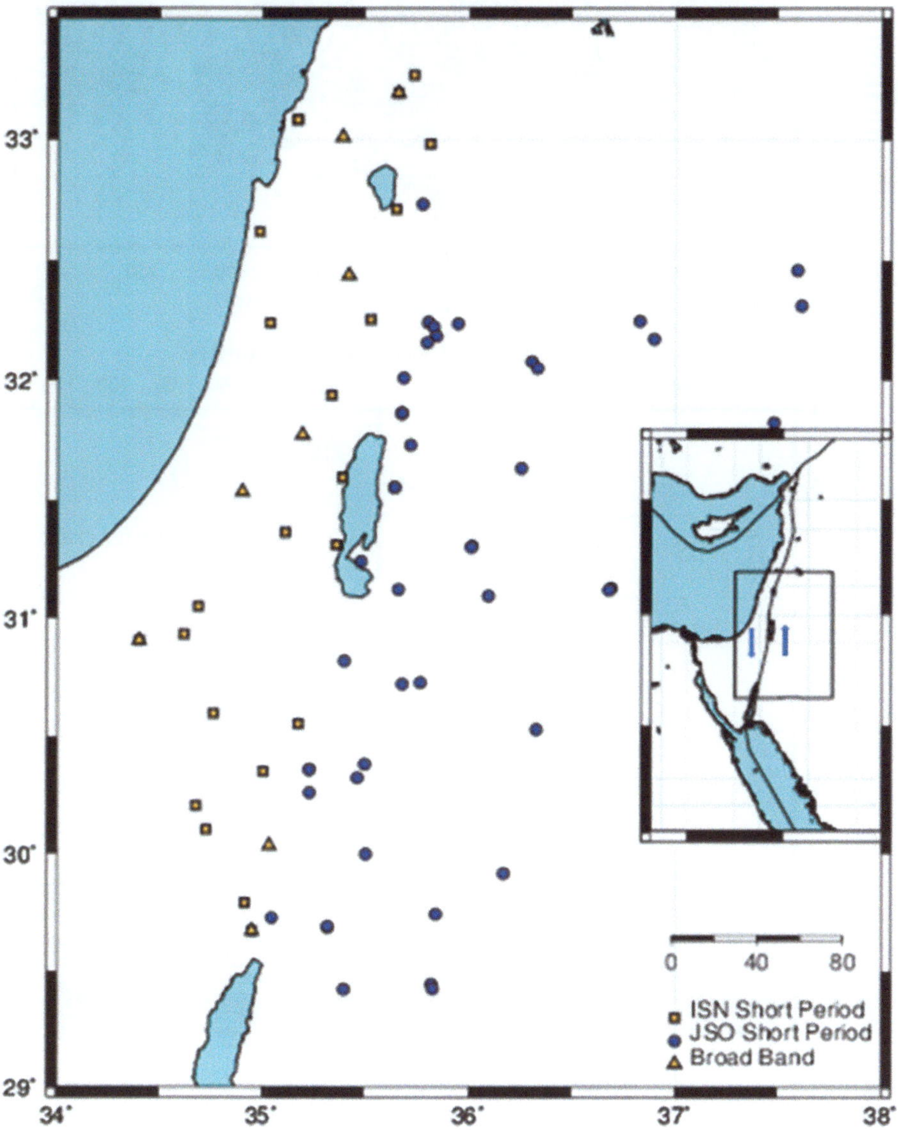

Fig. 9.2 Location of short-period seismic stations (*square*) and broadband stations (*triangle*) that were used for the location of earthquakes along the Dead Sea Transform. The proximity of some stations is due to the fact that the sites of some stations were changed due to technical issues during this period of time. The inset represents a schematic Middle East tectonics where the rectangle marks the study area of the Dead Sea Transform, and the blue arrows present the principal plate movement on both sides of the transform

Brink et al. 1993, 2006; Gitterman et al. 2006; Mechie et al. 2009). Most of the seismic activity, in the Sea of Galilee and Jordan Valley, occurs in the upper crust with two peaks at depths of 2–3 km and about 9–10 km (Fig. 9.3b). The activity in the HB is relatively constant at all depths. Most of the seismic activity in the Dead

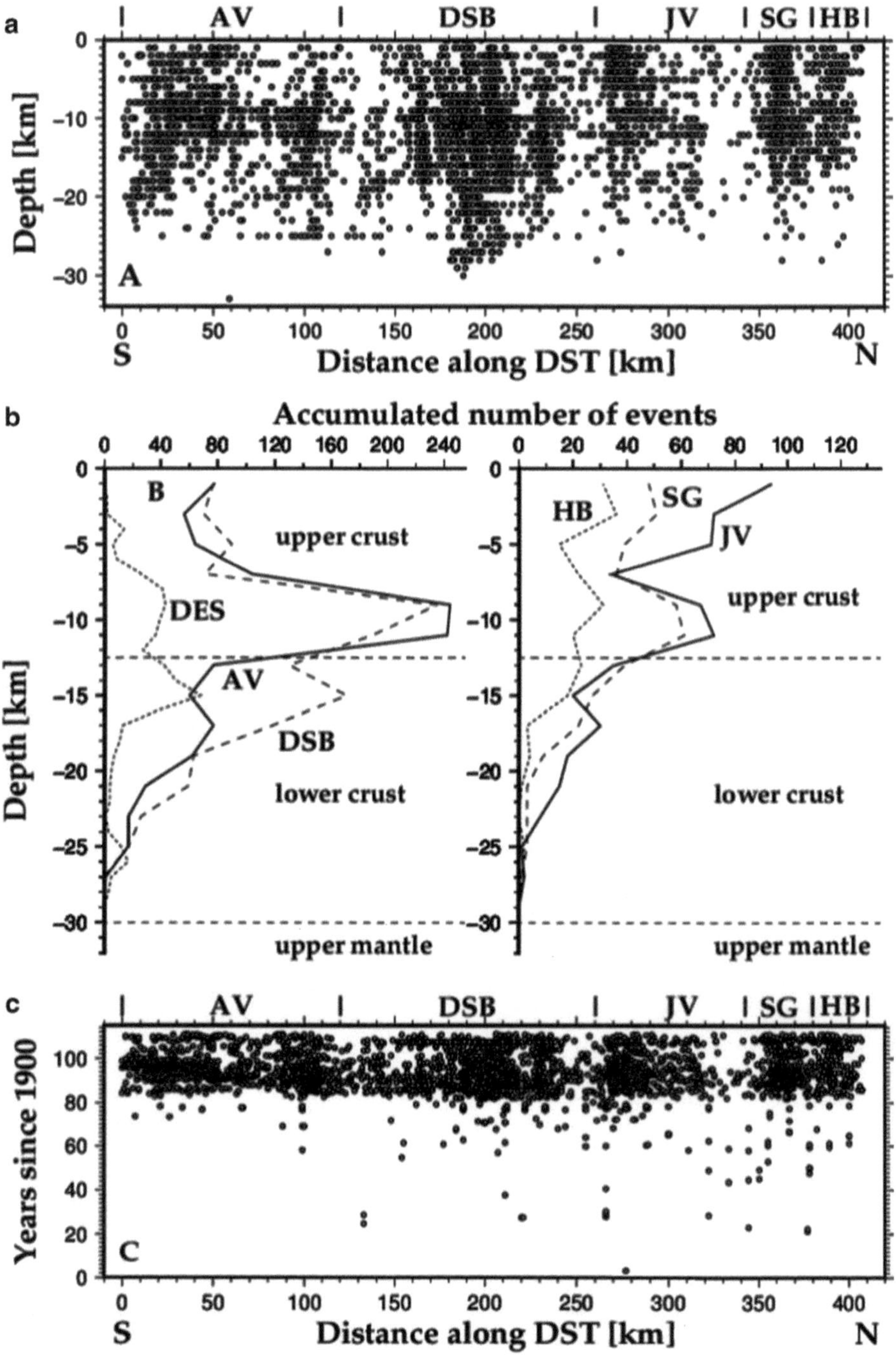

Fig. 9.3 (**a**) Distribution of earthquakes as a function of depth along the Dead Sea Transform based on recordings of permanent and temporary stations, where *AV* Arava Valley, *DSB* Dead Sea basin, *JV* Jordan valley, *SG* Sea of Galilee, *HB* Hula Basin (*lower part*); (**b**) Accumulated number of events as a function of depth in the crust and upper part of the upper mantle (*upper part*) in the five sections of the DST, where AV (solid line) DSB (*dashed line*) on the left side and JV (*solid line*), SG (*dashed line*) and HB (*dotted line*) are on the right side. The *horizontal dashed lines* present

Sea basin and Arava fault is deep with a pronounced peak at about 9–10 km and a secondary peak, in the case of DSB, at a depth of 15–16 km. The seismic activity at shallow depths in DSB and AV is rather minimal as compared to the deeper ones.

In all the sections along the transform there is significant amount of seismic activity in the lower crust, with a clear peak in the DSB at about 15–16 km. Seismic activity extends nonuniformly down to depths of 20 km and more in all sections of the Dead Sea Transform, mainly in AV, DSB, JV and SG. In total, in the Dead Sea Transform 1,220 out of 3,800 earthquakes (about 30 %) have deep hypocenters in the lower crust. Several heat flow studies reported on low values along the DST except for the Sea of Galilee basin (Ben-Avraham et al. 1978; Feinstein 1987). Recently, Förester et al. (2007) reported low values near the SG and AV and higher values in the eastern margin of the transform. The occurrences of hypocenters in the lower crust, as were also reported by Aldersons et al. (2003) and Hofstetter et al. (2012), requires a relatively thick brittle part of the crust. It can be are in a good agreement with the observations of the low heat flow measurements, suggesting relatively cold crust. Förester et al. (2010) and Petrunin et al. (2012) argued that the heat flow should be 50–60 mW m^{-2}, and in the Dead Sea basin the brittle layer is about 20–22 km and up to 27 km locally. The maximal depth of earthquakes throughout the Dead Sea Transform is about 25 km and even deeper in some parts, i.e. the Dead Sea basin (Fig. 9.3). The DESIRE campaign enriched the database of the Dead Sea basin, however, one can expect that comparable results could have been obtained in other parts of the Dead Sea Transform had similar campaigns been operating there.

Most of the seismic activity in the DST occurs in the middle of the sections while in the borders between two adjacent sections it is rather limited. This fact agrees with observations of Hofstetter et al. (2012) regarding the seismic activity at the northern and southern ends of the Dead Sea Basin. Out of seven relatively moderate to strong earthquakes ($M_d \geq 5.0$) that occurred along the DST only two are located in the section of the Jordan Valley, just north of the Dead Sea basin, in 1903 and 1928, and all the other earthquakes in 1927, 1956, 1970, 1979 and 2004 are located in the Dead Sea basin. Furthermore, the epicenters of all those earthquakes are located within the basin and relatively far from the southern or northern basin ends, i.e. the strong earthquake of 1927 with M 6.2.

The distribution of earthquakes along the DST as a function of time is presented in Fig. 9.3c. It is clear that the rate of occurrence from the early 1980s of last century is relatively uniform throughout the whole DST, with some temporal higher rate of activity from time to time. If we use only the observations of the permanent seismic networks of Jordan and Israel then for $M_d \geq 2.0$ the rate of "earthquake productivity" is rather uniform of about 7–9 earthquakes/km/year in the last 30 years in the five sections of the Dead Sea Transform. The fact that until the 1980s the number of reported earthquakes is rather small, can be explained by the sparse operating regional stations in the whole region, which hampers the possibility to identify any significant change in

Fig. 9.3 (continued) schematic inner division of the crust and upper mantle (After Ginzburg et al. 1979a, b; El-Isa et al. 1987; ten Brink et al. 1993, 2006; Gitterman et al. 2006; Mechie et al. 2009); (**c**) Distribution of earthquakes as a function of time along the Dead Sea Transform. The DESIRE recordings, marked as DES, are mainly in the Dead Sea basin

the seismic activity at any section of the DST. Furthermore, in those years the Arava Valley appears to be depleted from any seismic activity, as there were no operating local stations in its vicinity recording the microearthquake activity. Any strong earthquake that might occur in the Arava Valley would have been recorded by some distant regional stations, which is the case for several moderate to strong earthquakes that occurred in the other sections and were recorded by the sparse regional network in the Middle East.

The DESIRE campaign focused mainly on the southern Dead Sea basin, providing very detailed measurements in this region (Weber et al. 2009a, b; Braeuer et al. Braeuer et al. 2012a, b) during 18 months. The DESIRE recordings were added in Fig. 9.3. Most of the events were of small magnitudes and were not recorded by the permanent networks of either Jordan or Israel. In this study, as observed also by the permanent networks, there is the clear indication of seismic activity essentially at all depths down to about 30 km. The activity as a function of the depth as resulted from the 3-D inversion using the SIMULPS algorithm, done by Braeuer et al. (2012b; Fig. 9.3b therein), are presented also in Fig. 9.3b and marked as DES. There are two clear activity peaks at 9, 15 km and a small peak at 26 km. The shape of the event-depth distribution is similar to the equivalent distribution of DSB. We note that the second peak in the event-depth distribution of DES at 15 km is the largest one as compared to the peak at 9 km (by about 50 %), while in the DSB event-depth distribution we get the opposite order. However, we suppose that a longer period of recording time is needed in order to reach a more conclusive determination of the pattern of the seismic activity.

Catalog completeness was studied by Shapira and Hofstetter (2002; see also http://www.gii.co.il) for the period 1900–2000, taking into account the small number of local and regional seismic stations in the first half of the twentieth century anywhere in the Middle East, and the increasing number of seismic stations in the second half in Israel, Jordan and neighbouring countries. Through this period of time the detectability of seismic events has been improved and catalog completeness is considered in the following time steps: $M_d \geq 5.0$ from 1900 to 1939, $M_d \geq 4.0$ from 1940 to 1962, $M_d \geq 3.0$ from 1963 to 1982, and $M_d \geq 2.0$ from 1983 to recent times. The detectability of earthquakes in the Dead Sea Transform is rather uniform as a consequence of the operation of both seismological networks in Israel and Jordan since 1983, and the fact that there is a large number of operating stations on both sides of the transform.

Shapira and Hofstetter (2002) determined the *b*-value using a maximum likelihood method (Aki 1965; Wiechert 1980) for the time period 1900 through 2000, a total of 100 years, to be 0.96. There was no significant change in the seismic activity in the last 11 years and our results are very similar to those of Shapira and Hofstetter (2002). Overall there is a good agreement with the *b*-values obtained in former studies, i.e., by Salamon et al. (1996), $b = 1.00$ for the whole Dead Sea Transform (from southernmost part of Sinai to Eastern Turkey) for a 95 year period. The *b*-value is within the error bounds (about 0.05 to 0.1 units) of recent studies of Arieh (1967), $b = 0.8$ for a 60 year period, Ben-Menahem and Aboodi (1971), $b = 0.86$ for a 2,500 year period, and Shapira and Feldman (1987), $b = 0.8$ for a 10 month period, although different magnitude scales and regions were used in these studies.

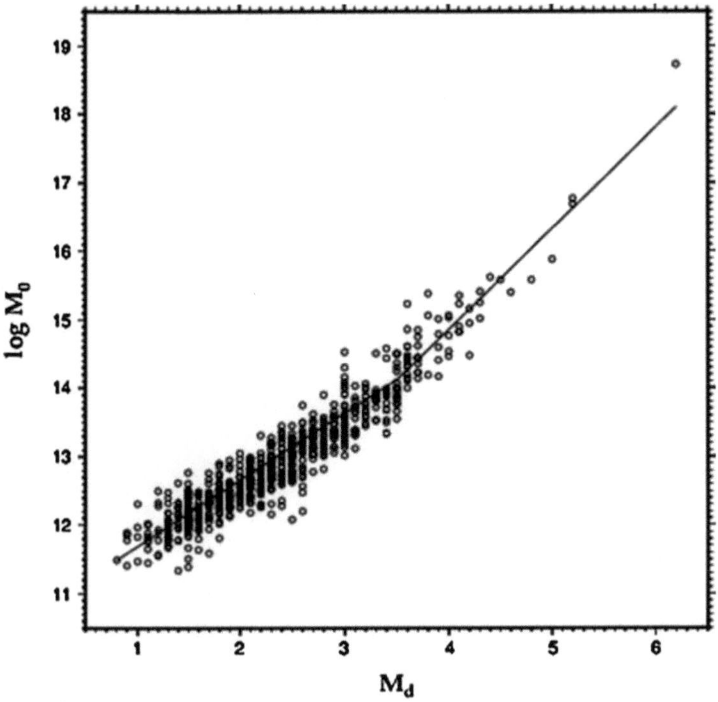

Fig. 9.4 The relation between the seismic moment M_0 and the magnitude M_d, including seismic moment of the earthquake of 1927 (After Ben-Menahem et al. 1976)

9.4 Seismic Moment and Stress Drop Estimation

We calculate the seismic moment, M_0, corner frequency, f_0, stress drop, $\Delta\sigma$, and source radius, r_0 based on the dislocation model of Brune (1970, 1971), using the spectra of S-waves recorded by the short period stations (three-components or vertical component) or the broadband stations in the case of relatively strong earthquakes. Here we describe briefly its application to the Israel Seismic Network following Shapira and Hofstetter (1993).

We manually analyzed more than 4,300 S-wave spectra, with M_0 values ranging from $3.1x10^{11}$ N·m to $5.4x10^{18}$ N·m. We assume the relationship $\log M_0 = aM_d + b$ between the mean seismic moment M_0 and the coda magnitude M_d of a given earthquake. The best fit is simultaneously obtained for two parts of the magnitude range and we get for the first part

$$\log M_0 = (0.9 \pm 0.1)\,M_d + (10.7 \pm 0.4),\ 0.8 \leq M_d < 3.5$$

and for the second part

$$\log M_0 = (1.5 \pm 0.1)\,M_d + (8.8 \pm 0.3),\ 3.5 \leq M_d < 6.2$$

where M_0 is in N·m and the magnitude range is $0.8 \leq M_d \leq 6.2$ (see Fig. 9.4). For the small magnitude earthquakes the magnitude coefficients in the first relation are

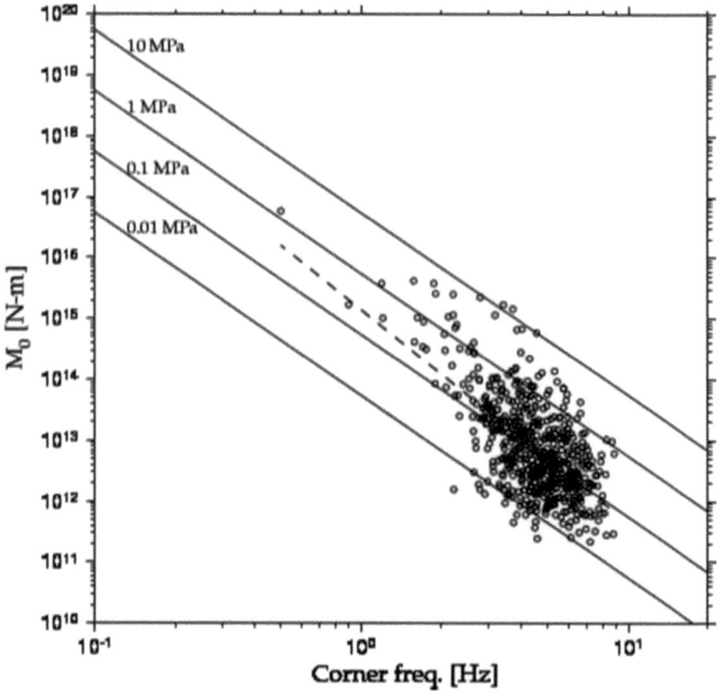

Fig. 9.5 The relation between the corner frequency f_0 and the seismic moment M_0, for the earthquakes in this study, where for reference the solid lines represent stress drop values of 0.01, 0.1, 1 and 10 MPa, based on f^{-3} scaling. The *dashed line* represent a least-squares fit of the observations with a slope of about −3.5

slightly lower than those in similar relations, as obtained by Bakun (1984) and Hanks and Boore (1984) for earthquakes in California, and by van Eck and Hofstetter (1989), Shapira and Hofstetter (1993), Hofstetter et al. (1996), Hofstetter (2003) and Hofstetter et al. (2008) for the Dead Sea Transform, the Carmel fault, the Eastern Mediterranean region, or the Gulf of Aqaba, with relatively similar magnitude range. The second relation is similar to the relation of Hofstetter and Shapira (2000) for earthquakes in the Eastern Mediterranean region.

The corner frequency decreases with increasing seismic moment, as can be seen in Fig. 9.5, where for reference the solid lines represent stress drop values of 1, 10 and 100 bars, based on f^{-3} scaling. We estimate the stress drop of the earthquakes $\Delta\sigma$, based on the dislocation model of Brune (1970, 1971), using the equation $\Delta\sigma = 8.47 M_0 f_0^3/\beta_s$ where f_0 is the corner frequency, and β_s is the rupture velocity. Shapira and Hofstetter (1993) provided details of the application of the method in the case of seismograms observed by the ISN (short period or broadband). It is similar to former values obtained for earthquakes occurring in the Dead Sea basin (van Eck and Hofstetter 1989) or other parts of the Dead Sea Transform, the Carmel fault and the Gulf of Aqaba (Shapira and Hofstetter 1993; Hofstetter et al. 1996; Hofstetter 2003). In general, the stress drop clearly increases with the increasing seismic

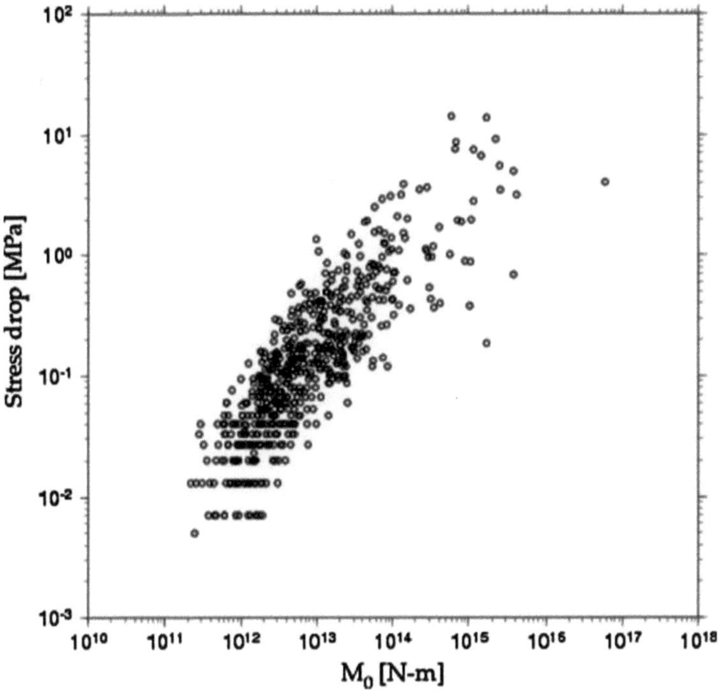

Fig. 9.6 The relation between Brune's stress drop $\Delta\sigma$ and the seismic moment M_0, for the earthquakes in this study

moment (Fig. 9.6), but there is no simple relationship between the stress drop and the seismic moment (or magnitude). This has already been noted for earthquakes in California by Hanks and Thatcher (1972), Thatcher and Hanks (1973), Kanamori and Anderson (1975), and for the Eastern Mediterranean region by Shapira and Hofstetter (1993). Using Brune's model (1970) we estimate the source radius using the relation $r_0 = a(\beta_s/f_0)$ where a has the value 0.37 (Fig. 9.7). Typical source radius values are between 0.2 and 0.6 km for the small magnitude earthquakes.

9.5 Discussion and Conclusions

Figure 9.5 illustrates the relationship between the corner frequency f_0 and the seismic moment M_0 , for the Dead Sea Transform. As can be seen our observations suggest a relationship of $\sim f^{-3.5}$. Van Eck and Hofstetter (1989) reported also a steeper fall than −3. It was interpreted by them as f_{max} for frequencies just below 10 Hz. In this case we observe the tendency already at frequencies of 3–4 Hz, which clearly suggests f_0. Mayeda et al. (2007) presented a similar relationship between the seismic moment and corner frequency for the Hector Mine sequence, California, where the scaling of seismic moment and corner frequency does not follow f^{-3} and is consistent with f^{-4}.

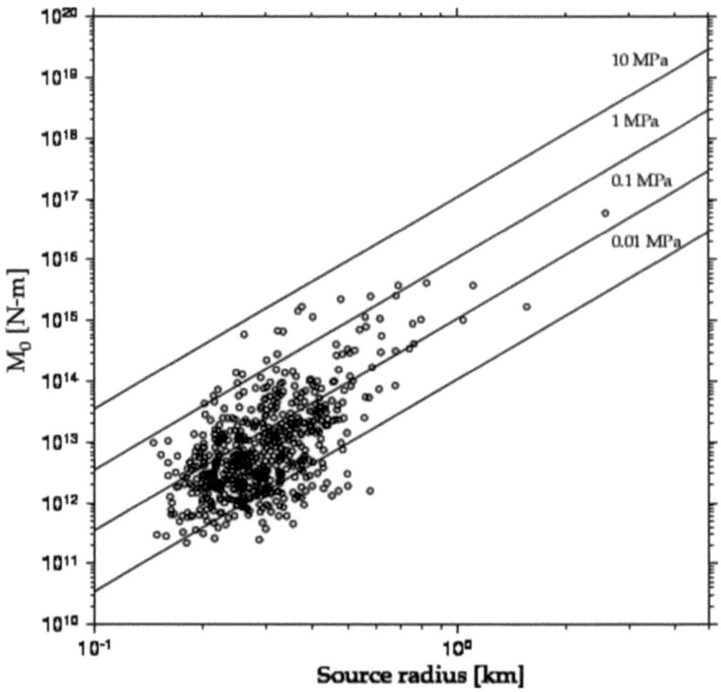

Fig. 9.7 The relation between the area r and the seismic moment M_0, for the earthquakes in this study, where for reference the *solid lines* represent stress drop values of 0.01, 0.1, 1 and 10 MPa, based on f^{-3} scaling

They used this evidence to support the idea of non-self-similarity, which means that a large earthquake is not a simple integer multiplication of a small one.

When studying the seismic efficiency of the Dead Sea Transform one should take into consideration the total length of the transform (Garfunkel et al. 1981), which is well beyond the section of the transform from the Gulf of Aqaba to Lebanon, shown in this study. The DST is essentially depleted from strong earthquakes except the one in 1927 with the seismic moment of 5.4×10^{18} N m (Ben-Menahem et al. 1976). The seismic moment contribution of the aftershocks of the 1995 sequence (Hofstetter 2003), the moderate activity in 2004 in the northern part of the Dead Sea basin (Hofstetter et al. 2008) and the recent activity in southern Lebanon in 2008–2010 (Meirova and Hofstetter 2012) is relatively minor. Even if we sum up the calculated seismic moment for all the earthquakes in the catalog, based on the above mentioned Mo-Md relationship, including the moderate earthquakes in the Dead Sea basin in 1956, 1970 and 1979, and 2 earthquakes in the Jordan Valley in 1903 and 1928, we still get a small contribution relative to that of the earthquake in 1927, where the latter contributes about 5 % to the total seismic moment, in good agreement with Garfunkel et al. (1981). Salamon et al. (2003) showed that the contribution of the Gulf of Aqaba earthquake in 1995 to the accumulated seismic moment is less than 40 % of the expected sum. They used a rate of relative motion of 5 mm/year.

Recent studies (Wdowinski et al. 2004; Le Beon et al. 2008; Sadeh et al. 2012) reported somewhat smaller slip rate values from 3.1 to 5.4 in various parts of the Dead Sea Transform, and a probable representative average value is 4.3 mm/year. If we use the latter value of slip rate and taking into account the contribution of the 1927 earthquake then the accumulated seismic moment is about 60 % of the expected sum, which means that the transformal motion along the Arabian-Sinai plate border is of major nature. Yet the large difference between the calculated seismic moment and expected moment due to the displacement along the transform is an open question that should be further explored.

The DESIRE campaign provided very detailed measurements in the southern Dead Sea basin and adjacent regions (Weber et al. 2009a, b; Braeuer et al. 2012a, b), which were taken in a period of 18 months. The accumulated seismic moment of these observations is equivalent to an earthquake of Md 4.2, essentially a small contribution. Similar campaigns, for longer periods of time using a large set of seismic stations, should be conducted in other parts of the Dead Sea Transform to better characterize the seismic activity of each part.

Acknowledgements The study was supported by EOST, Univ. of Strasbourg, Strasbourg, France, Chaire Gutenberg Strasbourg, France, and the Earth Sciences and Research Administration, Ministry of Energy and Water, Israel. Some figures in this report were prepared using the GMT program (Wessel and Smith 1991).

References

Aki K (1965) Maximum likelihood estimate of b in the formula log $N = a - bM$ and its confidence limits. Bull Earthq Res Inst 43:237–239

Aldersons F, Ben-Avraham Z, Hofstetter A, Kissling E, Al-Yazjeen T (2003) Lower-crustal strength under the Dead Sea basin from local earthquake and rheological modeling, Earth Planet. Sci Lett 214:129–142

Al-Tarazi E, Sandvol E, Gomez F (2006) The February 11, 2004 Dead Sea earthquake M_L=5.2 in Jordan and its tectonic implication. Tectonophysics 422:149–158

Amiran D, Arieh E, Turcotte T (1994) Earthquakes in Israel and adjacent areas: macroseismic observations since 100 B.C.E. Isr Explor J 44:76–91

Arieh E (1967) Seismicity of Israel and adjacent areas. Geol Surv Isr Bull 43:1–14

Arieh E, Rotstein Y, Peled U (1982) The Dead Sea earthquake of April 23, 1979. Seismol Soc Am Bull 72:1627–1634

Avni R (1998) The 1927 Jericho Earthquake, comprehensive macroseismic analysis based on contemporary sources. PhD thesis, University of Ben-Gurion, Israel, 211 pp. (in Hebrew with English abstract)

Bakun WH (1984) Seismic moments, local magnitudes, and coda-duration magnitudes for earthquakes in central California. Bull Seismol Soc Am 74:439–458

Ben-Avraham Z, Henel R, Villinger H (1978) Heat flow through the Dead Sea rift. Mar Geol 28:253–269

Ben-Avraham Z, Garfunkel Z, Lazar M (2008) Geology and evolution of the Southern Dead Sea fault with emphasis on subsurface structure. Annu Rev Earth Planet Sci 36:357–387

Ben-Menahem A, Aboodi E (1971) Tectonic patterns in the Red Sea region. J Geophys Res 76:2674–2689

Ben-Menahem A, Nur A, Vered M (1976) Tectonics, seismicity and structure of the Afro-Euroasian junction-the breaking of an incoherent plate. Phys Earth Planet Inter 12:1–50
Braeuer B, Asch G, Hofstetter A, Haberland C, Jaser D, El-Kelani R, Weber M (2012a) Microseimicity distribution in the southern Dead Sea area and its implications on the structure of the basin. Geophys J Int 188:873–878
Braeuer B, Asch G, Hofstetter A, Haberland CH, Jaser D, El-Kelani R, Weber M (2012b) High resolution local earthquake tomography of the southern Dead Sea area. Geophys J Int 191:881–897. doi:10.1111/j.1365-246X.2012.05668.x
Brune J (1970) Tectonic stress and the spectra of seismic shear waves from earthquakes. J Geophys Res 73:4997–5009
Brune J (1971) Correction. J Geophys Res 76:5002
El-Isa Z, Mechie J, Prodehl C, Makris J, Rihm R (1987) A crustal structure study of Jordan derived from seismic refraction data. Tectonophysics 138:235–253
Feinstein S (1987) Constraints on the thermal history of the Dead Sea graben as revealed by coal ranks in deep boreholes. Tectonophysics 141:135–150
Förster A, Förster H-J, Masarweh R, Masri A, Tarawneh K, DESERT Group (2007) The surface heat flow of the Arabian Shield in Jordan. J Asian Earth Sci 30:271–284. doi:10.1016/j.jseaes.2006.09.002
Förster H-J, Förster A, Oberhänsli R, Stromeyer D (2010) Lithospheric composition and thermal structure of the Arabian Shield in Jordan. Tectonophysics 481:29–37
Freund R, Garfunkel Z, Zak I, Goldberg M, Weissbrod T, Derin B (1970) The shear along the Dead Sea Rift. Philos Trans R Soc Lond Ser A 267:107–130
Garfunkel Z, Zak I, Freund R (1981) Active faulting in the Dead Sea rift. Tectonophysics 80:1–26
Ginzburg A, Makris J, Fuchs K, Prodehl C, Kaminski W, Amitai U (1979a) A seismic study of the crust and upper mantle of the Jordan-Dead Sea Rift and their transition toward the Mediterranean Sea. J Geophys Res 84:1569–1582
Ginzburg A, Makris J, Fuchs K, Perathoner B, Prodehl C (1979b) Detailed structure of the crust and upper mantle along the Jordan-Dead Sea Rift. J Geophys Res 84:5605–5612
Gitterman Y, Pinsky V, Amrat A-Q, Jaser D, Mayyas O, Nakanishi K, Hofstetter A (2006) Source features, scaling and location of calibration explosions in Israel and Jordan for CTBT monitoring. Isr J Earth Sci 54:199–217
Hanks TC, Boore DM (1984) Moment-magnitude relations in theory and practice. J Geophys Res 89:6229–6235
Hanks TC, Thatcher W (1972) A graphical representation of seismic source parameters. J Geophys Res 77:4393–4405
Hofstetter A (2003) Seismic observations of the 22/11/1995 Gulf of Aqaba earthquake sequence. Tectonophysics 369:21–36
Hofstetter A, Shapira A (2000) Determination of earthquake energy release in the Eastern Mediterranean region. Geophys J Int 143:1–16
Hofstetter A, van Eck T, Shapira A (1996) Seismic activity along fault branches of the Dead Sea-Jordan transform system: the Carmel-Tirtza fault system. Tectonophysics 267:317–330
Hofstetter A, Klinger Y, Amrat A, Rivera L, Dorbath L (2007) Stress inversion and focal mechanisms along the Levantine Fault from seismological data. Tectonophysics 429:165–181
Hofstetter R, Gitterman Y, Pinsky V, Kraeva N, Feldman L (2008) Seismological observations of the Northern Dead Sea Basin earthquake on 11/2/2004. Isr J Earth Sci 57:101–124
Hofstetter A, Dorbath C, Calò M (2012) Crustal structure of the Dead Sea basin from local earthquake tomography. Geophys J Int 189:554–568
Kanamori H, Anderson D (1975) Theoretical basis of some empirical relations in seismology. Bull Seismol Soc Am 65:1073–1095
Le Beon M, Klinger Y, Amrat A-Q, Agnon A, Dorbath L, Baer G, Ruegg J-C, Charade O, Mayyas O (2008) Slip rate and locking depth from GPS profiles across the southern Dead Sea Transform. J Geophys Res 113. doi:10.1029/2007JB005280

Mayeda K, Malagnini L, Walter W (2007) A new spectral ratio method using narrow band coda envelopes: evidence for non-self-similarity in the Hector Mine sequence. Geophys Res Lett 14:11303–11308

Mechie J, Abu-Ayyash K, Ben-Avraham Z, El-Kelani R, Qabbani I, Weber M, DESIRE Group (2009) Crustal structure of the southern Dead Sea basin derived from project DESIRE wide-angle seismic data. Geophys J Int 178:457–478

Meirova T, Hofstetter A (2012) Observations of seismic activity in Southern Lebanon. J Seism. doi:10.1007/s10950-012-9343-2

Petrunin A, Meneses Rioseco E, Sobolev S, Weber M (2012) Thermomechanical model reconciles contradictory geophysical observations at the Dead Sea Basin. Geochem Geophys Geosyst 13:Q04011. doi:10.1029/2011GC003929

Rotstein Y, Bartov Y, Hofstetter A (1991) Active compressional tectonics in the Jericho area, Dead Sea Rift. Tectonophysics 198:239–259

Sadeh M, Hamiel Y, Ziv A, Bock Y, Fang P, Wdowinski S (2012) Crustal deformation along the Dead Sea Transform and the Carmel Fault inferred from 12 years of GPS measurements. J Geophys Res 117:B08410. doi:10.1029/2012JB009241

Salamon A (2005) Natural seismogenic effects of the 11 February 2004, $M_L = 5.2$, Dead Sea earthquake. Isr J Earth Sci 54:145–169

Salamon A, Hofstetter A, Garfunkel Z, Ron H (1996) Seismicity of Eastern Mediterranean region: perspective of the Sinai subplate. Tectonophysics 263:293–307

Salamon A, Hofstetter A, Garfunkel Z, Ron H (2003) Seismotectonics of the Sinai Subplate – The Eastern Mediterranean Region. Geophys J Int 155:149–173

Shamir G, Eyal Y, Bruner I (2006) Localized versus distributed shear in transform plate boundary zones: the case of the Dead Sea transform in the Jericho Valley. Geochem Geophys Geosyst 6:1–21

Shapira A (1988) Magnitude scales for regional earthquakes monitored in Israel. Isr J Earth Sci 37:17–22

Shapira A (1997) On the seismicity of the Dead Sea basin. In: Niemi TM, Ben-Avraham Z, Gat JR (eds) The Dead Sea; the lake and its settings, Oxford monographs on geology and geophysics, no. 36. Oxford University Press, New York, pp 82–86

Shapira A, Feldman L (1987) Microseismicity of three locations along the Jordan rift. Tectonophysics 141:89–94

Shapira A, Hofstetter A (1993) Source parameters and scaling relationships of earthquakes in Israel. Tectonophysics 217:217–226

Shapira A, Hofstetter A (2002) Seismic parameters of seismogenic zones. Appendix C In: Shapira A 2002. An updated map of peak ground accelerations for Israel Building Code IS413, http://seis.gii.co.il/heb/hazards/docs/seismicity.pdf, Geophysical Institute of Israel Report 592/230/02, 74 p

Shapira A, Avni R, Nur A (1992) A new estimate for the epicenter of the Jericho earthquake of 11 July 1927. Isr J Earth Sci 42:93–96

ten Brink US, Ben–Avraham Z, Bell RE, Hassouneh M, Coleman DF, Andreasen G, Tibor G, Coakley B (1993) Structure of the Dead Sea pull–apart basin from gravity analysis. J Geophys Res 98:21877–21894

ten Brink US, Al-Zoubi AS, Flores CH, Rotstein Y, Qabbani I, Harder SH, Keller GR (2006) Seismic imaging of deep low-velocity zone beneath the Dead Sea basin and transform fault: implications for strain localization and crustal rigidity. Geophys Res Lett 33, L24314. doi:10.1029/2006GL027890

Thatcher W, Hanks T (1973) Source parameters of southern California earthquakes. J Geophys Res 78:8547–8576

Van Eck T, Hofstetter A (1989) Microearthquake activity in Dead Sea region. Geophys J Int 99:605–620

Van Eck T, Hofstetter A (1990) Fault geometry and spatial clustering of microearthquakes along the Dead Sea-Jordan fault zone. Tectonophysics 180:15–27

Wdowinski S, Bock Y, Baer G, Prawirodirdjo L, Bechor N, Naaman S, Knafo R, Forrai Y, Melzer Y (2004) GPS measurements of current crustal movements along the Dead Sea Fault. J Geophys Res 109:B05403

Weber M, DESERT Group (2004) The crustal structure of the Dead Sea Transform. Geophys J Int 156:655–681. doi:10.1111/j.1365-246X.2004.02143.x

Weber M, DESIRE Group (2009a) Anatomy of the Dead Sea Transform from lithospheric to microscopic scale. Rev Geophys 47:RGR2002, 44 pp

Weber M, DESIRE Group (2009b) Results of geophysical studies across the Dead Sea Transform: the Arava/Araba Valley and the Dead Sea Basin. Isr J Earth Sci 58:147–162

Wessel P, Smith W (1991) Free software helps maps and display data. EOS Trans AGU 72:441

Wiechert D (1980) Estimation of the earthquake recurrence parameters for unequal observation periods for different magnitudes. Bull Seismol Soc Am 70:1337–1346

Zhang H, Thurber CH (2003) Double-difference tomography: the method and its application to the Hayward fault, California. Bull Seismol Soc Am 93:1875–1889

Chapter 10
The Evolution of Neogene-Quaternary Water-Bodies in the Dead Sea Rift Valley

Mordechai Stein

Abstract During the late Neogene-Quaternary times the tectonic depressions along the Dead Sea transform accomodated several hypersaline to fresh water-bodies: the late Neogene Sedom lagoon and the Quaternary lakes of Amora, Samra, Lisan, Dead Sea and the Sea of Galilee. The sedimentary sections that were deposited at these water-bodies store the environmental conditions that prevailed in their watershed. The lakes are located between the Sahara-Arabia desert belt and the Mediterranean climate zone and their large watershed receives rains and detritus particles from these climatic regions. Thus, the sedimentary sections record the climate –hydrological conditions in the desert belt and Mediterranean climate zone during the late Neogene and the Quaternary Ice Ages. Over the past two decades extensive efforts were devoted to establish a precise and high-resolution chronology for the environmental history of the water-bodies. The chronologies are mainly based on U-series, radiocarbon and oxygen isotope stratigraphy methods and provide the base for detailed geochemical and sedimentological reconstruction of the limnological-hydrological conditions in the watershed. The solutions that filled the lakes are mixtures of Ca-chloride brines that were originally formed in the late Neogene Kinnarot-Sedom lagoon and freshwater from the watershed. The mixing between these solutions is dictated by regional hydrological regime that in turn reflects the changes in the global climate patterns. The water balance in the lakes is reflected in their surface level and salinity. The precise chronology of the lacustrine sedimentary sequences allows the comparison with other global climate archives such as the ice cores and ocean temperatures and opens the way for understanding the global climate connection of hydrological changes in the Levant region. It appears that most of the rains in the

M. Stein (✉)
Geological Survey of Israel, 30 Malkhe Israel St., Jerusalem 95501, Israel
e-mail: motistein@gsi.gov.il

Z. Garfunkel et al. (eds.), *Dead Sea Transform Fault System: Reviews*,
Modern Approaches in Solid Earth Sciences 6, DOI 10.1007/978-94-017-8872-4_10,

lakes watershed are derived from the Mediterranean and reflect the climate – oceanographic conditions in the north Atlantic. Glacial periods were colder and wetter and interglacials were warmer and drier. Episodes of extreme aridity in the Levant region were accompanied by massive salt deposition at the shrinking lakes, e.g. at ~400 ka (MIS11); 130–120 ka (last interglacial MIS5e) and at 14–13 ka (the post-glacial BÖlling-Allerod period). Nevertheless, various lines of evidence indicate on intrusions of wetness from southern sources into the Red Sea region –Arava valley during interglacials that are possibly related to the African monsoon. These wet intrusions could facilitate the migration of humans out of Africa along the rift valley. The lacustrine sedimentary sections indicate also on episodes of abrupt aridities that expanded over the sub-tropical regions from the Sahara to the Mediterranean (e.g. at ~3.5–3.3 ka, coinciding with the late Bronze cultural crisis). These events coincide with ice rafting episodes in the north Atlantic that were linked to solar minima events.

Keywords Dead Sea • Neogene-Quaternary • Paleoclimate • Paleohydrology • Paleolimnology

10.1 Introduction

The breakup of the African-Arabian continent in the late Cenozoic was expressed by extensive tectonic and magmatic activity that included the opening of the Red Sea, the formation of the Suez Rift, the development of Dead Sea Transform (DST) with deep tectonic basins along its route, as well as production of large fields of continental alkali basalts (cf. Garfunkel 1981; Stein and Hofmann 1992, and references therein). The DST extending from the Gulf of Aqaba to the Taurus Mountains (Fig. 10.1) accommodated during the past 20 Ma a left lateral movement of ~110 km along the Arava–Jordan valley. This movement affected the development of the deep tectonic depressions of the Dead Sea, Kinnarot basin, Hula basin, Bakaa and Gaab Valleys. Since the late Neogene the tectonic depressions were filled by marine lagoons and lakes that received brines and freshwater from marine lagoons, the regional aquifers and runoff (Zak 1967; Neev and Emery 1995; Stein 2001). Figure 10.2 illustrates a general stratigraphic section of the Neogene-Quaternary sedimentary sequences that were deposited in the above-mentioned water-bodies. The limnological-hydrological evolution of the late Neogene-Quaternary water-bodies and the relation to the regional and global climate history is the subject of this review. The chapter is organized in the following structure: Sect. 10.2 provides a brief overview on the late Neogene history of the region;

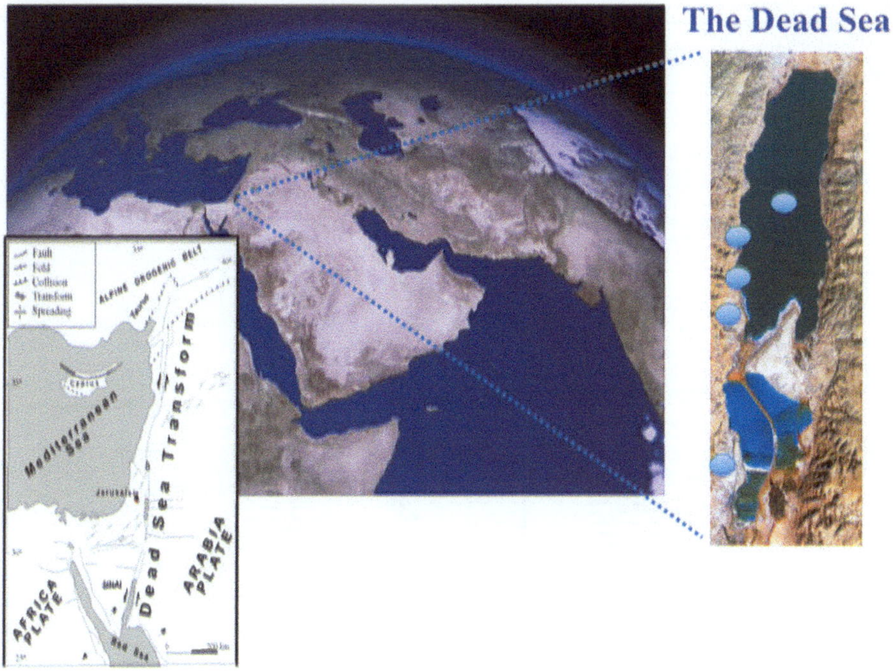

Fig. 10.1 Location map of the Dead Sea and sampling sites. The inset shows a regional tectonic map that outlines the locus of the Dead Sea Transform fault (= DST) extending between the Red Sea and the Taurus Mt. The African, Sinai and Arabia sub-plates and some other major faults. The Dead Sea basin accommodated several Neogene-Quaternary water-bodies (lagoons and lakes) that are the subject of this chapter. The basin is located between the Mediterranean and desert climate zones and the sediments deposited from the lakes reflect the climate-hydrological conditions in both regions. Circles in the Dead Sea and its margins mark the major study sites and sedimentary sections that are described in the chapter: from south to North: Perazim valley –west of the Mount Sedom salt diapir (exposures of the Lisan Fm.), Massada plain (exposures of the Lisan and Samra Fms at the foothill of the Massada archeological site); Ze'elim plain (where deep modern erosion exposed the Holocene section), Ein Gedi spa (where onshore shallow drilling recovered the Holocene section and offshore drilling recovered ~350 m of late Pleistocene and Holocene sediments), and the mid-lake ICDP deep drilling site (recovered 460 m of sediments deposited during the past ~220 ka, between MIS1 and MIS7)

Sect. 10.3 describes the limnological-hydrological history of the Quaternary lakes filling the basins; Sect. 10.4 deals with temporal changes in the regional climate-hydrological patterns and their relation to global climate; Sect. 10.5 refers to the topic of relation between human culture development in the Levant region and climate and hydrological regimes.

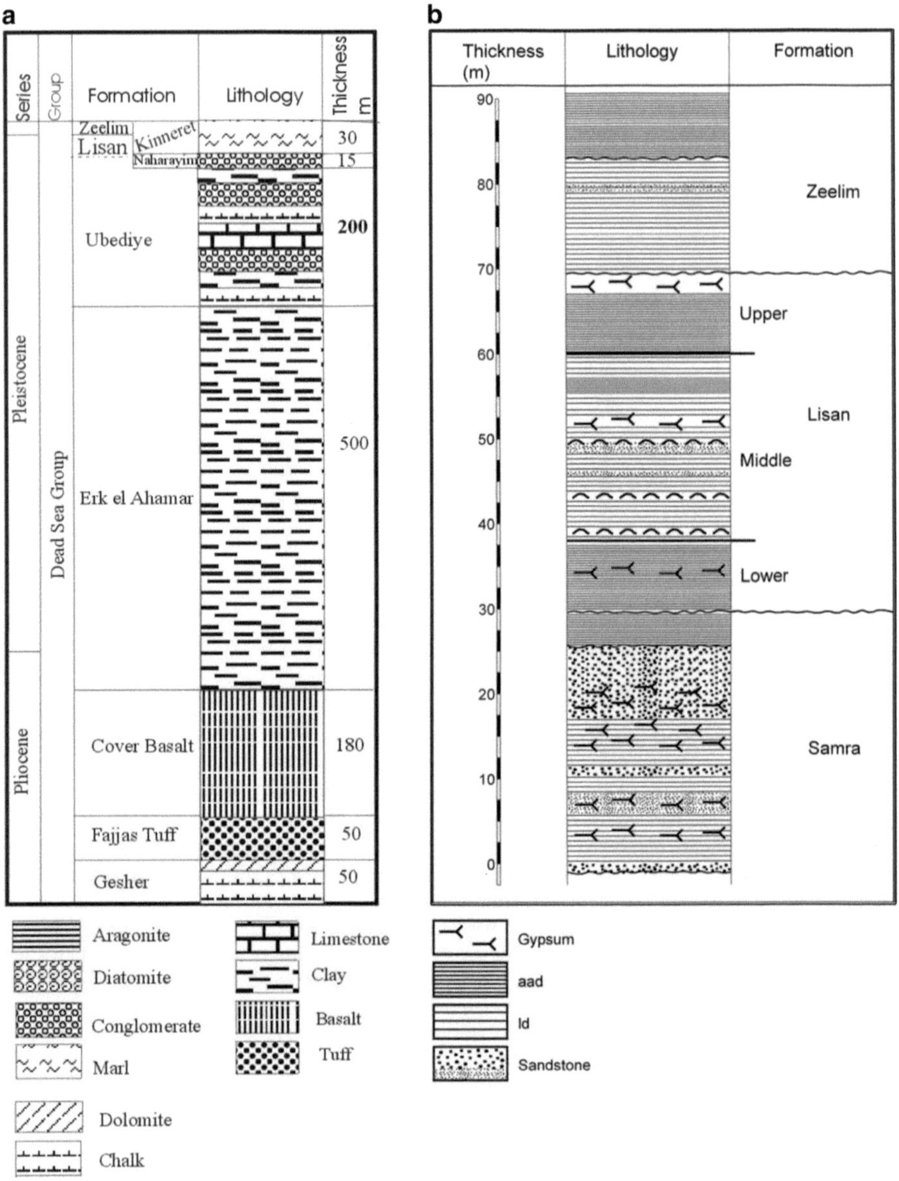

Fig. 10.2 General stratigraphic section of the Neogene-Quaternary formations in the Dead Sea-Jordan Valley and Kinnarot basin. The left panel shows the Dead Sea Group and the formations that are exposed at the Kinnarot basin. The late Pleistocene – Holocene periods are represented in the Kinnarot basin by the Kinneret Fm. that are stratigraphically correlated with the Lisan and Ze'elim Fms in the Jordan Valley and Dead Sea basin. The right panel shows the Samra, Lisan and Ze'elim Fms that are exposed at the vicinity of the Dead Sea basin and Jordan Valley, and were deposited during the last interglacial (MIS5) last glacial, (MIS4-2) and Holocene periods, respectively. It should be noted that a longer stratigraphic section that goes back to the mid Pleistocene is exposed at Mount Sedom salt diapir (Fig. 10.3)

10.2 Late Neogene Marine Transgressions and the Sedom Lagoon

10.2.1 The Late Neogene Marine Transgressions

During the early and middle Miocene a thick fluvio-lacustrine sequence – the Hazeva Formation was deposited all over the Negev desert, suggesting a subsidence of this region (Zilberman 1992; Calvo 2002), while in northern Israel, a low relief terrains prevailed, as suggested by the low-energy fluvial and lacustrine sediments, which compose the Hordos Fm. (Shaliv 1991). The tectonic activity along the DST during the Neogene was accompanied by uplift of the N-S oriented "mountain backbone" of Israel (Wdowinsky and Zilberman 1997). The "mountain backbone" was dissected by two structural lows: The Yizre'el and the Beer-Sheva Valleys, which were invaded during the Neogene by marine ingressions that intercalated with lacustrine sequences, e.g. the Bira and Gesher Formations in the lower Galilee and Yizre'el Valley that statigraphically lie between Lower Basalt (and the Um-Sabune Conglomerate) and the Cover Basalt (Gvirtzman and Buchbinder 1969; Martinoti et al. 1978; Shaliv 1991; Sandler et al. 2004; Rozenbaum et al. 2013). K-Ar and Ar-Ar dating of the basalts places the interval of deposition of the Bira and Gesher Formations between ~10 and 4 Ma (Heimann et al. 1996; Shaliv 1991; Rozenbaum et al. 2013). This interval includes the Messinian desiccation event at the Mediterranean Sea whose expression on-land is not entirely clear. The marine ingressions that intruded the Yizre'el and Jordan Valleys probably reached the Kinnarot and Dead Sea basins forming the "Sedom lagoon". In the Kinnarot basin thick sequence of salts was recovered by the Zemah borehole (Marcus and Slager 1985), and was deposited between ~9.5 and 5 Ma (Ar-Ar ages, Heimann and Stein in prep), within the range of ages of the Bira and Gesher marine ingressions. Thus, it is possible to argue that Kinnarot basin deposited salts from marine origin. The age of the Sedom lagoon, which possibly carried the evaporated seawater to the Dead Sea basin is not well constrained and it is not straightforward linking it with the Yizre'el valley Bira and Gesher ingressions (see below). The Sedom lagoon comprised Ca-chloride brine that was produced by the interaction of the evaporated marine water with the limestones of the basin wallrocks (Starinsky 1974; Stein et al. 2000). A thick sequence of salts, mainly halites and some dolomites was deposited from the Ca-chloride brine comprising the Sedom Formation (Zak 1967). Other salt-bodies that possibly were related to the Sedom lagoon (and the marine transgression) are salt diapirs that were detected in the central Jordan Valley (Shaliv 1991; Bilitzki and Mimran 1996; Gardosh and Bruner 1998), and the deposits of gypsum exposed at the margins of the central Jordan valley (Schulman 1962; Raab et al. 1997; Raab 1998). The deep canyon of the Beer-Sheva Valley (Neev 1960) comprised another route for the late Miocene-early Miocene marine transgressions (Martinoti et al. 1978; Buchbinder and Zilberman 1997).

10.2.2 The Sedom Formation

The Sedom Formation exposed at the Mount Sedom salt diapir (SW side of the Dead Sea, Figs. 10.1 and 10.3) and recovered from several boreholes drilled at the southern part of the Dead Sea comprises thick sequence of salts, marls and clastic units. At Mount Sedom the formation thickness reaches ~2,000 m (Zak 1967). In more details, the formation consists mainly of halite units cyclically interbedded with gypsum, anhydrite, dolomite and minor amounts of silt, marl and clay. In boreholes, the salt rocks overlie a thick late Neogene sequence of basin-fill sediments. Water-insoluble materials, which have been left on top of the rising diapir when their host halite beds dissolved, compose the cap- rock and form the upper part of the salt-diapir. It were the extensive thickness of halites in the Sedom Fm, and other geochemical observations that led Zak (1967) to suggest that the evaporites precipitated in a marine lagoon, probably connected to the Mediterranean Sea. Most notable of these geochemical indicators are the molar Br/Cl ratio in the halites and the $\delta^{34}S$ composition of the Ca-sulfates (19.8 ± 0.8 ‰: Raab et al. 1997; Stein et al. 2000), which attest to a marine origin of salts. It is consistent with the sulfur isotope composition of Neogene marine water (Stein et al. 2000). The proposed depositional environment for accumulation of such a thick sequence of halite is a tectonically subsiding marine lagoon separated from the open ocean by a sill that allows continuous flow of seawater (Fig. 10.4) (Zak 1967). A long-standing question is the age of the Sedom lagoon and its relation to the late Neogene marine transgressions in northern Israel (e.g. the Bira and Gesher Formations in the Yizre'el and Jordan Valleys) and to the Messinian desiccation event in the Mediterranean (Hsu 1973). Zak (1967) used general stratigraphic and some faunal considerations to suggest a late Pliocene-early Pleistocene for the age of the Sedom lagoon. He also estimated as ~1 Ma the duration of deposition of salts in the lagoon. Based on the chronology of the Neogene basalts at northern Israel and assuming that the Sedom lagoon is related to the marine transgressions within the Bira and Gesher Formations, Steinitz and Bartov (1991) estimated that the age of the Sedom lagoon lies between 6 and 3.4 Ma (see compilation of the various age estimates in Belmaker et al. 2013). Recently, Belmaker et al. (2013) applied atmospheric ^{10}Be to date directly the salts of the Sedom Fm. The data indicate that the salts comprising the Me'arat Sedom and Lot Members were deposited between ~5 and 3 Ma, similar to Steinitz and Bartov estimate. Interestingly, the lower part of the range of these ages coincides with ^{10}Be burial ages recovered from the lacustrine Erk-El Ahmer Formation that is exposed in the Jordan Valley south to the Kinnarot basin. The ^{10}Be burial ages lie in the range of ~4.3–3.6 Ma (Davis et al. 2011). This may infer that between ~4 and 3 ka freshwater and hypersaline water-bodies already co-existed in the Dead Sea-Jordan valley, similar to the late Quaternary and present day configuration. The implication of the atmospheric and burial ^{10}Be ages could be that the time of marine transgressions and the disconnection of the Sedom lagoon from the open sea occurred earlier, perhaps in the mid-Pliocene time around the "climax" of the Messinian salinity crisis (at ~5.7 Ma, Cornée et al. 2006).

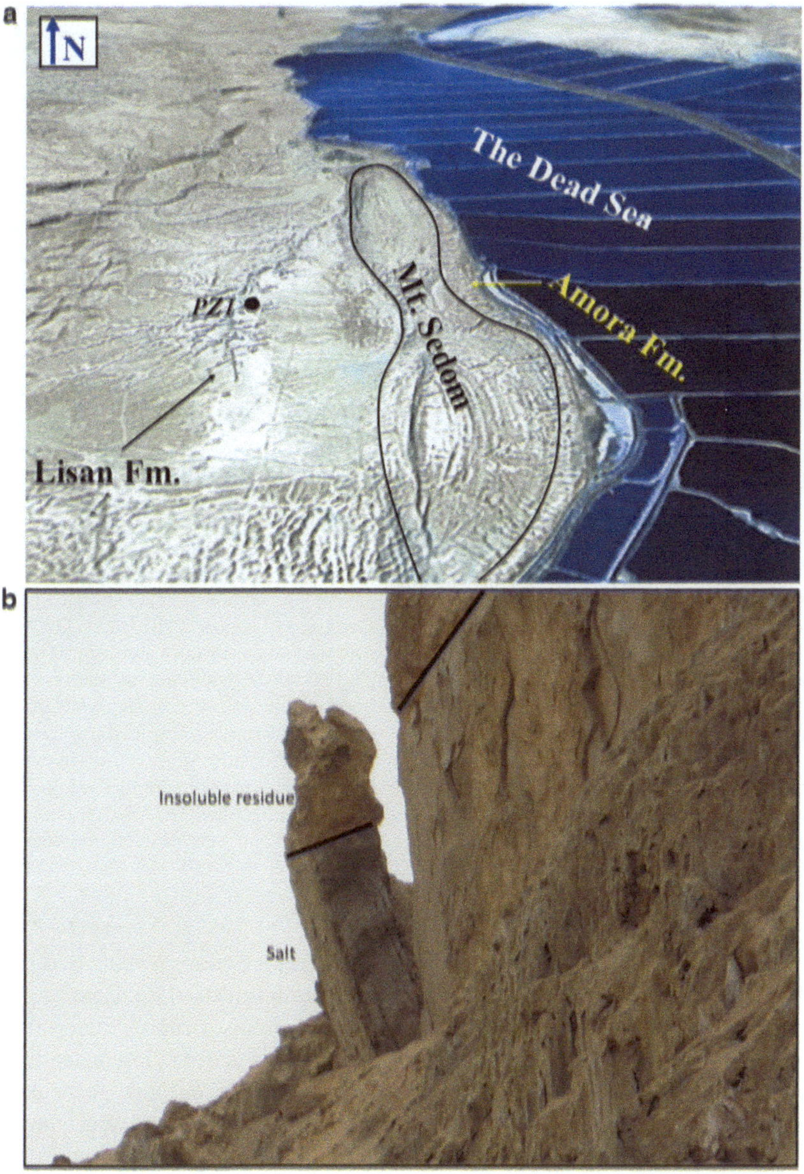

Fig. 10.3 (**a**) The Mount Sedom salt diapir and the Amiaz Plateau with the location of the PZ1 section at the Perazim Valley. The Amora Fm. is exposed along the eastern flanks of the diapir (e.g. the Arubotaim Cave section described by Torfstein et al. 2009). (**b**) The salt units (dipping vertically) comprising the Mt. Sedom salt diapir. The "Lot's Wife" pillar is composed of salt and insoluble residue that indicates on the activity of ancient dissolution plan

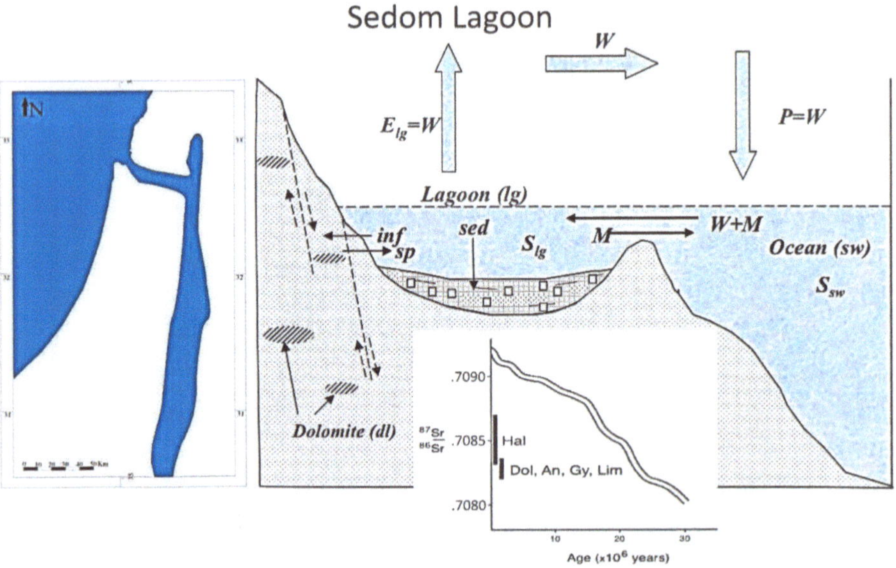

Fig. 10.4 The depicted geographical extension of the Sedom lagoon (*left hand side*), intruding from the Mediterranean via the Yisre'el Valley into Dead Sea – Kinnarot rift valley. On the *right hand side* – the lagoon model (see text) showing the flow of seawater (*sw*) and lagoon water (*lg*) over a sill, sedimentation of salts (*sed*) in the evaporating lagoon (E_{lg}) and the interaction of evaporated seawater that infiltered (*inf*) the aquifers in the Judea Cretaceous limestones walls of the rift, interacted with the limestones to produce epigenetic dolomites (*dl*) and back spilled (*sp*) Ca-chloride brine that mixed with fresh-coming evaporated seawater. $^{87}Sr/^{86}Sr$ ratios (*lower inset*) in evaporates of the Sedom Fm. at Mt. Sedom. In the *lower inset*: $^{87}Sr/^{86}Sr$ isotope ratios in the Sedom Fm salts (*HAL* halites, Dol dolomites, An Anhydrite, Gy Gypsum, Lim Limstone). The $^{87}Sr/^{86}Sr$ ratios are not consistent with values of late Neogene seawater (*upper curves*) and were possibly modified by the lagoon evaporated seawater-limestone interaction (After Stein et al. 2000)

Closer look at the exposed salt sequences at Mount Sedom reveals that the Lot and Me'arat Sedom Members comprise thick salts units but also laminated sequences of salt and marls (Fig. 10.5) that indicate seasonal input of runoff material to the Sedom lagoon. Similar laminated sequences comprise the overlying lacustrine sections of the Pleistocene – Holocene formations (e.g. sections that were recently described at the ICDP deep drill at the deepest bottom of the Dead Sea, Stein et al. 2011). The similarity may indicate that significant intervals of the Sedom Fm. were characterized by lacustrine-type deposition that followed the major phase(s) of marine transgression(s).

The evolution of the Ca – chloride brines comprising the Dead Sea solution and saline springs discharging at the Dead Sea margins and Lake Kinneret (e.g. the Ein Qedem springs) was investigated by Starinsky (1974). He suggested that the evaporated seawaters filling the lagoon penetrated the Cretaceous wall rocks of the basin via the mountain aquifers, dolomitized the limestone and returned back to the lagoon as Ca-chloride brine. This model was later supported by the strontium isotope composition of the Sedom Formation salts and epigenetic dolomites: Stein

Fig. 10.5 Laminated and folded salt and marly units of the Lot member of the Sedom Formation. The photo was taken with one of the dissolution chimneys of the Arubotaim Cave. On the left hand side of the photograph pebbly material marking old exit of the cave

et al. (2000) proposed that Sr in the Sedom evaporites represents a mixture between the evaporated seawater entering the lagoon and the Ca-chloride brine that was released to the lagoon from the Cretaceous wall rocks (Fig. 10.4). An independent assessment of this model was done by Stein et al. (2002) who demonstrated that epigenetic dolomites from the Judea desert to the west of Mt. Sedom yielded $^{87}Sr/^{86}Sr$ ratios that are significantly higher than the stratigraphic correlated limestone but are identical to the ratio of the some of Sedom salts. The Ca-chloride brines that were produced in the Sedom lagoon played a pivotal rule in the forthcoming geochemical and limnological evolution of the water-bodies that occupied the Dead Sea – Jordan basin (Katz et al. 1977; Stein et al. 1997; Gavrieli and Stein 2006; Katz and Starinsky 2009).

10.3 Quaternary Lakes in the Dead Sea Basin

10.3.1 General

After the disconnection of the Sedom lagoon from the open-sea several lakes filled the tectonic depressions along the Dead Sea basin-Jordan Valley depositing sequences of lacustrine sediments that comprise the Amora, Samra, Lisan and

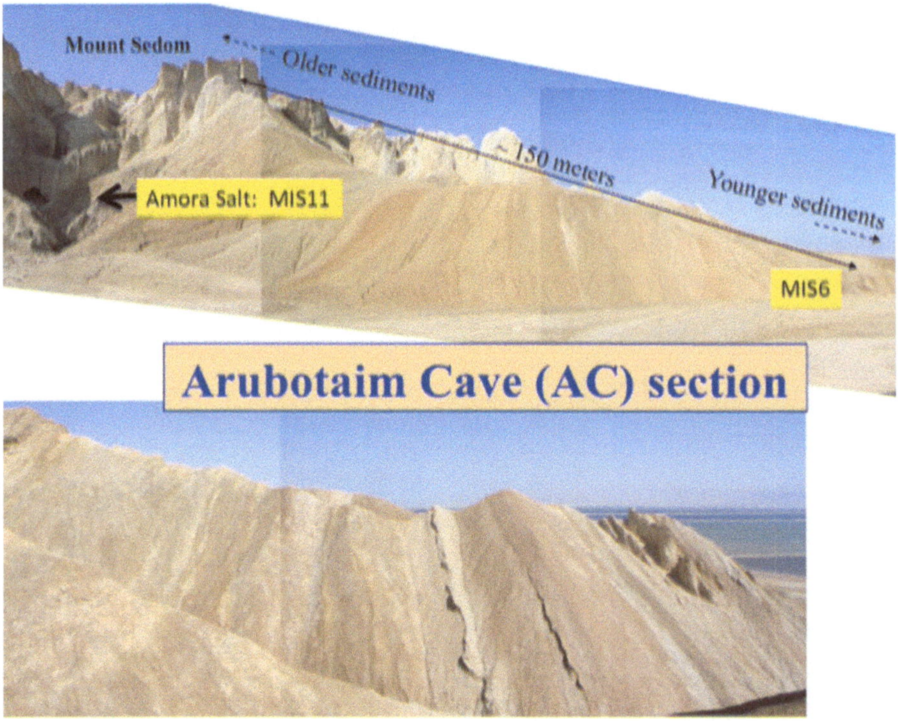

Fig. 10.6 The Arubotaim Cave section of the Amora Fm. on the eastern flank of Mt. Sedom. The chronology of the section was determined by U-Th dating and oxygen isotope stratigraphy (after Torfstein et al. 2009). A several meters thick massive salt unit (the Amora salt) is marked. The salt was deposited during MIS11 marking a significant drop in lake level reflecting regional aridity

Ze'elim Formations (Stein 2001; Fig. 2). The late Pleistocene pre-Lisan lacustrine sedimentary sequence exposed at the dipping flanks Mt. Sedom (Fig. 10.6) was termed by Zak (1967) as the Amora Formation. The time of origin of Lake Amora is not known, and it could well go back to the late Pliocene-early Pleistocene period (Torfstein et al. 2009). The important evolutionary step in the geochemical-sedimentological evolution of the Dead Sea water-bodies is the appearance of primary aragonite in Lake Amora that possibly reflects a major change in the behavior of the regional hydrological system that allowed the transport of bi-carbonate to the lakes and its mixing with the Ca-chloride brine. This would require enhanced dissolution of the limestone in the watershed and push out of the saline aquifers. Enhanced dissolution of the Jurassic-Cretaceous limestones would reflect soil accumulation and development of vegetation over the Judea-Galilee Mountains.

In the following chapter I summarize the sedimentological-limnological histories of the lakes that filled the Dead Sea rift tectonic depression after the disconnection of the Sedom lagoon. A significant part of the chapter deals with the description of the Lisan Formation that comprises the sediments deposited from the last glacial Lake Lisan. This reflects the comprehensive studies and data that were achieved for this Formation, e.g. definition of sedimentary facies, U-Th and radiocarbon chronology and limnological – hydrological reconstruction (e.g. Bartov et al. 2002; Haase-Schramm et al. 2004; Stein and Goldstein 2006; Stein et al. 1997; Torfstein et al. 2013a, b). The extensive investigation of the Lisan Formation paved the way to the exploration of the older Formations: the Samra and Amora Formations and the younger Ze'elim Formation that was deposited in the Holocene Dead Sea. Thus, I will discuss first the sedimentological- and limnological history of the last glacial Lake Lisan and then move to the other lakes.

10.3.2 Lake Lisan (70–14 ka BP)

Lake Lisan occupied the tectonic depressions along the Dead Sea basin-Jordan Valley between ~70 and 14 ka BP (Haase-Schramm et al. 2004). At its highest stand (160 m bsl) the lake extended from the Sea of Galilee in the north to the Hazeva area in the south (Fig. 10.7). The lake deposited mainly sequences of primary aragonite and silty- detritus material that comprise the *aad facies* (Machlus et al. 2000) and sequences of laminated silty-detritus that comprise the *ld facies* (Haliva-Cohen et al. 2012) (Fig. 10.8). In addition, the sections are interrupted by several sequences with prominent gypsum layers (Fig. 10.9). The *aad facies* dominates the Upper and Lower Members of the Lisan Formation. The silty-detritus material consists of typically ~8–10 μm size particles that originated from desert dust that was blown to the Judea and Galilee Mountains from the North Sahara deserts during major dust storms that are associated with the activity of the Cyprus Cyclone (Haliva-Cohen et al. 2012 and references there). The aragonite was precipitated directly from the surface water of the lake that represents a mixture between freshwater loaded with bi-carbonate and the Ca-chloride brine (Stein et al. 1997). Most of the freshwater originated in the northern and central parts of the Dead Sea watershed reflecting the Mediterranean winter rains activated by the Cyprus Cyclone (Bartov et al. 2003; Kushnir and Stein 2010). The *ld facies* dominated the formation sequences of low lake stands (mainly during MIS3 the interval of deposition of the Middle Member of the Lisan Formation). This facies represents flood material washed to the lake during times of limited supply of bi-carbonate and comprise surface cover such as the loess material in the northern Negev desert (Haliva-Cohen et al. 2012). I elaborate on this topic below.

10.3.2.1 U-Series Chronology of the Lisan Formation

The excellent preservation and high U concentrations (~3 ppm) of Lisan aragonite make it useful for U-Th dating. The pioneering attempts in applying this method to Lake Lisan sediments were made by Kaufman and colleagues (Kaufman 1971;

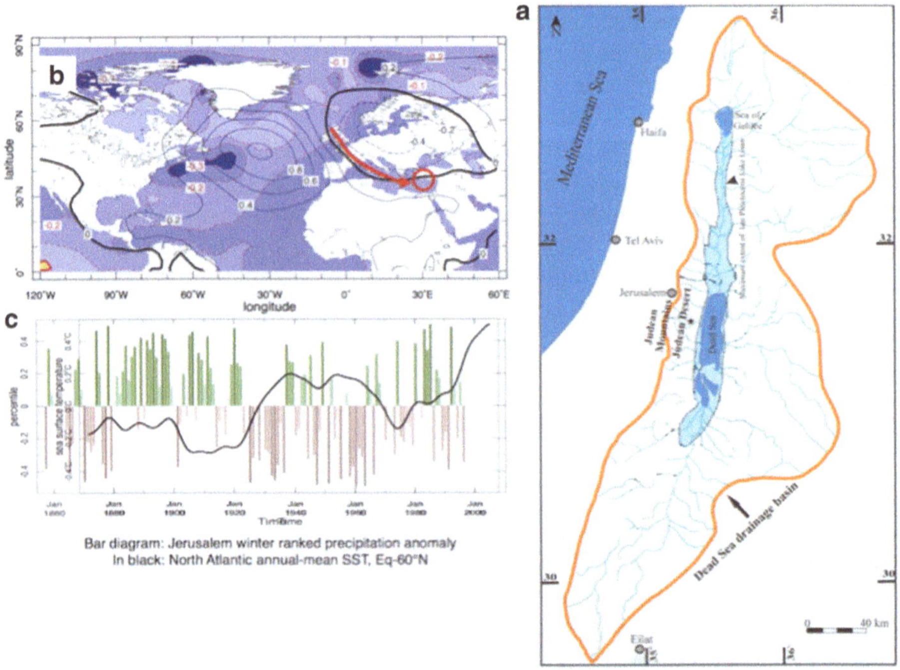

Fig. 10.7 (**a**) The Dead Sea drainage area extending from the Gulf of Aqaba to Mt. Hermon, encompassing the Arabian-Sahara desert and Mediterranean climate zones. Most of the rains in the watershed occur at wintertime due to the activity of the Mediterranean (Cyprus) Cyclone (**b**) High-pressure conditions in the central Atlantic causes the cold westerlies to move towards the relatively warm east Mediterranean where cyclogenesis occurs. (**c**) Instrumental rain patterns from a Jerusalem station during the past 120 years showing decadal changes in precipitations (e.g. higher than average between 1880–1920 (*green bars*) and lower than average between 1929–1960 (*brown bars*). The *black curve* shows North Atlantic (*NA*) annual-mean SST (60N), indicating correlation between NA colder/warmer SST and higher/lower Jerusalem precipitation (After Kushnir and Stein 2010)

Kaufman et al. 1992) applying α-counting technique. Using U-Th 'isochrons' to correct for detrital contributions to the ^{238}U, ^{234}U, and ^{230}Th, they showed that U-series dating can provide reliable ages and that it is possible to correlate sections of the Lisan Formation located in different sites in the Dead Sea region. Their ages were obtained on leached samples, which consist of different proportions of detritus and aragonite (allows for large spread on 'isochron' diagrams). Schramm et al. (2000) and Haase-Schramm et al. (2004) applied "Thermal Ionization Mass spectrometric (TIMS)" technique to determine precisely the ^{238}U, ^{234}U, and, ^{230}Th, ^{232}Th activities in the Lisan aragonite and detritus material and constructed a high-resolution ^{234}U-^{230}Th age-height model of a detailed stratigraphic section of Lisan Formation: the PZ-1 section at Perazim Valley, to the west of the Mt. Sedom (Fig. 10.1). This section was selected for geochemical, paleoseismic and paleomagnetic studies because of its location on an elevated terrace of the ancient lake at 306 m bmsl, 120 m above the present Dead Sea,

Fig. 10.8 The Samra-Lisan transition at ~70 ka. The late interglacial Samra Formation (~140–70 ka, Waldmann et al. 2009) comprises sequences of laminated detritus material (the *ld facies* composed mainly of calcite and quartz) that was washed to the Dead Sea basin by seasonal floods (Haliva-Cohen et al. 2012). At ~70 ka BP Lake Lisan rose above the landscape of the Samra Fm. depositing sequences of laminated primary aragonite and silty detritus (the *aad facies*) at intervals of positive freshwater supply to the lake and sequences of gypsum during lake retreat and overturn (Stein et al. 1997)

which makes it sensitive to level changes (Machlus et al. 2000). The section is not disturbed by recent faults and landslides, but contains many layers that were disturbed by ancient earthquakes (Marco et al. 1996). Thus, the precise chronology of this section provides an absolute time framework for comparing limnological, tectonic and paleomagnetic events in the region and the response of the lake to global climatic changes. The ages achieved by U-Th dating of primary aragonites range from ~70 to 14 ka BP. This age interval corresponds to the last glacial interval in the global climate stratigraphy. Machlus et al. (2000) divided the Lisan Formation at Perazim Valley into three Members and Schramm et al. (2000) and Haase-Schramm et al. (2004) dated these members to: ~70–50; ~49–31 and 30–14 ka, corresponding to the marine isotope stages MIS4, 3 and 2, respectively. The Lower and Upper Members consist mainly of alternating laminae of aragonite and detritus (the *aad* facies), which were deposited during high lake stands, and the Middle Member contains abundant clastic layers (the *ld* facies) deposited during relatively low lake stands. Recent dating of the Lisan Formation at the M1 section at Massada plain, beneath the Archeological site of Massada, revealed that the Middle Member evolved via several short-term intervals of high and low stand conditions (Torfstein et al. 2013a, b). At ~27–24 ka the lake rose to its highest elevation at of ~160 m bsl (Bartov et al. 2002,

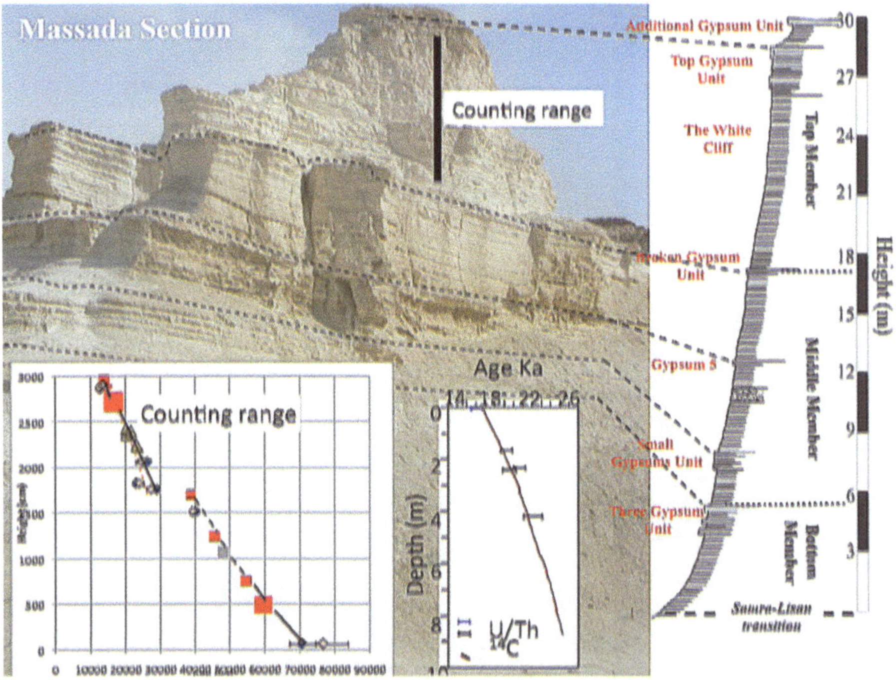

Fig. 10.9 The Lisan Formation exposure at the M1 section beneath the archeological site of Massada (After Torfstein et al. 2009). The section comprises sequences of alternating laminae of primary aragonite (that was deposited from the lake solution) and silty-detritus, and gypsum layers (that form benches). The gypsum was deposited at episodes of lake level decline and overturn. U-Th ages are from Torfstein et al. manuscript submitted. The Upper Member of the Lisan Fm. (termed the "White cliff") was studied in details by Prasad et al. (2004) who performed laminae counting (marked in the figure as "counting range") and time series analyses of the thickness of the aragonite and detrital laminae

2003). At that time it overflowed the sills separating between the Dead Sea basin and the Kinnarot basin (where the Sea of Galilee is located) and extended from the northern side of the Sea of Galilee (the Beteicha valley) to Hazeva in the south. Independent reconstruction of the elevation of the Sea of Galilee levels arrived to the same pattern of highest stand at ~26–24 ka (Fig. 10.10) (Hazan et al. 2005; Stein 2014). The lake's

Fig. 10.10 (continued) Torfstein et al. 2013a). (**b**) Zoom on the past 15 Ka. Note the different lake level scale in the two panels. During the Holocene the lake fluctuated between 430 and 370 m bmsl (with unknown drop at ~8.1 Ka) (After Bookman et al. 2004; Migowski et al. 2006; Stein et al. 2010); (**c**) During the last glacial period the lake fluctuated between 370 and 160 m bmsl. The panel shows the stratigraphy of the Lisan Fm. divided to three members that corresponds to MIS4, 3 and 2 (After Torfstein et al. 2013b). At ~14–13.5 Ka and possibly at ~120–130 Ka the lakes dropped below 450 m bmsl depositing thick sequences of salt (Stein et al. 2010; Goldstein et al. 2013). (**d**) Lake level curves for the hypersaline-terminal Lake Lisan and Dead Sea and the fresh (flow through) Lake Kinneret. There are several times of lakes convergence when the water level rose above the Wadi Malih and Yarmouk sills (e.g. at 1–5). At intervals 6–10 the lakes are separated but appears to fluctuate simultaneously, reflecting the precipitation in their watershed (lake Kinneret curve after Hazan et al. 2005 and Stein 2014)

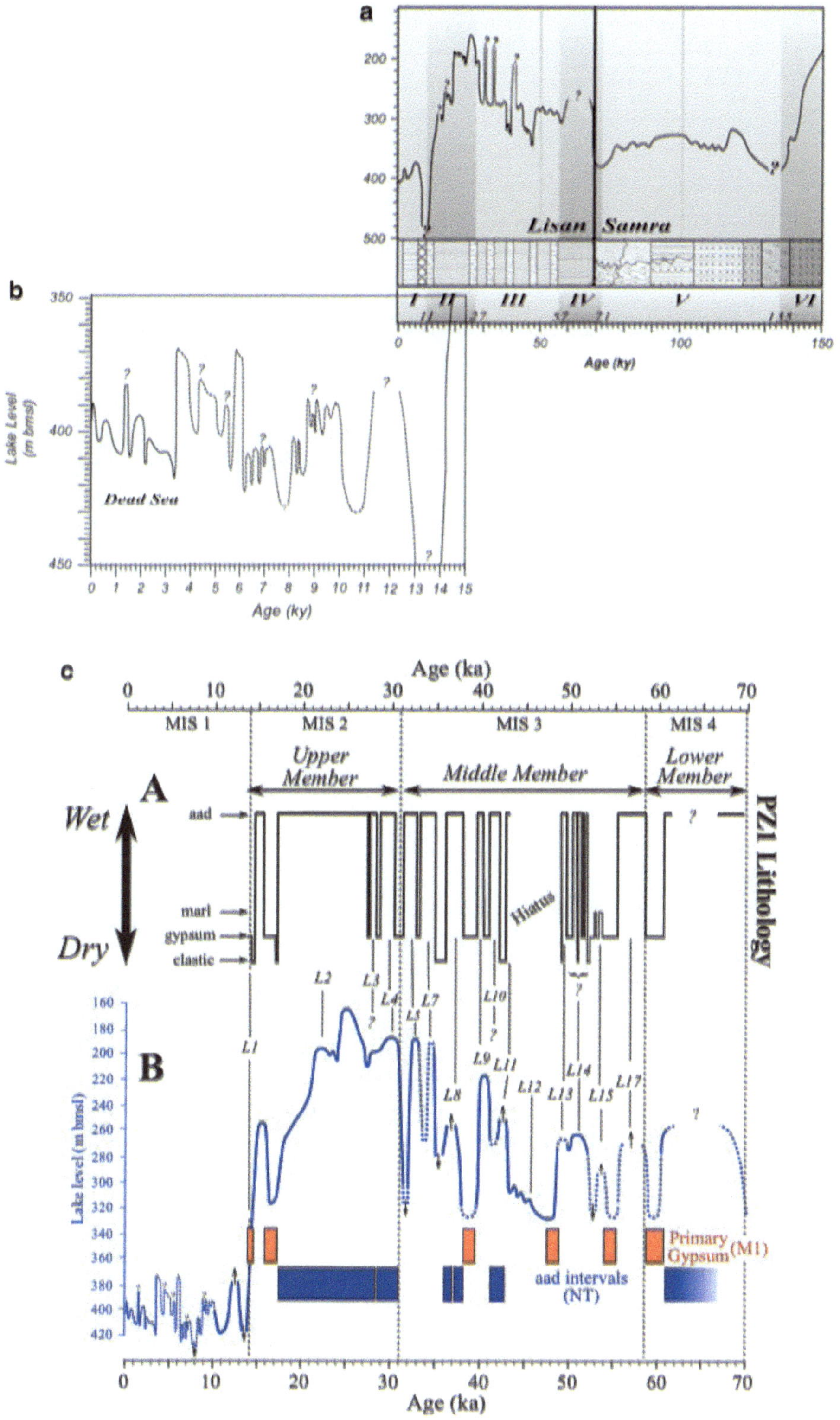

Fig. 10.10 Composite-level curve for the lakes filling the Dead Sea basin during: (**a**) The time interval present to 150 Ka (After Bartov et al. 2002, 2003, 2007; Waldmann et al. 2007;

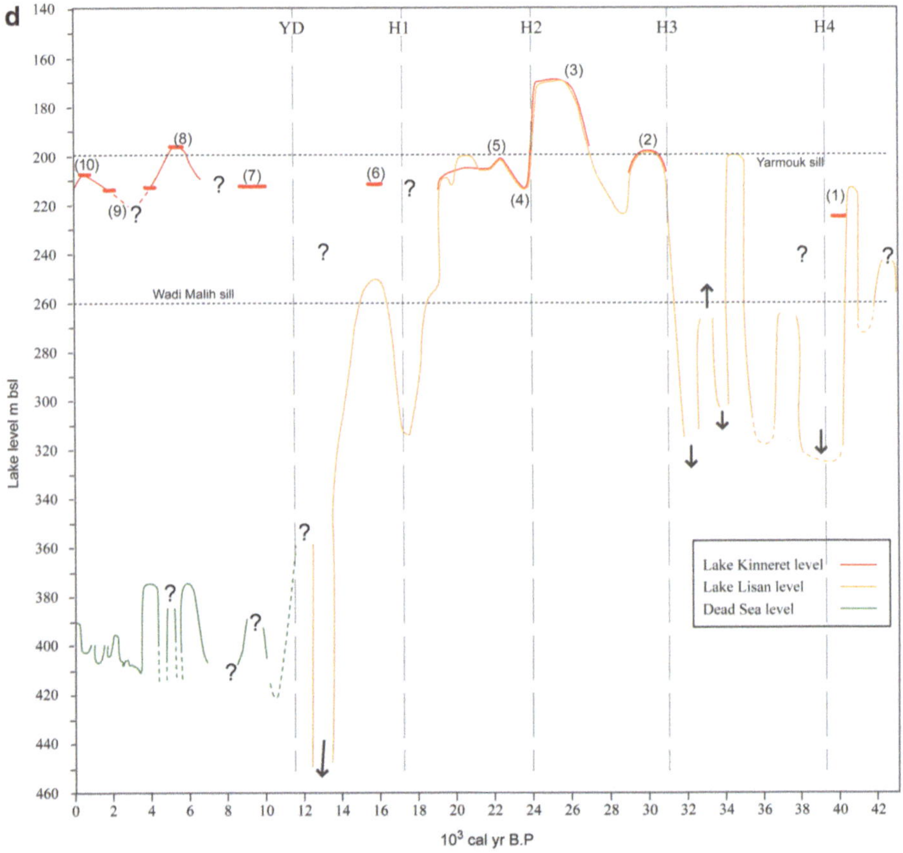

Fig. 10.10 (continued)

high stand coincides with the timing of maximum extent of glaciers in the northern hemisphere and minimum global sea level. Torfstein et al. (2013a) obtained U and Th isotope data from several sections of the Lisan Formation at the Dead Sea basin and Jordan valley (Fig. 10.10a) and combined the data with prominent stratigraphical markers (e.g. the major gypsum units) to produce an integrated multi-site U-Th age model for the Lisan Formation. The ages of gypsum marker units were correlated with the Heinrich events that were identified in the North Atlantic deep cores records and other global archives.

10.3.2.2 Long Term Freshening of Lake Lisan and Deposition of Gypsum

The recently conducted ICDP drill (see below) recovered the sedimentary sections of the ancient lakes down to the Amora Fm. (Stein et al. 2011). Porewaters that were extracted from the drilled cores provide unprecedented information on the composition of brine at deepest environment of the lakes. Analyses of chlorine concentration and oxygen isotopes in the porewaters revealed a long term freshening of the lower brine of the lake (the hypolimnium), possibly by turbulent mixing across the (fresher) epilimnium and the hypolimnium, which also affected the radiocarbon and strontium budgets in the lake (Lazar et al. 2014; Stein et al. 2013). This long-term freshening process arrived to an abrupt end at the very top of the Lisan Formation, when the lake dropped catastrophically from its high-stand elevation of ~200 m bmsl to below 500 m bmsl when it deposited a thick sequence of salts (see below). The freshening trend commenced after the previous dramatic lake retreat and deposition of massive salts that occurred during the last interglacial period (when Lake Samra occupied the basin).

10.3.3 The Post-glacial Retreat of Lake Lisan

At ~17.4 ka BP Lake Lisan commenced its retreat from the ~200 m bmsl to the minimum level of the post-glacial Lake Lisan and Holocene Dead Sea (Fig. 10.10). The lake retreat coincided with the global warming that generally imposed more arid conditions in the east Mediterranean – Levant region (Neev and Emery 1995; Stein 2001; Haase-Schramm et al. 2004). Between 17.4 and 16 ka, the lake deposited a sequence of gypsum layers intercalated with detritus (the Upper Gypsum Unit = UGU, Torfstein et al. 2008). In the North Atlantic the time interval between 17.4 and 16 ka BP is known as the Heinrich event H1. At Massada and Beit-Ha'Arava sections, the UGU is overlain by 1–2 m of *aad* sequence topped by gypsum. This sequence is termed the Additional Aragonite Gypsum Unit (AGU). At Beit-Ha'Arava section the AGU is covered by a 3 m thick silty unit termed the Fazael Fm. This formation comprises the surface cover in the central Jordan Valley from Beit-Ha'Arava to the Beit Shean comprising recycled mountain soils. The age of the AGU is not well established. In Beit-Ha'Arava section it was dated to 11.6 ka BP, which would place it at the Younger Dryas time interval. Accepting this age, the UGU and AGU were deposited during the post-glacial time interval of 17.4 to ~11.6 ka BP. This period, accommodated dramatic changes in lake level that possibly declined between ~14.5 and 13.5 ka BP to below 450 m bmsl and rose back during the Younger Dryas to ~330 m bmsl before declining again at 11–10 ka BP to below 430 m bmsl. The information on this period and these abrupt fluctuations in lake levels is derived mainly from sediment cores that were drilled along the retreating shores of the modern Dead Sea (Fig. 10.11). The drill-holes penetrated Holocene and pre–Holocene (Lisan) sections, recovering various sedimentary facies ranging from deep-water lake sediments to shore and fluvial sediments. A massive and thick salt unit was penetrated in most boreholes (Fig. 10.11), corresponding in age (~11–10 ka BP, Stein et al. 2010) to Migowski et al. (2004) who placed it at the base of the Holocene. The dissolution of this salt unit during the modern man-related lake retreat triggered the collapse of the Holocene sections above it and formation of sinkholes along the Dead Sea shores (e.g. Abelson et al. 2006).

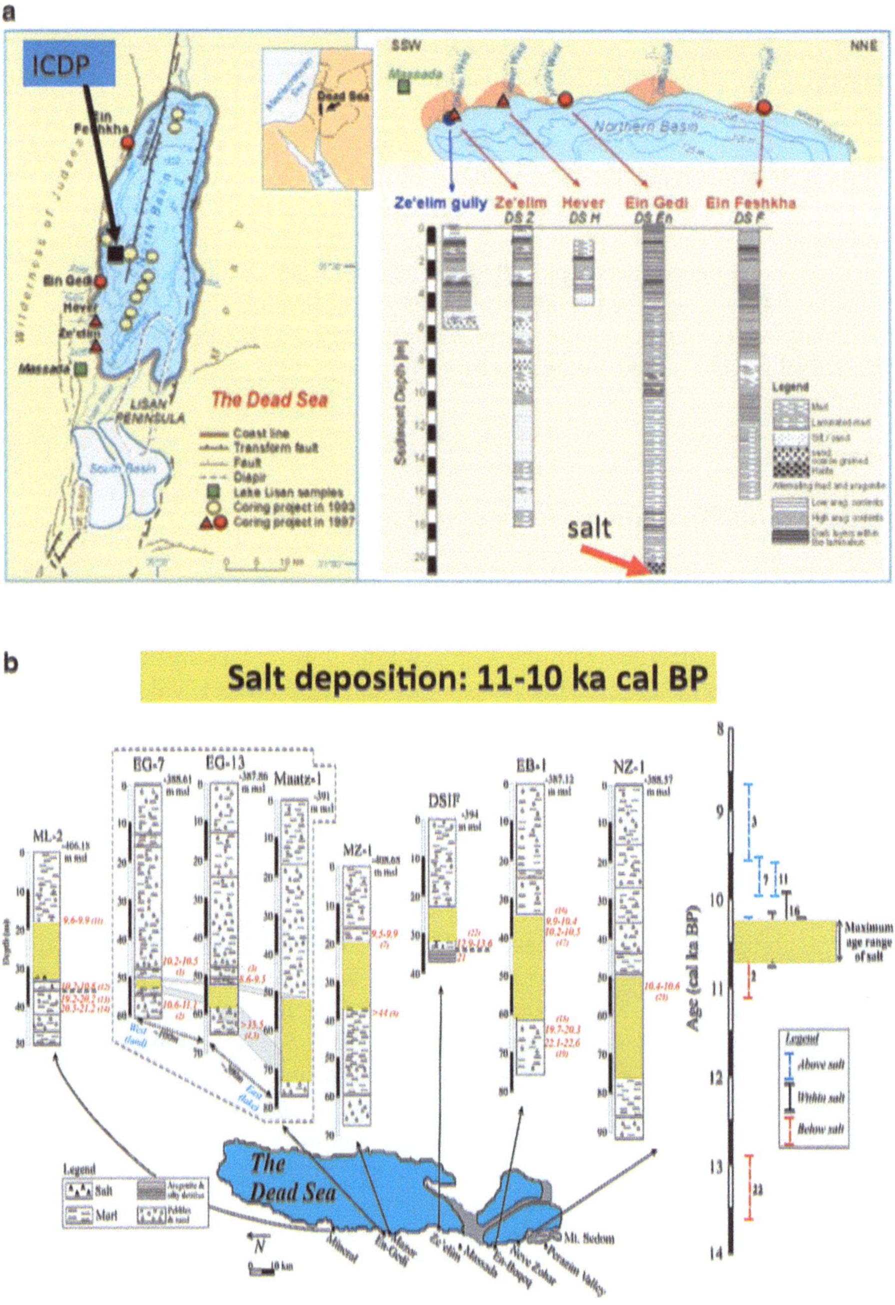

Fig. 10.11 (**a**) Scientific drilling projects at the Dead Sea. The figure illustrates the location of boreholes drilled in: the 1993 campaign (leaders: J. Negendank and Z. Ben-Avraham, see Heim et al. 1997); 1997 campaign (leaders J. Negendank and M. Stein and A. Agnon, see Migowski et al. 2004, 2006), and the deep site drilled in 2010/2011 under the ICDP-Dead Sea project, see Stein et al. (2011). (**b**) Boreholes recovering the post-glacial to Holocene Dead Sea sections containing the thick salt unit (in *yellow*) dated to ~11–10 ka (After Stein et al. 2010). (**c**) Thick halite units exposed at the Arubotaim Cave section (late-Pleistocene Amora Fm.), and recovered in boreholes drilled into the sections of the Holocene Dead Sea. The Amora salt was deposited during MIS11 (~400 Ka) and the early Holocene salt was deposited between 11–10 ka (Fig. b). The deposition of the salts indicates significant decline of the Dead Sea (below 400 bmsl) and aridity in the drainage area

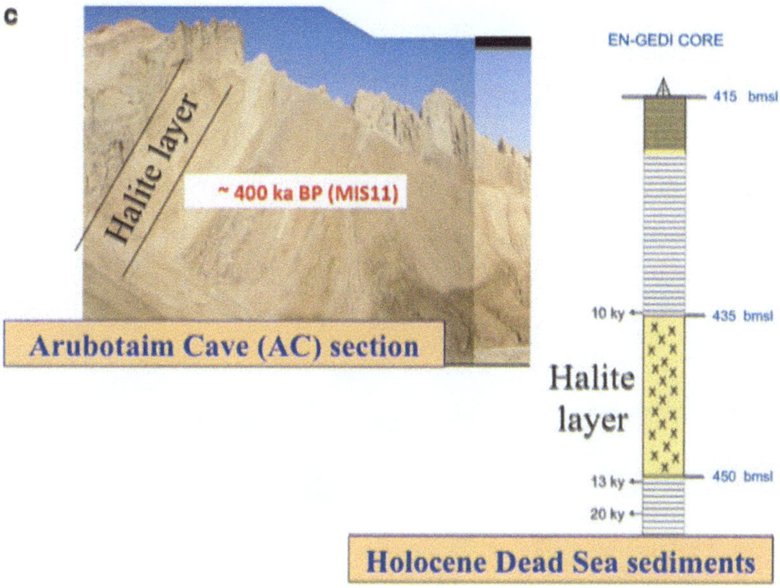

Fig. 10.11 (continued)

The salt unit was first recovered from the DSIF core drilled at the Ze'elim valley (Yechieli et al. 1993). The sedimentary sequence of the core consists of fan delta deposits, marls (*ld* facies type) and the massive salt unit. The drilling recovered ~7 m thick salt sequence, which overlies ~4 m of marly material (containing some cubic secondary halite and disseminated gypsum). These marly sediments unconformably overlie the laminated glacial Lisan aragonites, dated by U-Th age to 21 ± 2 ka BP (Yechieli et al. 1993). Wood debris recovered from this marly sequence yielded a radiocarbon age of 11.3 ± 1.6 ka BP (radiocarbon age), which is calibrated to ~13.2 (13.6–12.9 ka cal BP) within the Younger Dryas time interval. The evidence for an abrupt lake level drop between ~14.6–13.2 ka cal BP and the requirement for a lake level rise prior to the deposition of the salt layer at ~11 ka cal BP indicate that between 13.2 and 11 ka cal BP lake level rose from below 465 m bmsl to above 415 m bmsl (highest elevation of the salt unit at the Ze'elim DSIF site). Recently, Liu et al. (2013) reported on desert varnish that was deposited on the relief of the recessional shorelines of Lake Lisan, e.g. on the fan delta surfaces between 280 and 365 m bmsl along the western margins of the Dead Sea (e.g. Nahal Darga). The varnish comprises thin (a few micron in thickness) laminae characterized by a low Mn and Ba orange/yellow surface layer and a high Mn and Ba dark basal layer (e.g. Fig. 10.12). The deposition of the dark basal layers immediately after the lake recession represents a wet period coinciding with the Younger Dryas (YD) (12.9–11.6 ka). In contrast, varnish from the distal base of fan deltas contains only orange/yellow surface layers, diagnostic of the Holocene relatively dry climate. The absence of the dark basal layers in the varnish further indicates an YD high-stand at ~365 m msl and a lake level rise of at least 100 m from its Bølling/Ållerød low-stand. Liu et al. suggest that the wetter conditions during the YD could reflects a southward shift of the Atlantic meridional overturning

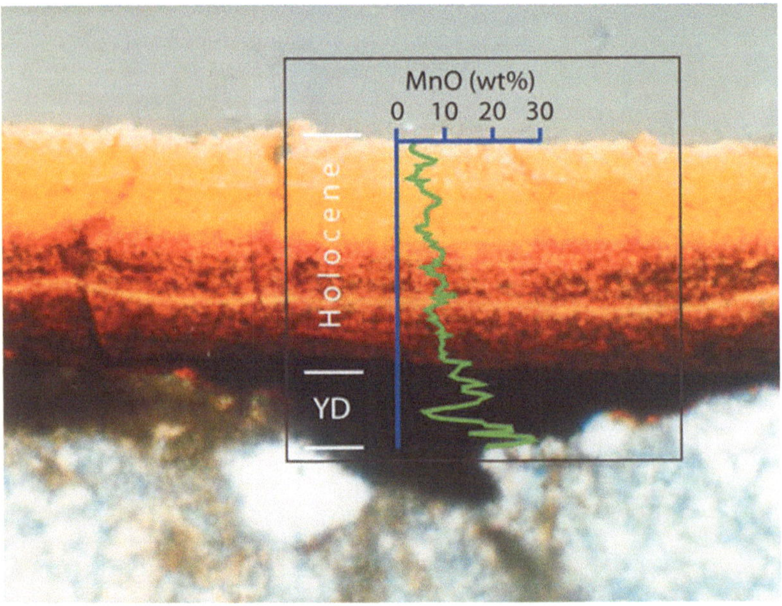

Fig. 10.12 Rock-varnish comprising thin (a few micron in thickness) laminae characterized by a low Mn and Ba *orange/yellow* surface layer and a high Mn and Ba dark basal layer (After Liu et al., 2013)

circulation, which in turns modulated the mid-latitude westerly winds in the east Mediterranean-central Levant region and led to enhanced rains.

10.3.4 Lakes Amora and Samra

The Amora and Samra Formations comprise the sedimentary sequences that were deposited in lakes Amora and Samra, respectively. The name Samra Formation was designated to the sedimentary sequence that was deposited in the water-body filling the Dead Sea tectonic depression during the last interglacial (MIS5), while the name Amora Formation accommodated a long sedimentary sequence that was deposited and accumulated in the tectonic depression since the time of disconnection of the Sedom lagoon from the open sea (~ 3 Ma, Torfstein et al. 2009), thus containing significant part of the mid to late Quaternary history of the basin. The stratigraphy of the Amora Fm. was first established by Zak (1967) in his canonical study of the Mount Sedom salt diapir. The rising diapir uplifted and tilted the lacustrine sequences that were deposited above the Sedom Fm. On the eastern flanks of Mount Sedom the lacustrine layers lie in dips of ~90° (Fig. 10.6) providing beautiful exposure of the formation and the possibility to investigate in details the entire mid to late Pleistocene history of the lacustrine system in the Dead Sea basin. Torfstein et al.

(2009) established the chronology of the Amora Fm at the Arubotaim Cave (AC) section by combining U-Th dating with oxygen isotope stratigraphy and paleomagnetic constraints. The idea of applying oxygen isotopes ratios as stratigraphic–chronologic tool is based on the simple relation and constant difference between the $\delta^{18}O$ in the source of rain water in the east Mediterranean and the primary aragonite deposited from the lake water (Kolodny et al. 2005). The East Mediterranean (EM) is considered to be the main source of water to the DSB drainage area. Accordingly, the Amora $\delta^{18}O$ record is tuned to the EM $\delta^{18}O$ record published by Almogi-Labin et al. (2009). The base of the exposed Amora Formation at the Arubotaim Cave (AC) section is younger than 780 ka, consistent with the normal paleomagnetic epoch. The uppermost part of the Arubotaim Cave section was dated to the last interglacial MIS5. In between, the formation comprises several sequences of glacial-interglacial ages (MIS 18 to 5). A prominent salt unit is constrained by U-Th dating to approximately 400 ka (corresponding to MIS11). This unit represents the only exposed occurrence of massive halite in the Quaternary DSB lake deposits. While massive halite is relatively abundant in the central and southern basins of the Dead Sea (Neev and Emery 1995) its precipitation or preservation on the shallow margins of the basin is rare. Massive salt units were recently recovered by the ICDP –Dead Sea deep drill (Stein et al. 2011). A prominent salt unit in the ICDP core was recently dated to MIS5e. Pebbles that could mark a proximate shoreline and thus a significant lake retreat overlie this unit (Goldstein et al. 2013).

10.3.5 The Holocene Dead Sea

The Holocene Dead Sea evolved from the last glacial Lake Lisan. As indicated above, Lake Lisan reached its highest elevations of ~160–200 m bmsl during Marine Isotope Stage 2 (MIS2) and at ~17.4 ka commenced its post-glacial retreat arriving to its minimum level (below 450 m bmsl) between 14 and 13 ka cal BP, rising back to ~330 m bmsl during the Younger Dryas and then declined again between 11 and 10 ka when a thick salt unit was deposited (Fig. 10.11, and see Stein et al. 2010).

The sedimentary section that was deposited from the Holocene Dead Sea comprises the Ze'elim Formation (Bookman et al. 2004). The lithology and chronology of the upper part of the Ze'elim Formation (past 3 kyear) were described and established in wall-exposures of recently formed gullies (e.g. the Ze'elim gully, Fig. 10.13) that dissect the exposed Dead Sea floor. The formation of these gullies reflects the anthropogenic retreat of the Dead Sea during the past few decades. The early part of the Holocene (~4 and 10 ka) is mainly recovered from sediment cores that were drilled along the retreating shores of the modern Dead Sea (Migowski et al. 2004, 2006). Yet, deep-cut exposures in the Ein Qedem shore and Darga Valley reveal units of the early Holocene (Kadan 1997; Stern 2010). Lake level reconstruction of the Holocene Dead Sea (Fig. 10.10) indicates moderate rise above the salt unit between ~10–9 ka followed by significant level declines during the time interval of 9–6 ka. At ~ 8.1; 5.6; 4.1 and 3.6 ka the lake declined abruptly to ~430 m bmsl and below.

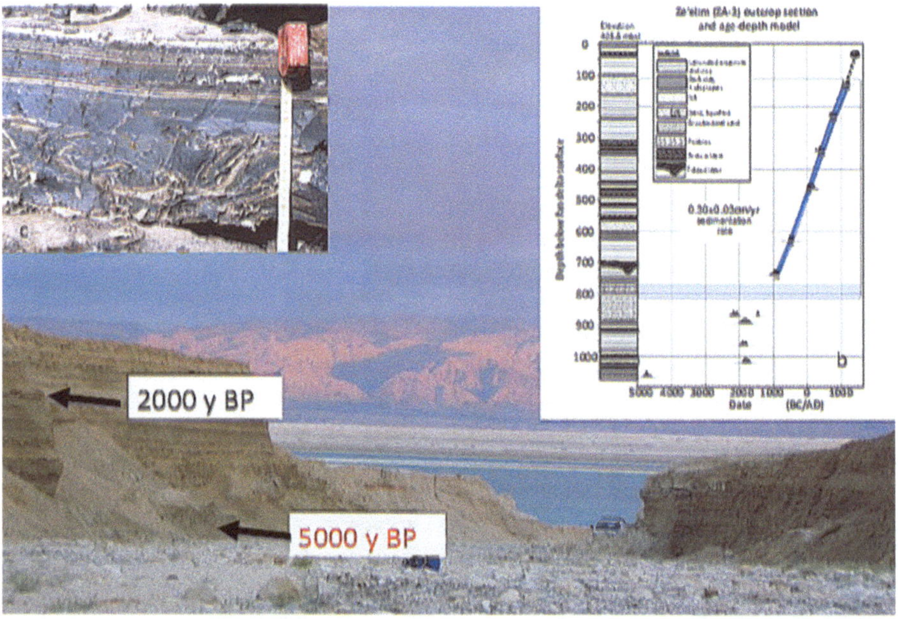

Fig. 10.13 The Ze'elim gully exposing the sedimentary sequence of the mid to late Holocene Dead Sea (the Ze'elim Formation). The Uppermost part of the section (past 2,500 years) comprises mainly sequences of *aad* couplets or detritus-aragonite–gypsum triplets. The lower part of the section is dominated by laminated detritus of the ld facies. The chronology of the Ze'elim Fm. was established by 14 C dating of organic debris that are abundant in the exposures (see example of an age – height model for the time interval of 1,000 BC to 1,000 AD, after Kagan et al. 2011). The Ze'elim gully exposures were used to establish high-resolution paleo-seismic chronology of the Dead Sea region (Ken-Tor et al. 2001; Kagan et al. 2011). The *upper left hand* side inset displays an example of "seismite" unit dated to 31 BC, which was reported by Flavious. This seismite comprises fragments of primary aragonite within silty-detritus matrix

These drops coincide with some global climate events (e.g. ice rafting in the North Atlantic) and were linked to solar activity (Kushnir and Stein 2010, and see below).

The term "Dead Sea" is related to the water-body that filled the Dead Sea basin after 10 ka. To a certain extent, this terminology is arbitrary since the lake with geochemical properties that are similar to the modern Dead Sea was formed only at ~3,000 years BP upon the transition from the late Bronze to the Iron Age. During the past 3,000 years the lake precipitated primary aragonite and fine detritus, while in the early and mid-Holocene almost no aragonite was deposited and the lithology of the lake's sediments was dominated by detrital calcites. Waldmann et al. (2007) attributed the change in the composition of the lake sediments and deposition of primary aragonite to reactivation of the Ca-chloride brines due to enhance precipitation on the Judea Mountain. The sediments deposited during the time interval of 10–3 ka BP resemble those of the last interglacial Samra Fm. Thus, from the viewpoint of the section lithology the conditions of the Holocene interglacial terminated at ~3–4 ka BP and the "neo – glacial" period commenced.

10.4 Paleo-Hydrological and Paleo- Climatic Patterns

10.4.1 General

The reconstruction of the paleo-hydrological conditions during the history of the lakes is based on the lithology, sedimentological facies and geochemistry (e.g. oxygen isotope compositions) of the sediments (Stein et al. 1997; Waldmann et al. 2007; Torfstein et al. 2009) as well as the lake level reconstruction (e.g. Bartov et al. 2002, 2003; Torfstein et al. 2013a, b; Stein 2014). The glacial intervals (e.g. MIS10, 8, 6, 4, 2) comprise mainly the *aad facies* indicating positive freshwater inputs and transport of bi-carbonate to the lake that allows the deposition of the primary aragonite (Stein et al. 1997). The interglacial and interstadial intervals (e.g. MIS11, 7, 3 5, 1) are characterized by sequences of the *ld*, gypsum (or anhydrite) and sand facies. The *ld* facies was attributed to floods carrying fine detritus from the surface cover (e.g. soils, settled dust, loess) to the lake (Haliva-Cohen et al. 2012). Delicate information on the paleo-hydrology is provided by the oxygen isotope compositions of the primary aragonites. Overall, the *long-term* lake oxygen isotopic composition was dominated by the composition of the coming freshwaters, evaporation and degree of mixing between the upper (fresher) lake layer and the lower brine (e.g. Kolodny et al. 2005; Lazar et al. 2014). Yet, short-term variations in the oxygen isotopic composition of the lake's upper water layer were controlled by the regional hydrological regime, relative humidity, water activity and evaporation conditions in the lakes (Gat 1984; Kolodny et al. 2005). Variations are reflected by discrete (possibly annual) $\delta^{18}O$ excursions towards lighter values showing shifts of approximately 2–3‰. Such excursions have been observed in the twentieth century Dead Sea as well as in a ~100 sequence of aragonite laminae in the Lisan Formation from ~35 ka BP (see examples in Kolodny et al. 2005 and Torfstein et al. 2009).

The following chapter deals with the implications of the lake limnology: composition, configuration and mainly lake levels to the understanding of regional hydrology and relation to global climate patterns. The lakes that have occupied the Dead Sea basin are considered as terminal lakes, whose levels have been sensitive to the amount of incoming water and evaporation. They thus provide a record of hydroclimatic variability in the Dead Sea drainage area and by inference in the entire Levant region. That said, several factors are combined in the shaping of the composition and limnological configuration of the lakes: e.g. layered or overturned lake, salinity and evaporation, hydrological conditions in the drainage area (e.g. Stein et al. 1997) and it is not expected that the lake response to regional climate and hydrology to be immediate or linear. The section deals first with a climatic model that explains the modern patterns of rains in the drainage area of the Dead Sea – and by inference the Levant region and then discusses the implications for the older lakes.

10.4.2 The Dead Sea as the Levant Rain Gauge

The Dead Sea basin consists of two sub-basins: a deep northern basin (~300 m deep), occupying about two thirds of the lake and a very shallow southern basin (which is currently essentially dry). The basins are separated by a sill at a level of ~402–403 m bmsl. The lake level curve (Fig. 10.10) shows the Holocene Dead Sea fluctuated between levels of ~430 and 370 m bmsl, and rose or declined beneath the sill.

Flooding the southern basin leads to an abrupt increase in the total lake evaporative flux and which buffers further increases in lake levels. In contrast, when lake levels drop below the sill, the surface level becomes more sensitive to changes in fresh water flux as the lake area is confined to the northern basin. Additional complexities in lake level response to freshwater influx are associated with changes in the surface area as the level changes (Abu Ghazle et al. 2009) and the dependence of the rate of evaporation to its surface salinity (Stanhill 1994).

Despite these complexities, Enzel et al. (2003) argued for a relatively simple relationship between lake level and regional precipitation using the instrumental precipitation record at Jerusalem and the measured lake levels recorded between 1930 and 1964 (before human intervention to the flow of the Jordan River and when lake levels were above the sill). They found that the Dead Sea level displays a multi-year rise when the annual precipitation amounts are distributed around a mean of 648 mm, with standard deviation of about 122 mm and a multi-year decline when the annual precipitation amounts drop to a mean of 445 mm, with a standard deviation of 117 mm. From their data, one could conclude that the rate of decline in lake level in the time interval of 1930–1960 is 8–10 cm year^{-1} consistent with the overall low annual precipitation amount during that period (average of 568 mm compare to the entire record average of 607 mm). In particular, between 1930 and 1945 the lake dropped by 2 m, yielding a rate of decline of 14 cm year^{-1}. This rapid decline came in the wake of the 10-year precipitation minimum of the entire instrumental record, averaging 390 mm year^{-1} between 1925 and 1934.

The Levant rainy season occurs during the months of October to April with most of the rainfall occurring in the winter months. During this season, cold-core upper-level, low-pressure troughs and ridges migrate from west to east over southern Europe and the Mediterranean (Ziv et al. 2006; Trigo 2006). These systems drive cold and often relatively dry air masses from Europe over the mountain ranges that circle the EM Basin to the north and over the relatively warm sea. The lifting of the air over topography and the encounter with the warm Mediterranean water lead to the formation of surface lows (cyclones) that provided with moisture from the water surface below produce clouds and precipitation. Of particular importance for the Levant are the cyclones that tend to deepen or form in the EM, known as Cyprus Lows (cf. Ziv et al. 2006). In the other half of the year (May to September) the Levant is dry due to the strong regional subsidence induced by the remote influence of the Indian summer monsoon system (Ziv et al. 2004).

Kushnir and Stein (2010) examined the possible effects of global climate phenomena on Levant precipitation variability. These are: the North Atlantic Oscillation

(**NAO**), the El Nino southern oscillation (**ENSO**) and Atlantic multidecadal variability (**AMV**). The NAO is associated with coherent latitudinal fluctuations in the Atlantic, wintertime eddy-driven jet stream. Connected with that are swings in the location of the Atlantic winter storm track from a northern path to a southern path and vice versa, depending on the NAO phase (Hurrell et al. 2003; Lee and Kim 2003). The downstream effect of this Atlantic-centric phenomenon "spills" into southern Europe and the northern Mediterranean countries all the way to the EM. Thus a negative NAO phase is associates with larger than normal rainfall in Turkey (particularly in the west) and a positive NAO phase leads to lower than normal precipitation there. Indeed, NAO influence was clearly identified in precipitation variability north of the Levant, over the Anatolian Peninsula and the Aegean Sea (Cullen and deMenocal 2000; Marshall et al. 2001). However, in the Levant (and along the North African coast of the Mediterranean) the influence of the NAO on precipitation is weak, and possibly in opposite phase to its effect to the north (Cullen and deMenocal 2000; Enzel et al. 2003).

Mariotti et al. (2005) found a clear ENSO linkage to Mediterranean and Levant precipitation but only in the relatively marginal fall season (September-November). Thus, one cannot rule out an overall weak influence of ENSO on the Levant wet season (October to April).

Kushnir and Stein (2010) proposed a new mechanism that explains the multi-decadal to millennial patterns of the Levant precipitation and moreover the observed antiphase climate-hydrology relation between the Levant and the sub-Sahara Sahel region. They argued that decadal and longer time-scale variations of Levant-Sahel hydroclimate are affected by the multi-year variability of North Atlantic SSTs and sea level pressure (SLP) over the east Atlantic (Fig. 10.7). Rainy winters in the Levant are associated with higher than normal SLP over the eastern North Atlantic extending into Western Europe and vice versa. Kushnir (1994) showed that a positive–negative AMV composite (warm-cold SST in the North Atlantic), exhibits a negative sea level pressure (SLP) anomaly over the mid-latitude North Atlantic, between 30 and 60°N. The implication is that SLP is higher than normal in the middle of the North Atlantic and over Western Europe when the basin is colder than normal. Higher than normal SLP in the mid-North Atlantic Basin, during the cold season, deflects the extratropical storms from their normal tracks increasing the chances of polar air intrusions directly into the Eastern Mediterranean and this, in turn, leads to higher likelihood for cyclogenesis and consequently higher precipitation in the Levant. Conversely, when Atlantic SSTs are higher than normal, lower than normal SLP in the mid-North Atlantic Basin brings rather warm air from subtropical origin to the Eastern Mediterranean reducing the chances of cyclogenesis and consequently leads to a dryer than normal winter in the Levant. The cold North- Atlantic scenario has consequences for the Sahel and North America regions. The EM winter environment, with the large temperature contrast between the cold Eurasian continent and the warm sea and the Anatolian topography to the north, create unique conditions that amplify the Atlantic impact compare to other parts of the Mediterranean.

10.4.3 Implications for the Hydrological-Climate Conditions During the Holocene and Late Pleistocene Periods

Could the North Atlantic – Levant – Sahel model be extended to the entire Holocene and late Pleistocene time intervals? Archives of Levant – Sahel paleo-climate exhibit large centennial to millennial fluctuations. Considering uncertainties in the chronologies and lake level elevations, particularly of the African lakes, the anti-phase behavior in lake level variability is apparent in the geological records. For example, during the Bölling-Allerod (B-A, ~14–13 ka BP) and the YD, when the Dead Sea level (DSL) dropped precipitously and then rose above the sill level, respectively, the African hydroclimate indicators (terrigenous dust record included) swung the other way – up and down. The subsequent millennial swings during the African humid period (AHP) and during the period of African lake decline also seem to maintain the anti-phase relationship. The relationship between centennial to millennial variability in the Levant and Sahel records and North Atlantic SST fluctuations on the same time scale is hard to discern with confidence from ocean sediment records, as these have a rather coarse and temporally uneven resolution. However, some support to our hypothesis emerges during the past few millennia. During the late Holocene, the Northern Hemisphere climate changed from the Medieval Climate Anomaly (MCA; 900–1400 AD), a relatively warm period in Europe and most likely the North Atlantic to the Little Ice Age (LIA, 1500–1900 AD), a cold period in these areas (see recent discussion by Denton and Broecker 2008). Based on records of the advance and retreat of mountain glaciers in Europe and North America and historical evidence on the extent of wintertime sea ice coverage in the North Atlantic, Denton and Broecker (2008) argued for associated centennial time-scale variations in the strength of the AMOC and related SSTs (AMV) in the Basin. According to their synthesis the North Atlantic was relatively warm during the MCA and cold during the Little Ice Age. A comparison between the record of advances and retreat of Swiss Alpine glaciers and the DSL in the last 4,000 years exhibits remarkable agreement between high lake stands and cold intervals in Europe. Consistent with the hypothesis of Denton and Broecker (2008) this supports the proposition regarding the role of the North Atlantic in Levant precipitation variability.

10.4.4 Catastrophic Lake Drops – The Drying of the Subtropical Zones

On the orbital time scale (e.g. the precession cycles of ~ 20 kyear) the Dead Sea lakes exhibit high-stands during north latitudes glacial periods and low stands during interglacials. On the millennial and multi-decadal scale they appear to follow the Kushnir and Stein North Atlantic AMV model. On both time scales it appears that the anti-phase hydrological correlation between the Levant and Sahel exists.

Recently, Torfstein et al. (2013a, b) re-evaluated the Lake Lisan level fluctuation during MIS3 (Fig. 10.10c). At: ~42.5, ~40.3, ~37, ~34, ~32.5, ~30.5, and ~28.1 ka the lake rose to elevations higher than 260 m bmsl and in between these episodes declined to elevations of ~280 m bmsl and below. These fluctuations appear to coincide with the millennial fluctuations in the ice cores records and by inference in the ice-cave-lake records that are referred to Dansgaard-Oeschger (D-O) stadial-interstadial cycles. The D-O cycles apparently reflect a global connection between the temperatures in the ice cores and patterns of monsoon and east Mediterranean rain fronts that are responsible for the speleothem growth in the Chinese cave or the rain precipitation in the Levant and Lake Lisan watershed. This topic is elaborated below. The sharp oscillations in lake levels during MIS3 appear to coincide with the oxygen isotope variations in the Greenland ice core. Within a D-O cycle the lake rose when Greenland temperature dropped and declined when Greenland temperature rose. The D-O type lake level fluctuations during MIS3 appear to be related to cold-warm North Atlantic temperatures and by inference to the multi-decadal millennial AMV model of Kushnir and Stein (2010). Yet, the D-O related lake fluctuations are punctuated by abrupt and large (>100 m) drops in lake levels that are accompanied by massive deposition of primary gypsum. The ages of these gypsum units are consistent with the timing of Heinrich (H) stadials in the North Atlantic and coincide with the H-related perturbations in the ice cores. The coincidence between the abrupt Lake Lisan drops and H-events was already noticed and discussed by Bartov et al. (2003). At a first glance this behavior looks paradoxical considering the "cold Atlantic SST – high Dead Sea level" Kushnir and Stein model. The association between the abrupt lake drops and Heinrich-stadials could be related to the impact of iceberg discharge into the North Atlantic. This process inhibited the production of North Atlantic Deep Water, resulting in the suppression of northward transport of heat from the tropics. Thus, lower sea surface temperatures in middle latitudes, including the Mediterranean, in combination with the cooling of air masses originating in the North Atlantic and bound for the Levant resulted in an overall weakening of the air-water thermal gradient, which inhibited vapor uptake from the sea, and in turn limited the delivery of precipitation to the DSB drainage (Bartov et al. 2003). The shutdown of Mediterranean rains during the H-events coincided with severe droughts and low lake stands in the sub-Sahara that were related to the weakening of the Indian monsoon. This could apparently be related to the spread of cooling into the Indian Ocean (Pausata et al. 2011). Modeling studies seem to support this hypothesis, showing that intense North Atlantic cooling, whether during the LGM or the Holocene, lead to cooling and drying of Europe and a southward shift in the Intertropical Convergence Zone (ITCZ) over the entire globe (Chiang 2009), thus yielding a more extreme pattern than the slow and relatively mild millennial cycles and insolation forced variability. The cooling of the Indian Ocean due to the sea-ice spread in the North Atlantic is also reflected in the Chinese speleothem records. There, sharp $\delta^{18}O$ excursions to heavier compositions coincided with the H-events. These $\delta^{18}O$ excursions were recently reinterpreted as reflecting reduction in the precipitation over the Indian Ocean and weakening of the Indian monsoon (Pausata et al. 2011). The shift in the global atmospheric systems during the H-events was reflected also in the patterns of atmospheric circulation and transport of desert dust. While the majority of dust storms that approached the northern Red Sea – east Mediterranean – Dead Sea originated during the late Quaternary at the Sahara deserts it appears that during the H-events dust was mainly mobilized from local sources, e.g. Red Sea margins, Negev desert, northern Sinai (Haliva-Cohen et al. 2012; Palchan et al. 2013; Roskin et al. 2011).

10.4.5 Wet Phases in the Negev Desert During Past Interglacials

The overall picture of the Levant paleoclimate on orbital time scale is of relatively wet and cold glacials and arid and warm interglacials. This pattern is overprinted by millennial and multi-decadal hydrological variations that reflect the SST in the north Atlantic and by abrupt catastrophic aridities that coincide with the H-events in the north Atlantic possibly reflecting cooling of the east Mediterranean and shut down of the cyclonic rain engine. The Sahara-Sahel deserts that are affected by the Indian monsoon system exhibit the opposite picture of wetter interglacials and arid intervals during cold North Atlantic SST. While the effects of the Mediterranean rains fade out currently towards the northern Negev desert (with ~200 mm rain at Beer Sheva) and the southern Negev desert, Arava valley and Gulf of Aqaba are considered as one of the most arid areas on earth with less than 20 mm of rain annually, there is a line of evidence for episodes with wetter conditions at the Negev desert during the interglacials that could cause significant floods in the major rivers that are draining to the Dead Sea from the south, mainly the Arava, Zin and Paran rivers. Enhanced deposition of travertines occurred at the Arava valley during MIS7 and 5 (Livnat and Kronfeld 1985; Enmar 1999) as well as precipitation of cave speleothems in the southern Negev Desert during MIS9, 7 and 5 (Vaks et al. 2007) might indicate a southern (relative to the DSB) source of humidity (Waldmann et al. 2010). This scenario is corroborated by the evidence for enhanced activity of fresh groundwater along the shore of the Gulf of Aqaba and the Red sea that caused a massive transformation of aragonitic skeletons of the coral reefs to calcite (Lazar and Stein 2010). The scenario of a southern humidity in the desert areas during interglacials stands in line with the evidence of filling of dry lakes in the Sahara (e.g. Gasse 2000). In this context, we point out that EM sapropel events, which are generally associated with enhanced summer monsoon precipitation over subtropical Africa and subsequent heavy Nile floods in the EM, usually corresponds to interglacial times (Rossignol-Strick 1985; Rossignol-Strick et al. 1998; Rossignol-Strick and Paterne 1999). Short humid episodes during MIS7 and other interglacial's could be related to enhanced activity of Red Sea Troughs (e.g. Greenbaum et al. 2006) or "Tropical Plumes" which pertain to the transport of moisture from the Tropics (i.e., West Africa) to extra-tropical latitudes (e.g. Ziv 2001; Waldmann et al. 2010). Such episodes have been recorded during the twentieth century and have occasionally triggered extremely heavy rain over short periods (hours, days) in the Negev Desert. A beautiful example for enhanced rains from southern Red Sea sources in the Negev desert occurred in January 2010, when strong floods washed via the major rivers to the southern Dead Sea carrying large amounts of detritus, mainly detrital calcites and quartz. The abundance of such flooding events and transport of detrital calcites can explain the sequences of these sediments in the interglacial formations, e.g. Samra and the early and mid part of the Holocene (10–6 kyear). Haliva –Cohen et al. (2012) classified the sedimentary facies in the Dead Sea lacustrine formations (Samra, Lisan, Ze'elim) into two types: the *aad facies* comprising sequences of aragonite and silty detritus laminae and the *ld facies* composed of laminated detritus. While sediments of the *aad* facies were deposited mainly during glacials when the lake was at high stand conditions and most of the rains were produced by the east Mediterranean cyclones, the sediments of the *ld facies* reflects occasional floods, mainly in the Negev desert that are abundant during interglacials. The major difference, however, between the *aad and ld facies* is in the appearance of aragonite that in turns requires supply of bi-carbonate to the Ca-chloride brine in the lakes (Stein et al. 1997). The bi-carbonate reflects massive dissolution of the carbonates surface cover of the northern and central mountain area of Israel. In this relation, I return to the Holocene section at the Dead Sea. The Ein Gedi core, which recovers the entire Holocene section shows that the lower the middle parts of the Holocene section (10–4 ka) are composed mainly the *ld* facies, while abundant aragonite laminae of the *aad* facies (here they appear in triplets with the addition of gypsum laminae) appear in the upper part of the section covering the past 3,000 years. As mentioned above Waldmann et al. (2007) explained the appearance of the aragonite by resumption of the activity and contribution of the Ca-chloride brines (similar to the modern saline springs at En Qedem). Yet, the supply of bi-carbonate is the initial requirement for deposition of aragonite.

10.5 The Climate – Human Development Connection in the Levant

The climate-hydrological evolution of the southern and central Levant – the region between the Red Sea and the Taurus Mt, were accompanied during the Neogene – Quaternary by the migration of early hominids out of Africa and the development of the pre-historical and historical cultures. A comprehensive review of this subject is beyond the scope of this manuscript. Nevertheless, I will outline several main developments in human culture evolution that are related to the evolution of the water-bodies and the hydrological conditions in their watershed as described in this chapter.

10.5.1 The Early Prehistoric Cultures

The Erq-el-Ahmer and Ubediya and Gesher B'not Yaakov archeological sites are located within succession of lacustrine sediments. As mentioned in Sect. 10.2, recent dating of the EEA formation suggests early age of ~4.5–3.6 Ma (Davis et al. 2011).

10.5.2 The Migration of the Homo-Sapience (the Anatomically Modern Hominids) Out of Africa

Several lines of evidence indicate on availability of freshwater along the shores of the currently hyperarid Red Sea, Gulf of Aqaba and the southern Arava Valley during past interglacials (Lazar and Stein 2010). The wetter conditions in the Red Sea–Arava valley is associated with the setting of "wetter" conditions during interglacials in the Sahel-Sahara regions as described in Sect. 4.5 above.

10.5.3 The Disappearance of the Neanderthal and the "Out of the Caves Movement"

During the glacial period when Lake Lisan extended over the Dead Sea basin and Jordan valley (Fig. 10.1), the area was characterized by wet and cold climate conditions that possibly pushed the Neanderthal and Homo-sapience to live in caves (e.g. the Tabun cave at Mt. Carmel). After the retreat of Lake Lisan and the Sea of Galilee from its high stand of ~160 m bmsl to ~200 m bmsl at ~24–23 ka BP (Fig. 10.10d) humans established settlements on the Sea of Galilee shore, e.g. the Ohalo-II site (Nadel 1995). The age of the Ohalo-II site coincides with the H2 event, which was expressed in the Levant area by expansion of the desert and enhanced aridity. I speculate that the Ohalo-II "people" were pushed to find a new source of food for their survival – fish in the Sea of Galilee. The extreme aridities at the Levant and the lake's watershed during the Heinrich events had an affect on the Neanderthal culture that eventually disappeared from the region. I think that the Neanderthals specified in "cave life" and could not survive the "out of the cave" development exemplified by the Ohalo-II people.

10.5.4 The Collapse of the Natufian Culture and the Neolithic Revolution

The Natufian people who lived in the central Levant region during the late glacial (~16–14 ka BP) when the region recovered from the aridity of the H1 time interval: 17.4–16 ka BP. They, however disappeared during the catastrophic aridity of

~14–13.5 ka (Stein et al. 2010). The emerging PPN cultures saw the recovery of the hydrological system during the Younger Dryas and the transition to the Holocene (~12–10 ka BP). The PPN-B people established the first agriculture settlements in the fertile valley and oasis's of Jericho, Gilgal and Fazael. Stein et al. (2010) suggested that the accumulation of the fertile terra rossa soil (the Fazael Formation) in these valleys combined with the resumption of the activity of the hydrological system at ~10 ka BP, allowed the "agriculture –Neolithic" revolution in the Jordan Valley.

10.5.5 The Rise and Fall Off the Bronze Cultures

During the fourth millennium BP the Mediterranean – Levant region saw the flourishing of the Bronze cultures when numerous cities and settlements were established across the Mycenaean kingdoms, the Hittite, and the Egyptian Empires. This prosperity arrived to a sudden termination at ~3,300–3,200 years cal BP when most of the cities were destroyed and settlements were abandoned. Several hypotheses were proposed to explain the sudden collapse of the mid-late Bronze cultures: deteriorating climate conditions, effects of the volcanic eruptions and devastating earthquakes. This chapter examines the chronology of events associated with the mid-Bronze crisis in light of the hydrological–climate history of the Levant-Nile land that is stored in the sedimentological-geochemical archives of the Dead Sea and Sahelian lakes. Between ~3,500 and 3,200 year BP the Dead Sea dropped abruptly by more than 60 m from its mid-Bronze elevation of ~370 m bmsl, marking catastrophic aridity in the regional watershed (Kushnir and Stein 2010; Kagan et al. (submitted); Langgut et al. 2014). The lake decline coincided with abrupt cooling of the North Atlantic, which we surmise led the shutdown of the regular process of Eastern Mediterranean cyclogenesis as well as monsoonal rains in the Sahel: the engines of Levant rains and Nile waters. The catastrophic aridity and the expansion of arid climates across the tropical-subtropical latitudes disrupted the long–term and multi-decadal antiphase Levant-Sahel hydrological pattern that possibly allowed the "out of Africa" migration of the prehistorical hominids and the frequent movements of the historical cultures along the Red Sea- Rift Jordan Valley corridor.

10.6 The ICDP Dead Sea Drilling Project and Future Perspectives

The studies of the sedimentary sequences that were deposited at the water-bodies that filled the Dead Sea basin focused mainly on the marginal terraces where the Lisan Formation is beautifully exposed and to a limited extent the Samra, Amora and Ze'elim Formations are exposed as well. Drilling projects at the shores of the currently retreating Dead Sea recovered the Holocene and segments of post-glacial

Fig. 10.14 Pictures from the ICDP Dead Sea drilling project (DSDDP) with DOSECC deep lake drilling facility. The displayed cores comprise sequences of marls and salt that were deposited during the last interglacial

sections (Migowski et al. 2004, 2006; Stein et al. 2010). The information on the behavior of the lake at time interval of low stands (below 400 m bsl) is stored in the deepest basin of the Dead Sea and required deep drilling. This was the main motivation behind the Dead Sea Deep Drilling Project (DSDDP) that was performed under the wings of ICDP. The drilling project took place between November 2010 and March 2011, with participation of scientists from Israel, Germany, USA, Switzerland, Japan and Norway. Drilling was conducted with the Large Lake Drilling Facility (Fig. 10.14) of DOSECC (Drilling Observation and Sampling of Earth's Continental Crust, Inc.). The project recovered core material reaching ~450 and ~350 m from a deep (~300 m below the lake level) and a shallow site (~3 m below lake level) respectively, plus additional shallow holes at both sites. The recovered cores are composed mainly of alternating intervals of marly units and salts that seem to represent glacial and interglacial cycles spanning an estimated interval of ~220,000 years (possibly to MIS 7). A pebble interval at a depth of ~235 m below the lake floor, below the deposits of the last ice age, lies atop a thick sequence of salt (Fig. 10.15), and implies an almost complete dry-out phase of the Paleo-Dead Sea. Recent, U-Th dating of this interval indicates that the severe drying occurred during the last interglacial 5e period (~ 130–120 ka BP, Goldstein

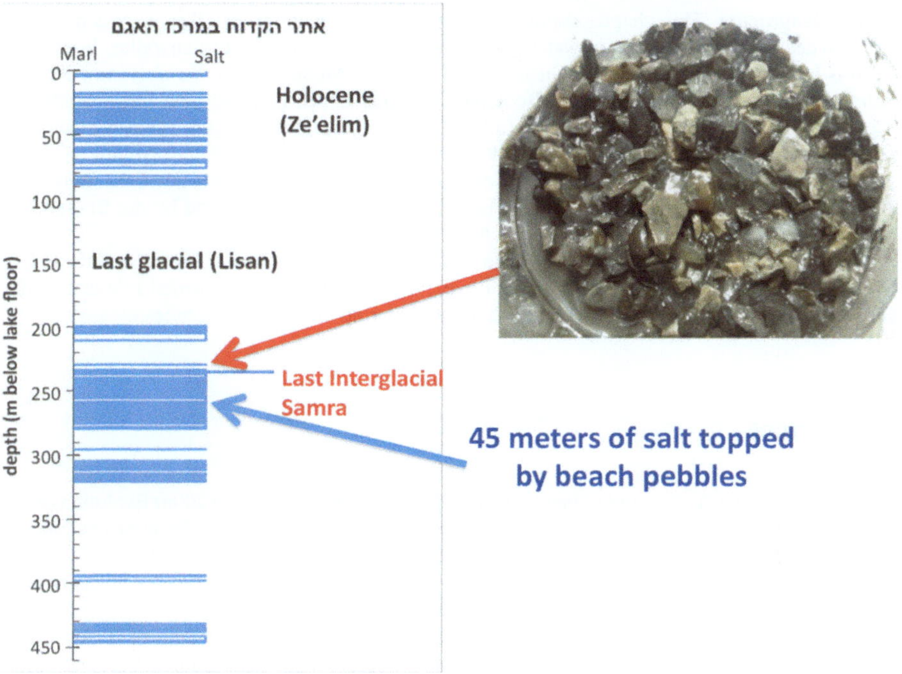

Fig. 10.15 A generalized lithology of the DSDDP core showing intervals of salt (*blue*) and marls (*white* sections). At the depth of ~235 m below the lake floor, a pebbly layer was discovered, atop a thick sequence of salt. This layer indicate the proximity of the shore line implying an significant dry-down of the Paleo-Dead Sea (during the last interglacial, Goldstein et al. 2013)

et al. 2013). This time of extreme aridity in the east Mediterranean – Levant region was accompanied by wet intrusions from the south via the Red Sea –Arava Valley (Waldmann et al. 2010; Lazar and Stein 2010). The next significant salt deposition event at the Dead Sea was that of the post-glacial BÖlling-Allerod period at ~14–13 ka BP (Stein et al. 2010), coinciding with the early phase of the African Wet Period at the Sahara-Sahel. The study of the contemporaneous extreme aridities in the Levant and the wet episodes in the Sahara-Sahel and they relation to the global climate patterns: the North Atlantic –Mediterranean "versus" the African monsoon systems, represents a major research theme that stands in the focus of current and future Dead Sea research.

The ICDP drill core provides the "Dead Sea research community" with unique opportunities to investigate in high-resolution both the changing hydrological-climate conditions and the tectonic history in the Levant region over time periods of great importance in human evolution and with implications to present and future environmental changes in this region.

Acknowledgments This chapter summarizes several of the many works that were done in the past two decades on the history of water-bodies in the Dead Sea rift and Jordan Valley. In particular, it is based on efforts of many colleagues and students: Amotz Agnon, Yuval Bartov, Reuven Belmaker, Zvi Ben-Avraham, Revital Bookman, Yehouda Enzel, Ittai Gavrieli, Steve Goldstein, Alexandra Haase-Schramm, Amit Haliva-Cohen, Nissim Hazan, Elisa Kagan, Yehushua Kolodny, Amitai Katz, Boaz Lazar, Thomas Litt, Malka Machlus, Shmuel Marco, Claudia Migowski, Joerg Negendank, Frank Neumann, Hagai Ron, Marcus Schwab, Abraham Starinsky, Ofra Stern, Adi Torfstein, Nicolas Waldmann and Yoseph Yechieli. The study was supported by the Bi-National US-Israel Scientific Foundation (BSF, grant # 2010.375).

I devote the chapter to memory of my close friend and colleague – the late Prof. Hagai Ron, with whom I shared early ideas and fieldwork at the Perazim Valley that had a pivotal role in creating and pushing the extensive study of Dead Sea research during the past two decades.

References

Abelson M, Yechieli Y, Crouvi O, Baer G, Wachs D, Bein A, Shtivelman V (2006) Evolution of the dead sea sinkholes. In: Enzel Y, Agnon A, Stein M (eds) New frontiers in Dead Sea paleoenvironmental research, Boulder, Colorado USA. GSA Specific paper 401, pp 241–253

Abu Ghazle S, Hartmann J, Jansen N, Kempe S (2009) Water input requirements of the rapidly shrinking Dead Sea. Naturwissenschaften 96:637–643

Almogi-Labin A, Bar-Matthews M, Shriki D, Kolosovsky E, Paterne M, Schilman B, Ayalon A, Aizenshtat Z, Matthews A (2009) Climatic variability during the last similar to 90 ka of the southern and northern Levantine basin as evident from marine records and speleothems. Quat Sci Rev 28:2882–2896

Bartov Y, Agnon A, Enzel Y, Reches Z, Stein M (2002) Sequence stratigraphy and reconstruction of Lake Lisan levels. Quat Res 57:9–21

Bartov Y, Goldstein SL, Stein M, Enzel Y (2003) Catastrophic arid episodes in the eastern Mediterranean linked with the North Atlantic Heinrich events. Geology 31:439–442

Bartov Y, Enzel Y, Porat N, Stein M (2007) Evolution of sedimentary deposition in the late Pleistocene and Holocene Dead Sea basin. J Sediment Res 77:680–692

Belmaker R, Lazar B, Beer J, Christl M, Tepelyakov N, Stein M (2013) ^{10}Be Dating of Neogene halite. Geochim Cosmochim Acta 122:418–429

Bilitzki S, Mimran Y (1996) Active salt diapirism at the Zaharat El-Qurei dome, lower Jordan valley (Jordan). Isr J Earth Sci 45:11–18

Bookman (Ken-Tor) R, Enzel Y, Agnon A, Stein M (2004) Late Holocene lake-levels of the Dead Sea. Geol Soc Am Bull 116(5/6): 555–571

Buchbinder B, Zilberman E (1997) Sequence stratigraphy of Miocene-Pliocene carbonates-siliciclastic shelf deposites in the eastern Mediterranean margins (Israel): effects of eustasy and tectonics. Sediment Geol 112:7–32

Calvo R (2002) Stratigraphy and petrology of the Hazeva Formation in the Arava and the Negev: implications for the development of sedimentary basins and the morphotectonics of the Dead Sea Rift Valley. PhD thesis, Hebrew University of Jerusalem, Jerusalem, 264 p

Chiang JCH (2009) The tropics in paleoclimate. Annu Rev Earth Planet Sci 37:263

Cornée JJ, Roger S, Münch P, Saint Martin JP, Féraud G, Conesa G, Pestrea-Saint Martin S (2006) Messinian events: new constraints from sedimentological investigations and new 40Ar/39Ar ages in the Melilla–Nador Basin (Morocco). Sediment Geol 151:127–147

Cullen HM, deMenocal PB (2000) North Atlantic influence on Tigris-Euphrates streamflow. Int J Climatol 20:853–863

Davis M, Matmon A, Fink D, Ron H, Niedermann S (2011) Dating Pliocene lacustrine sediments in the central Jordan Valley, Israel – Implications for cosmogenic burial dating. Earth Planet Sci Lett 305:317–327

Denton GH, Broecker WS (2008) Wobbly ocean conveyor circulation during the Holocene? Quat Sci Rev 27:1939–1950

Enmar L (1999) The travertines in the Northern and Central Arava: stratigraphy, petrology and geochemistry. Geological survey, Israel report, GSI/1/99

Enzel Y, Bookman R, Sharon D, Gvirtzman H, Dayan U, Ziv B, Stein M (2003) Late Holocene climates of the near east deduced from Dead Sea level variations and modem regional winter rainfall. Quat Res 60:263–273

Gardosh M, Bruner I (1998) A seismic survey in the Bet Shean region. The Geophysical Institute of Israel, Report 348/27/98, Lod, 13 pp

Garfunkel Z (1981) Internal structure of the Dead Sea leaky transform (rift) in relation to plate kinematics. Tectonophysics 80:81–108

Gasse F (2000) Hydrological changes in the African tropics since the Last Glacial Maximum. Quat Sci Rev 19:189–211

Gat JR (1984) The stable isotope composition of Dead Sea waters: Earth Planet. Sci Lett 71:361–376

Gavrieli I, Stein M (2006) On the source and fate of brines in the Dead Sea basin. In: Enzel Y, Agnon A, Stein M (eds) New frontiers in Dead Sea paleoenvironmental research, Boulder, Colorado USA. GSA specific paper 401, pp 183–194

Goldstein SL, Torfstein A, Stein M, Kitgawa H (2013) The ICDP Dead Sea deep drill core: chronology and implications for Levant climate change. Goldschmidt 2013 conference, Florence

Greenbaum N, Porat N, Rhodes E, Enzel Y (2006) Large floods during late oxygen isotope stage 3, southern Negev desert, Israel. Quat Sci Rev 25:704–719

Gvirtzman G, Buchbinder B (1969) Outcrops of Neogene formations in the central and southern coastal plain, Hashfela and Be'er Sheva region, Israel. Geol Surv Israel Bull 50:73 pp

Haase-Schramm A, Goldstein SL, Stein M (2004) U-Th dating of Lake Lisan (late Pleistocene dead sea) aragonite and implications for glacial east Mediterranean climate change. Geochim Cosmochim Acta 68:985–1005

Haliva-Cohen A, Stein M, Goldstein SL, Sandler A, Starinsky A (2012) Sources and transport routes of fine detritus material to the late Quaternary Dead Sea basin. Quat Sci Rev 49:55–70

Hazan N, Stein M, Agnon A, Marco S, Nadel D, Negendank JFW, Schwab M, Neev D (2005) The late Pleistocene-Holocene limnological history of Lake Kinneret (Sea of Galilee), Israel. Quat Res 63:60–77

Heim C, Nowaczyk NR, Negendank JFW, Leroy SAG, Ben-Avraham Z (1997) Near East desertification: evidence from the Dead Sea. Naturwissenschaften 84:398–401

Heimann A, Steinitz G, Mor D, Shaliv G (1996) The cover basalt formation, its age and its regional and tectonic setting: implications from K-Ar and $^{40}Ar/^{39}Ar$ geochronology. Isr J Earth Sci 45:55–69

Hsu KJ (1973) The desiccation deep-basin model for the Messinian event. In: Grooger CW (ed) Messinian events in the Mediterranean. North-Holland Publishing Company – Amsterdam, London, pp 60–67

Hurrell JW, Kushnir Y, Ottersen G, Visbeck M (2003) An overview of the North Atlantic Oscillation. In: Hurrell JW, Kushnir Y, Ottersen G, Visbeck M (eds) The North Atlantic oscillation: climatic significance and environmental impact. American Geophysical Union, Washington, DC, pp 1–35

Kadan G (1997) Evidence for Dead Sea lake-level fluctuations and recent tectonism from the Holocene fan delta of Nahal Darga. M.Sc. thesis, Ben-Gurion University of the Negev, Beer Sheva, Israel, 54 p. (in Hebrew, English abstract)

Kagan EJ, Langgut D, Boaretto E, Neumann F, Stein M. Dead Sea level changes during the Bronze-Iron Age transition. Radiocarbon (submitted)

Kagan E, Stein M, Agnon A, Neumann F (2011) Intrabasin paleoearthquake and quiescence correlation of the late Holocene Dead Sea. J Geophys Res 116:B04311

Katz A, Starinsky A (2009) Geochemical history of the Dead Sea. Aquat Geochem 15:159–194

Katz A, Kolodny Y, Nissenbaum A (1977) The geochemical evolution of the Pleistocene Lake Lisan-Dead Sea system. Geochim Cosmochim Acta 41:1609–1626

Kaufman A (1971) U-Series dating of Dead Sea Basin carbonates. Geochim Cosmochim Acta 35:1269–1281

Kaufman A, Yechieli Y, Gardosh M (1992) Reevaluation of the lake-sediment chronology in the Dead Sea basin, Israel, based on new 230Th/U dates. Quat Res 38:292–304

Ken-Tor R, Stein M, Enzel Y, Agnon A, Marco S, Negendank J (2001) Precision of calibrated radiocarbon ages of earthquakes in the Dead Sea basin. Radiocarbon 43:1371–1382

Kolodny Y, Stein M, Machlus M (2005) Sea-Rain–Lake relation in the last Glacial East Mediterranean revealed by $\delta^{18}O - \delta^{13}C$ in Lake Lisan aragonites. Geochim Cosmochim Acta 69:4045–4060

Kushnir Y (1994) Interdecadal variations in North-Atlantic sea-surface temperature and associated atmospheric conditions. J Clim 7:141–157

Kushnir Y, Stein M (2010) North Atlantic influence on 19 – 20th century rainfall in the Dead Sea watershed, teleconnections with the Sahel, and implication for Holocene climate fluctuations. Quat Sci Rev 29:1–18

Langgut D, Neumann FH, Stein M, Wagner A, Kagan EJ, Boaretto E, Finkelstein I (2014) Dead Sea Pollen record and history of human activity in the Judean Highlands (Israel) from the intermediate bronze into the iron ages (~2500–500 BCE). Palinology (in press)

Lazar B, Stein M (2010) Freshwater on the route of hominids out of Africa revealed by U-Th in Red Sea Corals. Geology 39:1067–1070

Lazar B, Sivan O, Yechieli Y, Gavrieli I, Levy E, Stein M (2014) Long-term freshening of the Dead Sea brine revealed by porewater Cl- and d18O in ICDP Dead Sea deep-drill. Earth Planet Sci Lett 400:94–101

Lee S, Kim HK (2003) The dynamic relationship between subtropical and eddy-driven jets. J Clim 60:490–1503

Liu T, Broecker WS, Stein M (2013) Rock varnish evidence for a Younger Dryas wet period in the Dead Sea basin of the Near East. Geophys Res Lett 40(10):2229–2235

Livnat A, Kronfeld J (1985) Paleoclimatic implications of U-series dates for lake sediments and travertines in the Arava Rift Valley, Israel. Quat Res 24:164–172

Machlus M, Enzel Y, Goldstein SL, Marco S, Stein M (2000) Reconstruction of low-levels of Lake Lisan between 55 and 35 kyr. Quat Inter 73/74:137–144

Marco S, Stein M, Agnon A, Ron H (1996) Long-term earthquake clustering: a 50,000-year paleoseismic record in the Dead Sea Graben. J Geophys Res 101(B3):6179–6191

Marcus E, Slager J (1985) The sedimentary-magmatic sequence of the Zemah 1 well (Jordan-Dead Sea Rift, Israel) and its emplacement in time and space. Isr J Earth Sci 34:1–10

Mariotti A, Ballabrera-Poy J, Zeng N (2005) Tropical influence on Euro-Asian autumn rainfall variability. Clim Dyn 24:511–521

Marshall J, Kushnir Y, Battisti D, Chang P, Czaja A, Dickson R, Hurrell J, McCartney M, Saravanan R, Visbeck M (2001) North Atlantic climate variability: phenomena, impacts and mechanisms. Int J Climatol 21:1863–1898

Martinoti GM, Gvirtzman G, Buchbinder B (1978) The late Miocene transgression in the Be'er Sheva area. Isr J Earth Sci 27:72–82

Migowski C, Agnon A, Bookman R, Negendank J, Stein M (2004) Recurrence pattern of Holocene earthquakes along the Dead Sea rift revealed by varve counting and radiocarbon dating of lacustrine sediments Earth Planet. Sci Lett 222:301–314

Migowski C, Stein M, Prasad S, Negendank JFW, Agnon A (2006) Holocene climate variability and cultural evolution in the Near East from the Dead Sea sedimentary record. Quat Res 66:421–431

Nadel D (1995) Radiocarbon dating of Ohalo II: archaeological and methodological implications. J Archaeol Sci 22:811–822

Neev D (1960) A pre-Neogene erosion channel in the southern coastal plain of Israel. Geol Surv Isr Bull 25:1–21

Neev D, Emery KO (1995) The destruction of Sodom, Gomorrah, and Jericho: geological, climatological, and archaeological background. Oxford University Press, New York, 175 pp

Palchan D, Stein M, Almogi-Labin A, Erel Y, Goldstein SL (2013) Dust transport and synoptic conditions over – Arabia deserts during the MIS6/5 and 2/1 transitions from grain-size, chemical and isotopic properties of Red Sea cores. Earth Planet Sci Lett 382:125–139
Pausata FSR, Battisti DS, Nisancioglu KH, Bitz CM (2011) Chinese stalagmite $\delta^{18}O$ controlled by changes in the Indian monsoon during a simulated Heinrich event. Nat Geosci 4:474–480
Prasad S, Vos H, Negendank JFW, Waldmann N, Goldstein SL, Stein M (2004) Evidence from Lake Lisan of solar influence on decadal-to centennial-scale climate variability during marine oxygen isotope stage 2. Geology 32:581–584
Raab M (1998) The origin of the evaporites in the Jordan-Dead Sea valley in view of the evolution of brines and evaporites during seawater evaporation. Geological Survey, Israel report, GSI/1/98, 138 pp
Raab M, Friedman GM, Spiro B, Starinsky A, Zak I (1997) The geological history of Messinian (upper Miocene) evaporites in the central Jordan Valley (Israel) and how strontium and sulfur isotopes relate to their origin. Carbonates Evaporites 12:296–324
Roskin J, Porat N, Tsoar H, Blumberg DG, Zander AM (2011) Age, origin and climatic controls on vegetated linear dunes in the northwestern Negev desert (Israel). Quat Sci Rev 30:1649–1674
Rossignol-Strick M (1985) Mediterranean Quaternary sapropels, an immediate response of the African monsoon to variation in insolation. Palaeogeogr Palaeoclimatol Palaeoecol 49:237–263
Rossignol-Strick M, Paterne M (1999) A synthetic pollen record of the eastern Mediterranean sapropels of the last 1 Ma: implications for the time-scale and formation of sapropels. Mar Geol 153(1–4):221–237
Rossignol-Strick M, Paterne M, Bassinot FC, Emeis KC, De Lange GJ (1998) An unusual mid-Pleistocene monsoon period over Africa and Asia. Nature 392:269–272
Rozenbaum AG, Shaked D, Zilberman E, Sandler A, Stein M, Fursterberg Y, Starinsky A (2013) Stratigraphy and lithology of Bira and Gesher formations in the Tabor Stream Area. Geologial Society Israel of annual meeting, Acco, Israel
Sandler A, Harlavan Y, Shaliv G (2004) The stratigraphy of Neogene conglomerates in the Yizre'el Valley. Isr J Earth Sci 53:77–86
Schramm A, Stein M, Goldstein SL (2000) Calibration of the14C timescale to 50 kyr by 234U-230Th dating of sediments from Lake Lisan (the paleo-Dead Sea). Earth Planet Sci Lett 175:27–40
Schulman N (1962) The Geology of the central Jordan Valley. PhD thesis, Hebrew University of Jerusalem, Jerusalem, Israel, 103 pp
Shaliv G (1991) Stages in the tectonic and volcanic history of the Neogene basin in the Lower Galilee and the valleys. PhD thesis, Hebrew University of Jerusalem, Jerusalem, Israel, 94 pp
Stanhill G (1994) Changes in the rate of evaporation from the Dead Sea. Int J Climatol 14:465–471
Starinsky A (1974) Relationship between Ca-chloride brines and sedimentary rocks in Israel. PhD thesis, The Hebrew University of Jerusalem, 213 pp
Stein M (2001) The history of Neogene-Quaternary water bodies in the Dead Sea Basin. J Paleolimnol 26:271–282
Stein M (2014) Late quaternary limnological history of Lake Kinneret. In: Zohary T, Sukenik A, Berman J, Nishri A (eds) Lake Kinneret: environments, biology and ecology, vol 6, Lake Kinneret ecology and management series: aquatic ecology series. Springer, 708 p (in press)
Stein M, Goldstein SL (2006) U-Th and radiocarbon chronologies of late Quaternary lacustrine records of the Dead Sea basin: methods and applications. In: Enzel Y, Agnon A, Stein M (eds) New frontiers in Dead Sea paleoenvironmental research, Boulder, Colorado USA, GSA specific paper 401, pp 141–154
Stein M, Hofmann AW (1992) Fossil plume-head beneath the Arabian lithosphere? Earth Planet Sci Lett 114:193–209
Stein M, Starinsky A, Katz A, Goldstein SL, Machlus M, Schramm A (1997) Strontium isotopic, chemical, and sedimentological evidence for the evolution of Lake Lisan and the Dead Sea. Geochim Cosmochim Acta 61:3975–3992

Stein M, Starinsky A, Agnon A, Katz A, Raab M, Spiro B, Zak I (2000) The impact of brine-rock interaction during marine evaporite formation on the isotopic Sr record in the oceans: evidence from Mt. Sedom, Israel. Geochim Cosmochim Acta 64:2039–2053
Stein M, Agnon A, Katz A, Starinsky A (2002) Strontium isotopes in discordant dolomite bodies of the Judea Group, Dead Sea basin. Isr J Earth Sci 51:219–224
Stein M, Torfstein A, Gavrieli I, Yechieli Y (2010) Abrupt aridities and salt deposition in the post-glacial Dead Sea and their North Atlantic connection. Quat Sci Rev 29:567–575
Stein M, Goldstein SL, Ben-Avraham Z (2011) Dead Sea cores: a window into past climate and seismicity. EOS 92(49):453–454
Stein M, Lazar B, Goldstein SL (2013) Radiocarbon reservoir ages as freshwater-brine monitors in Lake Lisan, Dead Sea system. Radiocarbon 55:1050–1057
Steinitz G, Bartov Y (1991) The Miocene-Pliocene history of the Dead Sea segment of the rift in light of K-Ar ages of basalts. Isr J Earth-Sci 40:199–208
Stern O (2010) Geochemistry, hydrology and paleo-hydrology of Ein Qedem spring system. Geological Survey of Israel report, GSI/17/2010, Jerusalem
Torfstein A, Gavrieli I, Katz A, Kolodny Y, Stein M (2008) Gypsum as a monitor of the paleo-limnological–hydrological conditions in Lake Lisan and the Dead Sea. Geochim Cosmochim Acta 70:2491–2579
Torfstein A, Haase-Schramm A, Waldmann N, Kolodny Y, Stein M (2009) U-series and oxygen isotope chronology of Lake Amora, Dead Sea basin Geochom. Cosmochim Acta 73:2603–2630
Torfstein A, Goldstein SL, Kagan E, Stein M (2013a) Multi-site integrated U-Th chronology of the last glacial Lake Lisan. Geochim Cosmochim Acta 104:210–234
Torfstein A, Goldstein SL, Enzel Y, Stein M (2013b) Impacts of abrupt climate changes in the Levant from last glacial Dead Sea levels. Quat Sci Rev 69:1–7
Trigo IF (2006) Climatology and interannual variability of storm-tracks in the Euro-Atlantic sector: a comparison between ERA-40 and NCEP/NCAR reanalyses. Clim Dyn 26(2–3):127–143
Vaks A, Bar-Matthews M, Ayalon A, Matthews A, Halicz L, Frumkin A (2007) Desert speleothems reveal climatic window for African exodus of early modern humans. Geology 35:831–834
Waldmann N, Starinsky A, Stein M (2007) Primary carbonates and Ca-chloride brines as monitors of a paleo-hydrological regime in the Dead Sea basin. Quat Sci Rev 26:2219–2228
Waldmann N, Stein M, Ariztegui D, Starinsky A (2009) Stratigraphy, depositional environments and level reconstruction of the last interglacial Lake Samra in the Dead Sea basin. Quat Res 72:1–15
Waldmann N, Torfstein A, Stein M (2010) Northward intrusions of low- and mid-latitude storms across the Saharo-Arabian belt during past interglacials. Geology 38:567–570
Wdowinsky S, Zilberman E (1997) Systematic analyses of large-scale topography and structure across the Dead Sea Rift. Tectonics 16(3):409–424
Yechieli Y, Magaritz M, Levy Y, Weber U, Kafri U, Woelfli W, Bonani G (1993) Late Quaternary geological history of the Dead Sea area. Isr Quat Res 39:59–67
Zak I (1967) The geology of Mount Sedom. PhD thesis, Hebrew University of Jerusalem, 208 pp
Zilberman E (1992) Remnants of Miocene landscape in the central and northern Negev and their paleogeographic implications. Geol Surv Isr Bull 83:54
Ziv B (2001) A subtropical rainstorm associated with a tropical plume over Africa and the Middle-East. Theor Appl Climatol 69:91–102
Ziv B, Saaroni H, Alpert P (2004) The factors governing the summer regime of the eastern Mediterranean. Int J Climatol 24(14):1859–1871
Ziv B, Dayan U, Kushnir Y, Roth C, Enzel Y (2006) Regional and global atmospheric patterns governing rainfall in the southern Levant. Int J Climatol 26:55–73

Chapter 11
The Story of Saline Water in the Dead Sea Rift – The Role of Runoff and Relative Humidity

Abraham Starinsky and Amitai Katz

Abstract Saline waters have been found in the subsurface of the Dead Sea basin along the area stretching from Lake Kinneret in the north to the Timna area in the south. The maximum salinity of the waters reaches 340 g L^{-1}, and their chemical composition is Ca-chloridic, like most subsurface brines in sedimentary basins elsewhere. It is now generally accepted that such waters evolved from ancient seawater via a three-stage mechanism, namely: (1) seawater evaporation in a marine lagoon; (2) modification of the resulting brine by water-rock interaction; (3) dilution of the brine by freshwater or mixing with other saline fluids.

The age of the parental seawater from which the salinity of the brines was derived is Late Miocene to Early Pliocene. All saline waters in the Dead Sea basin display Na/Cl ratios below that of seawater (0.86), indicating that their evolution should be tied to formation of rock salt bodies; indeed, such were actually found in the subsurface of the two tectonic depressions of the Dead Sea and Lake Kinneret (Sea of Galilee). It is likely that the ancient lagoon from which the seawater evaporated and in which the salt deposited was approximately 300 km long and 20 km wide.

The Na/Cl ratios in the brines and in fluid inclusions in halite crystals formed from these waters cover a wide range below the seawater value (0.86), reaching values as low as 0.1 eq/eq (All solute ratios in the paper are presented in equivalent units unless otherwise stated).

The maximum degree of evaporation that an aqueous solution may reach depends, amongst other things, on the relative humidity. Because evaporation may proceed only as long as the activity of water in the evaporating solution (a_{H2O}) is higher than the relative humidity above it, it is possible to estimate the *maximum* relative humidity that prevailed in the area during the deposition of the salt from the Na/Cl ratio in a given brine or in fluid inclusion enclosed in a halite sample.

A. Starinsky (✉) • A. Katz
The Fredy and Nadine Herrmann Institute of Earth Sciences, The Hebrew University of Jerusalem, Edmond J. Safra campus, 91904 Jerusalem, Israel
e-mail: stari@mail.huji.ac.il

Z. Garfunkel et al. (eds.), *Dead Sea Transform Fault System: Reviews*, Modern Approaches in Solid Earth Sciences 6, DOI 10.1007/978-94-017-8872-4_11,

The Pliocene lagoon waters that, at that stage, had high Mg/Ca ratios, started to migrate outwards from the basin towards the west, under the local hydraulic head, through the Judea group limestone layers of upper Cretaceous age which comprise in that area the margin of the basin. During their passage, the brines interacted with limestone beds resulting in discordant dolomite bodies. The resulting, brine accumulated at depth in the Northern Negev.

Upon decline of the regional hydraulic head, the waters that infiltrated out to the west reversed their flow backwards to the basin, now displaying a chemical composition significantly different than that on their way out. The newly acquired composition is thus characterized by low Mg/Ca and SO_4/Cl ratios, of ~0.5 and ~0.01, respectively, and is defined as R_1 water.

At some later time the lagoon was cut off from the sea, and its area was transformed to a lacustrine environment, allowing for other processes to take place which modified the composition of the lagoonal brines. As from the lagoon's cut off from the sea, the contribution of dissolved marine salts to the basin was substituted by freshwater solutes carried in by runoff.

The freshwater that started to feed the saline lake(s) was saturated with respect to $CaCO_3$ minerals (calcite and aragonite), deposited its entire load of dissolved Ca^{2+} as $CaCO_3$ and $CaSO_4$ in the lake. Additional Ca^{2+} to compensate for the excess SO_4+HCO_3 over Ca^{2+} was borrowed from the saline, Ca-chloridic brine in the lake, bringing about a marked increase in the Mg/Ca ratio therein. The present paper presents a model (Katz and Starinsky, Aquat Geochem 15:159–194, 2009) that describes the relation between the increase in the Mg/Ca ratio of the lake and the accumulated mass of $CaCO_3$ (calcite or aragonite) that was deposited in it. The evolving saline waters affected by this process, with Mg/Ca ratios >1, are defined as group R_2 waters.

An additional, significant modification of the brine inflicted by its passage from a marine lagoon to a lacustrine water body is reflected by the $^{87}Sr/^{86}Sr$ isotope ratio. We inspect two cases relevant to this question, the first in the Lake Kinneret area and the second in the Dead Sea basin.

The R_2 lacustrine waters on the eastern side of Lake Kinneret show $^{87}Sr/^{86}Sr$ isotope ratios around 0.706, contrasting in this regard with the R_1-type waters on the western side of the lake which are characterized by $^{87}Sr/^{86}Sr$ isotope ratios of ~0.708. We propose that the $^{87}Sr/^{86}Sr$ ratio transition 0.708 → 0.706 is driven by addition of freshwater with low $^{87}Sr/^{86}Sr$ ratios (0.704–0.707) originating in the runoff flowing to the lake over the basaltic terrain from the NE ($^{87}Sr/^{86}Sr \sim 0.704$).

In the southern, Dead Sea area, the lacustrine R_2 waters were fed by runoff with $^{87}Sr/^{86}Sr \sim 0.708$ and, therefore, remained unchanged in this respect.

The Timna water composition is a result of interaction between diluted subsurface Ca chloride brines that originated as group R_{S1} in the Dead Sea area with basic igneous rocks. The depletion of Mg^{2+} in the water is due to a reaction of destruction of Ca- plagioclase and formation of Mg^{2+} rich mixture of epidote and chlorite. The Sr isotope signature of ~0.706 was formed by exchange of high $^{87}Sr/^{86}Sr$ (~0.708) brines with low $^{87}Sr/^{86}Sr$ (~0.7045) igneous basic rocks, like olivine norite. The age of the water-rock interaction is estimated to be older than the R_s brine formation, i.e. <3–4 m.y.

Keywords Dead Sea Transform • Ca chloride brines • Climatic impact • Seawater evaporation • Water – rock-interaction

11.1 Introduction

Subsurface brines down to a depth of a few kilometers in Earth's crust, including sedimentary and basement environments, may reach salinities of more than 300 g/l. They usually belong to the Ca chloride group, which is defined by the relation $Ca > (SO_4 + HCO_3)$. There is a general debate about the origin of such saline water either from ancient seawater (by evaporation or freezing) or from fresh water by "internal processes" within basement rock environment (Bucher and Stober 2010).

It was proposed by Starinsky (1974) for subsurface brines in Israel, and by Carpenter (1978) for similar brines in the Illinois basin, that Ca chloride brines were formed from evaporated seawater (Table 11.1) with the same major solute chemistry of modern seawater. Later on, based on a series of studies conducted by Hardie (1996) that indicated a secular variation of seawater chemistry, Lowenstein et al. (2003) suggested that such brines, that are present in most Phanerozoic sedimentary basins, inherited their chemistries and salinities from evaporated paleoseawaters of a Ca chloridic composition ($CaCl_2$ seas). For the evolution of brines found in crystalline rocks in the northern hemisphere Frape and Fritz (1987) suggested a WRI[1] mechanism of fresh water while Herut et al. (1990) and Starinsky and Katz (2003) suggested a cryogenic marine origin during the Pleistocene glacial period. A cryogenic origin of Carboniferous-Permian sea was suggested by Katz et al. (2011) to saline water in the southern hemisphere in the Kaapvaal Craton (South Africa). Similarly, Frank et al. (2010) suggested a cryogenic origin to saline water in the McMurdo bay, Antarctica.

The Dead Sea rift is saturated with saline waters at least in the segment between Lake Kinneret and southern end of the Dead Sea. The waters appear in two main modes, either as deep seated brines or as springs. While the first group was found by drilling anywhere along the rift, the second one is restricted to the tectonic depressions of Kinneret – Beit Shean and the Dead Sea (Fig. 11.1). Hydrological mechanisms responsible for the emergence of the springs were suggested by Goldschmidt et al. (1967), Mazor and Mero (1969), and Gvirtzman et al. (1997) for the Lake Kinneret region, and by Matmon (1995) and Shalev and Yechieli (2007) for the Dead Sea area. The suggested processes are as follows: (1) Flow of meteoric water in the major regional aquifer; (2) Compaction of the strata due to tectonic movements (still ongoing); (3) Eastward flow of meteoric water which slowly forces deep

[1] Water Rock Interaction.

Table 11.1 Chemical composition of evaporated seawater in mg/L (Raab 1996). TDS in g/L

DE	Sample	Na	K	Mg	Ca	Sr	Cl	Br	SO_4	TDS	Density
1.0	MOW	11,050	416	1,326	422	8.5	19,870	68	2,784	35.95	1.023
1.1	W64	12,300	455	1,447	462	9.0	22,130	75	3,025	39.90	1.027
2.8	R1	30,000	1,062	3,606	1,150	21.0	53,580	190	7,600	97.21	1.069
2.8	Y1	29,690	1,001	3,623	1,112	21.0	53,550	190	7,520	96.71	1.069
4.7	R2	49,000	1,800	6,224	1,300	31.0	90,750	310	11,250	160.7	1.113
4.9	Y2	52,000	1,830	6,236	1,263	32.0	94,290	320	10,750	166.7	1.116
6.1	Y3	65,630	2,125	7,800	1,006	38.0	118,100	395	12,970	208.0	1.143
6.2	R3	64,360	2,250	8,169	1,000	38.0	119,000	400	13,530	208.7	1.143
7.9	R4	82,500	3,100	10,220	705	40.0	149,100	495	15,040	261.2	1.176
8.1	Y4	84,670	3,000	10,480	660	40.0	153,500	510	15,300	268.2	1.180
10.3	R5	101,600	4,219	13,570	413	46.0	185,800	630	18,900	325.1	1.217
11.5	R6	96,870	4,250	15,370	400	50.0	181,600	710	21,100	320.3	1.220
12.2	Y5	100,000	4,250	15,410	400	50.5	186,800	750	21,350	329.0	1.219
13.0	Y6	95,310	4,875	17,400	388	53.5	184,100	800	23,900	326.8	1.222
13.0	R7	95,310	4,750	16,780	400	51.5	185,400	800	23,300	326.8	1.221
13.8	R8	95,310	4,875	17,020	387	52.0	186,200	850	23,900	328.6	1.221
14.0	Y7	95,310	5,250	18,670	350	49.0	186,300	860	25,600	332.4	1.223
14.1	Y8	92,480	5,313	19,130	362	47.5	184,400	870	26,000	328.6	1.224
14.9	R9	93,930	5,367	18,480	333	42.8	188,600	917	24,410	332.1	1.224
16.6	R10	91,690	5,877	20,670	234	29.0	188,100	1,023	26,300	333.9	1.226

16.9	Y9	90,840	5,812	21,100	327	28.9	188,400	1,040	26,050	333.6	1.225
18.1	Y10	88,100	6,503	22,920	271	28.5	188,000	1,115	30,390	337.3	1.229
18.1	R11	88,170	6,345	22,250	284	28.7	186,200	1,117	30,020	334.4	1.229
19.8	R12	84,140	7,247	26,510	259	25.2	185,800	1,216	34,530	339.7	1.231
20.5	Y11	86,060	7,226	25,320	259	26.9	187,200	1,261	32,150	339.5	1.232
23.0	R13	79,890	8,360	28,840	238	24.4	186,800	1,417	35,910	341.5	1.234
23.2	Y12	80,220	8,447	29,100	227	23.4	187,400	1,425	37,370	344.2	1.237
25.8	R14	76,800	9,533	33,600	210	20.4	187,700	1,579	44,200	353.7	1.243
27.5	Y13	74,780	9,888	34,140	219	20.2	188,400	1,681	48,510	357.6	1.244
32.2	Y14	64,010	11,920	40,930	184	17.4	187,200	1,978	56,500	362.8	1.254
35.2	R15	60,190	12,820	43,740	167	24.0	187,500	2,169	58,110	364.7	1.258
41.3	R16	50,000	15,030	52,320	140	16.1	187,700	2,534	72,420	380.2	1.270
45.0	Y15	41,190	16,770	59,000	119	15.8	190,800	2,766	79,440	390.1	1.282
52.8	Y16	30,750	19,600	66,790	84	12.4	194,100	3,254	88,360	403.0	1.298
55.1	R17*	27,430	20,800	68,330	74	12.1	197,200	3,432	82,090	399.3	1.303
75.7	Y17*	15,980	27,800	81,630	46	9.9	232,300	4,634	67,460	429.8	1.318
138.0	R18*	6,555	8,918	102,400	69	5.9	273,100	8,333	50,600	449.3	1.326

DE Degree of evaporation – Br/Br_{sw} (molal)
MOW Mean Ocean Water (Riley and Chester 1965)
W64 Mediterranean sea water sample
*Br and DE corrected

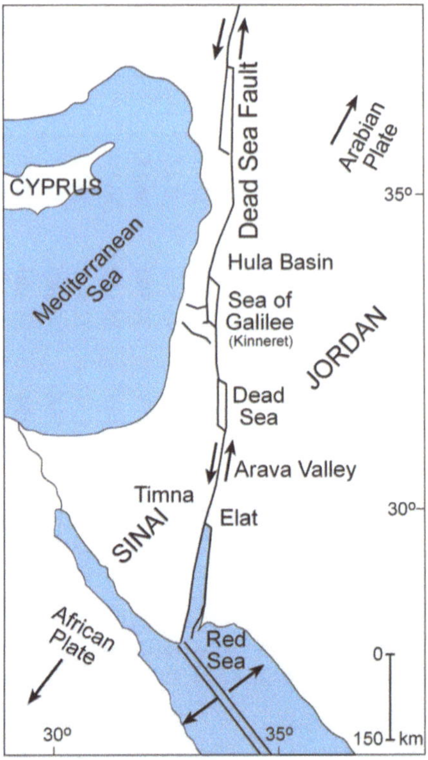

Fig. 11.1 Location map. The sites of saline waters along the Dead Sea transform fault include the tectonic depressions of Lake Kinneret (Sea of Galilee) and Dead Sea basin, as well as the Timna area (Modified after Ben Avraham et al. 2008)

brines towards the Dead Sea valley; (4) Forcing of Dead Sea deep brines to the surface by the pressure of the Dead Sea water body.

The area studied is a part of the active major fracture zone of the Dead Sea transform (DST), which extends along a distance of about 1,000 km from the northern Red Sea to the Taurus Mountains (Garfunkel 1997). The present study is focused on the southern part of the DST from Elat (Aqaba) to the north of Lake Kinneret (Figs. 11.1 and 11.2).

The sedimentary sequence of the study area is divided according to an earlier study in the Dead Sea basin fill by Zak and Freund (1981). Their division can be probably extended both northward and southward. Three main units are recognized: (a) A fluviatile-lacustrine sequence (Hatzeva Fm. in the south and Hordos Fm. in the north) consisting of conglomerates, sandstones and carbonates. (b) An evaporitic – lacustrine sequence composed mainly of rocksalt, less clastics, in the south (Fig. 11.2), and a unit of interlayered rocksalt, clastics and igneous rocks in the north. According to Horowitz (1987) salt deposition began at Upper Miocene – Lower Pliocene time. (c) A lacustrine sequence with fine clastics and some evaporites.

Major depressions that are interpreted as pull apart basins and are characterized by distinct structures and histories developed along this segment. The two major basins are the Dead Sea and the Kinneret – Beit Shean.

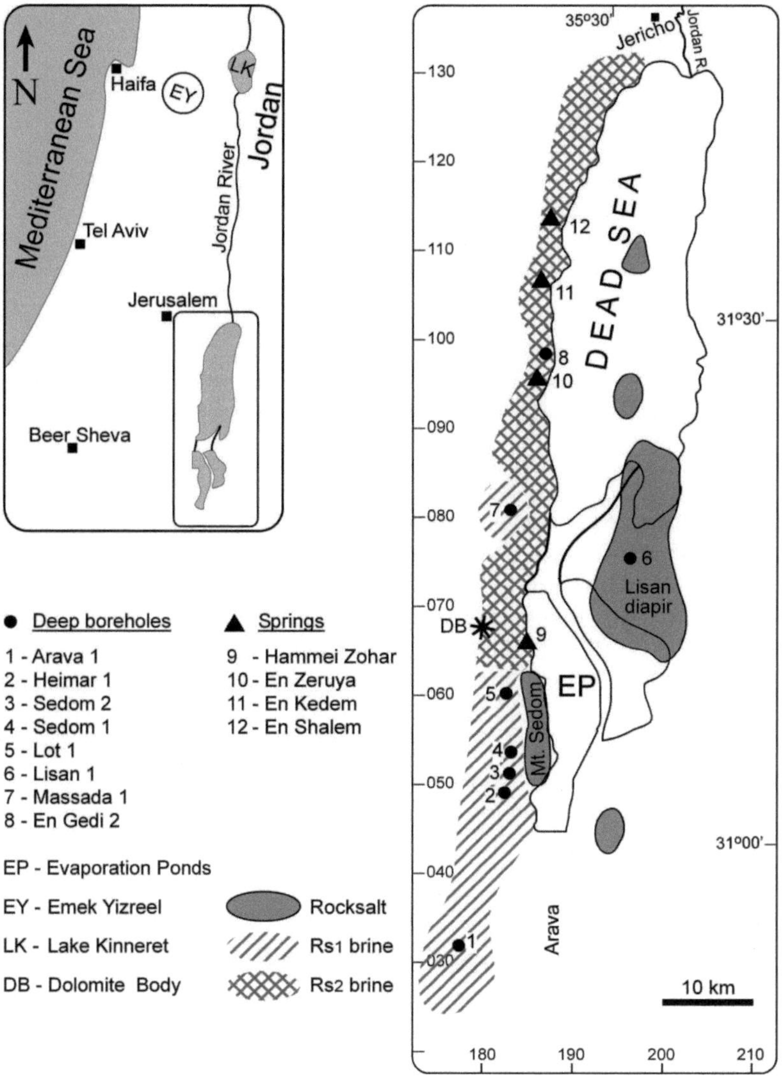

Fig. 11.2 A map showing the distribution of subsurface brines (R_{s1} and R_{s2}), deep boreholes, saline springs, epigenetic dolomite body (DB) and rock salt bodies in the Dead Sea basin. Standard universal coordinates are plotted along the top x axis and right ordinate. The abscissa and left hand ordinate represent the local coordinate system (From Katz and Starinsky 2009)

The Dead Sea basin extends from the center of the Arava valley to Jericho. It is about 150 km long and 15–17 km wide. A fill of 10 km thickness was discovered through geophysical exploration. The northern part of the basin consists of the water body of the Dead Sea, which is the lowest continental point on Earth (420 m below sea level). The Kinneret Bet – Shean basin includes in its northern part Lake

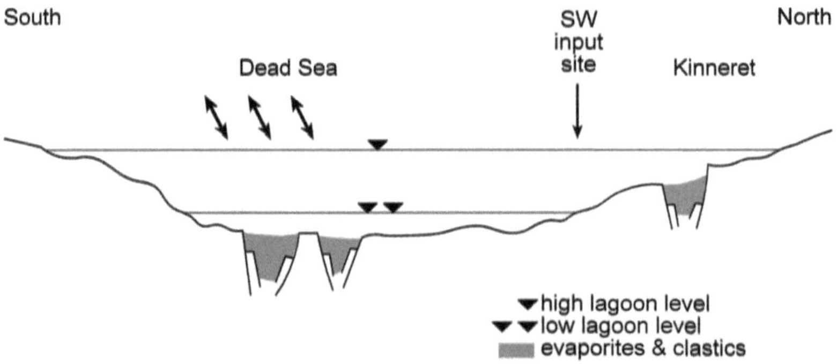

Fig. 11.3 Conceptual S-N cross section along the Sedom lagoon. Oblique, double-headed arrows denote the subsurface outward and backward migration areas of the brines

Kinneret, which extends along a distance of about 20 km, a maximum width of 12 km and is about 40 m deep, with a water level at about 200 m bsl.

At the time of deposition of the evaporites the Dead Sea rift was divided to several tectonic units. It appears that the salt was deposited in deep topographic depressions (Fig. 11.3). The thickness of rocksalt units varies significantly, being around 2 km in Mt. Sedom, 3–4 km in the Lisan diapir (Bender 1974) and over one km, in the north, close to Lake Kinneret (Zemah-1 borehole). The existence and emplacement of the deep tectonic basins and the evaporitic lithology along the DST play a most significant role in understanding the evolution, time of formation, appearance and migration of the brines, whether at depth or in spring discharge.

11.2 The Chemical Composition of the Brines

The geochemistry of saline waters in the Dead Sea rift was studied by many researchers. The most cited ones are Bentor (1961), Goldschmidt et al. (1967), Lerman (1967), Lerman and Shatkay (1968), Mazor et al. (1969), Mazor and Mero (1969), Bentor (1969), Starinsky (1974), Zak (1997) Kolodny et al. (1999), Gavrieli and Stein (2006), and Katz and Starinsky (2009).

The saline Dead Sea rift waters can be divided into two, *genetically distinct*, groups, R_1 (Mg/Ca<1) and R_2 (Mg/Ca>1), which were further divided into the following, *site-related* subgroups (Starinsky 1974):

<u>Group R_1: Characterized by Mg/Ca ratios ≤ 1</u>

- The Northern Negev (group R_0), in deep oil boreholes
- Western and northern sides of Lake Kinneret, usually in shallow boreholes (R_{n1})
- West and south of the Dead sea in deep oil boreholes (R_{s1})
- Timna area, with almost complete removal of Mg

Group R_2: Characterized by Mg/Ca ratios ≥1

- Liquid inclusions in the Sedom Salts (R_{s2})
- Dead Sea, in springs and adjacent boreholes (R_{s2})
- On the eastern side of Lake Kinneret (R_{n2})

11.2.1 Lake Kinneret

Freshwater Lake Kinneret is surrounded by a thermal saline water body with different salinities ranging up to 30 g L^{-1}. The source of the saline water was attributed by Goldschmidt et al. (1967), Starinsky (1974), Bergelson et al. (1999) and Gvirtzman et al. (1997) to the marine-evaporitic Sedom lagoon. An alternative interpretation was given by Mazor and Mero (1969) who explained the salinity of the waters by intrusion of Neogene seawater. The chemical composition and ionic ratios of the waters under regard is given in Tables 11.2a and 11.2b. The variation of salinity with depth is shown in the Kinneret 10b well (Fig. 11.4) and in the Rosh Pina borehole (Table 11.2a). The saline waters in the area are divided into an eastern and a western group, both of which are similar in their maximum salinity, major ion ratios such as K/Cl, Na/Cl and Sr/Ca, but differ in their Mg/Ca and $^{87}Sr/^{86}Sr$ isotope ratios. The eastern (R_{ne}) group waters are characterized by Mg/Ca > 1 ratios and $^{87}Sr/^{86}Sr \sim 0.706$ and are designated as R_{n2} or simply R_2 waters. The western group shows Mg/Ca < 1 and $^{87}Sr/^{86}Sr \sim 0.708$ ratios and are designated as R_{n1} or just R_1 waters.

A clear correlation exists between location and the chemical composition of the saline water (Mazor and Mero 1969). Each group has typical ionic ratios which have been almost constant during tens of years of measurements. These characteristic values in the main groups (Table 11.2b) were interpreted by Kolodny et al. (1999) as reflecting isolated brine pockets. Very saline waters (up to 200 gr L^{-1}) were reported from the deep Rosh Pina-1 borehole at a depth of ~3,800 m (Table 11.2a). The different ionic ratios between the deep and shallower waters in Rosh Pina rule out dilution as an explanation. Moreover, it is unclear whether the two waters have the same age.

Hence, it seems that the "shallow" saline water in the Kinneret area represents a diluted subsurface brine body which is buried at depth. The highest temperature (61 °C) was measured in the Tiberias hot springs. Starinsky et al. (1979) attributed the origin of thermal waters in the area to a normal thermal gradient. In such a case it seems that the Hammei Tiberias hot waters emerge from a depth of about 2 km. Very similar water in composition was found in Rosh Pina-1 borehole (Tables 11.2a and 11.2b) at a depth of around 2,500 m.

11.2.2 Dead Sea

Saline waters in the Dead Sea area appear as springs (Fig. 11.5) or were found at depth.

R_{s1} waters, which are characterized by Mg/Ca < 1 ratios occur only west and south of the Dead Sea basin (Fig. 11.2), and are known only from the deep

Table 11.2a Chemical composition in selected saline samples from the Lake Kinneret area. Composition in mg/L

Group	Sample	Source		Depth (m)	Na	K	Mg	Ca	Sr	Cl	Br	SO_4	HCO_3	TDS
East	TR268	Bet Zayda	R_{NE}	78–96	36,61	41.0	839.0	725.0	9.50	9,546		36	198	15,056
East	t18	Bet Zayda	R_{NE}	78–96	3,621	35.6	861.2	734.8	10.4	9,690	123	33	179	15,288
East	275	Ein Gofra	R_{NE}		1,466	86.3	254.5	258.4	12.4	3,088		218		5,384
East		1064	R_{NE}		6,922	380	1,576	814.0		16,570	216	395	172	27,045
East		1066	R_{NE}		8,397	384	1,915	1,112		20,030	253	864	122	33,077
East		1071	R_{NE}		9,093	296	2,258	1,285		23,050		4	186	36,172
East		1069	R_{NE}		6,483	337	1,822	1,052		16,680	196	782	219	27,571
East		1075	R_{NE}		7,127	195	1,734	944		17,540	176	5	469	28,190
East		CH5	R_{NE}		6,890	332	764	3,446		18,510	237	765	146	31,090
East		CH12	R_{NE}		7,129	433	1,196	1,933		17,730		741	43	29,205
East	4855	Ha'On-Chof	R_{NE}		2,689	371	544	287	22.1	5,800	61	759		10,533
East	4856	Ha'On-Yam	R_{NE}		4,073	178	896	587	53.4	9,883	114	244		16,028
Tiberias	140	Tiberias H. S. main	R_{NW}		7,096	335	651	3,498	62.2	18,190		849	149	30,829
Tiberias	TR255	Tiberias H. S. roman	R_{NW}		7,154	356	692	3,592	69.3	18,630	245	809		31,547
Tiberias	TR202	Tiberias H. S. roman	R_{NW}		7,029	348	675	3,524	67.7	18,480	247	833		31,204
Tabgha	271	Sartan Iver	R_{NW}		1,420	52.3	144	509	11.9	3,075		260		5,472
Tabgha	TR134	Sartan Iver	R_{NW}		1,293	51.0	134	413	10.9	2,765		233	342	5,242
Fuliya		Fuliya 6	R_{NW}		7,718	305	844	1,912		16,260	115	2,140	224	29,518
Fuliya	274	Fuliya new	R_{NW}		672	22.6	92	228	2.9	1,323		172		2,511
Fuliya	TR271	Kinneret 10 well	R_{NW}	600–882	8,364	293	748	2,081	38.4	17,410	126	1,940	226	31,230
Barbutim		1,020/3	R_{NW}		1,314	55.0	156	523		3,070	30	235	341	5,724
	D 295	Rosh Pina – 1	R_{NW}	2,486–586	8,000	359	725	4,804		22,790		12	6	36,696
	D 23	Rosh Pina – 1	R_{NW}	3,845–64	33,400	1,160	2,920	27,000		105,200	1,740	700	171	172,291

Table 11.2b Ion ratios in the main group of the Kinneret area (eq/eq)

Group	Na/Cl	Mg/Cl	K/Cl	100*Br/Cl
NE of Kinneret	0.59	0.26	0.0039	
Eastern shore	0.60–0.64	0.26–0.33	0.017–0.018	0.52–0.58
Tiberias	0.54–0.64	0.17–0.23	0.017–0.018	0.48–0.61
Tiberias hot springs	0.57–0.61	0.10–0.12	0.017–0.020	0.53–0.57
Tabgha	0.69–0.73	0.11–0.15	0.016–0.017	0.52–0.58
Barbutim	0.66–0.68	0.13–0.15	0.015–0.017	0.34–0.37
Fuliya	0.73–0.78	0.12–0.14	0.015–0.016	0.25–0.29

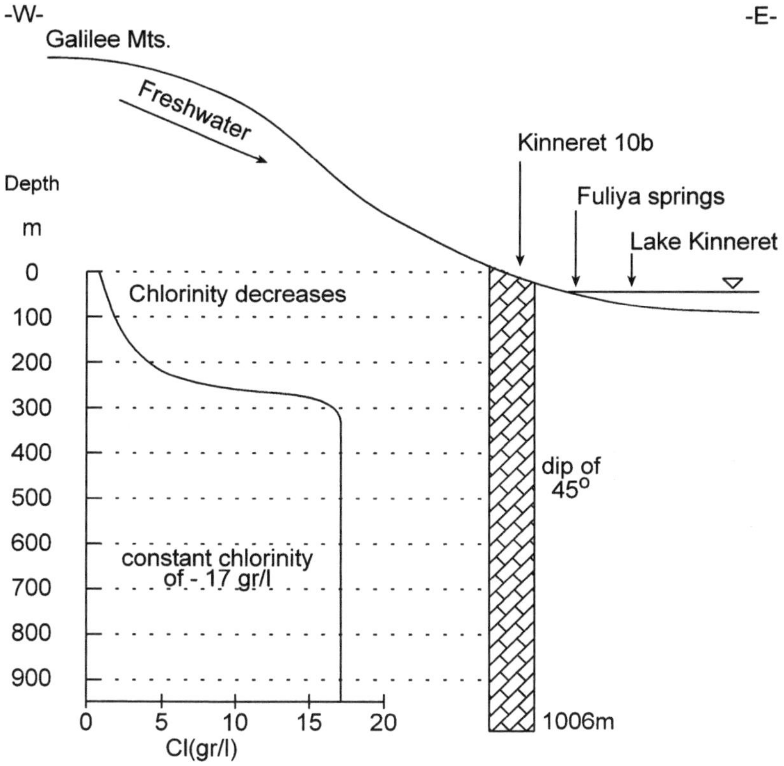

Fig. 11.4 W-E Cross section through the Fuliya spring area NW of Lake Kinneret, and the salinity profile in the Kinneret 10b borehole (From Moise et al. 2000)

subsurface (Tables 11.3a and 11.3b). The waters were found in different stratigraphic units, including Dead Sea graben fill (Arava-1, Sedom-1 and Sedom-2), Ramon group (Triassic), and Negev group (Paleozoic, in the Heimar-1 and Lot-1 bore holes). No specific linkage was found between the chemical composition of the saline waters and their stratigraphic position. The maximum salinity achieved is about 320 g L^{-1}. The Na/Cl, K/Cl, Br/Cl, and Mg/Ca ratios in the brines are

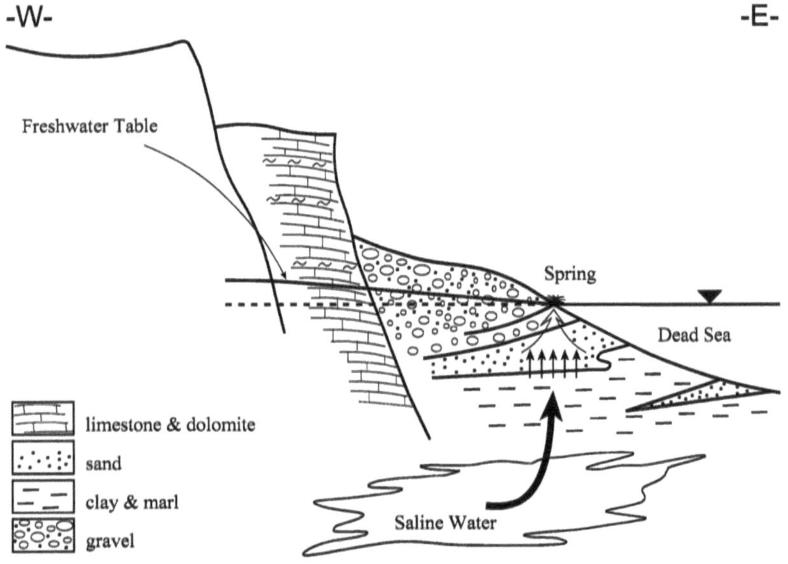

Fig. 11.5 W-E cross section showing the relation between subsurface brine, saline springs and the Dead Sea (From Katz and Starinsky 2009)

displayed in Figs. 11.6, 11.7 and 11.8, where the Na/Cl ratio in the Dead Sea brines is lower than in waters of group R_n in the Lake Kinneret area. Mg/Ca ratios in the R_{s1} waters are similar to the corresponding ratio in the waters on the western side of this lake.

Group R_{s2} waters show Mg/Ca > 1 ratios and includes the modern Dead Sea brine as well as the saline spring waters and shallow bore hole brines along the western shores of the Dead Sea. The same waters were also revealed in one deep bore hole (En Gedi-2) at a depth of about 1,600 m, within a Senonian rock formation (Table 11.3).

In their K/Cl, Br/Cl and Na/Cl ratios the R_{s2} waters are similar to R_{s1} waters. The R_{s2} group includes also the Temile springs in Mt. Sedom which differ from other waters in the same group by their K-rich composition and by a different Na-Cl-Br relationship. The $^{87}Sr/^{86}Sr$ isotope ratios in both groups are similar, being around 0.708.

11.2.3 Timna

Timna saline waters were studied previously by Kanfi (1972), Starinsky (1974) and by Beyth et al. (1981). The waters were found during copper exploitation works in the Timna area in a series of shallow boreholes and tunnels.

The saline waters have a salinity (TDS) of up to 16 g L^{-1} and were divided into four groups according to their chemical – geographic relations (Table 11.4).

Table 11.3a Chemical composition of selected brines from the Dead Sea area (in mg/L)

Group	Source	Sample	Depth (m)/Years	Na	K	Mg	Ca	Sr	Cl	Br	SO_4	HCO_3	TDS
R_{s1}	**Amiaz – 1**	**D 527**	**3737–3793**	13,670	8,190	17,800	113,400	1,200	281,200				435,460
	Arava – 1	**BT 84**	**1935–45**	50,480	1,030	12,040	50,640	1,205	202,200	3,670	122	61	321,450
	Heimar – 1	**BT 87**	**1459–73**	9,960	1,220	242	24,010		56,730	1,035	655		93,852
	Heimar – 1	**BT 88**	**1740–2050**	26,920	1,250	4,107	39,780		113,450		1,230		186,740
	Heimar – 1	**BT 89**	**2079–94**	24,000	1,600	3,630	43,010		137,800	2,750	1,280		214,070
	Lot – 1	**FG 25**	**1269–79**	15,120	1,680	13,960	34,680		131,000	3,040	1,000	61	200,540
	Lot – 1	**FG 26**	**1684–94**	9,896	800	9,196	42,930		119,100	2,465	816	24	185,230
	Massada – 1	**D 750**	**2388–2404**	33,800	1,000	5,034	44,340		144,400	2,760	352		231,690
	Massada – 1	**D 849**	**2535–2541**	35,000	1,100	4,888	46,700		153,800	2,780	247		244,520
	Sedom – 1	**D 35**	**Artesian**	20,160	28,580	19,990	83,190		269,960	3,120	90		425,090
	Sedom – 2	**D 59**	**Artesian**	33,200	999	8,510	17,330		106,600	1,790	905	121	169,460
	Zuk Tamrur 3	**D 3325**	**1995**	35,570	331	5,003	20,410	937	112,200	2,610	133		177,190
R_{s2}	**Dead Sea**	**D 411**	**1975**	39,250	7,050	39,200	16,000	295	207,800			261	309,860
	Dead Sea	**D 443**	**1976**	37,800	7,530	43,800	17,500	309	221,000	5,100	440	265	333,740
	Dead Sea	**D 442**	**1976**	39,500	7,590	42,600	16,500	241	218,500	4,900	510	265	330,610
	Dead Sea	**D 791**	**1983**	38,890	7,500	43,690	16,250	338	223,500				330,170
	Dead Sea	**D 803**	**1983**	39,050	8,180	44,000	16,500	330	224,400				332,460
	Dead Sea	**D 1369**	**1992**	29,910	6,220	37,040	15,020	285	200,600		2,080	273	291,430
	Dead Sea	**D 3164**	**1994**	39,710	9,060	52,420	19,430	374	226,200		1,320	264	348,780
	Dead Sea		**1997**	35,880	7,780	47,200	17,850		225,000	5,600	494	249	340,050
	Dead Sea	**D 1901**	**2003**	35,600	7,410	43,800	16,700	314	219,700		280	200	324,000
	Dead Sea		**2004**	36,100	8,400	50,200	19,400	360	235,000				349,460
	Dead Sea		**2005**	33,000	7,650	45,500	18,000	330	225,000	5,500	2,000		336,980

Table 11.3b Average concentrations and standard deviations of R_{S2} brines (mg/L)

Source	n	Date	Depth (m)		Na	K	Mg	Ca	Sr	Cl	Br	SO_4	HCO_3
En Gedi 2	2	1957	1928–71	Average	52,438	4,460	21,726	15897.5	323	176,500		412	283
En Gedi 3	10	1993–96	33	Average	24,602	2,400	12,379	8,468	190	90,687	1,750	1,278	154
				stdv.	1,051	187	623	437	62	2,673		155	34
En Kedem N	11	1991–96	Spring	Average	26,358	4,082	21,011	11,826	198	125,168	2,840	731	91
				stdv.	450	391	734	450	37	3,814	40	215	25
En Kedem S	10	1991–96	Spring	Average	25,757	4,005	20,619	11,434	209	123,610	2,820	794	83
				stdv.	1,423	411	1,469	707	11	6,788		66	21
En Shalem	2	1997	Spring	Average	24,895	3,875	20,575	11,500		119,195	2,465	968	70
Enot Zeruya	11	1992–94	Spring	Average	24,951	2,566	13,671	9,006	170	94,334		2,030	134
				stdv.	1,208	329	776	633	9	5,297		1,471	24
Hamei Zohar 3	2	1973; 92	43; 556	Average	11,504	1,228	7,243	4,273	76	47,100	1,078	730	154
Mt. Sedom	4		Springs	Average	22,419	15,792	42,576	40,427		250,736	3,024	95	127

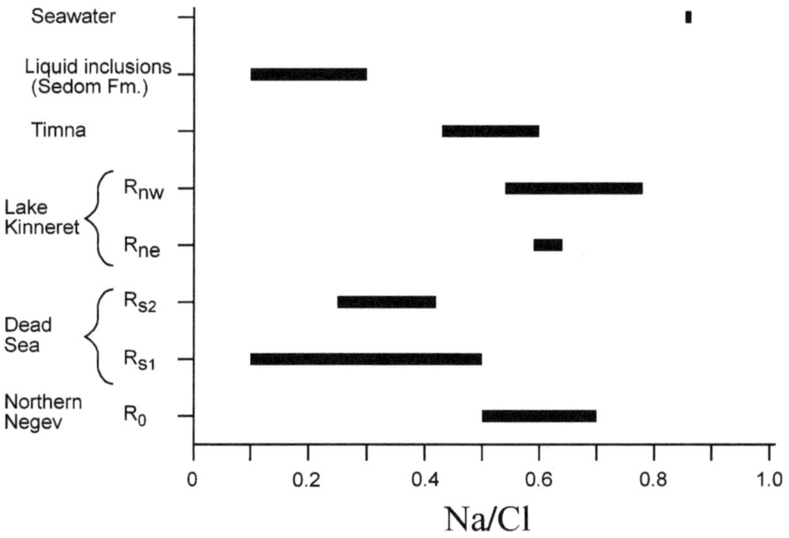

Fig. 11.6 The Na/Cl ratio (in equivalent units) in the water groups in the Dead Sea rift valley. All waters display ratios below that of modern seawater, indicating halite precipitation during their evolution. The R_o waters show higher Na/Cl ratios, attributed to lesser evaporation than do the R_{s1} brines. Waters sampled in the Dead Sea region are of lower ratios than those collected in the Lake Kinneret area. Na/Cl ratios in fluid inclusions in Sedom halite fall within the R_{s1} Na/Cl ratio range. The northern Negev waters show similar ratios to those in the Kinneret area

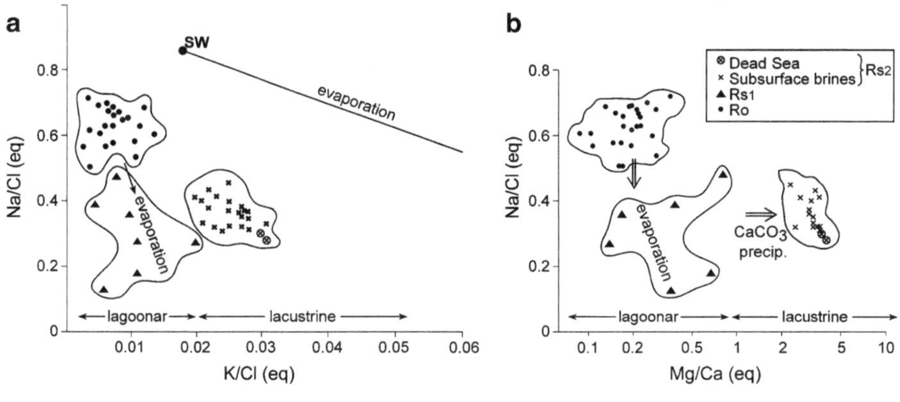

Fig. 11.7 The linkage between Na/Cl-K/Cl and Na/Cl-Mg/Ca ratios in saline waters in the northern Negev (R_o group) and in the Dead Sea area (R_{s1} and R_{s2} groups)

They were found in two sites, 3 km apart:

1. Pit N in the open mine east of Mt. Timna at 25–45 m msl (Fig. 11.9) – groups T_1, T_2, T_{2a}.
2. Mining tunnels in Mt. Hakhlil at 20–50 m bmsl (below mean sea level).

It is possible that these waters fill the entire area.

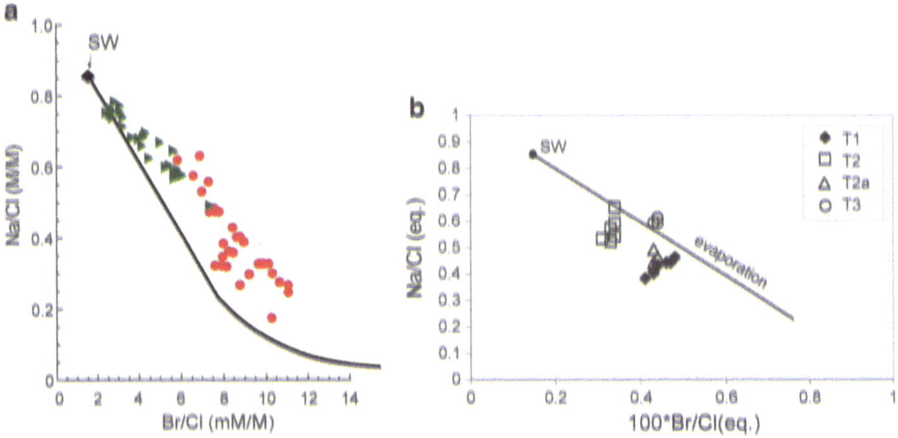

Fig. 11.8 Na/Cl-Br/Cl relationships in the Dead Sea rift waters and in Timna. (**a**) Dead Sea and Kinneret basins: most data points deviate upwards from the seawater evaporation path. *Green triangles*: Kinneret basin; *Red circles*: Dead Sea basin. (**b**) Timna area: A few data points fall close to the seawater evaporation path but most are located below it, which is attributed to albitization of silicate minerals

The Timna region is comprised of a Precambrian, magmatic-metamorphic core surrounded by sedimentary rocks of Paleozoic age. The magmatic rock complex is made up by a few intrusions of granite, monzodiorite and cumulates of olivine norite.

Cambrian sandstones, shale and clastics of the Yam Suf Group, which includes the Amude Shlomo, Timna, Shkhoret and Netafim Formations, make up the sedimentary rock column. The Timna Fm. represents a middle Cambrian marine episode.

The magmatic rocks display mineral replacement phenomena, where the olivine norite rocks are the most affected, as reflected by the following three reactions:

1. Olivine → serpentine
2. Amphibole/pyroxene → chlorite
3. Plagioclase → epidote and chlorite.

$$10CaAl_2Si_2O_8 + 4Mg^{+2} + 10H_2O \rightarrow 4Ca^{+2} + 6H^+ + Mg_4Al_2O_{10}(OH)_8 + 3Ca_2Al_6Si_6O_{22}(OH)_2 \tag{11.1}$$

The high concentrations of chlorite and epidote impart a greenish coloration to the replaced olivine norite rocks, explaining their local geological field designation as "lawn gabbro". On the basis of $\delta^{18}O$ analyses Beyth et al. (1997) proposed that the observed replacements were brought about by WR interaction of the gabbro rocks with a marine brine of Neogene age some 13–15 Ma. The maximum salinity of the Timna waters reaches 16 g L^{-1}. The waters are Ca-chloridic, almost completely deprived of Mg^{2+}. Each of the groups is characterized by fairly

Table 11.4 Chemical composition of selected samples from Timna area (mg/L)

Source	Group	Sample (D)	Na	K	Mg	Ca	Sr	Cl	Br	SO_4	HCO_3	TDS
Open mine (N)	T1	682	2,350	18.0	24.0	2,770	65.0	8,260	85.0	712	12	14,296
	T1	735	2,674	27.0	13.8	3,340	67.0	9,919	95.9	815		16,952
	T1	767	2,479	25.4	11.8	3,143	60.7	9,323	91.6	763		15,898
	T1	768	2,434	25.0	12.0	3,078	60.3	9,379	91.3	763		15,843
	T1	772	2,263	27.4	38.2	2,830	57.4	9,127	84.0	824		15,251
	T2	683	1,525	14.0	24.0	1,530	33.0	4,540	34.0	1,160	12	8,872
	T2	684	1,660	14.0	12.0	1,500	33.0	4,880	34.0	1,260	15	9,408
	T2	687	1,137	16.0	102.7	915	22.8	2,781	21.1	1,420		6,416
	T2	736	1,670	18.4	20.1	1,661	31.4	4,851	32.8	1,190		9,475
	T2	773	1,519	16.1	14.9	1,489	29.9	4,320	32.6	1,210		8,632
	T2	774	1,533	16.1	14.8	1,502	29.9	4,338	32.4	1,210		8,676
	T2a	740	1,886	28.0	97.0	2,060	53.0	5,900	57.0	830	37	10,948
	T2a	740	1,877	29.2	112.9	2,244	47.6	6,518	62.3	920		11,811
	T2a	769	1,729	28.6	93.3	2,078	44.8	6,063	58.9	924		11,020
	T2a	775	1,811	20.1	29.3	2,029	40.8	6,084	60.9	799		10,874
Underground mine	T3	686	1,816	16.1	15.2	1,433	30.9	4,863	44.1	905		9,123
	T3	741	1,678	14.8	9.5	1,342	30.0	4,455	41.4	835		8,406
	T3	742	1,316	13.0	6.0	960	25.0	3,323	33.0	690	24	6,390

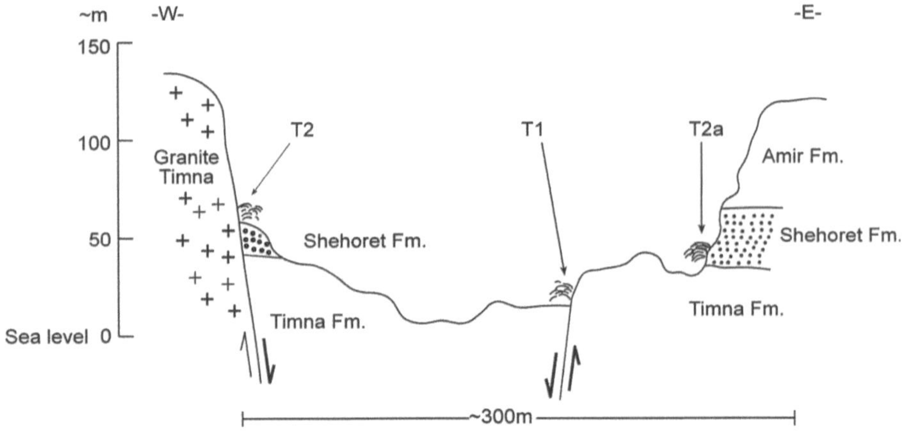

Fig. 11.9 E-W cross section through open pit N in Timna, from which saline waters were sampled. The three water groups T1, T2, and T2a are indicated near their corresponding discharge points (From Beyth et al. 1981)

constant, yet distinct K/Cl, Na/Cl, and SO_4/Cl ratios, and range in temperature between 16 and 32 °C.

The Sr isotope ratio of the Timna waters (<0.706) is lower than that of the saline waters in the Dead Sea region (~0.708).

In a comprehensive study, to be published soon, Spiro et al. evaluated the source of the Timna waters and proposed the mechanisms responsible for their chemical composition.

11.2.4 Saline Waters in the Northern Negev (R_o)

Saline waters in the Northern Negev were found in six deep bore holes (Boker-1, Kurnub-1, Makhtesh Katan-2, Keren-1, Sherif-1 and Zavoa-1). The chemical composition of these waters shows a marked similarity (Tables 11.5a and 11.5b) and their salinity varies between106-170 g L^{-1} (TDS). The waters were sampled from the deeper part of the geological section, of Paleozoic-Jurassic age. Some reside in limestone (Sherif-1 and Zavoa-1) while others were sampled from sandstone layers (Kurnub-1, Makhtesh Katan-2, Keren-1). The composition of the waters seems to be unrelated to the lithology of their host aquifers. Figure 11.10 details the locations of these R_0 waters and their Na/Cl-Br/Cl relationships. The latter fall along the seawater evaporation line, supporting their proposed ancient seawater origin. Their relation to the saline waters in the Dead Sea basin is discussed later in this paper.

Table 11.5a Chemical composition of subsurface brines from the Northern Negev (R_O) in mg/L

Source	Sample	Depth (m)	Na	K	Mg	Ca	Sr	Cl	Br	SO_4	HCO_3	TDS
Boqer – 1	D 517	2,730–46	31,080	314	1,660	15,870	427	81,060		1,170		131,581
Kurnub – 1	D 326	2,334–44	31,950	565	1,062	13,950	331	77,660		753		126,268
Kurnub – 1	BT 81	2,501–6	30,000	487	846	16,008		75,720		577		123,638
Makhtesh Qatan – 2	FG 1	2,101–25	25,950	901	2,550	14,800		73,580	725	895	76	119,477
Makhtesh Qatan – 2	FG 2	2,275–82	29,600	581	2,190	22,300		89,190	924	531	39	145,355
Qeren – 1	BT 73	2,708–19	41,260	1,010	1,585	15,407	284	95,760	733	497	158	156,694
Qeren – 1	BT 74	2,919–35	42,680	1,080	2,094	15,570		99,630	816	505	146	162,521
Sherif –1	D 15	2,843–82	35,000	300	2,128	17,900	684	94,150	782	644	110	151,698
Zavoa – 1	D 175	2,680–91	36,800	408	1,960	16,500	585	92,100	730	474	14	149,571

Table 11.5b Average ionic ratios of subsurface brines from the Northern Negev (R_O) (n=6, in eq/eq)

	Na/Cl	K/Cl	Br/Cl	SO_4/Cl	Mg/Ca	Sr/Ca
Average	0.62	0.0059	0.0036	0.0050	0.18	0.0130
Stdev	0.06	0.0020	0.0004	0.0014	0.01	0.0046

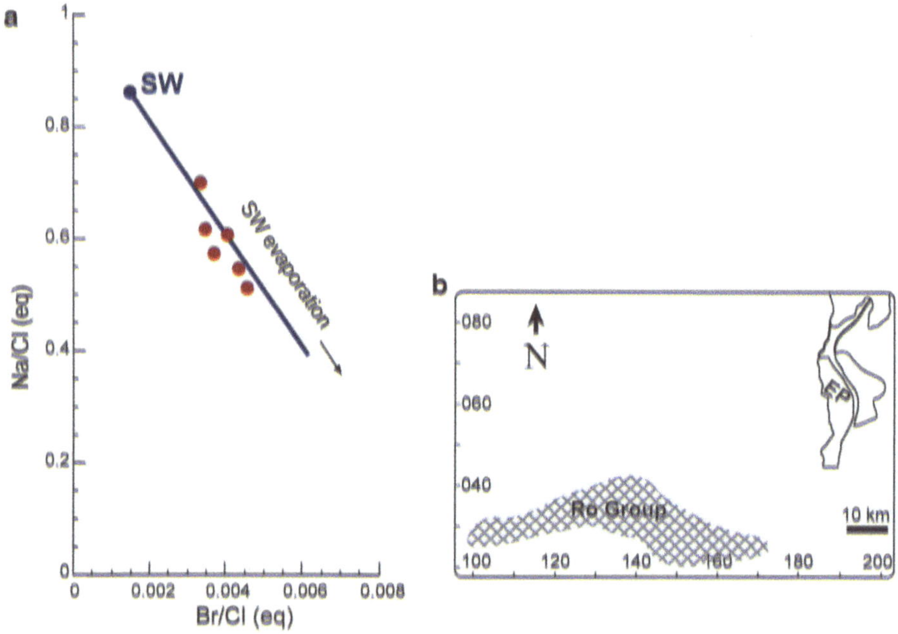

Fig. 11.10 Location and Na/Cl-Br/Cl relationships of saline waters in deep bore holes in the northern Negev. (**a**) Na/Cl-Br/Cl relationship. (**b**) Location map of the waters

11.3 The Transformation of Ancient Seawater to Dead Sea Rift Brines

Although the brines were derived from ancient seawater their major chemical composition is completely different from modern seawater. They are characterized by:

1. Being Ca chloride brines (Fig. 11.11)
2. A wide range of salinities, up to 340 gr/l
3. A wide range of ratios of Na, SO_4, K and Mg to Cl
4. High ratios of Ca, Sr, and Br to Cl
5. A wide range of $^{87}Sr/^{86}Sr$, from 0.7056 to 0.7082 (Table 11.6)

Based on rocksalt diapirs along the Dead Sea transform it is a consensus that the saline waters originated from ancient seawater of probably Pliocene age (Zak 1967; Starinsky 1974; Horowitz 1987).

The chemical transformation of lagoonar seawater into Dead Sea rift subsurface brines took place through three main stages:

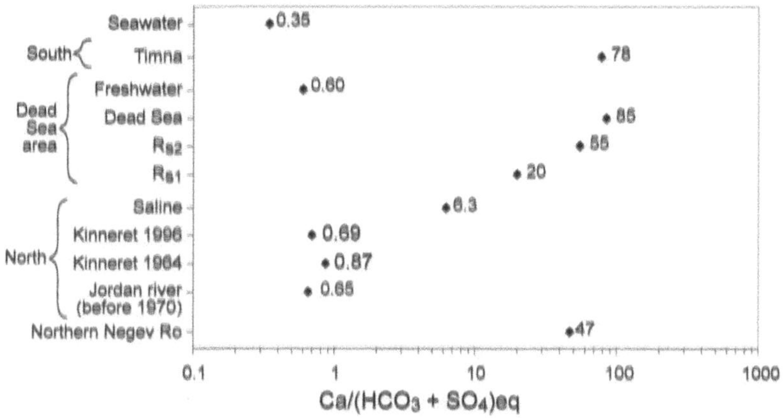

Fig. 11.11 The Ca/(HCO_3+SO_4) ratio (in equivalent units) in the various water groups in the Dead Sea rift valley and in seawater. The freshwaters and seawater ratios are distinctly lower than those of the saline, Ca-chloridic waters

1. Evaporation of seawater in a lagoon, connected to the Mediterranean Sea.
2. Water-rock interaction of the evaporated seawater during infiltration and migration into the surrounding rocks.
3. Mixing either in the lagoon or at depth with waters of various salinities and compositions.

Evaporation: Evaporation may take place via two different paths- equilibrium or fractional. In the first case the formed minerals stay in contact with the evaporated solution all along the process and a reaction between the solids and the liquid continues uninterruptedly. Minerals may redissolve and new ones form instead. In the second process the precipitating minerals are removed from the solution immediately after their formation. Once a mineral is formed it stays unaltered in the system until the end of the process.

Several experiments were carried out in the laboratory to follow the chemical evolution of seawater induced by evaporation (e.g. Braitsch 1971) while others, beginning with those of Harvie et al. (1980, 1982, 1984) use Pitzer's model calculations to predict the path of evaporation of seawater-like systems.

In the present study we use the data of Raab (1996) who carried out a series of evaporation experiments by fractional crystallization of Mediterranean seawater at a temperature of 25 °C. The chemical composition of the changing seawater at the different degrees of evaporation (DE) tested (up to 138) is shown in Table 11.1.

The first mineral precipitates very early as aragonite ($CaCO_3$), followed by gypsum ($CaSO_4.2H_2O$) at a degree of evaporation of 3. At a degree of 10, halite (NaCl) starts to crystallize. Thereafter, the new minerals to form consist of a K-Mg-SO_4 composition, starting to crystallize at various degrees of evaporation that depend on the system's operational mode, fractional or at equilibrium. The order of precipitation during seawater evaporation is given in Fig. 11.12 and the behavior of most used ionic ratios is displayed in Fig. 11.13 (Raab 1996).

Table 11.6 $^{87}Sr/^{86}Sr$ ratios in selected water samples from the Dead Sea rift

Region		Sample no.	Locality	Sampling date	$^{87}Sr/^{86}Sr$
Sea of Galilee	R_f	ER 147	En Jweiza sp.		0.70758
	R_f	ER 146	En Dankale sp.		0.70441
	R_f	ER 145	En Iron sp.		0.70501
	R_f	ER 165	N. Fara sp.		0.70628
	R_f	D 604	Jordan river – Arik Bridge	1973	0.70693
	R_{NE}	D 654	Bet Habek (well)		0.70566
	R_{NE}	TR 268	Bet Zayda	1991	0.70574
	R_{NW}	D 779	Hammam el Malih sp.	1978	0.70771
	R_{NW}	D 2601	Hammei Tiberias sp.	1990	0.70777
	R_{NW}	D 2600	Hammei Tiberias sp.	1990	0.70775; 0.70774
	R_{NW}	D 3096	Hammei Tiberias sp.	1990	0.70776; 0.70777
	R_{NW}	D 401	Hammei Tiberias sp.	1975	0.70787; 0.70784
	R_{NW}	TR 271	Kinneret 10 b (well)	1991	0.70787
	R_{NW}	D 2603	Kinneret 2 (well)	1990	0.70777
	R_{NW}	TR 183	Sartan Iver sp.	1991	0.70772
	R_{NW}	D 3092	Sartan Iver sp.	1990	0.70772
Dead Sea		ER 59	Jordan river – South		0.70802
	R_{S1}	D 527	Amiaz 1 (well)	1972	0.70805
	R_{S1}	D 173	Amiaz 1 (well)	1960	0.70791; 0.70804
	R_{S1}	NT 43	En Feshcha sp.		0.70803
	R_{S1}	NT 47	En Feshcha sp.		0.70799
	R_{S1}		En Hogla sp.		0.70797
	R_{S1}	D 750	Massada (well)	1976	0.70832
	R_{S1}	D 3562	Zerah 1 (well)	2004	0.70797
	R_{S2}	NT 34	Dead Sea water	1955	0.70802
	R_{S2}	TR 152	En Ashlag sp.	1990	0.70810
	R_{S2}	TR 153	En Ashlag sp.	1990	0.70810
	R_{S2}	TR 154	En Ashlag sp.	1990	0.70810
	R_{S2}	D 1280	En Gedi 3 (well)	1988	0.70793
	R_{S2}	D 3166	En Gedi-3 north (well)	1990	0.70795
	R_{S2}	D 1283	En Kedem sp.	1988	0.70791
	R_{S2}	D 1575	En Zruya sp.	1989	0.70796; 0.70796
	R_{S2}	D 1905	Hammei Zohar	1989	0.70794
Timna	T1	D 682	open mine pit (N)	1981	0.70598
	T1	D 735	open mine pit (N)	1981	0.70599
	T1	D 767	open mine pit (N)		0.70603
	T1	D 768	open mine pit (N)		0.70603
	T1	D 772	open mine pit (N)		0.70602
	T2	D 736	open mine pit (N)	1981	0.70583
	T2	D 773	open mine pit (N)		0.70587
	T2	D 774	open mine pit (N)		0.70589
	T2a	D 775	open mine pit (N)		0.70591
	T3	D 686	underground mine	1981	0.70589
	T3	D 741	underground mine	1981	0.70599
	T3	D 742	underground mine	1981	0.70586

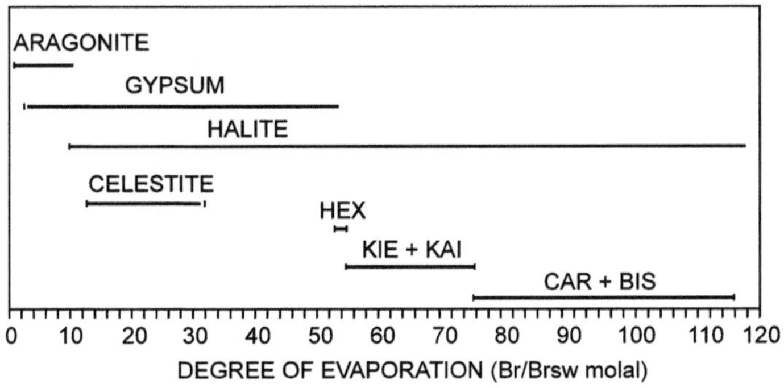

Fig. 11.12 The minerals that form during fractional evaporation of seawater at given degrees of evaporation: *KAI* kainite, *Kie* kieserite, *Hex* hexahydrite, *CAR* carnallite, *BIS* Bischofite (After Raab 1996)

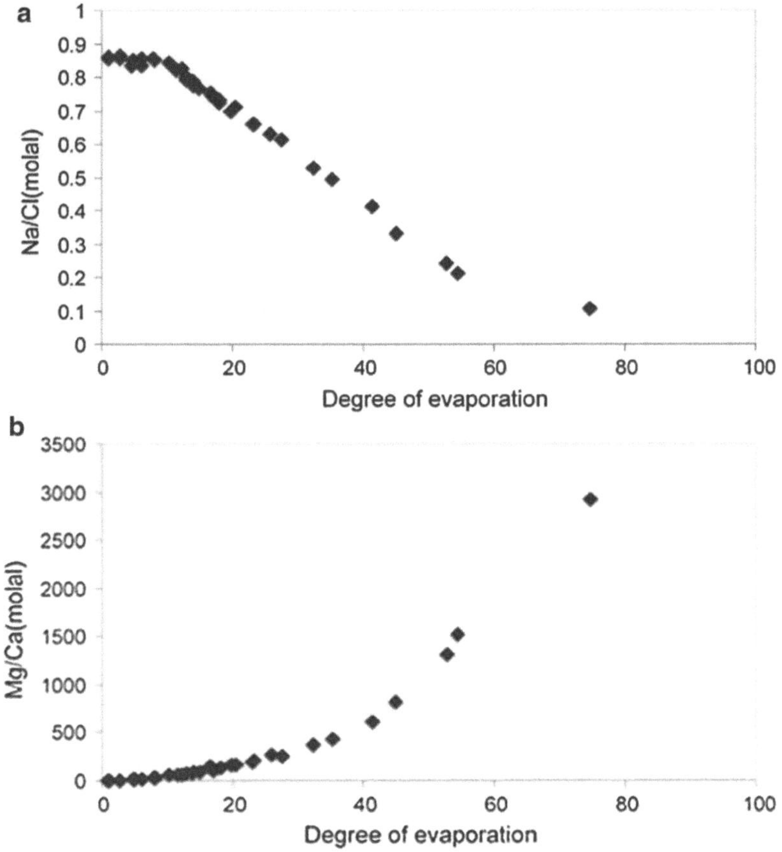

Fig. 11.13 Evolution of the Na/Cl and Mg/Cl ratios during the evaporation of seawater: (**a**) the Na/Cl ratio remains constant until DE ~ 10 where halite starts to crystallize. (**b**) The Mg/Ca ratio increases upon evaporation due to crystallization of gypsum and aragonite

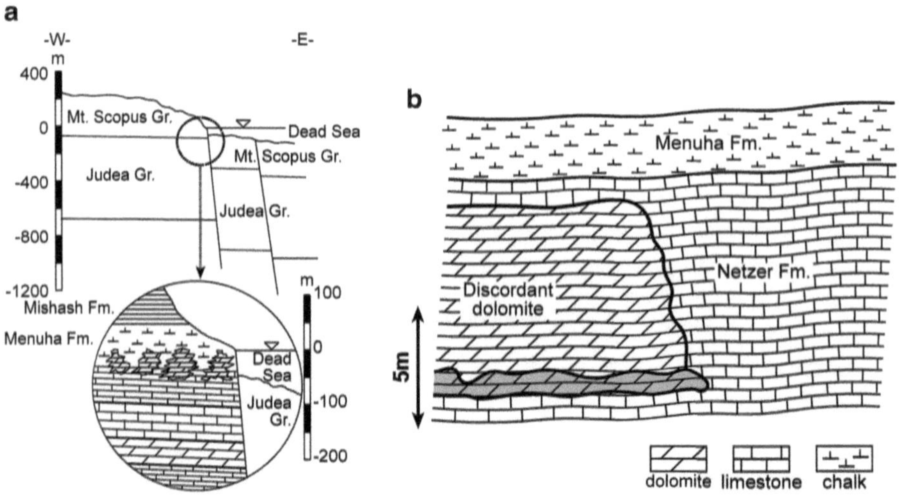

Fig. 11.14 Geological evidence for WRI between the Sedom lagoon brine and Upper Cretaceous (Turonian-Senonian) carbonate rocks west of the Dead Sea basin. (**a**) W-E cross section showing the linkage between the Sedom Lagoon and the geological rock sequence. Insert: detailed view of the discordant dolomite bodies products (After Raz 1983). (**b**) The spatial relationship between a discordant, epigenetic dolomite body that formed during the existence of the Sedom lagoon and the affected rocks of Turonian-Senonian age (After Stein et al. 2002)

Dolomitization: Because in seawater $Ca^{2+} < SO_4^{2+}$, only a small fraction of the SO_4^{2-} ions dissolved therein can be tied into gypsum during evaporation in a marine evaporitic lagoon. Therefore, the concentration of SO_4^{2-} uninterruptedly increases until the degree of evaporation reaches at least 50–60, in spite of gypsum precipitation (Raab 1996). At the same time, though, the concentration of Ca^{2+} decreases (Table 11.1), and the Mg/Ca ratio increases from about 5 (seawater value) to a few hundreds or a few thousands at high degrees of evaporation (Fig. 11.13). Brines that follow such a trail and get in touch with $CaCO_3$ minerals, within the lagoon or in the surrounding rocks, initiate dolomitization of these minerals, via the following reaction:

$$2CaCO_3 + Mg^{2+} \rightarrow CaMg(CO_3)_2 + Ca^{2+} \tag{11.2}$$

While Mg^{2+} is lost from the brine to form dolomite, the concentration of Ca^{2+} does not increase during the early stage of dolomitization because the brine is saturated with respect to gypsum, and Ca^{2+} ions released from $CaCO_3$ minerals via eq. [11.2] combine with dissolved SO_4^{2-} ions to form additional gypsum. As a result, the concentration of Mg^{2+} and the Mg/Ca ratio in the brine at first decrease, and the process takes place during the marine lagoon stage. A large number of papers describe the occurrences of epigenetic dolomite bodies in the carbonate rock sequence adjacent to the Dead Sea (Fig. 11.14) (e.g. Bentor and Vroman 1960; Starinsky 1974; Raz 1983; Stein et al. 2002).

11.4 The History of the Water

11.4.1 Na-Cl System

Large salt bodies in the DSB (Dead Sea Basin) testify to the existence of an ancient lagoon within which they were deposited. Such are known from two areas, the Dead Sea and the Lake Kinneret regions. Zak (1967) attributed the source of the lagoon's water to the ancient Mediterranean Sea which invaded the DSB via the Valley of Jezreel. The overflow point, or sill, over which the Mediterranean seawater invaded the DSB, was probably located in the western side of the Bet Shean valley. Although data required for exact delineation of the lagoon's dimensions are missing, the present location of the saline waters that originated therein as well as considerations dictated by the present topography allow the assumption that during its maximum expansion the lagoon occupied the area between Rosh Pina in the north and Idan – Hatzeva south of the Dead Sea. An N-S transect of the lagoon is presented in Fig. 11.3. Na/Cl ratios in the lagoon brine that was extracted from fluid inclusions from a few halite samples in the Sedom salt (Horita et al. 2002; García-Veigas et al. 2009) (Fig. 11.15) fall within 0.1–0.3, placing the degree of evaporation of the respective brines between 40 and 70, relative to the parent seawater. Seawater evaporation in the laboratory and thermodynamic calculations indicate that the major minerals that precipitate within that concentration range are halite, gypsum and anhydrite. As earlier stated in the introduction chapter, the Na/Cl seawater ratio (0.86) starts to decrease during evaporative concentration once halite crystallization sets in; indeed, all saline waters found so far in the DSB display Na/Cl ratios lower than that of seawater.

Mt. Sedom, which is the most thoroughly investigated body of all salt deposits in the DSB will be used for the explanation of the Sedom lagoon from which it was deposited. According to Zak (1967) about ¾ of the Mt. Sedom diapir section are comprised of evaporites, leaving one quarter of the section to clastics. Halite, with lesser amounts of gypsum and anhydrite, predominates among the evaporite minerals. Sand, clay and marl make up the clastic parts of the section. Hiatuses in the salt column indicate absence of seawater inflow, namely cutoff events of seawater supply from the Mediterranean Sea somewhere close to the bathymetric sill in the north, followed by drop in the lagoon's level.

The deposition rates of the salt and clastics differ by orders of magnitude. Applying the Lisan Formation as a crude gauge of clastic deposition rate in the Dead Sea area during late Pleistocene, one arrives at a 20–25 m of clastics per 40–50 kY, equal to some 0.5 mm year^{-1}. Dividing the integrated thickness of the clastic units in the Sedom Fm. (about 500 m) by the aforementioned deposition rate yields an accumulation period of about 1×10^6 years for these sediments. The accumulation rate of salt in the Sedom Fm must have been linked to the rate of evaporation of the Sedom lagoon. Complete evaporation of a 10 m thick layer of seawater would produce around 26.7 g halite cm^2, or a salt layer (ρ_{bulk} ~2 g cm^{-3}) of 13.4 cm thickness. Evaporation of brine from the Dead Sea has been estimated to be around 1.5 m year^{-1} (1.3–1.75 m year^{-1}, after Stanhill 1994), corresponding to about 2 cm year^{-1} salt. Hence, the deposition of the 1,500 m thick evaporite section of Mt Sedom would have required roughly

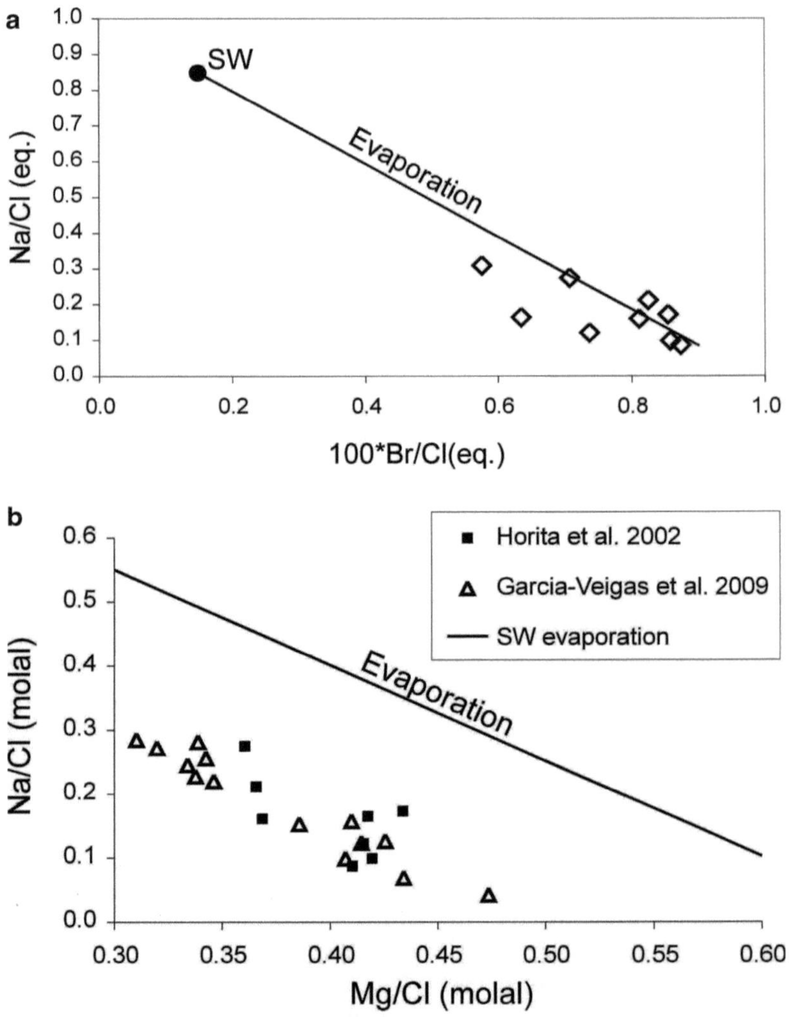

Fig. 11.15 (**a**) The relationship between the Na/Cl and Br/Cl ratios in fluid inclusions in halite crystals from the Mt. Sedom (From Horita et al. 2002). (**b**) The relationship between the Na/Cl and Mg/Cl ratios in Mt. Sedom halite samples. *Triangle markers* – Data from García-Veigas et al. 2009; *diamond markers* – Data from Horita et al. 2002)

75,000 years. Obviously, the accumulation periods of both materials are very crude estimates at best, but their ratio indicates that the accumulation of the clastics required significantly more time than did the accumulation of salt. In other words, one may assume that during the long periods of salt-free deposition of clastics, the lagoon area was completely or partially disconnected from the sea. Moreover, the absence of gypsum in the clastic layers indicates low salinity (i.e. less than 100 g L^{-1}) of the lagoon during these times (gypsum starts to precipitate from evaporating seawater at a threefold concentration). The significance of these observations is their hydrologic consequence, namely, that the lowering of the lakes water stand allows the backward

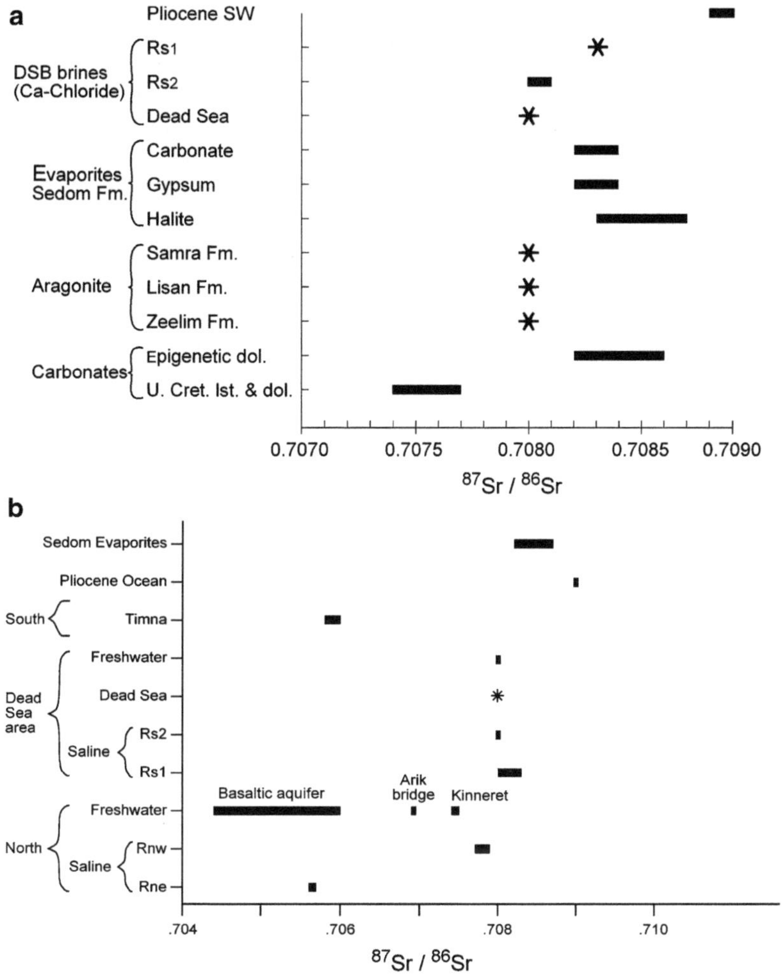

Fig. 11.16 (**a**)$^{87}Sr/^{86}Sr$ isotope ratios in Dead Sea basin brines, minerals and rocks (From Katz and Starinsky 2009). (**b**) $^{87}Sr/^{86}Sr$ isotope ratios in the various water groups in the Dead Sea basin, in Pliocene seawater and in the Sedom Fm. evaporites

(=eastwards) flow to the DSB of saline brines that were formed and wandered out (i.e. westwards) of the DSB lake during its former high stand periods. This conclusion is in full agreement with the current state, where numerous saline springs and seepages discharge into the low-standing Dead Sea. The returning fluid entered the lake where it mixed with the resident brine, evaporating together and depositing the salt and gypsum that comprise the fossil evaporitic section of the Mt. Sedom diapir.

The brine that returned into the DSB is Ca-chloridic in composition and displays $^{87}Sr/^{86}Sr$ ratios lower than that of its parental Pliocene seawater (Fig. 11.16). The lowering of the isotopic ratio is a result of the dolomitization by this brine of the surrounding carbonate Cretaceous country rocks during the westward voyage out and then back into the lagoon.

The conclusion from the forgoing discussion is that the Mt. Sedom diapir evaporites were deposited at relatively low-stand periods of the lagoon during which only a partial connection with the sea existed. The degrees of evaporation in the lagoon, of the seawater that invaded the DSB and precipitated the salt therein, were derived from analyses of fluid inclusions in the Sedom Fm. halite samples that were collected in the southern, topographically deepest part of the DSB, representing relatively high concentration factors.

It is reasonable to assume that the Sedom lagoon's volume changed during its lifetime along with its bottom. If so, in periods of maximum expansion, when the lagoon reached the northern coastline of Lake Kinneret, its brine must have been less evaporated than when it was constrained to the Mt. Sedom area only.

11.4.2 Ca-Mg System

Lowering of Mg/Ca ratio, as examlified in R_1 fluids, may have been caused by consumption of Mg^{2+}, addition of Ca^{2+}, or both, to the affected solution. The most reasonable process is dolomitization (decrease in Mg^{2+} along with a parallel increase in Ca^{2+}) which is directly supported by numerous dolomite bodies that discordantly replace layered carbonate rocks of Upper Cretaceous age (Fig. 11.14). While evaporation must have taken place on the surface, dolomitization could have taken place either by replacement of *contemporaneous* aragonite in the lagoon, and/or limestone in the subsurface, *after* infiltration of the concentrated seawater into the surrounding country rocks, on the western side of the lagoon. Dolomite layers do appear in the Sedom Fm. salt units, confirming dolomite formation also within the lagoon. The massive, discordant dolomite bodies in the Judea desert indicate that the dolomitization potential of the brine was not exhausted in the lagoon, allowing the continuation of this WR interaction during the brine's infiltration and migration in the subsurface.

Elevation of Mg/Ca ratios, reflected by the R_2 brines can be caused by consumption of Ca^{2+} without affecting the Mg^{2+} concentration in the water, such as precipitation of gypsum in a marine evaporitic lagoon, or by influx of freshwater with $(Ca+Mg)/HCO_3>1$ into a lacustrine, $CaCO_3$-precipitating, water body. Removal of Ca^{2+} in gypsum upon evaporation of seawater occurs at a DE of 3, well before the onset of halite crystallization (DE = 10), and must have increased the Mg/Ca ratio to values well above that of seawater (5.2), as actually observed in modern lagoons (Fig. 11.13.). The alternative process (supply of Mg^{2+} by freshwater) requires a lacustrine environment, which could have formed only *after* disconnection of the basin from the ancient Mediterranean Sea.

It is mandatory that removal of Ca^{2+} into gypsum did take place in the Sedom lagoon, like in any other marine evaporitic system, shifting the Mg/Ca ratio well above 5. Yet, the highest Mg/Ca ratio found in R_2 waters is around 5. Hence, rather than resorting to gypsum precipitation as responsible for the Mg/Ca ratios in R_2 waters (1–5) the elevation of this ratio should be attributed to addition of Mg^{2+} from freshwater to a *low* (<1) Mg/Ca brine, in a lacustrine environment. The obvious candidate brine for that purpose is R_1 water which evolved earlier, during the lagoon stage, migrated westwards while dolomitizing limestone en route. Eastward back-flow of R_1 brine from the northern Negev to the DST was made possible as a result of the intermittent uplift of the shoulders of the depression in which the brine has migrated and was stored. When that happened, the R_1 brine found its way either to the DSB in the south or to shallower parts of the DST north of the Sedom lagoon. Before the disconnection of the lagoon from the Mediterranean Sea, the returning brine admixed into the lagoon waters, significantly affecting its Sr isotope ratio and somewhat changing its chemical composition.[2] Once the lagoon became disconnected from its Mediterranean feeder, the returning R_1 brine started to fill the DSB depression, forming a (terminal) saline lake nourished by freshwater flowing in from its drainage basin.

The above scenario, which was initiated by the lagoon's cutoff from the seawater source, persisted until recent time and is manifested by the current situation in the Dead Sea. The freshwater is characterized by the ratio $Ca^{2+} < (HCO_3^- + SO_4^{2-})$. Assuming that the lakes maintain an approximately constant volume, the annual input of H_2O by the fresh water sources is lost by evaporation and the dissolved salts therein accumulate or precipitate. Crystallization of aragonite (or calcite[3]) and gypsum rapidly removes the Ca^{2+}, HCO_3^- and SO_4^{2-} imported by the freshwater plus some ambient Ca^{2+} from the resident brine to make up the charge balance. The Mg^{2+} remains in solution, gradually increasing the Mg/Ca ratio to its present value (~5). Hence, for lakes in which no other processes but freshwater influx took place one may regard the Mg/Ca in the water roughly as a qualitative time monitor. The youngest saline lake in the area is the modern Dead Sea in which Mg/Ca ~5, the highest amongst the saline waters that discharge into it, which show Mg/Ca ratios between 2 and 3, and were themselves saline lake waters in the past.

The R_1 brines appear in all areas mentioned above (Fig. 11.17). In the north they are found in the Tiberias spa, on the western shore of Lake Kinneret as well as east of the lake (in Hamat Gader). West of the Dead Sea the waters were sampled in deep oil drillings, quite far from the lake. The saline waters in Timna belong exclusively to this group. All these waters were originally more saline and were subsequently diluted. On the other hand, R_2 waters occur in shallow bore holes along the entire eastern shore of Lake Kinneret between Ha'On in the south and Bet Zaida in the north. The same is true for all the saline seepages along the western side of the Dead Sea, and for a few points near the eastern shores of that lake. Thus, the spatial distribution of the two water groups is in agreement with the geochemical evolution path proposed above namely, that the transformation of R_1 waters to R_2 waters took place on the surface and was arrested once these waters were trapped in the subsurface.

[2] Dolomitization of upper Cretaceous limestone by the brine resulted in a decrease of the $^{87}Sr/^{86}Sr$ ratio and the Mg/Ca ratio of the brine relative to the corresponding ratios in the fresh, Pliocene seawater that flowed through the lagoon (to be discussed later in this paper).

[3] Depending on the Mg/Ca ratio in the lake.

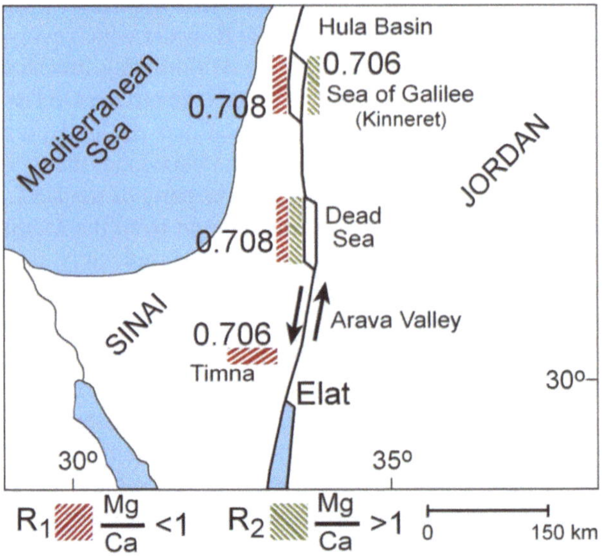

Fig. 11.17 A map showing the relationship between $^{87}Sr/^{86}Sr$ and the Mg/Ca ratios in the saline waters in the Dead Sea rift. The low $^{87}Sr/^{86}Sr$ ratio (0.706) in R_1 brines was determined by water-rock interaction while in waters of group R_2 it was set by the freshwaters that discharge into the saline lake during the lacustrine period

Table 11.7 Stages of evolution of saline water in the Dead Sea rift

Na/Cl	Mg/Ca	Location	Event
0.86	5.2	Sedom lagoon	SW evaporation
0.86	↑↑↑ 5.2	Sedom lagoon	Gypsum crystallization
0.86 ↓↓↓	No effect	Sedom lagoon	Halite crystallization
No effect	↓ ↓	Sedom lagoon	Aragonite dolomitization
No effect	No effect	Judea carbonate rocks	Brine infiltration & migration
No effect	↓↓↓ <1	Judea carbonate rocks	Limestone dolomitization
No effect	↓↓↓ <1	Judea carbonate rocks	Partial brine backflow to lagoon
R_1 brine formation & storage completed			
<0.86	<1	~ Bet Shean watershed	Disconnection from Mediterranean Sea
<0.86	<1	Dead Sea basin	Formation of an inland saline lake
<0.86	1–5 ↑↑↑ <1	Dead Sea basin	Freshwater inflow and evaporation
R_2 brine formation completed			

The orderly, relative time frame of events detailed above is schematically summarized in Table 11.7.

A quantitative model for the $R_0 \rightarrow R_2$ brine transformation was worked out by Katz and Starinsky (2009), and applied to explain the large difference in the Mg/Ca ratio between the Sedom (R_0-group) brines (average Mg/Ca = 0.16) and modern

Dead Sea water (Mg/Ca = 4.2–4.6). The model calculates the increase in the Mg/Ca ratio in the DSB lake, resulting from addition of freshwater and R_o brine into a saline lake. The chemical compositions of the fresh and saline waters are detailed in the paper, and it is assumed that the total H_2O input from the fresh and saline sources is balanced by evaporation.

The volumes of freshwater and R_1 solutions discharging into the lake are calculated by mass balancing Ca^{2+}, HCO_3^- and SO_4^{2-} contained in the primary aragonite and gypsum enclosed within the lacustrine sediments of the DSB lake with the same masses contained in the fresh and saline water inputs. The total mass of H_2O imported by the lake (normalized per unit time and unit area) is gleaned from modern evaporation rates prevailing in the area (1.5–2 m $year^{-1}$).

The model is summarized by Eq. 11.2 below (Eq. #13 in Katz and Starinsky's 2009):

$$\left(\frac{Mg}{Ca}\right)_L = \frac{\left([Mg]_i\, d_A \cdot 0.1 + V_b\,[Mg]_b + V_f\,[Mg]_f\right)}{\left([Ca]_i \cdot d_A \cdot 0.1 + V_b\left([Ca]_b - [HCO_3]_b - [SO_4]_b\right) + V_f\left([Ca]_f - [HCO_3]_f - [SO_4]_f\right)\right)}, \tag{11.3}$$

Where the solutes are designated by their chemical symbols, d_A is the average depth of the lake (in m), V_b and V_f are the respective volumes (in liters) per unit area of brine and freshwater that make up the water column of the lake, and 0.1 is a conversion factor translating depth (in m) to volume (in liters). The subscript i stands for the *starting* concentrations of the subscripted ions (Ca^{2+} and Mg^{2+}) in the lake, from which the Mg/Ca ratio evolved.

Figure 11.18 is an example of the application of Eq. 11.3 to simulate the evolution of the Mg/Ca ratio in a lake similar to the DSB Lake. The shape of the resulting curve is very sensitive to the ratio between gypsum and aragonite in the geological section, and to the chemical composition of the brine and freshwater.

The model emphasizes the significant role played by the removal of Ca^{2+} into evaporitic $CaCO_3$ and $CaSO_4$ minerals on the Mg/Ca ratio in the DSB Lake, and quantifies the contribution of the freshwater and brine that must have been imported into the lake to maintain its level and to balance the masses of the precipitating minerals.

The change in the Mg/Ca ratio in the lake's water displays a non linear relation to the accumulated masses of evaporitic minerals (aragonite and gypsum). The major factors that affect the shape of the Mg/Ca-elevation curve displayed in the Fig. 11.18 are the ratio between masses of aragonite and gypsum and their total concentration in the sediment as well as the chemical composition of the freshwater and brine that are selected as basic inputs to the model.

During the last 50 years or so, the Mg/Ca (equivalent) ratio in the Dead fluctuated between 4.2 and 4.6. This can be explained by the negligible deposition of aragonite and gypsum from the lake.

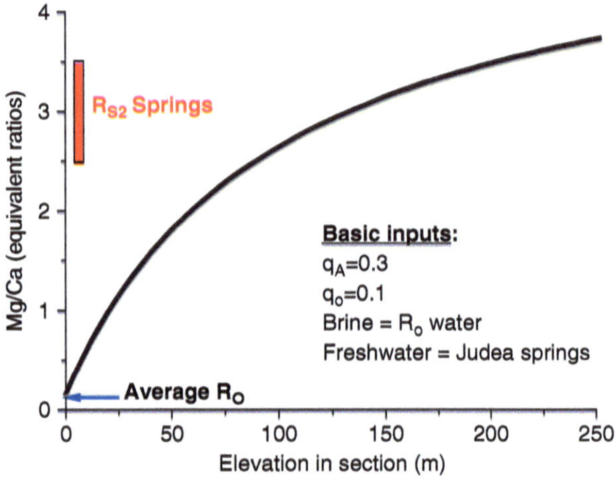

Fig. 11.18 Dependency of the Mg/Ca ratio in brine on the accumulated thickness of $CaCO_3$ and $CaSO_4$ minerals that precipitated in a saline lake in the Dead Sea basin. Under the specific inputs selected for this example the R_{s2} group waters could evolve from Ro brine after the deposition of a ~90–200 m thick sediment column (From Katz and Starinsky 2009)

11.4.3 $^{87}Sr/^{86}Sr$ Ratio in the Dead Sea Brines

The two saline, high Mg/Ca ratio subgroups are distinguished from each other by their isotopic, $^{87}Sr/^{86}Sr$ ratios. Around the Dead Sea the waters display $^{87}Sr/^{86}Sr$ ratios around 0.708, as opposed to the Lake Kinneret region, which is characterized, in the eastern side by ratios around 0.706 (Fig. 11.17). As shown earlier, the chemical composition of waters with high Mg/Ca ratios is dictated by freshwater, hence their $^{87}Sr/^{86}Sr$ ratios must differ between the two areas, reflecting the different lithology of their drainage systems. In the north, the incoming freshwater drain a basaltic terrain while the waters from the south drain a carbonate rock environment. Freshwater flowing nowadays into Lake Kinneret displays low $^{87}Sr/^{86}Sr$ ratios, around 0.704 (Table 11.6), in contrast to freshwater discharging into the Dead Sea with $^{87}Sr/^{86}Sr$ ratios of about 0.708.

11.4.4 The Evolution of Timna Saline Water

Three main features pertaining to the geological environment of the Timna waters, differ from those characterizing the saline waters in the Dead sea and in the Lake Kinneret basins:

1. The Dead Sea and Kinneret tectonic depressions contain salt bodies that very likely reflect ancient evaporitic activity responsible for the formation of the observed salt deposits. In contrast, the Timna area is not located in a tectonic basin, neither does it include any salt body that could have pointed to the origin of the local saline waters.
2. The saline spring waters in the Dead Sea and Kinneret areas are propelled to the surface under their own hydraulic heads, while the exposure of the Timna waters

was achieved by drilling and by artificial removal of the gangue overburden above the copper ores.

3. The waters in Timna reside within a sandy silicate aquifer, as opposed to the waters in the northern areas which are hosted within carbonate rock aquifers.

The Timna waters must have originated as saline Ca-chloridic waters that were *imported* to Timna, probably by southward migration from the Sedom lagoon in the Dead Sea region. Later, the waters interacted with the resident silicate rocks, mainly with the magmatic minerals plagioclase, olivine, pyroxene and amphibole, giving rise to the secondary minerals chlorite and epidote. The present low salinity of these waters is probably the result of (ongoing?) dilution.

The original $^{87}Sr/^{86}Sr$ and Sr/Ca ratios in the gabbro before interaction with the brine were 0.7045 and 0.013, respectively. The brine came in from the north with a $^{87}Sr/^{86}Sr$ ratio of ~0.708, such that the $^{87}Sr/^{86}Sr = 0.706$ ratio characterizing the Timna saline waters under regard reflects an intermediate value between the aforementioned solid and aqueous reactants in the WRI.

A question relevant to this explanation is which water type (from the north) participated in the interaction with the Timna gabbro. Were these R_1 brines that formed within the Sedom lagoon or R_2 waters that represent the later, lacustrine stage? A crude Ca-Mg mass balance may provide the answer. Chloritization of the gabbro minerals must have ended upon consumption of all available Mg^{2+} from the solution. The Mg^{2+} concentration in the Timna waters is almost zero, which allowed Spiro et al. (in preparation) to show that the original brine imported from the north must have had a Mg/Ca ratio between 0.3 and 0.5, attributing it to the R_1 group of brines. If so, it is possible that these waters arrived at Timna already during the lagoon stage.

11.5 The Climatic Impact on Brine Evolution

The possible maximum evaporation of a surface water body (lake or lagoon) is determined by the relative humidity above it. The control of relative humidity on evaporitic mineral facies formation was discussed earlier by Kinsman (1976) for normal evaporated seawater. Krumgalz et al. (2000), in their discussion of the Dead Sea future have shown the possible variation of the Dead Sea surface level and volume as a function of the relative humidity in the range of 10–70 %.

The relationship between the activity of water (a_{H2O}) and the Na/Cl ratio in the Dead Sea water and in evaporated seawater is displayed in Fig. 11.19.

A given mineral may crystallize from different brines under different relative humidity conditions, depending on the composition of the evaporating brine. Crystallization of halite from evaporating seawater, for example, may take place when RH falls below 76 %, while an RH as low as 67 % is required to start the same process in a Ca-chloridic brine such as the modern Dead Sea. The mineral bischofite may precipitate from seawater under an RH of 50–60 % as opposed to a RH of 30 % required for the same to happen in the Dead Sea (Fig. 11.19).

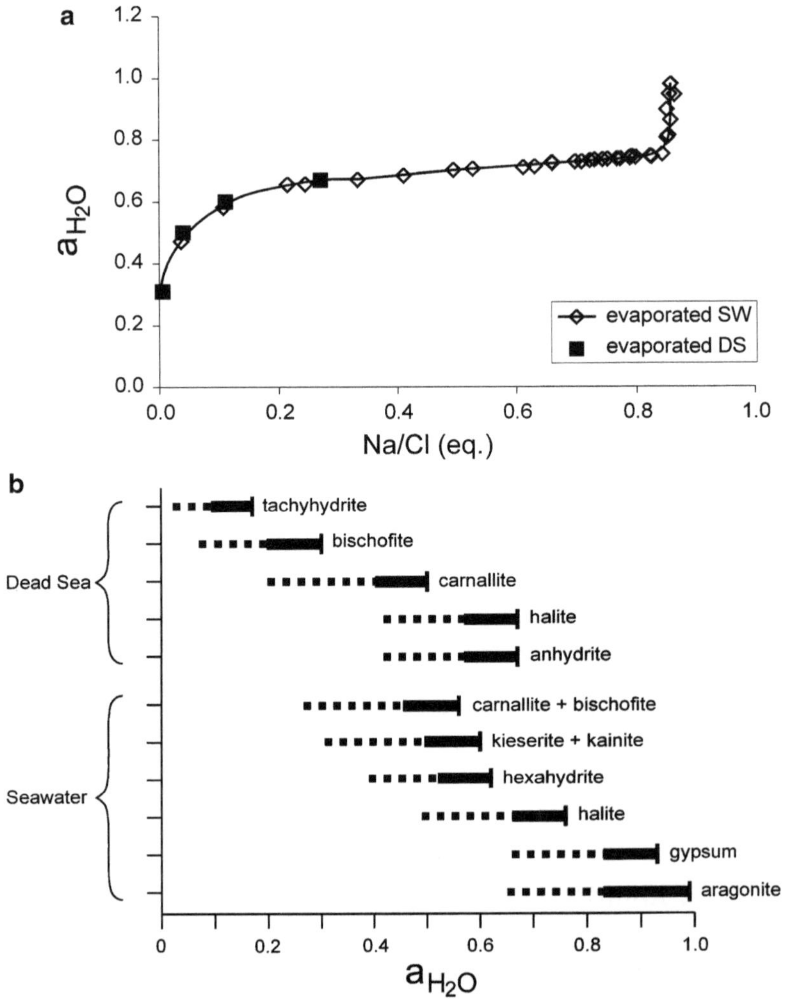

Fig. 11.19 (**a**) The relationship between the Na/Cl ratio and the activity of H_2O (a_{H2O}) in evaporating seawater and Dead Sea water. The plot allows determination of the maximum relative humidity that could exist during evaporation in an ancient saline lake. (**b**) The relationship between the crystallization points of minerals and the H_2O activity in evaporating seawater and Dead Sea water

It follows, that the Na/Cl ratio in brines can be used as a monitor of the ancient RH under which these brines formed, provided that it represents the *unaltered*, original Na/Cl ratio that prevailed in the brine during the last crystallization event of halite therein.[4]

[4]A common alteration of the original Na/Cl ratio in brine is dissolution of halite upon its dilution. Such would result in increase of the original (<1) ratio that is achieved by any marine-evaporitic brine once halite starts to crystallize from it.

Although the activity of water at crystallization onset points of a given mineral during the evaporation of seawater or Dead sea water (Fig. 11.19b) may differ, the same points fall on the same position on the corresponding a_{H2O} – Na/Cl plot (Fig. 11.19a).

Turning on to the Dead Sea rift saline waters, one may check whether or not the Na/Cl ratio in these brines indeed represents its original value that existed during halite deposition. The relevant Na/Cl ratio data is given in Fig. 11.6. The Lake Kinneret and Northern Negev waters possess Na/Cl ratios within the 0.5–0.7 range, corresponding to a maximum RH of 65–70 % during their formation. Fluid inclusions in the Sedom Fm. rock salt crystals have Na/Cl ratios between 0.1 and 0.3, indicating a maximum RH of 57–65 % in the atmosphere above the brine.

An alternative account for the wide scatter of the RH values calculated for the northern area brines is that the RH during the salt deposition period in Zemakh was rather high, disabling further evaporation. It should be noted that such waters (R_0 group) were formed also in the southern area of the Dead Sea rift at the same time. Once the RH decreased, evaporation resumed, accompanied by the lagoon shrinkage. Such a scenario may be reflected by the lower Na/Cl ratios in the Sedom salt fluid inclusions (Na/Cl=0.1–0.3) and in the R_{s1} waters (Na/Cl=0.1–0.5), which probably represent the shrinkage of the lagoon and salt deposition in the south. During that period the RH in the overhanging atmosphere must have been between 57 and 65 %.

Unfortunately, no fluid inclusion data is available for the northern salt (Zemah-1).

Acknowledgements The authors are thankful to Zvi Garfunkel, Michael Beyth and Baruch Spiro for fruitful discussions. Olga Polin and Carmel Gorni were very helpful in the preparation of the manuscript. The paper greatly benefitted from thorough reviewing by Joris Gieskes and Abraham Lerman.

References

Ben Avraham Z, Garfunkel Z, Lazar M (2008) Geology and evolution of the southern Dead Sea fault with emphasis on subsurface structure. Ann Rev Earth Planet Sci 36:357–387

Bender F (1974) Geology of Jordan, Supplementary edition of vol 7. Gebruder Borntraeger, Berlin/Stuttgart, 196p

Bentor YK (1961) Some geochemical aspects of the Dead Sea and the question of its age. Geochim Cosmochim Acta 25:239–260

Bentor YK (1969) On the evolution of subsurface brines in Israel. Chem Geol 4:83–110

Bentor YK, Vroman A (1960) The geological map of Israel 1:100,000, sheet 16, Mount Sedom, Geological Survey, Israel

Bergelson G, Nativ R, Bein A (1999) Salinization and dilation history of groundwater discharging into the Sea of Galilee, the Dead Sea Transform. Isr Appl Geoch 14:91–118

Beyth M, Starinsky A, Lazar B (1981) Low temperature saline waters, Timna, Southern Israel (Preliminary results). In: Ministry of Energy and Infrastructure, GSI. Report no. 606/81: 20 pp

Beyth M, Ayalon A, Longstaffe F, Matthews A (1997) Epigenetic alteration in the Precambrian igneous complex of mount Timna, Southern Israel: oxygen isotope studies. Isr J Earth Sci 46:1–12

Braitsch O (1971) Salt deposits: their origin and composition. Springer, Berlin/Heidelberg/New York, 279 pp
Bucher K, Stober I (2010) Fluids in the upper continental crust. Geofluids 10:241–253
Carpenter AB (1978) Origin and chemical evolution of brines in sedimentary basins. Okla Geol Surv Circ 79:60–67
Frank TD, Gui Z, ANDRILL Science Team (2010) Cryogenic origin for brine in the subsurface of southern McMurdo Sound, Antarctica. Geology 38:587–590
Frape SK, Fritz P (1987) Geochemical trends for ground waters from the Canadian Shield. In: Fritz P, Frape SK (eds) Saline water and gases in crystalline rocks, Geological Association of Canada special paper 33. Geological Association of Canada, St. John's, pp 19–38
García-Veigas J, Rosell L, Zak I, Playà E, Ayora C, Starinsky A (2009) Evidence of potash salt formation in the Pliocene Sedom Lagoon (Dead Sea Rift, Israel). Chem Geol 265:499–511
Garfunkel Z (1997) The history and formation of Dead Sea basin. In: Niemi TM, Ben-Avraham Z, Gat JR (eds) The Dead Sea: the lake and its setting, Oxford monographs in geology and geophysics, 36. Oxford University Press, Oxford, 298 pp
Gavrieli I, Stein M (2006) On the origin and fate of the brines in the Dead Sea basin. In: Enzel Y, Agnon A, Stein M (eds) New frontiers in the Dead Sea paleoenvironmental research, Geological Society of America Special paper 401. Geological Society of America, Boulder
Goldschmidt MJ, Arad A, Neev D (1967) The mechanism of saline springs in the Lake Tiberias depression. Geol Surv Isr Bull Hydrol Serv Part 11:1–19
Gvirtzman H, Garven G, Gvirtzman G (1997) Hydrogeological modeling of the saline hot springs at the Sea of Galilee, Israel. Water Resour Res 33:913–926
Hardie LA (1996) Secular variation in seawater chemistry: an explanation for the coupled secular variation in the mineralogies of marine limestones and potash evaporites over the past 600 m.y. Geology 24:279–283
Harvie CE, Eugster HP, Wear JH (1982) Mineral equilibria in the six-component seawater system, Na-K-Mg-Ca-SO_4-Cl-H_2O at 25 Composition of the saturated solutions. Geochim Cosmochim Acta 46:1603–1618
Harvie CE, Moller N, Weare H (1984) The prediction of mineral solubilities in natural waters: the Na-K-Mg-Cl-SO_4-OH-CO_3-CO_2-H_2O system to high ionic strengths at 25°C. Geochim Cosmochim Acta 48:723–751
Harvie CE, Wear JH, Hardie LA, Eugster HP (1980) Evaporation of seawater: calculated mineral sequences. Science 208:498–500
Herut B, Starinsky A, Katz A, Bein A (1990) The role of seawater freezing in the formation of subsurface brines. Geochim Cosmochim Acta 54:13–21
Horita J, Zimmermann H, Holland HD (2002) Chemical evolution of seawater during the phanerozoic: implications from the record of marine evaporites. Geochim Cosmochim Acta 66:3733–3756
Horowitz A (1987) Palynological evidence for the age and rate of sedimentation along the Dead Sea rift and structural implications. Tectonophysics 141:107–115
Kanfi Y (1972) Flow regime in the Southern Arava and its affect on the water geochemistry. M.Sc. thesis, The Hebrew University, Jerusalem (in Hebrew)
Katz A, Starinsky A (2009) Geochemical history of the Dead Sea. Aquat Geochem 15:159–194
Katz A, Starinsky A, Marion GM (2011) Saline waters in basement rocks of the Kaapvaal Craton, South Africa. Chem Geol 289:163–170
Kinsman DJJ (1976) Evaporites: relative humidity control on primary mineral facies. J Sed Petrol 46:273–279
Kolodny Y, Katz A, Starinsky A, Moise T, Simon E (1999) Chemical tracing of salinity sources in Lake Kinneret (Sea of Galilee). Isr Limnol Oceanogr 44(4):1035–1044
Krumgalz BS, Hecht A, Starinsky A, Katz A (2000) Thermodynamic constraints on Dead Sea evaporation: can the Dead Sea dry up? Chem Geol 165:1–11
Lerman A (1967) Model of chemical evolution of a chloride lake – The Dead Sea. Geochim Cosmochim Acta 31:2309–2330

Lerman A, Shatkay A (1968) Dead Sea brines: degree of halite saturation by electrode measurements. Earth Planet Sci Lett 5:63–66

Lowenstein TK, Hardie LA, Timofeeff MN, Demicco RV (2003) Secular variation in seawater chemistry and the origin of calcium chloride basinal brines. Geology 31:857–860

Matmon D (1995) Simulations of groundwater flow between the Mediterranean Sea and the Jordan Rift Valley. M.Sc. thesis, The Hebrew University of Jerusalem

Mazor E, Mero F (1969) Geochemical tracing of mineral and fresh water sources in the Lake Tiberias basin. Isr J Hydrol 7:276–317

Mazor E, Rosenthal E, Ekstein J (1969) Geochemical tracing of mineral water sources in the southwestern Dead Sea basin. Isr J Hydrol 7:246–248

Moise T, Starinsky A, Katz A, Kolodny Y (2000) Ra isotopes and Rn in brines and groundwaters of the Jordan – Dead Sea Rift Valley: enrichment, retardation, and mixing. Geochim Cosmochim Acta 64(14):2371–2388

Raab M (1996) The origin of the evaporites in the Jordan-Arava valley in view of the evolution of brines and evaporites during seawater evaporation. Ph.D. thesis, The Hebrew University of Jerusalem (in Hebrew, English summary), 114 pp

Raz E (1983) The geology of the Judea desert. Geol Survey Israel, Jerusalem 83/3: 130 p

Riley JP, Chester R (1965) Introduction to marine chemistry. Academic Press, London, 465pp

Shalev E, Yechieli Y (2007) The effect of Dead Sea level fluctuations on the discharge of thermal springs. Isr J Earth Sci 56:19–27

Stanhill G (1994) Changes in the rate of evaporation from the Dead Sea. Int J Climatol 14:465–471

Starinsky A (1974) Relationship between Ca-chloride brines and sedimentary rocks in Israel. PhD thesis, The Hebrew University of Jerusalem (in Hebrew, English summary), 177 pp

Starinsky A, Katz A (2003) The formation of natural cryogenic brines. Geochim Cosmochim Acta 67:1475–1484

Starinsky A, Katz A, Levitte D (1979) Temperature-composition-depth relationship in rift valley hot springs: Hammat Gader, Northern Israel. Chem Geol 27:233–244

Stein M, Agnon A, Katz A, Starinsky A (2002) Strontium isotopes in discordant dolomite bodies of the Judea Group, Dead Sea basin. Isr J Earth Sci 51:219–224

Zak I (1967) The geology of Mount Sedom. PhD thesis, The Hebrew University of Jerusalem (in Hebrew, English summary)

Zak I (1997) Evolution of the Dead Sea brines. In: Niemi TM, Ben-Avraham Z, Gat JR (eds) The Dead Sea: the lake and its setting. Oxford monographs in geology and geophysics, 36. Oxford University Press, Oxford, 298 pp

Zak I, Freund R (1981) Asymmetry and basin migration in the Dead Sea rift. Tectonophysics 80:27–38

Index

Z. Garfunkel et al. (eds.), *Dead Sea Transform Fault System: Reviews*, Modern Approaches in Solid Earth Sciences 6, DOI 10.1007/978-94-017-8872-4,

MIX
Papier aus verantwortungsvollen Quellen
Paper from responsible sources
FSC® C105338

If you have any concerns about our products,
you can contact us on
ProductSafety@springernature.com

In case Publisher is established outside the EU,
the EU authorized representative is:
Springer Nature Customer Service Center GmbH
Europaplatz 3, 69115 Heidelberg, Germany

Printed by Libri Plureos GmbH
in Hamburg, Germany